CHEMISTRY

Students' book

Third edition

Nuffield Advanced Science

Published for the Nuffield–Chelsea Curriculum Trust by Longman Group Ltd

Longman Group Limited
Longman House, Burnt Mill, Harlow, Essex CM20 2JE, England
and Associated Companies throughout the world.

First published 1970
Revised edition published 1984
Third edition published 1994
Copyright © 1970, 1984, 1994 The Nuffield–Chelsea Curriculum Trust.

Set in 10/12 pt Times Roman and Univers Condensed
Produced by Keyspools Ltd.
Printed in Great Britain by Butler & Tanner Ltd.

ISBN 0582 233461

All rights reserved. No part of this publication may be reproduced, stored in a retrieval system, or transmitted in any form or by any means, electronic, mechanical, photocopying, recording, or otherwise without either the prior permission of the publishers or a licence permitting restricted copying in the United Kingdom issued by the Copyright Licensing Agency Ltd, 90 Tottenham Court Road, London W1P 9HE

The publishers' policy is to use paper manufactured from sustainable forests.

Note
All references to the **Book of data** are to the **revised** edition which is part of the present series.

Cover picture
Laboratory glassware
Will and Deni McIntyre/Science Photo Library

General Editor:
Michael Vokins

Contributors:
John Apsey
David Craggs
Alastair Fleming
Alan Furse
Frances Hawkesford
Brian Hitchen
Andrew Hunt
Glyn James
Roger Norris
Bill Price
Bryan Stokes
Michael Vokins

Adviser:
Peter Burrows

Contents

Introduction *1*

Topic 1	Iron compounds: an introduction to inorganic chemistry	*3*
Topic 2	Alcohols: an introduction to organic chemistry	*29*
Reading task 1	The carbon cycle	*52*
Topic 3	Atoms, ions and acids	*56*
Topic 4	Energy and reactions	*85*
Topic 5	The halogens and redox reactions	*103*
Reading task 2	Origin of the chemical elements	*128*
Topic 6	Covalent bonding	*132*
Topic 7	Hydrocarbons and halogenoalkanes	*155*
Topic 8	How fast? Rates of reaction	*212*
Reading task 3	Photochemistry	*240*
Topic 9	Intermolecular forces and solubility	*245*
Topic 10	Entropy	*272*
Topic 11	How far? Reversible reactions	*289*
Topic 12	Carbon compounds with acidic and basic properties	*323*
Topic 13	Redox equilibria and electrochemical cells	*358*
Topic 14	Natural products and polymers	*391*
Reading task 4	Diabetes	*431*
Topic 15	The transition elements	*435*
Topic 16	Organic synthesis	*473*
Reading task 5	Drugs	*501*
Topic 17	Nitrogen compounds	*507*
Reading task 6	Nitrates in agriculture	*532*
Topic 18	Instrumental methods	*537*
Appendix 1	Help with mathematics	*553*
Appendix 2	Laboratory safety	*559*

Acknowledgements *562*

Index *563*

Introduction

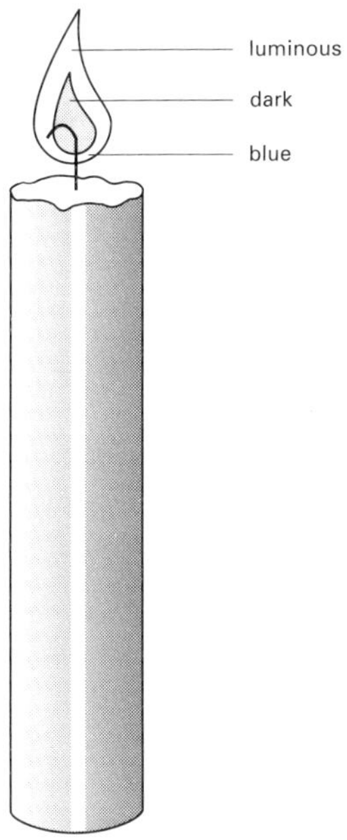

luminous
dark
blue

Figure 1 A candle burning

This course

You are now beginning an advanced course in chemistry. What is chemistry about at this level? What will you learn? These are questions that cannot be answered fully until the end of the course. But if you consider the essential nature of chemistry – what chemists do and how they think – then you will get some idea of what to expect.

We might start with a candle burning. That was how Michael Faraday started a series of lectures (the last time being Christmas 1860). He was quite firm about one point: that the explanations he offered his audience about how a candle burnt would need to be altered as new experiments produced new observations and ideas.

For Faraday the study of chemical changes was the study of the conversion of a substance to new different substances. When Faraday burnt his candle he could demonstrate to his audience, in his own words, that:

- Fresh air is necessary for combustion.
- The new substances formed are carbon dioxide and water.
- The heat of the flame is exactly where the chemical action is.
- Particles are responsible for the luminous part of the flame.
- The dark part of a candle flame consists of vaporized candle wax.
- A candle waits until it is hot enough before it burns.

This was about as far as Faraday could go; he could not write formulae or equations because the values of atomic masses were disputed. He certainly had no idea that future chemists would describe how in a candle flame:

- Wax molecules break up in a set of steps to form carbon dioxide molecules.
- Energy is involved as bonds break and form.
- Light is emitted as electrons lose energy.
- The form of the flame depends on the rate of reaction.
- A definite temperature is needed before combustion becomes spontaneous.

In this course you will be using experiments, as Faraday did in his lectures, to help you understand how and why chemical reactions occur; and by the end of the course you should be able to understand a twentieth-century description of a candle flame. You should also expect that in another hundred years the chemists' candle will be described in quite different ways. You may even make a contribution to that change.

EXPERIMENT
Light a candle and write down everything you notice about the candle and its flame as it burns. A careful chemist can make about *forty* observations!

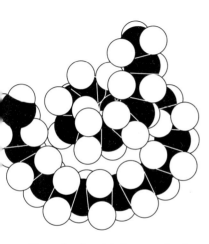

Figure 2 A model of a wax molecule

This book

This book is the third edition of Nuffield Advanced Chemistry. It is supported by a *Book of data* that you will need to consult regularly.

Later in the course you will choose one of several *Special Studies*: these are designed to help you appreciate the links between chemistry and its applications because chemistry is useful knowledge – useful in the mining industry, useful to dieticians, useful in the production of the multitude of products that we find so valuable every day.

If you have not already done so, start by flicking through this book. Look for features that are different from other science books you have used.

As you work through Topic 1 you will meet the main features that we use during the course to help you organize your learning.

> **REVIEW TASK**
> Using your knowledge of science, write down how you think Faraday's ideas might be demonstrated.

Review task We suggest you work on these in a group. You will be asked to try to recall facts and ideas that you should already have met.

Experiments You will be expected to try to interpret your experiments – to look for patterns and to explain what you observe. Merely describing what you see happening in an experiment will not improve your understanding.

> **SAFETY** ⚠
> In boxes like this you will find important safety instructions.

Safety You will be safe and successful in your laboratory work provided:
- You plan your work taking note of the safety information provided.
- You wear eye protection and whatever else is recommended.
- You carry out all instructions thoughtfully and correctly.
- You follow the general guide-lines in Appendix 2 on 'Laboratory safety'.

Comments These give extra information, sometimes curious facts, sometimes more advanced explanations which you may find helpful but which are not part of the main development of the course.

> **COMMENT**
> The blue colour in the flame is due to the electrons in a C_2 fragment rearranging to a lower energy level.

Questions and Study tasks The Questions are usually short and aim to guide you through some aspect of the work. The Study tasks are more extensive, designed particularly to develop your ability to extract information from your reading and report your findings clearly.

Investigations In the Investigations you are expected to work out your own experimental procedures to answer the problem posed.

Summary The Summaries list the facts and ideas you should know by the end of a Topic. Sometimes we suggest ways to organize an outline summary of a Topic, so that you can concentrate on learning the main ideas before adding the detail.

Review questions and Examination questions The Review questions are short items to help you practise your use of the ideas in the Topics. The Examination questions are taken from past Advanced Chemistry papers. You may find them rather difficult if you attempt them without a period of review and revision of the whole Topic.

The main point that we made in the account of Faraday's candle is that chemists deal in both ideas and experiments. Ideas and experiments go hand in hand: experiments lead to ideas and ideas to further experiments. We have organized this book so that you can follow the same pattern, in the hope that your study of chemistry will be successful and enjoyable.

TOPIC 1

Iron compounds: an introduction to inorganic chemistry

IRON

*Fe fi fo fum
As hard as nails
As tough as they come*

*I'm the most important
Metal known to man
(though aluminium
is more common
do we need another can?)*

*Five per cent of the earth's crust
I am also the stone at its centre
Iron fist in iron glove
Adding weight to the system
I am the firma in the terra*

*Fe fi fo
Don't drop me on your toe*

*My hobbies are space travel
And changing the course of history
(they even named an Age after me
– eat your heart out Gold)*

*And changing shape of course
From axe heads and plough shares
To masks maidens and missiles
I am malleable
I bend to your will
I am both the sword and the shield
The bullet and the forceps*

*I am all around you
And more much more
You are all around me 2, 3, 4 . . .*

♬

*You've got me
Under your skin
I'm in your blood
What a spin that I'm in
Haemoglobin
You've got me
Under your skin*

♬

*So strike while I'm hot
For if I'm not there
What are you?
Anaemic that's what*

*Fe fi
High and mighty
Iron*

*Gregarious and fancy free
Easy going that's me
No hidden depths
I'm not elusive
To be conclusive
You get what you see
fe   Fe*

Roger McGough

We are going to begin with the chemistry of iron and some of its compounds. Iron is the metal which has most affected the way we live over the last 250 years. Iron, and its most important alloy, steel, are at the heart of technologically sophisticated societies.

Iron is everywhere. The Earth's core behaves as if it is iron; the Earth as a whole is 35% iron while the Earth's crust contains 5% of iron; meteorites are composed mostly of iron; ploughed fields are red to brown in colour because of the iron compounds in the soil; red pottery contains iron; a red-brick wall contains iron; red paints are usually based on iron compounds; haemoglobin contains iron and you have about 3.5 g of iron in your body and need 10 mg of iron in your food each day.

Learning how to manufacture and use iron in large quantities was the key to the industrial revolution in the eighteenth and nineteenth centuries, just as understanding semiconductors is the key to the development of computers and information technology today.

When you have finished work on this Topic you should be confident about the ideas listed in the **Summary** at the end of the Topic (on page 26) and be able to tackle the **Review Questions** provided (starting on page 26). Take a look at those sections now.

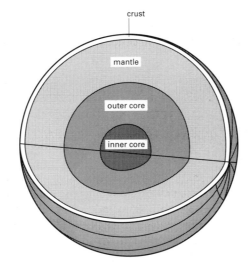

Element	Approximate composition of the Earth as a whole / g kg^{-1}	Composition of the Earth's crust / g kg^{-1}
Fe	300	70
O	295	470
Mg	150	30
Si	145	270
Al	15	85

Figure 1.1 Iron in the Earth and the Earth's crust

Figure 1.2 The impact of an iron meteor made this crater in Arizona

Figure 1.3 This cast-iron bridge over the River Severn at Coalbrookdale is one of the triumphs of the Industrial Revolution

Figure 1.4 Stansted airport terminal uses iron in its structure

1.1 The chemistry of iron

You may remember from experiments that you did in your previous science course that iron forms two series of compounds, known as iron(II) compounds and iron(III) compounds. Thus there are two iron oxides, iron(II) oxide with the formula FeO, and iron(III) oxide with the formula Fe_2O_3; two iron chlorides with the formulae $FeCl_2$ and $FeCl_3$, and so on. (See the Comment box opposite for more information about the oxidation numbers (I), (II), etc.)

REVIEW TASK
Working in groups, make a list of the different types of reaction that you have already met. Consider processes such as combustion, precipitation, electrolysis and any other types you can think of. Find an example of each type.

EXPERIMENT 1.1

Iron compounds

In this introductory experiment you are asked to make a general survey of the types of reaction that can take place with iron compounds.

There are hazards associated with several of the compounds you will be using or producing: provided you take care and follow the procedures the experiments are safe.

SAFETY
Hydrochloric, nitric and sulphuric acids, sodium hydroxide and hydrogen peroxide are corrosive even when dilute but be especially careful when using concentrated solutions; barium compounds are harmful; nitrogen oxides and sulphur dioxide are toxic materials and should be treated with care.

Copy the main hazard warning signs into your notes and make sure you are familiar with them.

Procedure
1 Appearance

Examine the collection of iron and its compounds put out for you to inspect.

- *In your notes:* design a table in which you can record the names, appearance, formulae and any hazardous properties of the compounds you are examining.

2 Solubility

Shake small amounts of iron compounds with pure water and decide whether the compounds are soluble or insoluble. Measure the pH of your solutions using Full-range Indicator paper or solution. Also measure the pH of the water you used to make the solutions (as a control).

- *In your notes:* present your results in a table.

3 Precipitation reactions

a Mix small volumes of a fresh solution of iron(II) sulphate in water with the following solutions, noting carefully what you see:
- Sodium hydroxide solution
- Dilute sulphuric acid
- Barium chloride solution.

b Repeat **3a** using a fresh solution in water of iron(III) sulphate.

- *In your notes:* record your results for **3a** and **3b** together in a table so that you can easily compare them. Name as many products as you can. Try to explain why there is no reaction with dilute sulphuric acid.

A compound that dissolves is called a solute, the liquid used is called a solvent.

1 Iron compounds

> **COMMENT**
> The Roman numerals (I, II, III, etc.) in a name indicate the degree of oxidation of the iron. In iron(II) oxide there are two oxygen atoms to every two atoms of iron; in iron(III) oxide there are three oxygen atoms to every two atoms of iron. The uncombined element is described as iron(0) because the iron atoms are not oxidized at all.
> Iron(II) compounds used to be known as 'ferrous compounds', and iron(III) as 'ferric', and you can still see these names on medicines.
> You will be studying oxidation and reduction in Topic 5.

4 Oxidation

You are going to try to **oxidize** iron(II) to iron(III). Use a fresh dilute solution of iron(II) sulphate to which an equal volume of dilute sulphuric acid has been added. Some reactions take place more readily in acidic conditions, and the acid will also limit reaction with oxygen and water. The difference in colour of iron(II) hydroxide and iron(III) hydroxide should help you decide whether a reaction has taken place.

Mix small volumes of acidic iron(II) sulphate solution with equal volumes of the following oxidizing agents, followed by sodium hydroxide solution, until a precipitate is obtained:
- Hydrogen peroxide solution
- Potassium manganate(VII) solution
- 5 drops of concentrated nitric acid; and warm.

■ *In your notes:* design a table in which to record the results from this experiment and the next. Write the formulae of as many compounds as you can, and name as many products as you can. Write down in your table all the colour changes: in each case describe the colour before and after reaction.

5 Reduction

You are now going to try to **reduce** iron(III) to iron(II). Use a fresh dilute solution of acidified iron(III) sulphate.

Mix small volumes of acidic iron(III) sulphate solution with equal volumes of the following reducing agents followed by sodium hydroxide solution until a precipitate is obtained:
- A small piece of zinc; warm and leave for 2 minutes before adding the sodium hydroxide solution
- Sodium sulphite solution; and warm.

6 Thermal decomposition

a Do this experiment in a fume cupboard as toxic fumes are produced. Heat a small amount of solid iron(III) nitrate in an ignition tube. Test the gases produced with a glowing splint, and moist pH paper.

b Repeat **a** using solid iron(II) sulphate.

■ *In your notes:* describe what happens and name as many products as you can for **6a** and **b**.

7 Tests for the presence of iron compounds

a Add a few drops of potassium thiocyanate solution, KCNS(aq), to a few crystals of:
- Iron(II) sulphate
- Iron(III) sulphate.

b Add a few drops of potassium hexacyanoferrate(III) solution, $K_3Fe(CN)_6$(aq), to a few crystals of:
- Iron(II) sulphate
- Iron(III) sulphate.

■ *In your notes:* record the colours produced.

Interpretation of the experiments

Appearance of the compounds

You should have noticed that most iron compounds are coloured. Iron(II) compounds are generally green, iron(III) compounds are yellow or brown.

Precipitation reactions

The reactions with sodium hydroxide solution produce distinctively coloured precipitates:

$$FeSO_4(aq) + 2NaOH(aq) \longrightarrow \underset{\text{pale green}}{Fe(OH)_2(s)} + Na_2SO_4(aq)$$

$$Fe_2(SO_4)_3(aq) + 6NaOH(aq) \longrightarrow \underset{\text{dark brown}}{2Fe(OH)_3(s)} + 3Na_2SO_4(aq)$$

The reaction with barium chloride solution shows that both iron solutions contain sulphate ions:

$$Ba^{2+}(aq) + SO_4^{2-}(aq) \longrightarrow Ba^{2+}SO_4^{2-}(s)$$

There is no reaction with sulphuric acid: the solutions already contain sulphate ions.

> **COMMENT**
>
> The colours of iron compounds have been used in attempts at archaeological fraud. Teeth soaked in iron(II) sulphate solution followed by tannic acid have the superficial appearance of prehistoric teeth, as do flints dipped in iron(III) chloride solution followed by exposure to ammonia fumes. Both these frauds were used in the Piltdown 'finds' of 1912–13, but not detected until 1954 after a thorough chemical study of the finds.

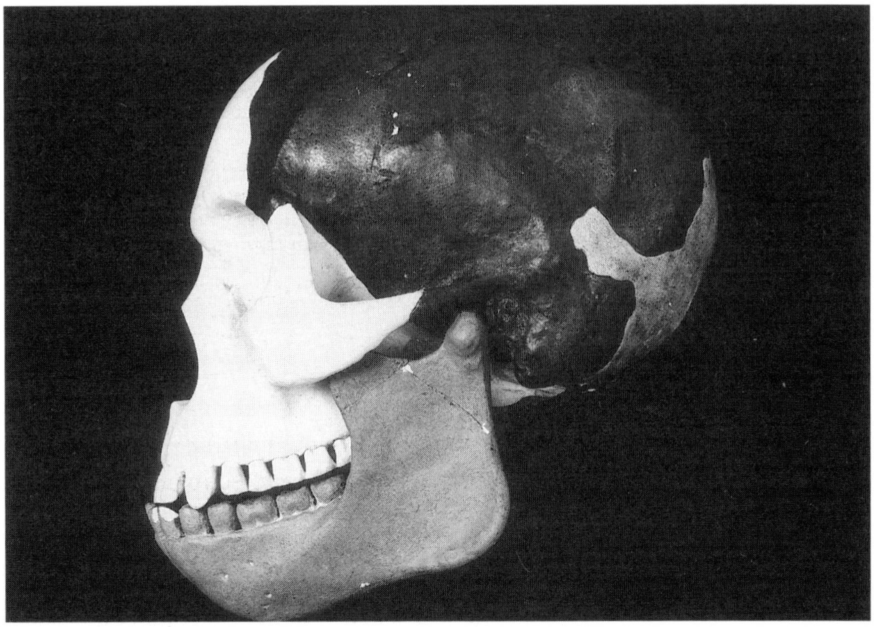

Figure 1.5 The 'missing link': an archaeological fraud involving the use of iron compounds

Oxidation–reduction reactions

You should have found that the oxidation of iron(II) to iron(III) is relatively easy. Although oxidizing agents are often 'oxygen-rich' it is unhelpful to limit oxidation reactions to those in which oxygen is actually added to a compound.

Changing iron(II) oxide, FeO, to iron(III) oxide, Fe_2O_3, is a simple case of oxidation, so in parallel we can classify any reaction that changes an iron(II) compound to an iron(III) compound as an oxidation reaction. For example

$$2FeSO_4(aq) + H_2O_2(aq) + H_2SO_4(aq) \longrightarrow Fe_2(SO_4)_3(aq) + 2H_2O(l)$$

Similarly the reduction of iron(III) to iron(II) is also easy in the right conditions; and any reaction in which iron(III) changes to iron(II) is classified as a reduction. For example,

$$Fe_2(SO_4)_3(aq) + Zn(s) \longrightarrow 2FeSO_4(aq) + ZnSO_4(aq)$$

Oxidation–reduction reactions will be described in full in Topic 5.

Thermal decomposition

Heating decomposes many iron compounds in oxidation–reduction reactions. For example, iron(II) sulphate:

$$2FeSO_4.7H_2O(s) \longrightarrow \underset{\text{brick red}}{Fe_2O_3(s)} + SO_3(g) + SO_2(g) + 14H_2O(g)$$

The equation for the decomposition of iron(III) nitrate is

$$4Fe(NO_3)_3.9H_2O(s) \longrightarrow 2Fe_2O_3(s) + 12NO_2(g) + 3O_2(g) + 36H_2O(g)$$

> **QUESTION**
> Iron oxide made in this way is known as Jewellers' Rouge. Can you think why?

COMMENT

Because of their colours iron compounds are used in several artists' pigments including:

Pigment	Colour	Compound
Indian red	bluish-red	manufactured Fe_2O_3
Mars black	black	manufactured Fe_3O_4
Mars violet	purplish-red	manufactured Fe_2O_3
Venetian red	scarlet	impure natural Fe_2O_3
Yellow ochre	yellow	clay coloured by iron oxide
Prussian blue	deep greenish-blue	manufactured $Fe_3[Fe(CN)_6]_2$

You are not expected to learn the information included in COMMENTS.

Complex ion formation

Only iron(III) compounds react with potassium thiocyanate solution:

$$Fe^{3+}(aq) + CNS^-(aq) \longrightarrow \underset{\text{deep red}}{Fe(CNS)^{2+}(aq)}$$

This type of reaction will be considered in Topic 15. The deep colour serves as a sensitive test for the presence of iron(III) as a concentration greater than 10^{-4} mol dm^{-3} will give a visible colour.

QUESTION
Old solutions of iron(II) compounds also give reddish colours with thiocyanate ions. Why do you suppose old solutions of iron(II) compounds may contain traces of iron(III)?

The deep blue colour with potassium hexacyanoferrate(III) is also a sensitive test, this time for iron(II).

1.2 Introduction to the chemists' toolkit

Over the years chemists have developed special procedures to help them describe and explain their discoveries. We are going to call these procedures the **'chemists' toolkit'**. You will need to learn these aspects of chemistry with care otherwise chemical reactions will pass before you as no more than a colourful parade. So take careful note of these sections: they are all listed together in the index.

We will start with an account of how chemists measure and record the amounts of substances involved in quantitative experiments.

One *mole* of any substance is the amount of substance which contains as many elementary entities as there are carbon atoms in 12 grams (exactly) of pure carbon-12.

By an 'entity' we mean a single particle that can be exactly described. When using the mole as a unit the entity involved must be specified. It can be an atom, a molecule, or an ion of an element or of a compound, or even electrons.

So:

(Br)   (Br_2)   (Br^-)   (H_2SO_4)   and   (SO_4^{2-})

are all examples of entities that can be measured in moles.

The unit is the **mole** and its symbol is **mol**. Note that the unit 'mole', the symbol 'mol', and the adjective 'molar' are NOT connected to the terms 'molecule' or 'molecular'.

Figure 1.6 Amedeo Avogadro was an Italian nobleman, Count of Quaregna and Cerreto, who trained as a lawyer before turning to science. His recognition that gaseous elements might occur as molecules rather than atoms was published in 1811 but was both rejected and neglected for fifty years before being slowly accepted.

> **COMMENT**
> The mole is one of the seven base units of measurement of the International System of Units (SI units) from which all our other units are derived. A list of the base units, together with their definitions, is given in table 1.1 in the *Book of data*.

The **Avogadro constant** is the constant of proportionality between the amount of substance and number of specified entities of that substance.

amount of substance × Avogadro constant = number of specified entities
(in mol)   (mol^{-1})

The symbol for the Avogadro constant is L and its unit is mol^{-1}. The experimentally determined value is 6.02×10^{23} mol^{-1}. A value was originally calculated by an Austrian schoolteacher, Loschmidt, and the symbol L comes from his name.

Just as banks can count coins by weighing them, so chemists measure out their moles by weighing. 1 mole of carbon can be measured out as 12 g, which is called the **molar mass** of carbon (C).

The *molar mass* of a substance is defined as the mass of one mole of the substance; the entities must be specified.

The symbol for the molar mass of an element is A, and for a molecule is M; the unit is g mol^{-1}. Molar masses of the elements are given in the Periodic Table on the inside cover of this book: the specified entities for tables of elements are **single atoms**.

These traditional measures are still used:
Grocers pack eggs in boxes by the 'dozen', 12 eggs.
Ironmongers measure out by the 'gross', 144 nails.
Papermakers measure out by the 'ream', 480 sheets of paper.
And, of course, chemists use a special amount in their work: they measure out by the **mole**.

A table of molar masses of the elements enables us to work out how much to weigh out to get a mole of any substance.

Element	Entity	Molar mass/g mol^{-1}
Iron	Fe	56
	Fe^{2+}	56
	Fe^{3+}	56
Sulphur	S	32
Hydrogen	H	1
	H^+	1
	H_2	2
Oxygen	O	16
	O^{2-}	16
	O_2	32
Ozone	O_3	48

Notice that you have to decide exactly what entity you are working with. You must always think about and write down exactly which entities you are working with to get correct solutions to problems.

To work out the molar masses of compounds you add up the molar masses of the elements in the compound.

Substance	Entity	Molar mass/g mol^{-1}
Iron(II) sulphate	$FeSO_4$	152
Iron(III) sulphate	$Fe_2(SO_4)_3$	400
Hydrated iron(III) sulphate	$Fe_2(SO_4)_3.9H_2O$	562

Check that you can calculate these molar masses correctly.

Figure 1.7 One mole of iron, one mole of iron(II) sulphate, and one mole of iron(II) sulphate in solution

Now attempt the first three questions in the Review Questions at the end of this Topic, and write your answers in your notes.

1 Iron compounds

EXPERIMENT 1.2

The Thermit reaction

We are going to use a variety of metals from different positions in the **reactivity series** to try to produce iron from iron(III) oxide.

If iron(III) oxide and aluminium react, the balanced equation should be

$$Fe_2O_3(s) + 2Al(s) \longrightarrow 2Fe(s) + Al_2O_3(s)$$

and the amounts to mix will be

1 mol of Fe_2O_3		2 mol of Al	
2 Fe	112 g	2 Al	54 g
3 O	48 g		
	160 g		54 g

For an ordinary laboratory experiment we must scale down the quantities and use no more than $\frac{1}{20}$ mol of Fe_2O_3.

REVIEW TASK
In small groups, write down what you can remember about the reactivity series of the metals. Make a list of the types of reaction that can be used to place metals in the series and list as many metals as you can in decreasing order of reactivity.

Entity	Molar mass/g mol^{-1}
O	16
Al	27
Fe	56
Cu	63.5
Zn	65.4

QUESTION
Write the equations and calculate the reacting masses for the reactions that might occur between iron(III) oxide and copper and zinc. What amounts of copper and zinc are needed for 1 mol of Fe_2O_3?

SAFETY ⚠
Do not mix these compounds by grinding them together. Wear eye protection and keep behind the safety screen. Stand well back.

Procedure
These reactions will be demonstrated to you.

0.05 mol of dry iron(III) oxide, Fe_2O_3, is weighed out and mixed with 0.1 mol of aluminium, Al, in the form of fine powder. They are mixed by stirring them together on a sheet of paper and then poured into a small fireclay pot to form a conical pile. The pot needs to be stood in a bucket of sand because of the energy evolved.

We need magnesium as a 'fuse'. A depression at the top of the pile is filled with a little magnesium powder and a 10 cm length of magnesium ribbon inserted into it. When the fuse has been lit stand back; do not approach if the ignition seems to have failed as it can be delayed.

The mixtures of iron(III) oxide with copper and zinc can be tested by the same technique, using the appropriate molar ratios.

■   Which mixtures react? Is much energy given out?

Do not attempt to investigate any other mixtures using aluminium powder.

Interpretation of the experiments

These reactions are oxidation–reduction reactions:
- The aluminium has gained oxygen and has therefore been oxidized.
- The iron(III) oxide has lost oxygen and therefore the iron has been reduced.

Notice that the two processes have gone on simultaneously: whenever oxidation occurs there will be a balancing reduction.

COMMENT

When chemicals react and give off energy to their surroundings we say an **exothermic reaction** has taken place. Some reactions 'pull in' energy from their surroundings and in these cases we say an **endothermic reaction** has taken place.

You should have seen that aluminium and zinc are high enough in the reactivity series to reduce iron(III) oxide to iron with the release of a considerable amount of energy.

For the reduction of 1 mole of iron(III) oxide, the equations and matching amounts of energy are

$$Fe_2O_3(s) + 2Al(s) \longrightarrow 2Fe(s) + Al_2O_3(s) \quad -851.5 \text{ kJ (highly exothermic)}$$
$$Fe_2O_3(s) + 3Zn(s) \longrightarrow 2Fe(s) + 3ZnO(s) \quad -220.7 \text{ kJ (exothermic)}$$
$$Fe_2O_3(s) + 3Cu(s) \longrightarrow 2Fe(s) + 3CuO(s) \quad +352.3 \text{ kJ (endothermic)}$$

By convention the amount of energy in an exothermic change is given a negative value, and in an endothermic change is given a positive value.

We can see from these values why there was no dramatic flare-up with the copper–iron(III) oxide mixture: the reaction is endothermic so we would have to heat the mixture if there is to be any chance of a reaction. In practice even at high temperature the reaction does not take place.

In Topic 4 you will learn how to measure the energy given off or absorbed by reactions.

The Thermit process

The aluminium–iron(III) oxide reaction is known as the **Thermit reaction** and is used in industry in situations where it is not practicable to weld iron by the use of gas or electric heating. For example, the Thermit process is regularly used on the railways to produce the continuous rail system. To weld normal grade rail a significant amount of molten iron is needed and the rails to be welded have to be heated to a high temperature as part of the process.

The Thermit welding process uses 10 kilograms of Thermit mixture plus about 2 kilograms of alloying mixture, which varies with the type of rail being welded. The reaction is all over in 15 seconds and produces 7 kilograms of iron. Enough energy is released to produce molten iron at a temperature of 2050 °C.

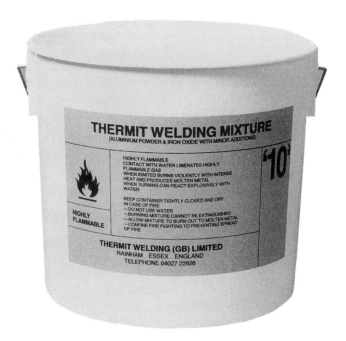

Figure 1.8 Thermit welding mixture

Figure 1.9 Welding on the railway using the Thermit reaction

1.3 More about the chemists' toolkit

Pharmacists, food scientists and chemical engineers all use special ways of planning their work and solving their problems. Similarly, for hundreds of years chemists have been devising names for their discoveries and methods. Some of the names and methods you will meet in your study of chemistry will seem straightforward and others oddly named. For example, you need to be able to construct and analyse chemical formulae with complete accuracy; most symbols for elements are a straightforward two-letter abbreviation of their names but some are unexpected.

COMMENT
Sodium   Na, from the Arabic for soda, natrun.
Potassium   K, from an Arabic word for 'burnt ashes'.
Silver   Ag, from the Latin name for the metal.
Copper   Cu, from the Latin name for the metal.
Iron   Fe, from the Latin name for the metal.

Writing formulae

The composition of compounds is recorded in their formulae, using procedures developed over many years. Thus for iron(II) chloride the chemical formula records:

- The elements present   Fe and Cl
- The number of atoms of each element   1 Fe and 2 Cl
- Their arrangement into cations and anions $Fe^{2+}2Cl^-$, i.e. $FeCl_2$

The proportion of one iron atom to two chlorine atoms is fixed; when you need to record two entities of iron(II) chloride you write **2FeCl₂**, and not Fe₂Cl₄.

Additional information can be added to the formula of a compound. When the solid contains 'water of crystallization' in a fixed proportion the number of molecules of water is recorded after the formula of the salt using a **dot** as in

hydrated iron(II) chloride $FeCl_2.4H_2O$

compared to

anhydrous iron(II) chloride $FeCl_2$

The general rule about writing the formulae of ionic compounds is that the numbers of each ion have to be chosen so that the positive and negative charges balance. When you are familiar with the Periodic Table you will find it helps you recall the charges on the various ions. But essentially you have to memorize this information:

Charge on the ion	Periodic Table group	Other examples
1+	Group 1 Li^+ Na^+ K^+	H^+ NH_4^+ Ag^+
2+	Group 2 Mg^{2+} Ca^{2+} Ba^{2+}	Cu^{2+} Fe^{2+}
3+	Group 3 Al^{3+}	Fe^{3+}
2−	Group 6 O^{2-} S^{2-}	SO_4^{2-} CO_3^{2-}
1−	Group 7 Cl^- Br^- I^-	OH^- NO_3^-

Copy this list into your notes and add other examples as you meet them.

Balancing equations

To write the **balanced equation** of a reaction you need to go through a careful sequence of steps. You must make sure that there is no change in the elements and their number of atoms as you change them from reactants into products. And you have to make sure you get the formulae right.

Consider the reaction of iron with dilute hydrochloric acid

Step 1 Identify the reactants and products by name
iron and **hydrochloric acid** ⟶ **iron(II) chloride** and **hydrogen**

Step 2 Write down their correct formulae
Fe and **HCl** ⟶ **FeCl₂** and **H₂**

Step 3 Balance the numbers of atoms of each element by adjusting the number of entities of each compound
$Fe + 2HCl \longrightarrow FeCl_2 + H_2$

Step 4 Arrive at a complete balanced equation by adding information about the physical state of the compounds
$Fe(s) + 2HCl(aq) \longrightarrow FeCl_2(aq) + H_2(g)$

An equation either records an experiment we have done or can be used to predict the amounts to mix in an investigation.

COMMENT
The state symbols are: (s) solid, (g) gas, (l) liquid, (aq) aqueous (solution in water, from Latin *aqua*)

INVESTIGATION 1.3 — The reaction of iron with copper(II) sulphate

When iron reacts with a salt of a metal lower in the reactivity series such as copper(II) sulphate the reaction could be either

$$Fe(s) + Cu^{2+}(aq) \longrightarrow Fe^{2+}(aq) + Cu(s)$$

or

$$2Fe(s) + 3Cu^{2+}(aq) \longrightarrow 2Fe^{3+}(aq) + 3Cu(s)$$

Procedure
Devise a quantitative experiment to find out which of the two reactions takes place.

Make a risk assessment before starting any experiments.

SAFETY ⚠
Copper(II) sulphate is harmful.

1.4 Preparation of some iron compounds

In this section you are going to learn more about how to calculate the reacting amounts needed to prepare iron compounds. You will also be developing your laboratory skills, and finding out more about the chemistry of metals.

EXPERIMENT 1.4a — Preparation of ammonium iron(II) sulphate (Mohr's salt), $(NH_4)_2SO_4 \cdot FeSO_4 \cdot 6H_2O$

When solutions of ammonium sulphate and iron(II) sulphate are mixed and allowed to evaporate, the crystals which form contain the two salts in an exact one to one ratio by moles. The two sets of ions can pack together in a regular pattern that results in crystals of constant composition The crystalline product is known as a 'double salt'. Its solution behaves just like a mixture of ammonium sulphate and iron(II) sulphate.

$$(NH_4)_2SO_4(aq) + FeSO_4(aq) \longrightarrow (NH_4)_2SO_4 \cdot FeSO_4 \cdot 6H_2O(s)$$

This double salt was introduced into general use by the chemist Mohr because it is less reactive with oxygen in the air than iron(II) sulphate.

You are first going to prepare iron(II) sulphate by reacting iron with dilute sulphuric acid:

$$Fe(s) + H_2SO_4(aq) \longrightarrow FeSO_4(aq) + H_2(g)$$

Use this equation to calculate the amount of iron that will react with 25 cm³ of dilute sulphuric acid ($\frac{1}{20}$ mole). Increase the amount of sulphuric acid by 10% when doing the experiment to make sure all the iron reacts. The dilute sulphuric acid solution (and the dilute ammonia) should contain 2 moles per cubic decimetre of solution.

You are then going to prepare ammonium sulphate by reacting dilute ammonia with dilute sulphuric acid:

$$2NH_3(aq) + H_2SO_4(aq) \longrightarrow (NH_4)_2SO_4(aq)$$

Use this equation to calculate the volume of ammonia solution that will react with 25 cm³ of dilute sulphuric acid ($\frac{1}{20}$ mole). Measure out 10% more ammonia to allow for loss due to evaporation while stored.

1 Iron compounds

> **COMMENT**
> In laboratories the preferred units for volume are the cubic centimetre, cm^3, and the cubic decimetre, dm^3. Shops sell liquids in packages marked with different units:
> 1 litre = 1000 cm^3 = **1 dm^3**   1 cl = **100 cm^3**   1 ml = **1 cm^3**
> The cubic centimetre and the cubic decimetre are preferred because they are related to the metre, one of the 'base' units. See page 2 in the *Book of data* for a description of the base units used in Science.

Procedure

Measure out 50 cm^3 of dilute sulphuric acid and divide into two equal portions.

Making the iron(II) sulphate

Heat the first 25 cm^3 portion of acid to boiling in a conical flask at least 250 cm^3 in size, remove the source of heat and stand the flask on a heat resisting mat.

Add your calculated amount of iron filings in **small** portions. The energy given out by the reaction will keep the solution close to boiling. Keep a plug of cotton wool in the mouth of the conical flask as much as possible **to reduce the escape of acid spray**.

When all the iron filings have been added leave the mixture to continue reacting slowly. When the reaction slows down add an extra 10% of acid (2.5 cm^3). Meanwhile prepare the ammonium sulphate.

Making the ammonium sulphate

Put the other 25 cm^3 portion of dilute sulphuric acid into a beaker and add sufficient dilute ammonia to neutralize the acid. Add the final portions of your calculated volume of ammonia solution in 5 cm^3 portions until a drop of the mixture turns red litmus paper to blue. Boil the solution briskly to drive off the excess of ammonia as a gas (TAKE CARE: ammonia gas is harmful). Concentrate the solution by leaving it boiling.

Separate the iron(II) sulphate solution from undissolved impurities and excess of iron by filtering; collect the solution in a beaker containing 5 cm^3 of dilute sulphuric acid to keep the solution acidic. Wash the filter paper with a small portion of water in order to collect all your iron(II) sulphate.

Making the double salt

Now add the iron(II) sulphate solution to the ammonium sulphate solution. If the ammonia was not neutralized completely the mixture will go cloudy and you will need to add a little more acid. Boil until the volume is reduced to about 40 cm^3, then remove the source of heat and wait until the beaker is cool enough to handle.

Pour your concentrated solution of Mohr's salt into a dust-free flat-bottomed crystallizing dish, cover with a watch glass, label the apparatus and set aside. Crystals should appear within an hour.

- What shape and colour are the crystals you obtain? Leave some crystals exposed to see whether they are stable in the air.
  What ions are present in a solution of Mohr's salt?
  Classify the reactions used in this preparation as neutralization or oxidation-reduction.

EXPERIMENT 1.4b

Preparation of ammonium iron(III) sulphate (iron alum), $(NH_4)_2SO_4.Fe_2(SO_4)_3.24H_2O$.

Iron alum is a 'double salt' of ammonium sulphate and iron(III) sulphate: mixed solutions of the two salts crystallize in a one to one mole ratio even though no new compound has been formed.

You are going to prepare the necessary amount of iron(III) sulphate by oxidizing iron(II) sulphate. Use these equations to calculate the amounts of iron(II) sulphate and ammonium sulphate you need in order to prepare $\frac{1}{200}$ mol of iron alum:

$$2FeSO_4(aq) + 2HNO_3(aq) + H_2SO_4(aq) \longrightarrow Fe_2(SO_4)_3(aq) + 2NO_2(g) + 2H_2O(l)$$

$$(NH_4)_2SO_4(aq) + Fe_2(SO_4)_3(aq) \longrightarrow (NH_4)_2SO_4.Fe_2(SO_4)_3.24H_2O(s)$$

You should find that the correct amount of iron(II) sulphate is 2.8 g. Check your calculation of the amount of ammonium sulphate with other students.

> **HINT**
> Molar mass of $FeSO_4.7H_2O$ is 278 g mol^{-1}.
> Molar mass of $(NH_4)_2SO_4$ is 132 g mol^{-1}.
> For every two moles of iron(II) sulphate you will need one mole of ammonium sulphate.

SAFETY
Concentrated nitric acid is corrosive; nitrogen dioxide gas is toxic. Use a fume cupboard if possible.

Procedure

Warm a mixture of 20 cm³ of pure water and 5 cm³ of dilute sulphuric acid in a boiling tube, add $\frac{1}{100}$ mol of hydrated iron(II) sulphate and shake to dissolve. Stand the boiling tube in a rack and place in a fume cupboard.

When the solution is cool oxidize the Fe^{2+} ions to Fe^{3+} by adding concentrated nitric acid (TAKE CARE) one drop at a time: the solution will go deep brown and then lighten to yellow-brown when the oxidation is complete. Avoid breathing the brown gas nitrogen dioxide which is given off.

Dissolve your calculated amount of ammonium sulphate in 10 cm³ of water, add to your iron(III) sulphate solution and transfer to a small beaker.

Boil the mixture until it is dark brown, then label and leave to crystallize in a warm dry place, protected from dust.

- What shape and colour are the crystals you obtain? Leave some crystals exposed to see whether they are stable in the air.
  What ions are present in a solution of iron alum?
  Which compounds have been oxidized and which reduced in this preparation?

EXPERIMENT 1.4c Preparation of a 'complex ion'

Iron is usually present in its compounds as a cation. But sometimes iron ions will form covalent bonds to anions: the resulting 'complex ion' in this case is an anion.

$$FeCl_3(aq) + 3K_2(C_2O_4)(aq) \longrightarrow K_3[Fe(C_2O_4)_3](aq) + 3KCl(aq)$$

Calculate the amounts of hydrated iron(III) chloride, $FeCl_3.6H_2O$, and hydrated potassium ethanedioate, $K_2(C_2O_4).H_2O$, that you need to mix in order to obtain $\frac{1}{100}$ mole of the product, $K_3[Fe(C_2O_4)_3]$. Have your calculation checked before you proceed.

Procedure

Dissolve the calculated amount of iron(III) chloride in 5 cm³ of pure water in a test-tube.

In another test-tube dissolve the calculated amount of potassium ethanedioate in 10 cm³ of pure water, warming gently.

Mix your solutions in a small beaker, cover and leave to crystallize. The product is called potassium triethandioatoferrate(III).

SAFETY
Potassium ethanedioate is poisonous; iron(III) chloride is irritant.

- Write an account of the experiment, including all your calculations; record the colour changes you observe.
  Leave a sample exposed to sunlight or a powerful light.

EXPERIMENT 1.4d The corrosion of iron

The rusting of iron can be studied using a corrosion indicator solution. Its ingredients are phenolphthalein, potassium hexacyanoferrate(III) and sodium chloride, all dissolved in agar gel.

The gel is made by dispersing 5 g of powdered agar in 250 cm³ of boiling water and adding the indicators: 5 cm³ of 5% potassium hexacyanoferrate(III) solution, 1 cm³ of 1% phenolphthalein solution, plus 7.5 g of sodium chloride.

Carry out a study of the corrosion of iron in a variety of conditions. Use iron nails placed in Petri dishes and covered with corrosion indicator gel before it sets. Make a risk assessment before starting any experiment.

> **COMMENT**
> The gel slows down the diffusion of products from the site of reaction.
> Sodium chloride speeds up corrosion.
> Phenolphthalein will detect any changes to alkalinity by turning pink.
> Potassium hexacyanoferrate(III) will detect any iron that dissolves by turning blue and also reacts with other cations.

1.5 Iron in the Periodic Table

When the elements are listed in order of increasing atomic number, elements having similar properties appear at periodic intervals in the list. These patterns are particularly well seen when the elements are arranged in the form known as the **Periodic Table**.

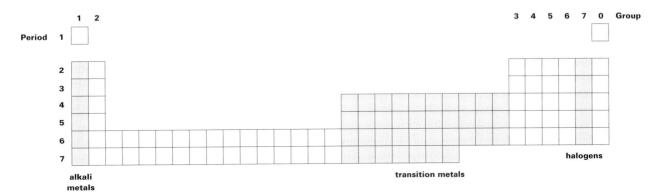

Figure 1.10 Periodic Table of the elements (in outline)

Look at the outline Periodic Table. **A vertical column is called a Group.** Groups are numbered from 1 to 7, with the last column called Group 0 (zero). Besides numbers, several groups have names.

Group number	Name
1	Alkali metals
2	Alkaline earth metals
7	Halogens
0	Noble (inert) gases

A horizontal row is called a Period. Periods are numbered from the top downwards.

Period number	Elements
1	H and He
2	Li to Ne
3	Na to Ar
4	K to Kr
...	...

Transition elements
Titanium (Ti) to copper (Cu)
Lanthanides
Cerium (Ce) to lutetium (Lu) – the 'rare earths'
Actinides
Thorium (Th) to lawrencium (Lw)

Three horizontal regions of the Table have names, and these are the transition elements, the lanthanides and the actinides.

The data included in the Periodic Table about each element are typically their atomic numbers and their molar masses. The number of protons in an atom is its **atomic number**, Z. The number of protons plus the number of neutrons in an atom is its **mass number** and the stable mass numbers tell us how many different types of an element's atoms occur naturally.

Atoms with the same number of protons but different numbers of neutrons are called **isotopes** of an element and are written as $^{54}_{26}Fe$ and $^{56}_{26}Fe$, etc.

The **molar mass** refers to the naturally occurring mixture of isotopes and is determined by an instrument known as a mass spectrometer. The mass spectrometer is described in Topic 18.

To produce electrically uncharged atoms the number of negatively charged electrons will have to equal the number of positively charged protons, so the commonest iron atom consists of

26 protons, p^+
30 neutrons, n
26 electrons, e^-

From the normal behaviour of electric charge we would expect the atoms to consist of a tightly compacted ball of the three types of particle. We now accept a model of the atom as a nucleus of protons and neutrons, with an external 'cloud' of electrons – a model in which neither the nucleus flies apart nor the electrons collapse into the nucleus!

Particle	Symbol	Mass/kg	Charge/C
Proton	p	1.7×10^{-27}	$+1.6 \times 10^{-19}$
Neutron	n	1.7×10^{-27}	zero
Electron	e	0.91×10^{-30}	-1.6×10^{-19}

Periodicity of physical properties

Periodicity is the name given to the regular recurrence of similar features. A school timetable exhibits periodicity; the same lessons recur at periodic intervals, usually every week.

Some interesting information about periodicity amongst the elements can be obtained by comparing the change in various physical properties with change in atomic number. These comparisons can be seen by assembling the data in tables or more readily by plotting charts using a computer spreadsheet.

Figure 1.11 shows the variation of atomic volumes of the elements. The atomic volume of an element is the volume occupied by one mole of atoms of the element.

$$\text{Atomic volume} = \frac{\text{Molar mass (g mol}^{-1})}{\text{Density (g cm}^{-3})}$$

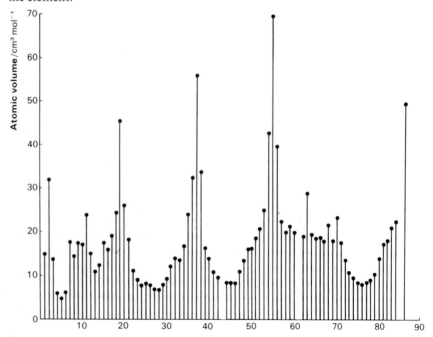

Figure 1.11 Atomic volumes of the elements

STUDY TASK
Use the *Book of data* to work out the atomic volumes of the Group 1 and 2 metals. If a computer is available, use a spreadsheet program to produce a bar chart automatically from the data.
What patterns can you detect in these atomic volumes?

The properties of iron compounds

You are now expected to bring together in your notes the essential ideas that you have met in this Topic. Use the outline below to prepare a rough draft of what you want to record: you may find that putting the material in the form of a coloured poster will help you to learn the ideas. Use the *Book of data* to look up any data that are needed.

REVIEW TASK
Working in groups, try to answer these questions which are about topics you should have met in your previous science course.
1 List the main differences between metallic and non-metallic elements.
2 List the main differences between ionic and covalent compounds.

1 Iron in the Periodic Table
Record the position of iron in the Periodic Table. List basic information about the iron atom such as atomic number, Z, molar mass, A, and the isotopes that occur naturally (are any of them radioactive?).
In what ways is iron a typically metallic rather than a non-metallic element?

2 Iron in the reactivity series
Where is iron placed in the reactivity series? What evidence is available to justify its position?

3 Iron compounds
Record the formulae, solubility, melting points, and colour of some typical iron compounds; does your information justify you classifying iron compounds as usually ionic or usually covalent?

4 Oxidation and reduction
Write a short account of what you understand by oxidation and reduction. Record some equations to justify your explanation.

5 Double salts
What do you understand by the term 'double salt'?
Can you find examples of double salts similar to the ones formed by iron? You will have to use reference books to answer this question.

1.6 Study task: History of the Periodic Table

Hint: *muria* is the Latin for brine.

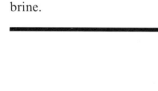

QUESTIONS
Read the following passage, then answer these questions based on it.

1. Which substances listed by Lavoisier as elements are no longer accepted as such?
2. What are the chemical names we use nowadays for the 'earthy substances'?
3. Try to work out what our names are for the radicals: muriatic, fluoric, boracic.
4. Since oxygen is called dephlogisticated air, what properties would you expect in phlogisticated air, and what is the English name for 'azote'?
5. How should light and caloric be classified nowadays?
6. Four elements in our Periodic Table are preceded by an element of heavier molar mass. Which are they?
7. Estimate astatine's boiling point and density, and sodium astatide's melting point and solubility in water from the data available on chlorine, bromine, iodine and their sodium salts.

The Periodic Table is one of the great achievements of chemical science, as it brings order and system to the enormous amount of information which is available about the chemical elements and their compounds. The first successful table to group elements according to their chemical behaviour was devised by Mendeleev in 1869. Mendeleev based his table on the sixty-odd elements then known. Since that time the table has grown to accommodate over one hundred elements. It has also been rearranged to take account of the electronic structures of the atoms, which were quite unknown to Mendeleev. That the original idea has proved capable of absorbing this new knowledge shows the value of Mendeleev's original proposal.

Before the Periodic Table was suggested, several scientists had made attempts to classify the elements according to their properties. One such attempt was that of Lavoisier.

Order in Groups: Lavoisier

Antoine Lavoisier, a French nobleman, had many scientific interests. In 1790 he was a member of the commission that introduced the metric system, but he is most famous for his recognition of oxygen as the component of air involved in combustion, leading to the downfall of the 'phlogiston theory'. In 1789 Lavoisier published one of the most influential books on chemistry ever written. It was called *Traité Elémentaire de Chimie* (Elements of Chemistry), and in it he gave a list of 'simple substances not decomposed by any known process of analysis', or, as we would say, a list of 'the elements'. He divided this list into several groups, based on the similar chemical behaviour of the elements in each group. As you can see from figure 1.13, his groups were elements of bodies, non-metallic substances, metallic bodies, and earthy substances. In Lavoisier's time, this last group was believed to be composed of elements because the substances had not then been broken down into anything simpler.

Figure 1.12 Lavoisier and his wife, who actively helped him in his laboratory

TABLE OF SIMPLE SUBSTANCES.

Simple substances belonging to all the kingdoms of nature, which may be considered as the elements of bodies.

New Names.	Correspondent old Names.
Light	Light.
Caloric	Heat. Principle or element of heat. Fire. Igneous fluid. Matter of fire and of heat.
Oxygen	Dephlogisticated air. Empyreal air. Vital air, or Base of vital air.
Azote	Phlogisticated air or gas. Mephitis, or its base.
Hydrogen	Inflammable air or gas, or the base of inflammable air.

Oxydable and Acidifiable simple Substances not Metallic.

New Names.	Correspondent old names.
Sulphur	
Phosphorus	The same names.
Charcoal	
Muriatic radical	
Fluoric radical	Still unknown.
Boracic radical	

Oxydable and Acidifiable simple Metallic Bodies.

New Names.		Correspondent Old Names.
Antimony		Antimony.
Arsenic		Arsenic.
Bismuth		Bismuth.
Cobalt		Cobalt.
Copper		Copper.
Gold		Gold.
Iron		Iron.
Lead	Regulus of	Lead.
Manganese		Manganese.
Mercury		Mercury.
Molybdena		Molybdena.
Nickel		Nickel.
Platina		Platina.
Silver		Silver.
Tin		Tin.
Tungstein		Tungstein.
Zinc		Zinc.

Salifiable simple Earthy Substances.

New Names.	Correspondent old Names.
Lime	Chalk, calcareous earth. Quicklime.
Magnesia	Magnesia, base of Epsom salt. Calcined or caustic magnesia.
Barytes	Barytes, or heavy earth.
Argill	Clay, earth of alum.
Silex	Siliceous or vitrifiable earth.

Figure 1.13 Lavoisier's classification of substances (1790) from the first English translation

Mendeleev's Table

Chemical knowledge advanced greatly in the eighty years after Lavoisier's attempt to find a pattern in the behaviour of elements. In particular the earthy substances were recognised as oxides, rather than elements, and chemists had learnt how to work out relative atomic masses from the analysis of compounds.

Dimitri Mendeleev, who published his work in 1869, was Professor of Chemistry at St Petersburg. He arranged the elements according to their atomic masses but he left gaps for elements which, he said, had not yet been discovered; and he listed separately some 'odd' elements (for example, cobalt and nickel) whose properties did not fit in with those of the main groups. Apart from the fact that it contained only about sixty elements, Mendeleev's Periodic Table is in principle much the same as that which we use today. That is, the outline of the jigsaw was complete, although a number of the pieces were still missing.

Figure 1.14 Dmitri Mendeleev, who produced a Periodic Table in 1869 similar to those used today

Figure 1.15 Mendeleev's Periodic Table. This memorial is opposite the entrance to the Technologičeskij Institut metro station in St Petersburg

Perhaps the most important feature of Mendeleev's work was that he left gaps in his table where he thought the 'missing' elements should be. This was important because, if a theoretical idea in science is to be really useful, it should not only explain the known facts but also enable new things to be predicted from it. In this way the theory can be tested by seeing whether or not the predictions prove to be correct, and also the theory can lead to scientific advance from following up the new ideas. With Mendeleev – and this drew attention to his table in the first place – not only were elements discovered which fitted the gaps in the table that he had left for them, but also their properties agreed remarkably well with those that Mendeleev had said they should have.

Take one example. When Mendeleev was arranging his table, he left a gap for an element between silicon and tin. He predicted that the molar mass of this element would be 72 and its density 5.5 g cm^{-3} – basing his predictions on the properties of other known elements which surrounded the gap. Fifteen years later the element was discovered. It had a molar mass of 72.6 g mol^{-1} and a density of 5.35 g cm^{-3}. It was given the name **germanium**. Mendeleev made other predictions about it too. The table shows how closely he was able to predict the properties of this new element, and provides confirmation of the correctness of his ideas.

Mendeleev's predictions	Observed properties
Colour will be light grey	Colour is dark grey
Will combine with two atoms of oxygen to form a white powder (the oxide) with a high melting point	Combines with two atoms of oxygen to form a white powder (the oxide) with a melting point above 1000 °C
The oxide will have a density of 4.7 g cm^{-3}	Density of the oxide is 4.228 g cm^{-3}
The chloride will have a boiling point of less than 100 °C	The chloride boils at 84 °C
The density of the chloride will be 1.9 g cm^{-3}	The density of the chloride is 1.844 g cm^{-3}

Summary

At the end of this Topic you should be able to:

a demonstrate understanding of atom, element, symbols for elements, atomic number, mass number, and isotopes

b calculate the composition of the nucleus of a particular atom, in terms of protons and neutrons, and the matching number of electrons, given its mass number and atomic number

c demonstrate understanding of molecule, compound, chemical formulae

d interpret and construct balanced chemical equations including the use of state symbols

e demonstrate understanding of and perform calculations using the concepts of amount of substance (the mole), molar mass in $g\,mol^{-1}$

f recall the meaning of the terms: solution, solvent, solute, saturated solution

g recall the meaning of the terms: concentrated and dilute, acid, base, alkali, salt

h demonstrate understanding of oxidation and reduction in terms of changes in ionic charge as well as reaction with oxygen

i recall the major properties associated with ionic bonding

j recall the meaning of the terms: endothermic and exothermic reactions

k recall the layout of the Periodic Table as Groups and Periods

l evaluate information by extraction from text and the *Book of data* about the uses and properties of metals.

Review questions

* Indicates that the *Periodic Table* is needed.

*1.1 Instructions for practical work often give the amounts of reactants in terms of moles. But for the actual measurement of these amounts you need to convert them into the units of the measuring instrument: for example, grams or cm^3. The following questions require you to convert molar quantities in this manner with the aid of the *Periodic Table*.

   a What is the mass of 0.1 mole of zinc atoms, Zn?
   b What is the mass of 0.02 mole of argon atoms, Ar?
   c What is the mass of 2 moles of nickel atoms, Ni?
   d What is the mass of 1 mole of phosphorus molecules, P_4?
   e What is the mass of 0.5 mole of sodium chloride, NaCl?
   f What mass of calcium chloride, $CaCl_2$, contains 2 moles of chloride ions, Cl^-?
   g What mass of aluminium sulphate, $Al_2(SO_4)_3$, contains 1 mole of aluminium ions, Al^{3+}?

*1.2 Calculate the mass of each of the following:

   a 1 mole of hydrogen molecules, H_2
   b 1 mole of hydrogen atoms, H
   c 1 mole of silica, SiO_2
   d 0.5 mole of carbon dioxide, CO_2
   e 0.25 mole of hydrated sodium carbonate, $Na_2CO_3.10H_2O$.

1 Iron compounds

***1.3** How many moles or part of a mole are each of the following?

- **a** 32 g of oxygen molecules, O_2
- **b** 32 g of oxygen atoms, O
- **c** 31 g of phosphorus molecules, P_4
- **d** 32 g of sulphur molecules, S_8
- **e** 50 g of calcium carbonate, $CaCO_3$.

1.4 Write down the formula of these compounds:

- **a** Barium chloride
- **b** Copper(II) nitrate
- **c** Sodium carbonate
- **d** Potassium hydroxide
- **e** Aluminium sulphate.

Anything you need to look up you should memorize.

1.5 Complete these equations by adding the most appropriate state symbols:

- **a** $NaOH + HCl(aq) \longrightarrow NaCl + H_2O$
- **b** $2NaOH + CuSO_4(aq) \longrightarrow CuO + Na_2SO_4 + H_2O$
- **c** $Na_2CO_3 + H_2SO_4(aq) \longrightarrow Na_2SO_4 + CO_2 + H_2O$
- **d** $NH_3 + HCl \longrightarrow NH_4Cl(s)$

1.6 Balance these equations using chemical formulae and state symbols:

- **a** Iron(II) nitrate + sodium hydroxide $\longrightarrow$ iron(II) hydroxide + sodium nitrate
- **b** Aluminium hydroxide $\longrightarrow$ aluminium oxide + steam
- **c** Sodium carbonate + zinc sulphate $\longrightarrow$ zinc carbonate + sodium sulphate
- **d** Copper(II) oxide + hydrochloric acid $\longrightarrow$ copper(II) chloride + water

1.7 Natural silicon consists of a mixture of three isotopes and its atomic number is 14.

Isotope	Isotopic mass /g mol^{-1}	Percentage abundance by numbers of atoms
A	28.0	92.2
B	29.0	4.7
C	30.0	3.1

- **a** In each of the isotopes, how many neutrons are there in each atom?
- **b** Calculate the molar mass of natural silicon. Show how you arrive at your answer.

Examination questions

1.8 The element gallium (symbol Ga) occurs in Group 3 of the Periodic Table, one row below aluminium and one row above the element indium. The existence of gallium was predicted by Mendeleev a few years before its discovery in 1875. Gallium has similar properties to aluminium, reacting with potassium hydroxide releasing hydrogen, forming a hydroxide and burning in chlorine to form a chloride.

a The value accepted in 1870 for the molar mass of aluminium was 27 and that for indium was 113 g mol^{-1}. Suggest, giving your reasons, a value for the molar mass of gallium.

b i Deduce a formula for gallium hydroxide.

ii What would be the products and equation of the reaction of gallium hydroxide with dilute aqueous nitric acid?

c Suggest an experimental procedure by which gallium potassium sulphate, 'gallium alum', might be prepared from gallium metal.

1.9 Copper has an atomic number of 29 and occurs naturally as two isotopes:

Isotopic mass/g mol^{-1}	Percentage abundance
63	69.1
65	30.9

a i Write down the number of electrons, protons and neutrons in one copper atom whose isotopic mass is 63 g mol^{-1}.

ii Calculate the molar mass of naturally occurring copper, Cu.

b The composition of alloys of gold and copper is often expressed in carats. Pure gold is described as 24 carat; the proportion of gold in an alloy in carats is given by the number of grams of gold in 24 grams of the alloy.
Calculate the molar ratio, to the nearest whole number, of gold to copper in an 18 carat gold ring. (Molar mass of gold = 197 g mol^{-1})

1.10 Choose at least **two** examples of important metallic elements and outline some of the effects that any shortage of these metals might have on our lives if reserves of their ores were to run out within the next twenty years. You need not restrict your answer to those metals whose ores are expected to be in short supply.
For each metal, suggest what alternative materials might be used, indicating how their physical and/or chemical properties make them suitable for the applications you describe.

TOPIC 2

Alcohols: an introduction to organic chemistry

Carbon atoms have an amazing ability to join together in chains, rings, balls and networks. So many of the molecules in living organisms are carbon compounds that we call the study of carbon chemistry 'organic chemistry'.

Figure 2.1

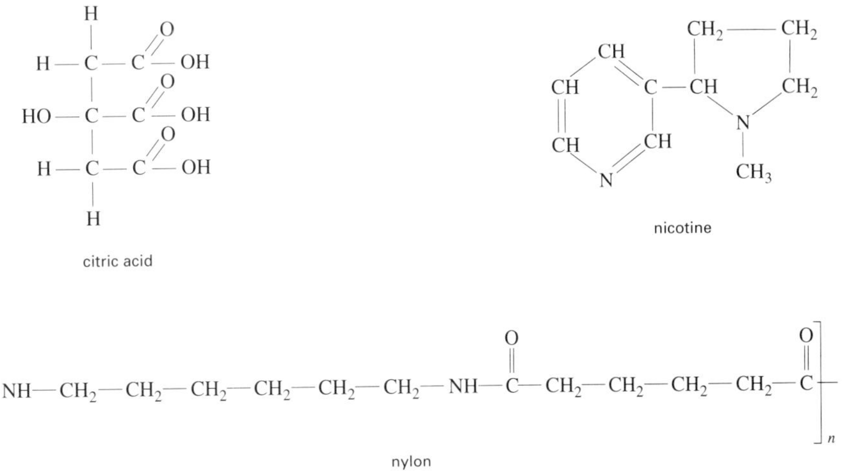

citric acid

nicotine

nylon

In the past people found almost all the carbon compounds they needed in living things. They found ways to extract drugs, perfumes and dyes from plants and animal tissues. Nowadays we get most of our carbon chemicals from oil, gas and coal. The petrochemical industry starts with crude oil, refines it, processes it and produces the huge variety of chemicals needed to manufacture plastics, fibres, drugs, pesticides, and so on.

We can make the study of such a huge number of carbon compounds more manageable by grouping them in families. You may already be familiar with some of the simpler members of the hydrogen-carbon family, which we call the **alkanes**.

The simplest alkane is methane, CH_4, which makes up most of natural gas (see figure 2.3). You will find an explanation of the rules for naming carbon compounds starting in section 2.2 on page 33. The naming of many compounds is based on the names of the alkanes.

Figure 2.2a An early aspirin bottle

Figure 2.2b Polystyrene foam blocks being used in Norway as the foundation of a road

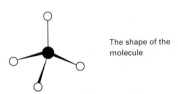

Figure 2.3 Drawings representing methane

2.1 Extraction of a natural product

Plants are a valuable source of chemicals. For thousands of years we have turned to plants for food flavours, perfumes, drugs and dyes. Cooks and chemists have discovered ways to use the different parts of plants as a source of flavour and chemicals: peppermint from leaves, ginger from roots, mustard from seeds, nutmeg from fruits and cloves from buds.

Figure 2.4 A wild yam is the source of a compound needed in the manufacture of contraceptive pills. This discovery was made by Russell Marker.

2 Alcohols

EXPERIMENT 2.1a

Extracting limonene from oranges by steam distillation

Heating plant material directly to distil off the chemicals does not work. The compounds either decompose or burn. Distilling in steam, however, makes it possible to separate plant chemicals without destroying them.

Steam distillation drives off from plant materials oily chemicals which do not mix with water. The oils will distil over in the steam just below 100 °C – well below the boiling point of the oil.

Figure 2.5 Growing oranges

Procedure

Put into a 250 cm^3 flask the finely ground or chopped **outer rind** of two oranges and 100 cm^3 of water. This is the minimum amount to use but the extraction can be scaled up to suit any size of container. Use only the **outer orange-coloured rind** which needs to be fresh.

Arrange the flask for distillation and heat on a wire gauze. Collect about 50 cm^3 of distillate in a measuring cylinder. You should be able to see an oily layer of limonene on top of the water. Use a dropping pipette to remove portions of the oily layer for Experiment 2.1b.

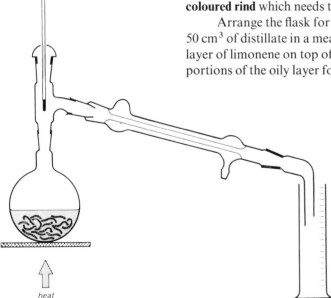

Figure 2.6 Steam distillation apparatus

Figure 2.7 a Growing lavender commercially b Extracting oil of lavender

- *In your notes:* describe how you carried out the steam distillation and what happened.

One aroma extracted by steam distillation on a commercial scale is oil of lavender. The flower heads are harvested in July and the oil extracted from 0.25 tonne of flower heads at a time by steam distillation.

EXPERIMENT 2.1b

Testing limonene

You are now going to compare the limonene from orange peel with two other carbon compounds with similar structures: cyclohexane and cyclohexene.

COMMENT
Remember that you are using pure liquids and not dilute solutions: a few drops is a large amount for a chemical test.

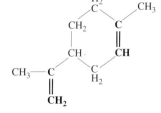

Figure 2.8

Cyclohex**ane**, obtained from petroleum

Cyclohex**ene**, obtained from petroleum

Limon**ene**, obtained from citrus fruits

SAFETY ⚠
Handle cyclohexane and cyclohexene with care as they are highly flammable (like most organic compounds); every bromine solution is harmful, the fumes are dangerous to your eyes, nose and lungs. Remember to wear eye protection throughout the experiment.

Procedure

1 Odour

Smell the three liquids cautiously by wafting the fumes towards your nose. Do not breathe in directly from their containers.

- *In your notes:* try to describe the odours.

2 Action of bromine

Measure out 1 cm³ of bromine solution in water into separate test-tubes using a dropping pipette. Add a few drops of each hydrocarbon to each tube. Shake the tubes gently from side to side.

- *In your notes:* what happens to the colour of the bromine each time?

3 Oxidation

To 1 cm³ of a very dilute, acidified solution of potassium manganate(VII), add a few drops of each hydrocarbon.

- *In your notes:* what happens to the colour of the potassium manganate(VII) each time? Does limonene act like cyclohexane or cyclohexene? What do you think is the reason for the similarity, the ring structure or the C=C bond?

Interpretation of the experiments

We will discuss the chemistry of these reactions in Topic 7, but you should have found that the compounds with C=C bonds in their structure are much more reactive than cyclohexane which only has C—C bonds. The compounds with C=C bonds are called **unsaturated** and the compounds with C—C bonds are called **saturated**.

> **COMMENT**
> When elements react together to form uncharged molecules the number of bonds, —, formed between the atoms depends on the number of pairs of electrons available. This can be deduced from their positions in the Periodic Table, at least for simple molecules.

Group	Element	Number of covalent bonds
—	H	1
4	C	4
5	N	3
6	O	2
7	Cl, Br, I	1

2.2 Naming carbon compounds

Chemists have identified or prepared over five million carbon compounds in laboratories around the world. In this course, we shall meet numerous compounds with up to six carbon atoms and several with two hundred or more (not that you have to learn these formulae). For instance:

$C_6H_3OCl_3$ TCP, an antiseptic
$C_{254}H_{377}N_{65}O_{75}S_6$ insulin, a hormone

In order to build up a picture of how this variety occurs, consider four organic compounds that are used as fuels:

CH_4 methane, found in natural gas
C_2H_6 ethane
C_3H_8 propane, in Calor gas
C_4H_{10} butane, in Camping Gaz

The carbon atoms in the molecules of these compounds form four bonds. The hydrogen atoms each form one bond.

REVIEW TASK

Working in groups, make a list of what you can recall about covalent compounds and the properties that distinguish them from ionic compounds. Then answer these questions:

1. Are the bonds in hydrocarbon molecules ionic or covalent?
2. Draw a diagram to show the number and arrangement of electrons in carbon and hydrogen atoms.
3. Draw a diagram to show how the electrons in carbon and hydrogen atoms form the bonds in methane.

The carbon atoms are arranged in chains, and each molecular formula differs from the one next to it on the list by a CH_2 unit, as shown below. These are the ways of representing the formulae of organic compounds:

- **Molecular formulae** show the number of atoms in a single molecule:

    CH_4 C_2H_6 C_3H_8 C_4H_{10}

- **Structural formulae** show how the atoms are grouped in the molecule:

    CH_4 CH_3-CH_3 $CH_3-CH_2-CH_3$ $CH_3-CH_2-CH_2-CH_3$

- **Displayed formulae** show all the atoms and all the bonds:

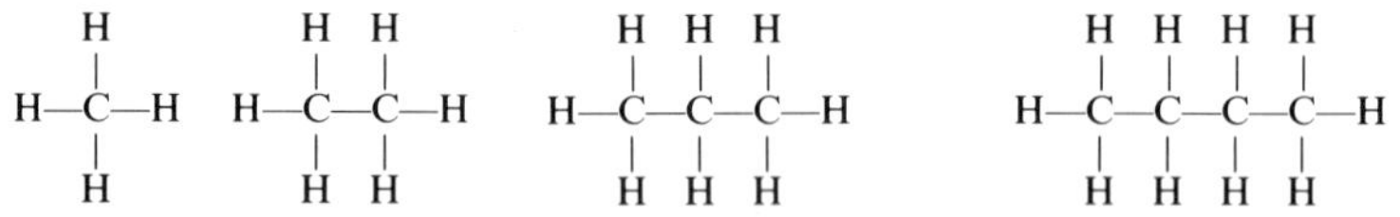

```
     H            H  H          H  H  H            H  H  H  H
     |            |  |          |  |  |            |  |  |  |
  H—C—H       H—C—C—H        H—C—C—C—H         H—C—C—C—C—H
     |            |  |          |  |  |            |  |  |  |
     H            H  H          H  H  H            H  H  H  H
  methane       ethane         propane            butane
```

Isomers are compounds that have the same molecular formula but different structures. For example, chemists have found **two** compounds with the formula C_4H_{10}. One boils at $-1\,°C$ while the other boils at $-12\,°C$. The molecular structures of the two compounds are shown below.

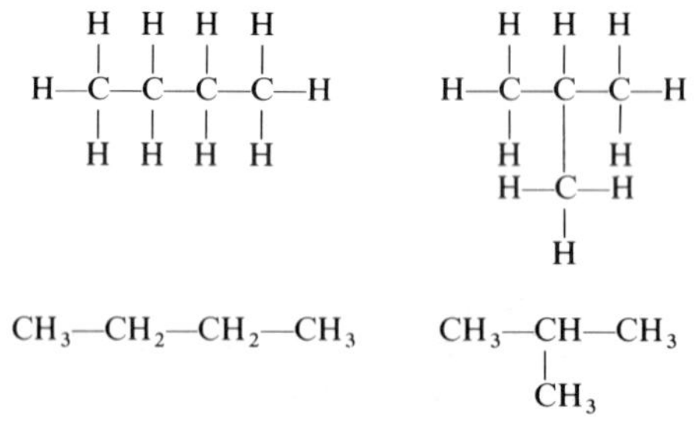

```
     H  H  H  H              H  H  H
     |  |  |  |              |  |  |
  H—C—C—C—C—H             H—C—C—C—H
     |  |  |  |              |  |  |
     H  H  H  H              H  |  H
                              H—C—H
                                 |
                                 H

  CH3—CH2—CH2—CH3          CH3—CH—CH3
                                 |
                                 CH3
      butane              2-methylpropane
```

The rules for naming compounds have been settled by international agreement amongst chemists. We need only a few rules to start with, so here we will deal only with some simple compounds of carbon and hydrogen, and some alcohols.

Names for compounds containing carbon atom chains

Compounds in which the molecules are made up of straight chains of carbon atoms with C—C bonds and combined with hydrogen only have the general name **alkanes.** Names for individual compounds all have the ending '-ane'. For example,

$$CH_3—CH_2—CH_2—CH_3 \quad \text{butane}$$

The names of the first four hydrocarbons in the series, containing 1, 2, 3 and 4 carbon atoms respectively, are methane, ethane, propane and butane. These do not follow any logical system and must be learned. The rest of the hydrocarbons in the series are named by using a Greek numeral root and the ending '-ane', for example, pentane (five carbon atoms in an unbranched chain), hexane (six carbon atoms). The roots are the same as those used in naming geometric figures (pentagon, hexagon, etc.).

Number of carbon atoms in chain	Molecular formula	Name
1	CH_4	methane
2	C_2H_6	ethane
3	C_3H_8	propane
4	C_4H_{10}	butane
5	C_5H_{12}	pentane
6	C_6H_{14}	hexane

Names for alkanes containing a ring of carbon atoms

These are named from the corresponding straight-chain hydrocarbon by adding the prefix 'cyclo-'. An example is cyclohexane, which can be represented as:

A special symbol is used for the very important unsaturated cyclic hydrocarbon named **benzene**:

QUESTIONS

For this activity you will need a set of molecular models or a computer with software for molecular modelling.

1 Make models of the molecules of the first five members of the alkane family.
2 Draw up a table with four columns showing the names of the alkanes, their formulae, drawings of their molecular shapes, and their physical states at room temperature. You will find their physical properties listed in table 5.5 of the *Book of data*.
3 Make models of the isomers with the formula C_5H_{12}. You should be able to make three different models. Write down the displayed formulae corresponding to the structures.

Names for alcohols

Most people think of drinks of various kinds when they see or hear the word alcohol. But for chemists alcohol, or ethanol, is only one of a family of similar compounds.

Alcohols have a hydroxyl group, —OH, attached to a carbon atom. The —OH group is an example of what is called a 'functional group'. The first three members of the alcohol series are methanol, ethanol and propanol. We name them by changing the end of the name of the corresponding alkane to **ol**.

CH_4 methane $\qquad$ CH_3—OH methanol
CH_3—CH_3 ethane $\qquad$ CH_3—CH_2—OH ethanol
CH_3—CH_2—CH_3 propane $\qquad$ CH_3—CH_2—CH_2—OH propan-1-ol

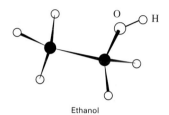
Ethanol

H_2O or H—O—H
Water

QUESTIONS

1 Look up the boiling points of the first four alkanes and the first four alcohols, from methanol to butan-1-ol, in table 5.5 in the *Book of data*. Do the values vary in a regular pattern?

2 Ethane is insoluble in water but ethanol, as in all alcoholic drinks, dissolves. Would you expect the other three alcohols to dissolve in water?

3a How does sodium react with water? Write an equation for the reaction.

b Predict what will happen on adding sodium to ethanol assuming that the —OH group behaves in a similar way in both compounds. Try to write an equation for your predicted change.

2.3 Reactions of alcohols

Alcohols are more reactive than alkanes because C—O and O—H bonds break more easily than C—H and C—C bonds in reactions with aqueous reagents. The alcohols share similar properties because they all have the **C—OH** group of atoms in their molecules.

In organic chemistry we use the term **functional group** to describe the atom, or group of atoms, which gives a compound its characteristic properties.

> **COMMENT**
> You will find a table of functional groups in Topic 7, on page 159. These are the functional groups which you must be able to recognize and recall.

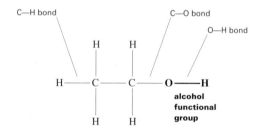

EXPERIMENT 2.3a   Experiments with alcohols

These reactions will introduce you to the reactions of the alcohol functional group and you will be able to see how the whole family of alcohols behaves.

Procedure

1   Solubility in water

To 1 cm³ of ethanol in a test-tube add 1 cm³ of water. Do the two liquids mix? Test the mixture with Full-range Indicator.

Repeat with a range of alcohols, increasing the volume of water used if mixing does not occur.

- *In your notes:* how does the solubility of alcohols compare with the solubility of alkanes? What is the trend in solubility in water in the series methanol to pentan-1-ol? Suggest an explanation for the trend.

2   Reaction with sodium

To 1 cm³ of ethanol in an evaporating basin add one small piece of freshly cut sodium the size of a rice grain (TAKE CARE). Is there any sign of reaction?

Repeat the experiment with a range of alcohols. Add ethanol to dissolve all traces of sodium before throwing away the reaction mixture.

- *In your notes:* what do you see on adding sodium to an alcohol? What are the products? Is this what you would predict by comparing alcohols to water? Which bond in the alcohol breaks?
  What is the trend in rate of reaction from methanol to pentan-1-ol? Can you suggest an explanation for the trend?

3   Oxidation

To 5 cm³ of dilute sulphuric acid in a boiling tube add a few drops of sodium dichromate(VI) solution. Next add 2 drops of ethanol and heat the reaction mixture until it **just** boils. Is there any sign of reaction? Is there any change of smell suggestive of a new organic compound?

Repeat the experiment with a range of alcohols.

- *In your notes:* record the colour changes and smells. Refer to the colours of chromium compounds in the *Book of data*, table 5.3 page 71, to see if this gives you any clues to what may be happening.

> **SAFETY**
> Remember that all alcohols are flammable; sodium dichromate(VI) is a skin irritant; concentrated sulphuric acid is corrosive; sodium is flammable and corrosive. You should wear your eye protection all the time.

EXPERIMENT 2.3b   Preparations using propan-1-ol

In these experiments you will be collecting the products of the reactions in order to study their properties. Carry out **1** and **either 2 or 3** and share your product with someone who carried out the alternative experiment.

Procedure

1   The dehydration of propan-1-ol

Put propan-1-ol in a test-tube to a depth of 1 cm. Push in some loosely packed ceramic fibre until all the propan-1-ol has been soaked up. Now add a 2 cm depth of aluminium oxide granules and arrange the apparatus for the collection of a gas, as shown in figure 2.9.

SAFETY ⚠️
Both propan-1-ol and the products are flammable; bromine water is harmful; sulphuric acid is corrosive; sodium dichromate(VI) is a skin irritant. Be ready for possible spillages and wear gloves when using sodium dichromate(VI).

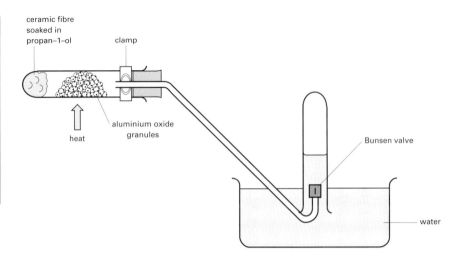

Figure 2.9

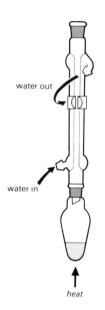

Figure 2.10 Reflux apparatus

SAFETY ⚠️
If your reaction mixture does not boil spontaneously stop adding the sodium dichromate(VI) otherwise the energy released may cause sudden and violent boiling. Seek advice.

Heat the granules gently and collect three or four test-tubes of gas, discarding the first one.

Carry out the following tests on the gas:

a Add 1 cm³ of bromine water. Shake the test-tube and look for any colour changes that suggest a reaction.

b Add 1 cm³ of a very dilute, acidified solution of potassium manganate(VII).

■ *In your notes:* note what you can see at each stage.
What do the tests **a** and **b** tell you about the nature of the product from the dehydration of propan-1-ol?

2 The oxidation of propan-1-ol

This experiment uses approximately equal quantities (0.02 mole) of propan-1-ol and an oxidizing agent, and they are refluxed together in order to oxidize the propan-1-ol as fully as possible under these conditions.

Measure 5 cm³ of water into a boiling tube. Add 6 g of sodium dichromate(VI) (WEAR GLOVES), shake and set aside to dissolve.

Put about 1.5 cm³ of propan-1-ol into a 50 cm³ pear-shaped flask and add about 5 cm³ of water and two or three anti-bumping granules. Put a condenser in the flask for reflux, as shown in figure 2.10.

Add 2 cm³ of concentrated sulphuric acid (TAKE CARE) down the condenser **in drops** from a dropping pipette. While the mixture is still warm start to add your sodium dichromate(VI) solution down the condenser **in drops** from a dropping pipette. The energy released from the reaction should make the mixture boil. **Add the solution a drop at a time** so that the mixture continues to boil without any external heating.

When all the sodium dichromate(VI) solution has been added use a low Bunsen burner flame to keep the mixture boiling for 10 minutes, not allowing any vapour to escape. At the end of that time remove the Bunsen burner and arrange the apparatus for distillation, as shown in figure 2.11. Gently distil 2–3 cm³ of liquid into a test-tube.

COMMENT
Benedict's solution contains a copper(II) compound and is used to test for organic reducing agents. An alternative reagent is Fehling's solution.

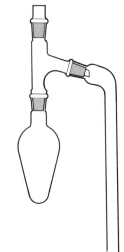

Figure 2.11

The liquid that collects is an aqueous solution of the product. Carry out the following tests on it, recording your results in a table. Leave two columns in your table in which to record the results of each test when performed on propan-1-ol and the product of the next reaction.

a Note the smell of the product.
b Will it neutralize an appreciable volume of sodium carbonate solution?
c Add a few drops of the product to 2 cm³ of Benedict's solution and boil gently.

■ *In your notes:* compare these results with those obtained using propan-1-ol, and the product made in the next experiment.

3 The partial oxidation of propan-1-ol

This experiment uses only half the quantity of oxidizing agent (0.01 mol) that the previous experiment used and the product is distilled from the reaction mixture immediately it is formed. In this way we hope to achieve a partial oxidation of propan-1-ol.

Place about 10 cm³ of dilute sulphuric acid in a flask and add about 3 g of sodium dichromate(VI) and 2 or 3 anti-bumping granules. Shake the contents of the flask until solution is complete (do not warm).

Add 1.5 cm³ of propan-1-ol in drops from a dropping pipette, shaking the flask so as to mix the contents, and then assemble the apparatus as shown in figure 2.11.

Gently and slowly distil 2 cm³ of liquid into a test-tube, taking care that none of the reaction mixture splashes over.

Carry out the tests **2a**, **b** and **c** above, comparing the results.

Interpretation of the reactions of alcohols

Behaviour with water

Alcohols will mix with water but as the hydrocarbon chain gets longer the solubility gets less. The solutions are neutral: this shows that the functional group does not form either hydrogen ions or hydroxide ions with water.

Reaction with sodium

The reaction of sodium with alcohols is very similar to the reaction of the metal with water. Alcohols react less vigorously than water.

$$2Na + 2CH_3{-}CH_2{-}CH_2OH \longrightarrow 2CH_3{-}CH_2{-}\mathbf{CH_2O}^-Na^+ + H_2$$
sodium propoxide

$$2Na + 2HOH \longrightarrow 2HO^-Na^+ + H_2$$
sodium hydroxide
usually written NaOH

In both these reactions each sodium atom loses an electron forming a positive ion while the hydrogen atoms in the —OH groups combine to form hydrogen gas. The organic product is called an **alkoxide**.

Oxidation

Sodium dichromate(VI) is an oxidizing agent. It oxidizes alcohols such as propan-1-ol first to compounds called **aldehydes** and then to acids, called **carboxylic acids**.

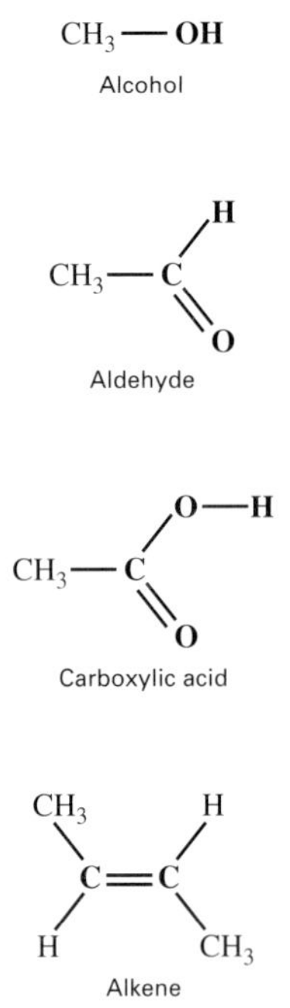

CH$_3$—OH
Alcohol

CH$_3$—C(H)=O
Aldehyde

CH$_3$—C(O—H)=O
Carboxylic acid

CH$_3$\C=C/H, H/ \CH$_3$
Alkene

Figure 2.12 Some functional groups

This reaction can be used to make an aldehyde if the product is separated from the reaction mixture as it forms. This prevents further oxidation to the acid.

$$CH_3-CH_2-CH_2OH \xrightarrow{\text{warm with acidified sodium dichromate(VI)}} CH_3-CH_2-CHO$$
propanal (an aldehyde)

Note the way this reaction is written. Sometimes it is convenient to write unbalanced chemical equations which show the main reactants and products and state the conditions for the change above the reaction arrow. State symbols are usually omitted.

Heating the alcohol with excess acidified sodium dichromate(VI) takes the process a stage further. Heating in a flask fitted with a reflux condenser helps to make sure there is no loss of volatile reactants or products.

$$CH_3-CH_2-CH_2OH \xrightarrow{\text{heat with excess acidified sodium dichromate(VI)}} CH_3-CH_2-CO_2H$$
propanoic acid
(a carboxylic acid)

Dehydration

Propan-1-ol loses water when its vapour passes over a catalyst such as aluminium oxide at about 400 °C. This is an example of an **elimination reaction**.

$$CH_3-CH_2-CH_2OH \longrightarrow CH_3-CH=CH_2 + H_2O$$
propene (an alkene)

The product is propene, a molecule with a double bond. Propene belongs to the family of hydrocarbons called **alkenes**.

Other alcohols behave in a similar way.

Another way to dehydrate an alcohol is to heat the liquid alcohol with concentrated phosphoric acid or sulphuric acid. You will have an opportunity to try this reaction in experiment 2.4.

QUESTIONS

1 Write an equation to show what happens when butan-1-ol reacts with sodium. Name the organic product and write out its displayed formula.
2 Make molecular models of ethene, ethanal and ethanoic acid.
3 Use a set of molecular models to show what happens in each of the following reactions. For each example name the main organic product and write down its displayed formula.
  a Passing ethanol vapour over aluminium oxide at 400 °C
  b Warming ethanol with acidified sodium dichromate(VI) and allowing the product to distil off as it forms.
  c Refluxing ethanol with excess acidified sodium dichromate(VI).

Naming more complicated alcohols

When the —OH group of an alcohol is attached to a carbon atom which is attached directly to only **one** other carbon atom, the compound is known as a **primary** alcohol.

When the —OH group is attached to a carbon atom which is attached directly to **two** other carbon atoms, the compound is a **secondary** alcohol, and when to **three** other carbon atoms, it is a **tertiary** alcohol. Make sure that you understand this naming by looking carefully at the following structural formulae, which are all isomers of $C_4H_{10}O$.

CH_3—CH_2—CH_2—CH_2—OH   butan-1-ol, a primary alcohol

CH_3—CH_2—CH—CH_3
                  |
                  OH   butan-2-ol, a secondary alcohol

           CH_3
             |
CH_3—C—CH_3 2-methylpropan-2-ol, a tertiary alcohol
             |
             OH

When alcohols contain more than one hydroxyl group they are known as **di**ols or **tri**ols, etc., after the number of hydroxyl groups they contain.

CH_2—CH_2
  | | ethane-1,2-diol (glycol)
  OH   OH

CH_2—CH—CH_2
  | | | propane-1,2,3-triol (glycerol)
  OH   OH   OH

> **COMMENT**
>
> The freezing point of an insect is the temperature at which some of its body water can be frozen. For most insects this would normally be about −1 °C but it depends upon the concentration and nature of the solute in the tissues. Just as wise motorists protect their car radiators by the addition of antifreeze in which the important component is ethane-1,2-diol (glycol), so it appears that some insects accumulate propane-1,2,3-triol (glycerol) in their body water during the autumn (figure 2.13). In this way the insects can withstand low temperatures and avoid the danger of cell damage by the formation of ice crystals. If they contain 15% propane-1,2,3-triol the eggs of the moth *Alsophila pometaria* can be cooled to −45 °C before ice crystals form.

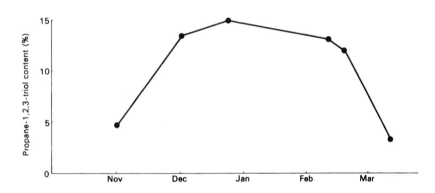

Figure 2.13 Propane-1,2,3,-triol content in the moth, *Alsophila pometaria*

QUESTIONS
A model kit will help you here.
1 Write the displayed formula for an isomer of propan-1-ol.
2 Write the displayed formula for one other isomer with the formula $C_4H_{10}O$. Three structural formulae are shown opposite.

2.4 How much?

Read a cookery recipe and you will see that it tells you how much you need of each ingredient. It will probably also tell you how many people the meal will feed. The recipe answers the question 'How much?'.

Chemists also need to answer this question when making new chemicals both on a laboratory scale and in industry. In this section you are going to find out how to calculate chemical recipes with the help of the ideas about chemical amounts which you met in Topic 1.

The chemists' toolkit: molar masses of organic compounds

Working out the molar masses of organic compounds should not be a problem provided you organize your work neatly. This also helps when you want to check a calculation. You could try organizing your calculations like these two examples:

Molar masses of the elements: $H = 1$; $C = 12$; $O = 16$; $Cl = 35.5 \text{ g mol}^{-1}$

Molar mass of ethanol	Molar mass of TCP
C_2H_5OH	$C_6H_3OCl_3$
$2C = 24$	$6C = 72$
$6H = 6$	$3H = 3$
$1O = \underline{16}$	$1O = 16$
46 g mol^{-1}	$3Cl = \underline{106.5}$
	197.5 g mol^{-1}

When an organic compound is liquid it is often convenient to measure it by volume rather than mass, using the relationship

mass = density × volume

QUESTIONS
1 Try working out the molar masses of limonene and citric acid (or insulin!) from the formulae given earlier in this Topic.
2 Work out the volumes of the liquids in figure 2.14. Densities are listed in table 5.5 in the *Book of data*.

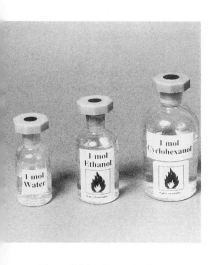

Figure 2.14 One mole of each liquid

Stages in a laboratory preparation

When setting out to make an organic compound you have to go through several stages.
1. Find out the equation for the reaction and calculate the amounts of reactants to use.
2. Measure out the reactants in the right proportions.
3. Mix the reagents in a suitable apparatus and keep them in the right conditions for long enough to complete the reaction; this often means heating and perhaps adding a catalyst.
4. Separate the product from the reaction mixture.
5. Purify the product.
6. Weigh the purified product and calculate the yield.
7. Carry out tests to check that the process has produced the required product.

EXPERIMENT 2.4 How much cyclohexene can you get from cyclohexanol?

We can illustrate these stages with the preparation of cyclohexene from cyclohexanol. Instead of passing the vapour over hot aluminium oxide, as in Experiment 2.3b, we here use concentrated phosphoric acid as the dehydrating agent.

QUESTIONS
1. What do the words 'dehydration' and 'dehydrating agent' mean?
2. Write the displayed formulae of cyclohexanol and cyclohexene. Show how dehydration converts the alcohol to the cycloalkene.
3. What are your predictions about the solubilities of cyclohexene and cyclohexanol in water?
4. Which compound would you expect to have the lower boiling point: cyclohexene or cyclohexanol? Check with the *Book of data*, table 5.5.
5. Work out the volume of 0.1 mol of cyclohexanol. You will find its density and molar mass in table 5.5 in the *Book of data*.

SAFETY
Cyclohexanol and cyclohexene are both flammable, do not store your product as it will eventually form unstable by-products; phosphoric acid is corrosive.

Procedure
Place 0.1 mol of cyclohexanol in a flask and add, dropwise, while shaking the flask, 4 cm^3 of concentrated phosphoric acid from a dropping pipette. Assemble the apparatus as shown in Figure 2.15.

Heat the flask gently and distil very slowly, collecting the liquid which comes over between 70 and 90 °C.

Pour the distillate into a separating funnel and add an equal volume of a saturated solution of sodium chloride. Shake the funnel and allow the two layers to separate. Run off the lower aqueous layer and then run the top layer (cyclohexene, density 0.81 g cm^{-3}) into a small conical flask.

Figure 2.15 Apparatus for distillation at a known temperature

> **COMMENT**
> Wash and dry your distillation apparatus as you will need it again.

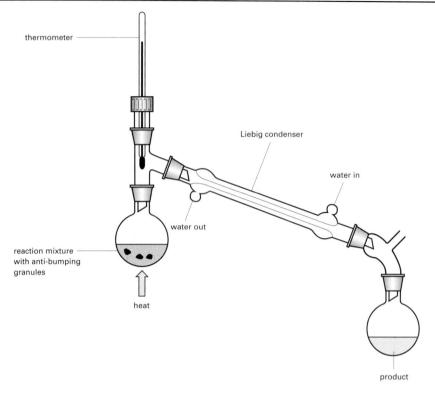

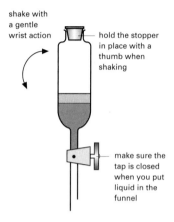

Figure 2.16 Using a separating funnel

To the crude alkene, add two or three pieces of anhydrous calcium chloride and stopper the flask. Shake for a few minutes until the liquid is clear.

Decant the alkene into a clean flask and redistil it, collecting the liquid distilling between 81–85 °C in a preweighed sample tube. Weigh the sample tube with your product.

Carry out test-tube tests on your product to show that it is an alkene and not an alcohol. Look at your results for Experiment 2.1b for some ideas.

- *In your notes:* describe this preparation using the seven stages given above as a guide to what to include.
    - What are the conditions for converting cyclohexanol to cyclohexene? Write an equation for the reaction.
    - How is the product separated from the reaction mixture and why does the method work?
    - There are three stages to the purification of the product. What are they and how do they work?
    - Calculate the theoretical yield from 0.1 mol cyclohexanol, based on 100% yield according to the equation. What percentage of the theoretical yield did you actually get?
    - What test did you carry out on the product and with what results?

2.5 Oxidation products from alcohols

When you oxidized samples of alcohols you obtained aldehydes with the carbonyl functional group $\diagdown\!\!\!C\!=\!O$. There are two closely related types of compound with this functional group, the **aldehydes** and the **ketones**.

Compounds with one alkyl group and one hydrogen atom attached to the carbonyl group are known as *aldehydes*. They are named after their parent alkane, with the terminal **e** replaced by **al**.

Compounds with two alkyl groups attached to the carbonyl group are known as *ketones*. They are named in the same manner, but with the terminal **e** replaced by **one**.

Examples of these two types of compound are:

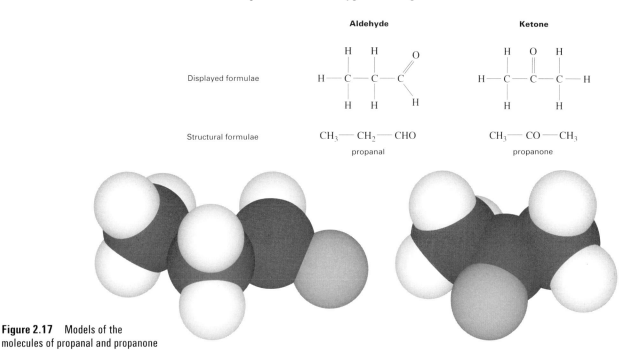

Figure 2.17 Models of the molecules of propanal and propanone

EXPERIMENT 2.5 Oxidation of carbonyl compounds

It is suggested you use propan**al** (an aldehyde) and propan**one** (a ketone) as examples of carbonyl compounds. Be careful to use the correct compound.

Procedure
1 Oxidation with sodium dichromate(VI)

a To a few cm³ of dilute sulphuric acid in a boiling tube add a few drops of sodium dichromate(VI) solution. Next add 2 drops of propan**al** and heat the mixture until it *just* boils. Try to tell from any colour change of the dichromate(VI) whether the propanal has been oxidized.

b Repeat the test using propan**one** instead of propanal.

■ *In your notes:* is there any difference in the behaviour of aldehydes and ketones?

> **SAFETY** ⚠
> Propanal and propanone are both highly flammable and exposure to their vapours will irritate your eyes.

2 Oxidation with Benedict's solution

a Add a few drops of propan**al** to 2.5 cm^3 of water in a boiling tube and add 2.5 cm^3 of Benedict's solution. Bring to the boil and allow to stand. Note the colour of any precipitate.

b Repeat using propan**one** instead of propanal.

■ *In your notes:* note the colour of any precipitate and any substance which does not react. Does Benedict's solution distinguish between aldehydes and ketones?

> **COMMENT**
> Liquid propanal is the pure compound so two drops is a large amount compared to a dilute solution

Oxidation products from alcohols

The oxidation of a primary alcohol produces an aldehyde in the first stage by removal of hydrogen:

$$CH_3-CH_2OH \xrightarrow{\text{oxidation}} CH_3-CHO$$

and aldehydes themselves can be oxidized quite readily to carboxylic acids in a second stage by addition of oxygen:

$$CH_3-CHO \xrightarrow{\text{oxidation}} CH_3-CO_2H$$

Secondary alcohols on the other hand are oxidized to ketones only. Ketones are not usually oxidized as the carbon chain would have to be broken up.

$$CH_3-CH(OH)-CH_3 \xrightarrow{\text{oxidation}} CH_3-CO-CH_3$$

All the reactions can be reversed by the use of suitable reducing agents.

2.6 Fermentation

The process of fermentation has been discovered and developed independently all over the world. All you need is vegetable material that contains sugar, plus yeast. Since yeast moulds occur naturally on fruit skins it is not difficult to get started: the skill is in matching the type of yeast to the type of fruit.

The process occurs in all living cells in the absence of oxygen and depends on the enzymes in the cells. The enzymes act as catalysts breaking a specific C—C bond at each step, releasing energy that is useful to the cell, and finishing with ethanol and carbon dioxide.

$$C_6H_{12}O_6 \longrightarrow 2CH_3CH_2OH + 2CO_2 \qquad -200 \text{ kJ (exothermic)}$$

> **COMMENT**
> The stages in fermentation are:
> Glucose, $C_6H_{12}O_6 \longrightarrow$ fructose, $C_6H_{12}O_6 \longrightarrow$ glyceraldehyde, $CH_2OH-CHOH-CHO \longrightarrow$ pyruvic acid, $CH_3-CO-CO_2H \longrightarrow$ ethanal, CH_3-CHO, and carbon dioxide $\longrightarrow$ ethanol, CH_3-CH_2OH
>
> The names are those used by biochemists and are not the systematic names based on alkanes. You are not expected to learn this information.

Figure 2.18 Beer-making in the 6th dynasty; Egyptian funerary model

EXPERIMENT 2.6a Making ginger beer

Here is an old recipe for making ginger beer which you might try at home. If you do, you will produce a slightly alcoholic drink with about 2% ethanol by volume. Regulations do not allow this experiment to be carried out in a chemistry laboratory if we want to taste the product.

Procedure

Squeeze the juice from the two lemons, pour it into your fermenting vessel, then add the sugar, cream of tartar and bruised ginger root. Pour six pints of boiling water onto the ingredients, stirring well. Make up to two gallons with cool water.

When it has cooled to 60–70 °F, take a little of the mixture in a teacup and cream the yeast. Float the piece of toast on top of the liquid in the fermenting vessel and pour the yeast mixture onto it.

Ferment at an even temperature of 60–70 °F. After 24 hours skim and bottle in plastic bottles, using a fine strainer. Use corks, not screw caps. It will be ready for drinking in three days. Check the corks and release any excess pressure building up in the first two days.

Ingredients
2 gallons of water
1½ pounds sugar
1 oz bruised root ginger
2 lemons
½ oz cream of tartar
½ oz dried yeast
1 slice toast

SAFETY ⚠

Use corks, not screw caps. Screw caps on fizzy drinks bottles have caused serious injuries when they have shot off the bottle. **Always point the top away from your face when opening a bottle, especially when the cork is tight.**

QUESTIONS

1a Convert the units used in the recipe into the usual laboratory units (grams, dm^3 and degrees Celsius). Use the conversion table on pages 12–15 in the *Book of data*.

b Rewrite the list of ingredients to include the systematic chemical names for table sugar, and cream of tartar. You will need a reference book.

2a Calculate the maximum concentration (in mol per dm^3) you could expect if all the sugar fermented to alcohol (1 mol sucrose ferments to 4 mol ethanol).

b Convert your answer from **2a** into a concentration in cm^3 of ethanol per 100 cm^3 and compare this with the expected 2% ginger beer which this procedure yields.

HINTS FOR QUESTION 2

Suppose that the mass of sucrose is W g in a volume V dm^3 of solution and that the molar mass of sucrose is M g mol^{-1}.

$$\text{Amount of sucrose (in mol)} = \frac{W \text{ (in g)}}{M \text{ (in g mol}^{-1})}$$

$$\text{Concentration of sucrose (in mol dm}^{-3}) = \frac{\text{amount of sucrose (in mol)}}{\text{volume of solution (in dm}^3)}$$

$$\text{Volume of ethanol (in cm}^3) = \frac{\text{mass of ethanol (in g)}}{\text{density of ethanol (in g cm}^{-3})}$$

EXPERIMENT 2.6b

Remember that regulations do not allow this experiment to be done in a chemistry laboratory.

The flavour of ginger beer

Almost every book with a recipe for ginger beer lists different proportions of ingredients. When you are making a compound such variation is not sensible but if you prefer your ginger beer sweeter or spicier then changing the recipe is a good idea.

Plan and carry out a study of the effect of varying the amounts of the ingredients on the flavour of ginger beer. Decide which ingredients should be kept constant and which varied. 1.2 dm^3 of water is suitable for filling a 1.5 dm^3 plastic fizzy drinks bottle.

A clean cooking bowl, covered with a cloth, is suitable for the initial fermentation but the piece of toast in the previous recipe appears to be unnecessary.

Summary

Now you have finished this Topic you need to organize the basic ideas into a form that will make it easy to learn them; when you have done that you will be ready to add extra details.

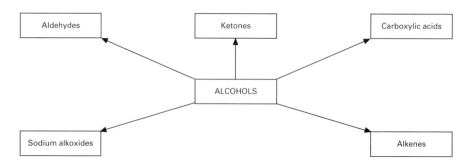

We suggest you draw a chart in the form of an illustrated poster (with plenty of colour) to show the reactions of ethanol, the products and their formulae. Leave the method and conditions used to a second more detailed chart.

At the end of this Topic you should be able to:
a demonstrate understanding of:
   i the nomenclature and corresponding structural and displayed formulae for alcohols
   ii the following terms as associated with the structure of organic molecules: molecular formula; structural formula; displayed formula; structural isomer; functional group
   iii the following terms as associated with organic reactions: redox and elimination reactions
b recall the major properties associated with covalent bonding
c interpret the physical properties of alcohols as characteristic of covalent compounds

(*continued*)

d recall the typical behaviour of alcohols, limited to:
combustion
solubility in water
treatment with sodium
oxidation to carboxylic acids, aldehydes and ketones
dehydration to alkenes
production by fermentation
e recall the processes of simple and steam distillation
f recall the typical behaviour of aldehydes and ketones, limited to:
combustion
production from alcohols by oxidation
oxidation of aldehydes to carboxylic acids
results of testing with Benedict's solution
g evaluate information by extraction from the text and the *Book of data* about the properties and uses of alcohols.

Review questions

2.1 Draw structural formulae for the following:

a Propan-2-ol
b Hexan-1-ol
c Propanal
d Pentan-3-one.

2.2 State the meaning of the following terms:

a Alkane
b Functional group
c Displayed formula
d Elimination reaction.

2.3 Calculate the molar mass of the following compounds:

a $CH_3CH_2CH_2CH(OH)CH_2CH_3$
b $CH_3CH=CHCHClCH_2OH$
c $CH_3COCH_2CH_3$
d $CH_3CH_2CH_2CH_2CHO$.

2.4 The following substances all have the same molecular formula $C_7H_{15}OH$.

A $CH_3-\underset{\underset{CH_3}{|}}{\overset{\overset{H}{|}}{C}}-\underset{\underset{H}{|}}{\overset{\overset{CH_3}{|}}{C}}-\underset{\underset{H}{|}}{\overset{\overset{H}{|}}{C}}-\underset{\underset{H}{|}}{\overset{\overset{H}{|}}{C}}-OH$

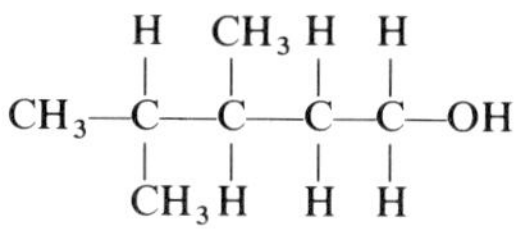

B $H-\underset{\underset{CH_3}{|}}{\overset{\overset{CH_3}{|}}{C}}-\underset{\underset{H}{|}}{\overset{\overset{H}{|}}{C}}-\underset{\underset{CH_3}{|}}{\overset{\overset{OH}{|}}{C}}-CH_3$

C $CH_3-\underset{\underset{CH_3}{|}}{\overset{\overset{H}{|}}{C}}-\underset{\underset{CH_3}{|}}{\overset{\overset{OH}{|}}{C}}-C_2H_5$

D $CH_3-\underset{\underset{CH_2CH_3}{|}}{\overset{\overset{OH}{|}}{C}}-CH(CH_3)_2$

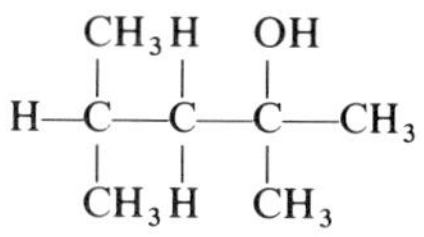

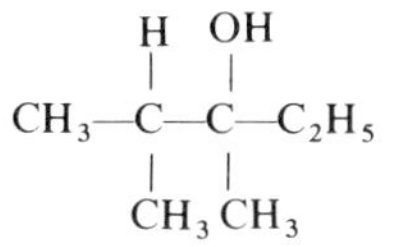

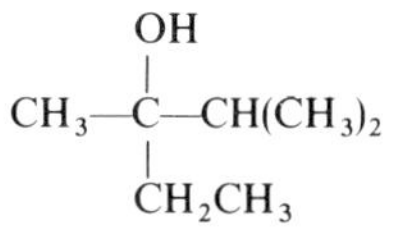

a Which substance is identical with **C**?
b Which substances are tertiary alcohols?
c Which substance is 2,4-dimethylpentan-2-ol?
d Which substance could be oxidized to an aldehyde?
e Which substance could be oxidized to a carboxylic acid containing the same number of carbon atoms?

2.5 In an experiment 3.7 g of butan-1-ol were heated with excess concentrated phosphoric acid. The main product of the reaction was obtained in a yield of 1.8 g after purification.

a Name the product and write an equation for the reaction.
b Calculate the percentage yield of the product by comparing the actual yield with the maximum theoretically obtainable.
c Name the function of the concentrated phosphoric acid in the reaction.

2.6 When a substance of the formula

$$CH_3-CH_2-CH(OH)-CH_2-CH_2-CH_3$$

is vaporized and passed over heated aluminium oxide granules, one mole of water is eliminated from each mole of the original substance.

a Name the original substance.
b Deduce the formula of each of the two possible isomers which result from the reaction given above.
c One mole of each of the isomers in **b** reacts with one mole of bromine molecules; give the formula of each of the products.

Examination questions

2.7 Complete the following table of reactions of ethanol, filling in the blank spaces.

Formulae of reagents added to ethanol	Reaction type	Formula of organic product	Name of organic product
			sodium ethoxide
		$CH_2=CH_2$	
	mild oxidation		

2.8 Lavender has been grown commercially in Norfolk for the past sixty years to produce perfume. After harvesting, the blooms are dried and the lavender oil extracted by steam distillation.

a Draw a labelled diagram of the apparatus you would use to carry out a steam distillation in the laboratory.

b The distillate is transferred to a separating funnel and shaken with a hydrocarbon. After separation, the hydrocarbon layer is run into a conical flask and a few spatula measures of anhydrous sodium sulphate are added. The mixture is then filtered.
Explain the reasons for using **i** a hydrocarbon solvent
 ii anhydrous sodium sulphate

c One of the substances present in lavender oil is linalool, $C_{10}H_{18}O$.

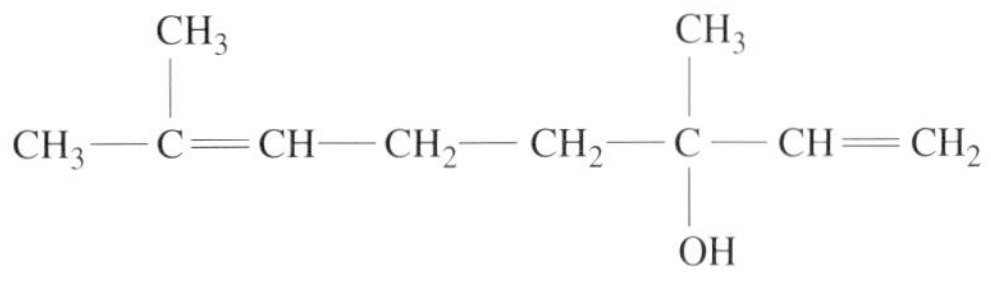

$$CH_3-C(CH_3)=CH-CH_2-CH_2-C(CH_3)(OH)-CH=CH_2$$

linalool

Draw the structural formula of the main organic product formed from linalool in each of the following reactions:

i With sodium, Na
ii With bromine, Br_2
iii With alumina, Al_2O_3, at 400 °C.

d Dried lavender flowers only contain a very small proportion of lavender oil. Chemists are often required to find a suitable substitute for naturally occurring plant oils. Suggest three factors which must be taken into account when seeking a suitable substitute.

2.9 Substances known as thio-compounds occur in both inorganic and organic chemistry, the prefix thio- implying the presence of a sulphur atom in place of an oxygen atom in the compounds. One example of this is seen in the thio-alcohols, many of whose reactions are similar to those of the corresponding alcohols.
The questions which follow are about thioethanol, C_2H_5SH.
Assume that its reactions are similiar to those of ethanol.

a Write a balanced equation for the reaction of thioethanol with sodium.
b Give the formula of the anion formed when thioethanol reacts with sodium.
c Name the products from the complete combustion of thioethanol.

READING TASK 1

THE CARBON CYCLE

Carbon makes up less than one per cent of our planet, but it is the key element for life on Earth. Plants, animals, microorganisms, our food and our bodies are all based on compounds of this versatile element – and carbon compounds in the atmosphere made the planet warm enough for life to evolve. Today carbon-based fossil fuels provide three-quarters of our energy needs, but this may change in future. Scientists are concerned that our dependence on fossil fuels is distorting the natural movement of carbon through the world's ecosystems – known as the **carbon cycle**. The common currency of the carbon cycle is the gas carbon dioxide. Although it makes up only 0.03 per cent of the atmosphere, this gas is the source of most carbon for living things as well as the product of burning fuel or decomposing organic matter.

The idea of a global carbon cycle was first proposed by the German chemist Justus von Liebig in 1840. During the nineteenth century, scientists such as the great French mathematician Jean–Baptiste Fourier and the British physicist John Tyndall contributed greatly to our knowledge of how carbon dioxide keeps the Earth warm. The Swedish chemist Svante Arrhenius predicted today's concern about the 'greenhouse effect' as long ago as 1896.

The carbon cycle can be divided into three parts, the biological part, the geochemical part and carbon in soil. Each part consists of carbon pools, that store carbon for different 'residence times'. The **biological part** of the cycle has the shortest storage time. Plants, the atmosphere and the surface layers of the oceans each contain approximately equal amounts of carbon – between about 500 and 700 billion tonnes. Animals, ourselves included, hold far less – between 1 and 2 billion tonnes. As organisms grow, die and decompose, over days, years, or even hundreds of years, carbon moves between these pools. Pools are known as sinks or sources, depending on whether they are net 'uptakers' or 'providers' of carbon. But in spite of these large

Figure R1.1 In the carbon cycle carbon moves between the atmosphere, plants and animals, the oceans and rocks

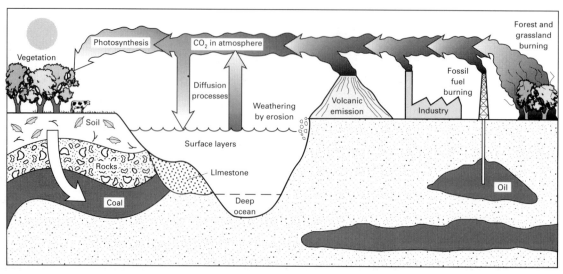

and fairly rapid movements of carbon, on a global scale the system is in dynamic equilibrium, so the pools remain about the same size.

The **geochemical part** of the carbon cycle stores carbon for far longer. There are two main pools: the deep ocean water, with an estimated 36 000 billion tonnes, and rocks, especially limestone, which hold an estimated 75 000 000 billion tonnes. Carbon moves to other parts of the cycle very gradually when deep ocean currents surface again after hundreds or thousands of years. Volcanic eruptions, weathering of rocks and burning fossil fuels such as oil, coal and gas all release carbon to the biological cycle, usually by way of the atmosphere, after many millions of years underground.

Intermediate between biological and geochemical carbon in terms of its 'cycling time' is the **carbon in soil**. Worldwide this amounts to about 1500 billion tonnes, roughly twice as much as is held in plants. Soil carbon is fairly stable and does not normally interchange very quickly with the other pools.

Plants are the major force behind the global carbon cycle. Through photosynthesis they convert carbon dioxide into their stems, trunks, leaves and roots. By fixing carbon in this way, the plants themselves grow, and carbon then enters the food chain as animals eat the plants. About twice as much carbon dioxide is taken up by photosynthesis on land as in the oceans and rivers. Land plants contain 250 times as much carbon as aquatic plants, most of it in the form of trees.

Both plants and blue-green algae are able to make their own organic food by photosynthesis, converting solar energy into chemical energy in carbohydrate molecules. Most organisms that can make their own food use photosynthesis, but a few live on sulphur or nitrogen compounds. All other organisms rely on finding their food ready-made, as sugars or other carbon compounds. However they feed, once plants and animals die, the carbon they have stored is normally released back into the atmosphere by decomposition by other organisms that feed on their remains.

Figure R1.2 The arrows show how many billons of tonnes of carbon move between these pools every year

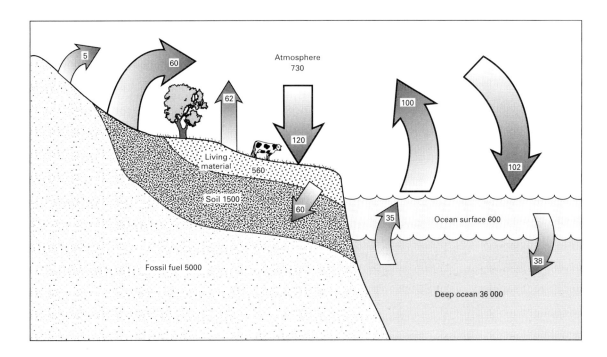

Carbon turnover

Carbon is constantly moving from one pool to another. Carbon dioxide from the atmosphere is taken in by plants to form carbohydrates, and some of this carbon then passes to animals. In the oceans, most of the carbon is in the debris left behind by living creatures, and as dissolved carbon dioxide. Plankton, coral and other creatures grow shells and skeletons of calcite (calcium carbonate) using the hydrogencarbonate ions HCO_3^- in sea-water. When they die, their shells and skeletons accumulate on the ocean floor, eventually to become limestone. This removes carbon from the sea until over millions of years the processes of mountain building expose the rocks on land.

Other rocks also lock away carbon, notably those that produce fossil fuels. This carbon too originates in living things – plants provide the carbon for coal, whereas oil and gas come mainly from marine animals.

The flow of carbon is not just into the geochemical part of the cycle. Weathering, the slow breakdown of rocks by organic acids in soil, results in release of carbon dioxide to the atmosphere from carbonates. Carbon also comes into the atmosphere when volcanoes erupt.

The greenhouse effect

Over the past 150 years, since the Industrial Revolution in the West, we have added carbon to the atmosphere from two main sources. First by burning fossil fuels, and, secondly, by changes in land use, especially the destruction of forests. The rate of carbon dioxide emission from both these sources has increased dramatically since 1950. The growth in demand for energy in both developed and developing countries has had the most effect. Coal is a mixture of complex hydrocarbons with a high content of carbon. Oil contains less carbon and more hydrogen, and natural gas (methane) has four hydrogen atoms for every carbon atom. Therefore combustion of coal produces more carbon dioxide for a given amount of energy released than the burning of oil, and natural gas produces the least carbon dioxide released into the atmosphere. The burning of wood and biomass fuels, which are mainly carbohydrates, releases carbon dioxide, but if more trees are planted to replace those burned, combustion of wood need not add to the carbon dioxide in the atmosphere.

We know how much carbon is released from the burning of fossil fuel because the coal, oil and gas industries keep records; more than five billion tonnes of carbon goes up in smoke each year, and the total is increasing. Because oceans and plants take in carbon naturally, the increase in carbon in the atmosphere is in fact only about half the value predicted from these figures. But we need to understand much more about the role of the world's many ecosystems in the global carbon cycle, as well as the effects of human disturbance. Scientists remain uncertain about the turnover of carbon world-wide, especially the contribution from changes to forests, grasslands and agriculture. When farmers convert forest to cropland, the carbon in the vegetation is oxidized through burning and decomposition, adding carbon dioxide to the atmosphere. Carbon in organic matter in soil may also be oxidized and lost to the atmosphere as cultivation continues, especially if the soil erodes at the same time. The effect is reversed when trees and forest grow again on abandoned land, but no-one is sure how much regeneration is going on.

Burning vegetation plays an important part in the global carbon cycle, although fire often merely speeds up the release of carbon that would otherwise be produced later by decomposition. Many people in developing countries use wood or agricultural waste as fuel, and both forests and grasslands are burned for farming. Vast areas of grasslands are cleared and burned annually, especially in the tropics, giving a large seasonal flow of carbon to the atmosphere. Some scientists claim that it is as large as the amount of carbon released every year by felling and burning tropical forests. But unlike forests, grasslands can recover quickly provided they are carefully managed.

If we carry on burning fossil fuel and destroying vegetation on the Earth's surface at current rates, the carbon dioxide in the atmosphere could reach three times today's levels by the

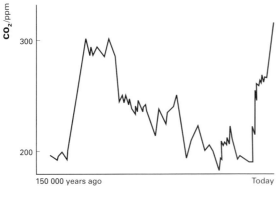

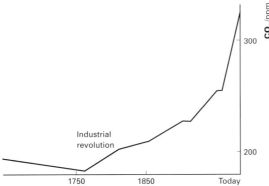

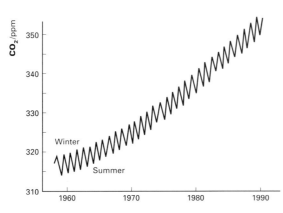

Figure R1.3 Change in carbon dioxide levels in the atmosphere in the past 150 000, 350 and 30 years, in parts per million. The annual fluctations are mainly due to photosynthesis in the Northern Hemisphere.

year 2100. Whenever politicians agree to act, the concentration of carbon dioxide may still double. It will take major changes in the use of energy world-wide, and in land management, to keep the carbon in the atmosphere at a steady and acceptable level. Every kilowatt-hour of electricity used in Britain today releases 1 kilogram of carbon dioxide, and each litre of petrol and diesel burned by private and public transport adds another 2.5 kilograms. And every hectare of tropical forest burned gives between 350 and 700 tonnes of carbon dioxide.

Questions

Reread the passage and answer these questions.

1 Find the paragraph that describes how carbon dioxide in the oceans eventually becomes part of the deposits on the ocean floor. Identify the stages involved in changing dissolved carbon dioxide into solid calcium carbonate.

What does the carbon dioxide react with in the first stage? What ions are formed as a result?

Consult the *Book of data* to check which elements are present in sea-water. Which substances must be involved in the reactions that produce the shells and skeletons of sea-creatures?

2 The article states that about half of the carbon dioxide produced by burning fossil fuels stays in the atmosphere. Assume that fossil fuels continue to be burned at the present rate for the next 100 years. Estimate the extra mass of carbon added to the atmosphere in this time.

The carbon in the atmosphere today is estimated to be 730 billion tonnes. How does the value you have calculated compare with that figure?

3 The article gives estimates of the sizes of the carbon pools, and of the movement of carbon between the pools. How reliable do you think these estimates are? How might these estimates have been made?

4 Do you think the 'greenhouse effect' is a serious problem? What changes are you prepared to accept in the way you live to prevent the problem getting worse?

TOPIC 3

Atoms, ions and acids

Many of the properties of compounds can be interpreted by simple models of how the electrons are distributed in these compounds.

In this Topic we shall examine simple models of electron distribution, and the ways in which they can account for the formulae and properties of compounds, and the forces which hold them together. Forces of attraction and repulsion are involved when oppositely charged particles form giant crystal lattices. We will be considering what ions are, why we believe they exist, and how they are formed.

Finally we will look at a special ion: the proton. Protons are of particular importance because they are responsible for acid-base reactions.

3.1 The story of the atom

The idea of atoms has a long history, and many people have contributed to the development of the idea. In the past hundred years, our model of the atom has developed rapidly as our need to explain more phenomena has demanded an ever more detailed model.

There is no one correct model of the atom, and there never can be. Matter behaves 'as if' it was made of our model atoms, but our mental models only represent matter and correspond to it in the way a map corresponds to the countryside it represents. This idea is older than science, being part of the philosophies of Plato and of the Buddha; the 'apparent' world of phenomena as we experience them and the 'real' world of actual things ('the thing in itself' as the philosopher Kant put it).

Figure 3.1 Maps are redrawn as changes occur; and scientific theories have to be rewritten as our knowledge changes.

3 Atoms, ions and acids

We make use of models to help us understand why things behave as they do. A simple model of the atom helps us understand the differences between solids, liquids and gases, but we need a more detailed model to explain why chemical changes take place. The model we need for this was developed mainly in the years between 1895 and 1940. More detailed models have been developed since then, but on the whole they contribute to our understanding of the nucleus rather than chemical reactions.

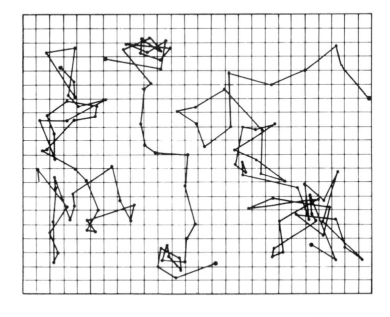

Figure 3.2 Brownian motion is good evidence for atoms. The position of the particle is marked every 5 seconds.

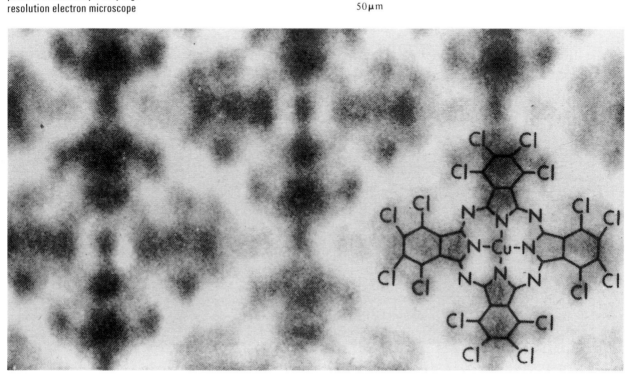

Figure 3.3 The position of individual atoms can be deduced from pictures obtained by a very high resolution electron microscope

STUDY TASK
As a member of a small group, use a range of books and other resources to study one of the stages in the development of our model of the atom. In your group find out about the people involved, their experiments and the key ideas they proposed.

Summarize your findings as a poster or OHP transparency to present to the other groups in a talk of about five minutes.
1. Democritus to Dalton, the first ideas about atoms
2. Faraday to J. J. Thomson, evidence for electrons
3. Marie Curie to Rutherford, radioactive particles and the proton
4. Geiger and Rutherford, the idea of the nuclear atom
5. Moseley, the idea of atomic number
6. Chadwick to Lise Meitner, the neutron and atomic fission

You will find collected biographies helpful, like Asimov's *Biographical Encyclopaedia of Science and Technology*. And check whether your library has a CD-ROM that is appropriate.

3.2 Flame colours and emission spectra

One way to study the arrangement of electrons in atoms is to disturb the electrons and then observe what happens as they go back to their original arrangement. A good way to do this is to heat compounds in a Bunsen burner flame and study the characteristic colours produced. These colours are caused by electrons losing the energy that they have gained from the Bunsen burner flame. The next experiment will enable you to gather some information about these colours.

EXPERIMENT 3.2 The flame colours of the elements of Groups 1 and 2

All compounds of a particular element give the same flame colour, but the chlorides are the best to use because they usually vaporize relatively easily in a Bunsen burner flame.

You can see the patterns in the spectra by looking at flame colours with a hand spectroscope.

SAFETY ⚠
Hydrochloric acid is corrosive, barium salts are harmful.

Procedure
Clean a platinum or nichrome wire by heating it in a non-luminous Bunsen burner flame, dipping it into a little concentrated hydrochloric acid on a watchglass (TAKE CARE) and heating it again. Continue this until the flame is not coloured by the wire.

Pour the impure acid away and take a fresh portion. Dip the clean wire into the acid and then into a small portion of powdered compound on a watchglass. Use chlorides where possible, otherwise use nitrates or carbonates. Hold the wire so that the powdered solid is in the edge of the flame and note any colour in the flame which results. The colour disappears fairly quickly but can be renewed by dipping the wire into acid and chloride again and reheating.

Observe the flame through a diffraction grating or direct vision spectroscope and look for the coloured lines that make up the spectrum.

You may need to take a fresh sample of solid for this and to do the experiment in a darkened corner of the laboratory.

When you want to examine the flame colours of other elements it is easier to exchange apparatus with other students. Cleaning your own wire thoroughly for each new sample is time consuming.

- *In your notes:* record the coloured lines that you see for each element which you examine. Use the colour plate C.2 in the *Book of data* to compare with your observations. Note the frequencies of the main spectral lines you observe for each element.
Are there any other trends you have noticed?

Interpretation of the emission spectra of elements

When atoms of an element are supplied with sufficient energy they will emit light. This energy may be provided in several ways. If the element is a gas it may be placed in an electric discharge tube at low pressure; neon signs work on this principle. Certain easily vaporized metals also emit light under these conditions; examples are the bluish-white street lamps which are mercury discharge tubes, and the yellow street lamps which are sodium discharge tubes. When the light which is emitted is examined through a spectroscope, it is found not to consist of a continuous range of colours like part of a rainbow, but to be made up of separate coloured lines.

Each element has its own characteristic set of lines, and these enable elements to be identified by examination of their spectra. Indeed, spectroscopic examination of the Sun revealed the existence of an element which at that time had not been discovered on Earth; it was named helium, from the Greek word *helios*, meaning the Sun.

This type of spectrum is known as a **line emission spectrum**. We can explain this by assuming that electrons in an atom can exist only in definite energy levels; they cannot possess energies of intermediate size. It is like a set of steps rather than a slope.

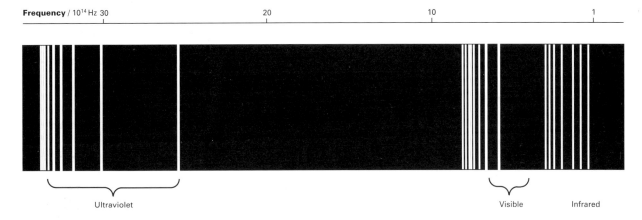

Figure 3.4 The line emission spectrum of hydrogen

Each spectral line is caused by electrons losing as light the energy they gained by being heated. The electrons fall back from a higher energy level to a lower energy level. The energy is lost as light of a frequency that depends on the amount of energy given out. The greater the difference between the energy levels, the higher will be the frequency of the corresponding spectral line.

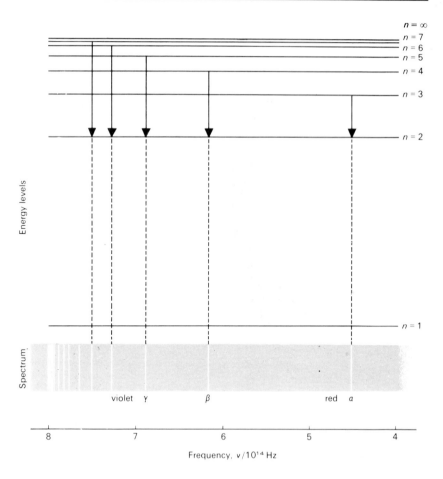

Figure 3.5 The energy levels of the hydrogen atom

In order for atoms to form positive ions, electrons must be removed completely from the atoms. This **ionization** requires the input of energy, which is known as the **ionization energy** and is given the symbol E_{mj}. It is important to know the amount of energy which is needed to remove an electron from an atom if we are to understand the energy changes involved in chemical bonding.

The first ionization energy of an element is the energy required to remove one mole of electrons from one mole of atoms of the element in the gaseous state, to form gaseous ions, and is given the symbol E_{m1}

$M(g) \longrightarrow M^+(g) + e^-$ First ionization energy E_{m1}

The values of ionization energies are given in kilojoules per mole, kJ mol^{-1}. Some values are:

$Na(g) \longrightarrow Na^+(g) + e^-$ First ionization energy = 496 kJ mol^{-1}

$Mg(g) \longrightarrow Mg^+(g) + e^-$ First ionization energy = 738 kJ mol^{-1}

$Al(g) \longrightarrow Al^+(g) + e^-$ First ionization energy = 578 kJ mol^{-1}

3.3 The arrangement of electrons in atoms

A knowledge of ionization energies provides valuable information about the arrangement of electrons within atoms.

The discussion of the ionization of an atom has so far considered the removal of one electron only; but if an atom containing several electrons is treated with sufficient vigour, then more than one electron may be removed from it. A succession of ionization energies is therefore possible. These may be determined, principally from spectroscopic measurements; a table of successive ionization energies for a number of elements is given in table 4.1 in the *Book of data*.

STUDY TASK

1 Use a spreadsheet to plot for sodium a graph of ionization energy, on the vertical axis, against the corresponding number of electrons removed, on the horizontal axis. The data you need is in table 4.1 of the *Book of data*.
   - What do you notice, first about the general trend in values, and second about their magnitude?
   - Why do you think a general increase in values occurs?

2 Now plot a graph of the logarithm (to base 10) of the ionization energy against number of electrons removed. Use the log button on your calculator or the log function in the spreadsheet.
   - Does this give any information about groups of electrons which can be removed more readily than others?
   - How many electrons are there in each group?

3 Using the same table in the *Book of data*, study the change in the first ionization energy of the elements for the first twenty elements. Plot the value of their first ionization energy, on the vertical axis, against their atomic number, on the horizontal axis. When you have plotted the points, draw lines between them to show the pattern, and label each point with the symbol for the element.
   - Where do the alkali metals lithium, sodium, and potassium appear?
   - Where do the noble gases occur?
   - Do you notice any groups of points in the pattern? How many elements are there in each section?
   - Do the numbers of elements in each section bear any relation to the numbers of electrons in the pattern you found in the successive ionization energies for sodium?

Interpretation of the patterns in the ionization energies

You should be able to see that your graph of the successive ionization energies of sodium shows that one electron needs much less energy for its removal than do the others. It must therefore be in a higher energy level relative to the nucleus; eight electrons required much more energy, and must be in a lower energy level; the last two electrons must be in a still lower energy level.

The lowest energy level is called the $n = 1$ level; the next level up is called the $n = 2$ level, and so on. Thus, for sodium there would appear to be:

2 electrons in the lowest energy level, $n = 1$
8 electrons in an intermediate energy level, $n = 2$
1 electron in the highest energy level, $n = 3$

We can represent this on an energy level diagram, as in figure 3.6.

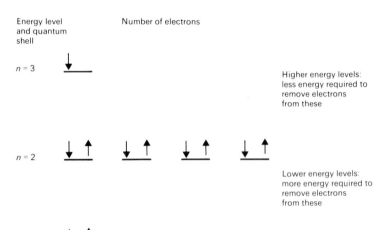

Figure 3.6 Energy levels of electrons in a sodium atom

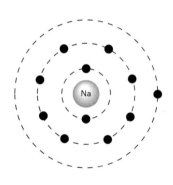

sodium atom, Na 2,8,1

Figure 3.7 Quantum shells of electrons in a sodium atom

The two electrons in the $n = 1$ energy level behave as if they are situated most of the time closer to the nucleus than the other electrons, and they are said to be in the **first quantum shell**. The eight electrons in the $n = 2$ energy level spend much of their time further from the nucleus, and are said to be in the **second quantum shell**. The single electron in sodium spends much of its time further still from the nucleus and is said to be in the **third quantum shell**. This is represented in figure 3.7.

Thus there are two ways of looking at electrons in atoms: from the point of view of their energy level, $n = 1, 2, 3, 4$, etc., and from the point of view of how far from the nucleus they are on average, that is, in the first, second, third, or fourth, etc., quantum shell.

The electrons in figure 3.6 have been represented by arrows. When an energy level is half full the next electrons pair up with existing ones. Electrons behave as though they have the property of spin, and paired electrons have their spins in opposite directions; this is represented by up and down arrows. Each electron pair occupies a separate part of the quantum shell known as an 'orbital'. The evidence for this comes from a more detailed examination of line spectra.

The successive ionization energies for potassium have a pattern in which 1 electron is most easily removed, followed by 8 which are more difficult, followed by 8 which are even more difficult, followed by 2 which are extremely difficult to remove.

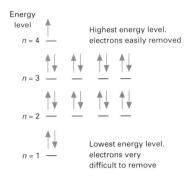

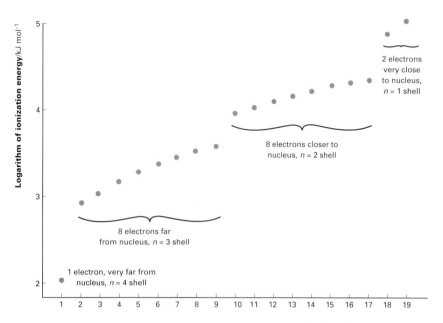

Figure 3.8 Ionization energy and energy levels of electrons in a potassium atom

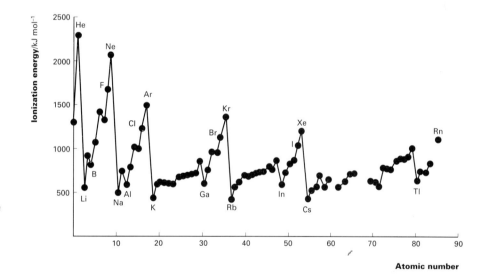

Figure 3.9 Periodic patterns in the first ionization energies of the elements

QUESTION

How many electrons do the $n = 1$, $n = 2$, $n = 3$, and $n = 4$ energy levels and quantum shells hold in potassium? Does a similar pattern show in the graph of first ionization energies which you plotted for twenty elements?

You will notice from the graph of the first ionization energies of the elements that the groups of eight are made up of groups of 2, 3, and 3 points on the curve. This indicates that the eight electrons are not all exactly the same as far as their energies are concerned. From this type of evidence, and also from studies of spectral lines, it has been concluded that the energy levels are split so that the $n = 2$ level has two electrons in a sub-level known as 2s (slightly more difficult to remove) and six electrons in a sub-level known as 2p (slightly less difficult to remove). This is shown in figure 3.10.

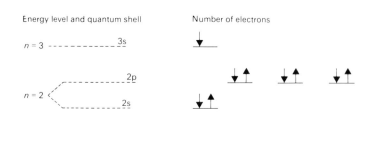

Figure 3.10 Energy levels of electrons in a sodium atom, showing sub-levels

From similar evidence, it has been concluded that the $n = 3$ level is split into s, p, and d. The $n = 4$ level adds an f sub-level and so on.

**All the s sub-levels can contain up to two electrons,
the p sub-levels six,
the d sub-levels ten,
and the f sub-levels fourteen.**

The arrangement of the $n = 1$ to $n = 4$ energy levels is shown in figure 3.11.
The atomic number of neon is 10, and its atom therefore contains 10 electrons.

2 are in the $n = 1$ level, and are in the s sub-level $1s^2$
2 are in the $n = 2$ level, and are in the s sub-level $2s^2$
6 are in the $n = 2$ level, and are in the p sub-level $2p^6$

And the **electronic structure of neon** is written: $1s^2 2s^2 2p^6$.

The next element, sodium, atomic number 11, has one electron in the $n = 3$ level, and it is in an s sub-level, since this is the lowest $n = 3$ level. The electronic structure of sodium is therefore $1s^2 2s^2 2p^6 3s^1$.

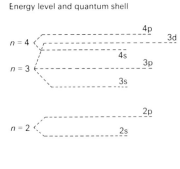

Figure 3.11 Energy levels of electrons in atoms

QUESTION
From figure 3.9 you will see that after calcium, atomic number 20, the 2, 3, 3 grouping is broken. How many elements produce the break?
- What might this indicate in terms of electrons?
- What is the name of this grouping of elements in the Periodic Table?

After calcium, a new energy level belonging to the $n = 3$ quantum shell becomes occupied; the electrons in it are known as d electrons. A d sub-level can hold 10 electrons when full. You will notice from figure 3.11 that the 3d level is just above the 4s level and is just below the 4p level; this is of great importance in the chemistry of the transition elements.

Electrons occupy the orbital with the LOWEST energy, so when the 4s orbital is full electrons are added to the 3d sub-level. The electronic structure for scandium is $1s^2 2s^2 2p^6 3s^2 3p^6 3d^1 4s^2$. Note that this is the numerical order and the sub-levels are not filled in that order in some cases.

Each major peak in the ionization energy curve in figure 3.9 represents an element with a completed quantum shell; the element concerned is a noble gas. The pattern of electronic structure is mirrored in the Periodic Table.

QUESTION
Work out the electronic structure of iron.

3 Atoms, ions and acids

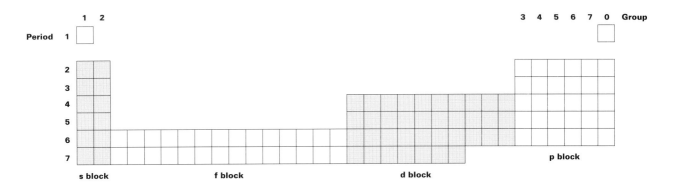

Figure 3.12 Part of the Periodic Table

STUDY TASK

Draw up a table in your notes with the headings shown below:

Elements	Atomic radius /nm	Ionic radius /nm	First ionization energy /kJ mol^{-1}

Complete the table for a short sequence of elements down Groups 1 and 2, and Groups 6 and 7. Collect enough data to be able to recognize any patterns. You will find the values you need in tables 4.4 and 4.1 in the *Book of data*. For the atomic radius, you should use values of the *metallic radius*, r_m, for Groups 1 and 2. For Groups 6 and 7 use the values quoted for the *van der Waals radius*, r_v. The van der Waals radius is half the distance between the centres of two atoms in adjacent molecules and is discussed further in Topic 9.

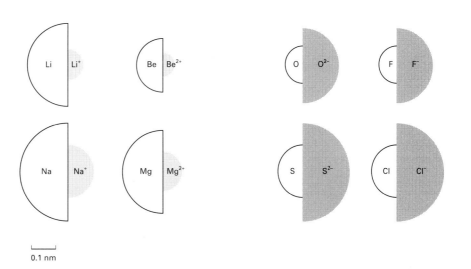

Figure 3.13 Atoms and ions drawn to scale

You may find it helpful to draw diagrams similar to those in figure 3.13. When you have completed your table, refer to it as you read the next two paragraphs and answer the questions. Write your answers in your notes after the table, in such a way as to make clear what you are answering.

3 Atoms, ions and acids 66

1 Trends in ionization energies

How does the first ionization energy change on going down a group in the Periodic Table? Suggest a reason for the patterns you find.

The chemical similarities existing among members of a group of elements arise because of the similar configurations of the outer electron shells of their atoms. Does your data support the theory that the elements in a group will have similar outer electronic shells?

2 Trends in ionic radii

How does atomic radius change down groups of the Periodic Table? How does ionic radius change in comparison to atomic radius down a group of the Periodic Table?

Suggest a reason for the patterns you have found.

3.4 Evidence for the ionic model

COMMENT

In sea water [NaCl(sea)] = 0.48 mol per dm^3 of solution

In our blood [NaCl(blood)] = 0.14 mol per dm^3 of solution

Sodium chloride is a very familiar compound but its properties reveal the importance of ion formation and ionic bonds. A violently reactive metal combines with a pungent, toxic gas and they are transformed into a safe material: fish live in its solution and our blood contains it in solution.

REVIEW TASK

From your previous science course you should be familiar with the effects of electrolysis and the evidence it provides for the existence of charged particles in solutions. In groups, list all you can recall of the properties of ionic compounds.

EXPERIMENT 3.4

SAFETY ⚠️
Copper salts and potassium manganate(VII) are harmful

COMMENT

In chemistry it is usual to quote conductance rather than resistance:

 = conductance

Resistance is measured in ohms and conductance in siemens.

Properties of ionic compounds

Since we cannot see ions and watch their behaviour, we have to rely on experimental observations to build a theoretical model to explain their properties. This is typical of chemistry – essentially an experimental science.

Procedure

1 The electrical resistance of solutions

In this experiment we shall be attempting to examine the properties of some solutions without decomposing them.

You are going to look at the electrical properties of solutions using a conductivity meter. This instrument is described in Topic 18. Use the dip cell to measure the conductance of the following solutions whose concentrations should all be $\frac{1}{10}$ mole per dm^3 of solution. Between each measurement it is important to rinse the dip cell in two separate fresh portions of **pure** water.

a Hydrochloric acid   b Ethanoic acid   c Sodium hydroxide
d Sodium chloride   e Sodium ethanoate.

■   What pattern can you see in the conductance values? Molar conductivities of individual ions are listed in table 6.4 in the *Book of data*.

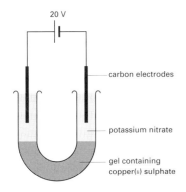

Figure 3.14

STUDY TASK

Summarize all the evidence which you have encountered for the existence of ions. Look up the melting and boiling points of some of the ionic compounds you have been working with.

2 Migration of ions

You may try either or both of the following experiments.

a Using copper(II) sulphate

Measure the volume of your U-tube and divide by four. Prepare a 'quarter volume' of 10 per cent solution of gelatin by adding gelatin to boiling water. While the gel is still hot add a 'quarter volume' of dilute copper(II) sulphate and pour the mixture into the U-tube. Cool the U-tube in ice, and when the gel has set add a concentrated solution of potassium nitrate to each limb of the U-tube. Place carbon electrodes in each limb of the U-tube and connect to a 20 V dc power supply for about 30 minutes.

- Note which electrode the coloured ion moves towards. Which ion is responsible for this colour? What does this tell you about the nature of the charge on the ion?

b Using potassium manganate(VII)

Cut a strip of chromatography paper to the width but slightly longer than a microscope slide, mark the middle with a pencil line, moisten with tap water and place on a slide. Use tweezers to place a small crystal of potassium manganate(VII) in the centre of the paper, and cover with another slide to reduce evaporation. Connect the paper to a 20 V dc power supply using crocodile clips and leave for about 20 minutes.

- Note which electrode the coloured ion moves towards. Which ion is responsible for this colour? What does this tell you about the charge on the ion?

3 Examination of ionic crystals

If you have time you could grow your own ionic crystals of salts like sodium chloride, ammonium chloride and potassium nitrate.

Examine your crystals under a light microscope and draw what you see. Try splitting some crystals of rock salt (sodium chloride) or calcite (calcium carbonate) using a penknife and tapping with a small hammer. Be sure to wear eye protection in case fragments fly up. You should find that the crystals split to regular shapes. This is due to the regular packing of ions which may be seen most clearly in a model of crystal structure.

Figure 3.15 a Ammonium chloride crystals

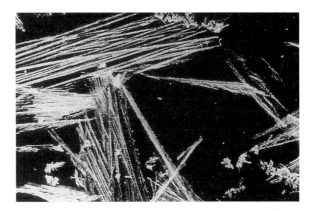

b Potassium nitrate crystals

3 **Atoms, ions and acids** 68

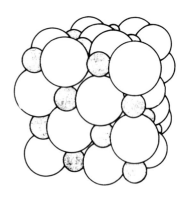

Figure 3.16 A sodium chloride crystal and a model of its ionic structure

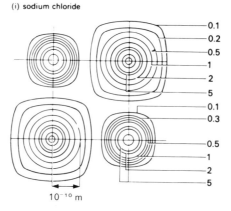

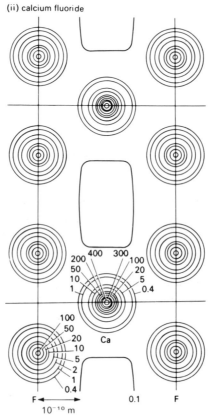

Figure 3.17 Electron density maps for (i) sodium chloride and (ii) calcium fluoride. Contours are at electrons per 10^{-30} m³.

3 Atoms, ions and acids

The shape of ions

Within a crystal, the ions are held in place by strong forces of attraction between oppositely charged ions, and by the repulsions of the like-charged ions. These forces are non-directional, unlike the forces in the covalent bond which are directional. So the ions are arranged in giant crystal lattices, in which attractive and repulsive forces are balanced.

When building models of ionic structures, those ions which are formed from single atoms are usually represented as spheres. You might think that as ions possess a complete outer electron shell, the electron distribution is spherical. There is some experimental evidence for this view, based on a study of electron density maps, obtained by X-ray diffraction measurements. X-ray diffraction is described in Topic 18.

Figure 3.17 shows electron density maps for sodium chloride and calcium fluoride. Do the maps suggest that these ions are discrete entities? Are the ions spherical?

In Topic 4 we will look at how the spherical model does not apply to large anions combining with small, densely charged cations.

STUDY TASK

Try to measure the radius of a sodium ion and of a chloride ion using figure 3.17. What difficulty is involved? Suggest one method of overcoming it.
What do you think limits the accuracy of ionic radii found from electron density maps?
What distance can you measure more accurately using the electron density map?
This task is designed to introduce you to the problem of deciding exactly how large ions are.

Electron arrangements in ions

Typical cations
Li^+ Na^+ K^+
Mg^{2+} Ca^{2+} Ba^{2+}
Al^{3+}

Typical anions
F^- Cl^- Br^- I^-
O^{2-} S^{2-}
N^{3-} (uncommon)

We know from their typical properties of electrical conductance and high melting and boiling points that most compounds formed between metallic and non-metallic elements are ionic, with the metals forming positively charged cations and the non-metals forming negatively charged anions.

The diagrams in figure 3.18 show how the **transfer of electrons** from one atom to another gives ions. In these diagrams the nucleus of each atom is represented by its symbol, and the shells of electrons are represented by groups of dots and crosses around the nucleus. The shell of lowest energy is nearest to the nucleus, and successively higher energy levels are shown at increasing distances.

Notice that the electronic structures of the ions that are formed are identical to those of a noble gas. Copy the diagrams into your notes and write in the names of the noble gases whose structures are formed in these examples.

3 Atoms, ions and acids

COMMENT
Although dots and crosses are used for the electrons in different atoms, it should not be thought that these electrons are distinguishable; this is merely a device used in the diagrams to enable their transfers to be followed.

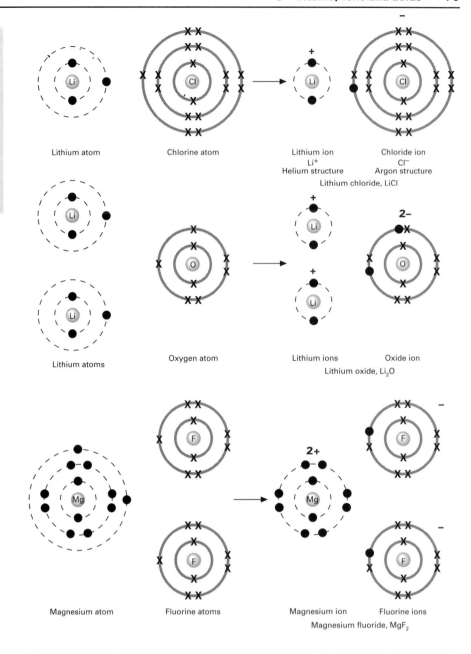

Figure 3.18 Transferring electrons to form ions

QUESTIONS
Draw similar diagrams to show how the following compounds might be formed by electron transfer: sodium fluoride, magnesium oxide, calcium chloride, aluminium oxide. Try to draw barium iodide. Add the names of the noble gases whose structures are formed in your structures.

You will notice that making up the **noble gas structure** leads to the experimentally determined empirical formula.

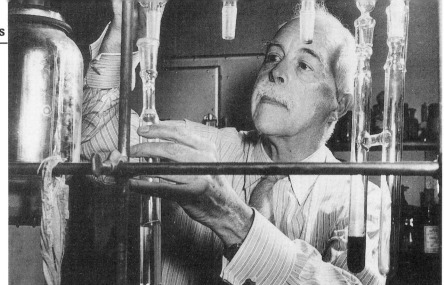

Figure 3.19 The proposal that ions had a 'noble gas' electronic structure was put forward by G. N. Lewis in 1916

You should remember when writing 'dot-and-cross' diagrams that the dots and crosses are a means of counting electrons, and showing the number present; they do not show the positions of the electrons. The electrons are distributed in space as diffuse negative charge-clouds.

Note also that not all ions have a noble gas structure, for example, the ions formed by d-block elements: the Cu^{2+} ion has an electronic structure

$1s^2 2s^2 2p^6 3s^2 3p^6 3d^9$

In conclusion, the electron transfer model, leading to the formation of ions with electronic structures the same as those of noble gases, accounts for the formulae of ionic compounds, the charges on them, and the non-directional electrostatic forces which hold them together in giant lattices.

Figure 3.20 Producing sulphuric trioxide for the manufacture of sulphuric acid (Rocksavage)

3.5 Acids and bases

REVIEW TASK
Working in groups, make a list of the typical properties of acids and bases, for example hydrochloric acid and sodium hydroxide.

Acids and bases account for eight of the top fifteen chemicals that are manufactured in large amounts. You should already have some knowledge of the behaviour of acids and bases and in this section we will be looking at a definition of wide application.

EXPERIMENT 3.5

What is an acid?

These experiments are intended to raise questions about the use of the term 'acid' and lead to a useful method of interpreting acid-base reactions.

Procedure

1 Use a selection of indicators to check the neutralization of acids by bases. Suitable **dilute** solutions to test are acids such as hydrochloric acid, sulphuric acid and ethanoic acid, and bases such as sodium hydroxide, calcium hydroxide and ammonia. Remember that even when dilute some acids are corrosive or irritant to your skin.

- Write equations for at least three of the reactions. Then convert the formulae into ionic formulae and rewrite the equations leaving out all the ions that are not affected by the neutralization process. Can you reduce your equations to a standard pattern?

> **COMMENT**
> Ions that are present in reagents but are not altered by a reaction are known as 'spectator ions'. In this reaction
>
> $$Na_2CO_3(aq) + 2HCl(aq) \longrightarrow 2NaCl(aq) + H_2O(l) + CO_2(g)$$
>
> the ionic form is (leaving out the state symbols for simplicity)
>
> $$\mathbf{2Na^+} + CO_3^{2-} + 2H^+ + \mathbf{2Cl^-} \longrightarrow \mathbf{2Na^+} + \mathbf{2Cl^-} + H_2O + CO_2$$
>
> with the spectator ions in bold type.

2 Test samples of a solution of hydrogen chloride in a dry hydrocarbon solvent with magnesium metal, anhydrous sodium carbonate, and dry indicator paper; test for electrical conductivity.
 Add the same small amount of water to each sample and notice any differences in reactivity.

3 Allow ammonia gas to drift across a small basin containing hydrogen chloride in hydrocarbon solvent (TAKE CARE – irritant gases).

- Write an equation for the reaction between the two gases. How can hydrogen ions be involved in this reaction?

> **HINT**
> To obtain carbonic acid solution exhale through a sample of water using a drinking straw. Throw away the straw as soon as you have used it.

4 Measure the pH of some aqueous solutions of acids and bases of concentration $\frac{1}{10}$ mole per dm^3 of solution, using indicators or a pH meter. Suitable acids include carbonic, ethanoic, phosphoric, and hydrochloric acids. Suitable bases include sodium hydroxide, calcium hydroxide, ammonia and urea.

- The solutions of acids and bases have the same amount of substance in solution, so why do these solutions have such varied pH values?

> **QUESTION**
> Arrhenius proposed that 'acids' are substances that dissolve in water forming hydrogen ions. Does this description cover all we know about acids?

Interpretation of acid–base reactions

You should have found that most of the neutralization reactions between acids and bases in aqueous solution follow a common pattern

acid + base ⟶ salt + water

and can be summarized in one ionic equation

$$H^+(aq) + OH^-(aq) \longrightarrow H_2O(l)$$

This common pattern of behaviour was recognised by Arrhenius and his definition of acids and bases was

> An *Arrhenius acid* is a compound containing hydrogen which will form hydrogen ions in water.
>
> An *Arrhenius base* is a compound which will form hydroxide ions in water.

> **COMMENT**
> Not all the hydrogen atoms in a compound need be acidic: they are in H_2CO_3, H_3PO_4, but only one is acidic in CH_3CO_2H and none are acidic in ethanol, CH_3CH_2OH.
> Bases need not contain hydroxide groups, but they do produce hydroxide ions when they react with water:
>
> $CaO(s) + H_2O(l) \longrightarrow Ca^{2+}(aq) + 2OH^-(aq)$
> $NH_3(g) + H_2O(l) \longrightarrow NH_4^+(aq) + OH^-(aq)$

However the experiments have shown that the Arrhenius definitions are limited to behaviour in water. A better definition would be based on features that are common to reactions between gases and reactions in solvents other than water, for example,

$$\overset{\longrightarrow H^+ \longrightarrow}{H{-}Cl(g) + NH_3(g)} \longrightarrow NH_4^+ . Cl^-(s)$$

and even reactions we have not thought of as acid–base reactions:

$$\overset{\longrightarrow H^+ \longrightarrow}{H{-}OH(l) + NH_3(g)} \longrightarrow NH_4^+(aq) + OH^-(aq)$$

Figure 3.21 The Swedish chemist Svante Arrhenius proposed in 1884 that acids are substances that form hydrogen ions in water.

This approach led to a theory of acid–base behaviour that was put forward independently in 1923 by the Danish chemist Brønsted and the British chemist Lowry. In their theory, an acid is a substance which can provide protons (hydrogen ions) in a reaction; a base is a substance which can combine with protons.

An acid is a proton donor; a base is a proton acceptor.

In the reaction of hydrogen chloride gas with water,

$$\underset{\text{acid}}{HCl(g)} + \underset{\text{base}}{H_2O(l)} \xrightarrow{H^+} Cl^-(aq) + H_3O^+(aq)$$

the water behaves as a base by accepting a proton. The H_3O^+ ion is called the **hydroxonium ion**.

But water can also act as an acid. In aqueous ammonia, for example, the water is donating a proton in the acid–base reaction.

$$\underset{\text{acid}}{H_2O(l)} + \underset{\text{base}}{NH_3(aq)} \xrightarrow{H^+} OH^-(aq) + NH_4^+(aq)$$

It is this property that gives water its special importance in acid–base behaviour.

We can say that acid–base reactions are essentially a competition for protons between two bases.

Strong acids and bases

You will have seen in the experiments that the pH of solutions is not directly related to the number of protons that the molecules can accept or donate; if that was the case H_3PO_4 would have a smaller value pH than HCl.

The pH of a solution depends on the extent to which the available protons are actually transferred to water forming hydroxonium ions

$$HCl(g) + H_2O(l) \xrightarrow{100\%} Cl^-(aq) + H_3O^+(aq)$$

or

$$CH_3CH_2CO_2H(l) + H_2O(l) \xrightarrow{\text{less than 1\%}} CH_3CH_2CO_2^-(aq) + H_3O^+(aq)$$

Acids that donate their acidic protons almost completely to water are classified as **strong acids**; acids that interact very little with water are classified as **weak acids**. There is no sharp dividing line between strong and weak acids but there is a continuous spectrum of acid strength (figure 3.22).

In Topic 11 we shall look at the quantitative basis of the Brønsted–Lowry theory.

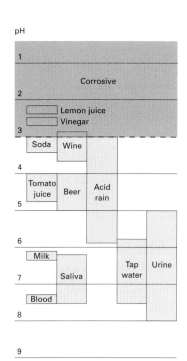

Figure 3.22 pH values of strong and weak acids and alkalis

3.6 The chemists' toolkit: solution concentration

In Topic 1 we showed you how to work out the molar masses of compounds and how to calculate the quantities to mix when doing an experiment. However many reactions are best carried out by mixing solutions and some compounds, like ammonia and hydrogen chloride, are only readily available as solutions.

For chemists the most useful information about a solution is how many moles of a compound there are per cubic decimetre of solution.

For sodium chloride a concentration of 1 mole in 1 cubic decimetre of solution is written as

$$[NaCl] = 1 \text{ mol dm}^{-3} \text{ or } 1 \text{ M}$$

Notice that square brackets, [ ], are used to identify the entity whose concentration is being recorded.

When we measure the sodium chloride very accurately as 58.44 g, dissolve the crystals in water and then add water until we have exactly 1.000 dm³ of solution the concentration can be written

$$[NaCl] = 1.000 \text{ mol dm}^{-3} \text{ or } 1.000 \text{ M}$$

To prepare solutions of other concentrations, you work out the amounts in proportion:

For   1 dm³ of solution of 1 M NaCl use 1 mol   ≡ 58.5 g
for   5 dm³   1 M   use 5 mol   ≡ (5 × 58.5) g
and for 100 cm³   1 M   use $\frac{1}{10}$ mol ≡ ($\frac{1}{10}$ × 58.5) g

The general expression is

$$\text{amount of substance (mol)} = \frac{\text{volume (cm}^3\text{)}}{1000} \times \text{concentration (mol dm}^{-3}\text{)}$$

Because sodium chloride consists of the entities Na^+ and Cl^-, when

$$[NaCl] = 1 \text{ M}$$

it is also correct to write

$$[Na^+] = [Cl^-] = 1 \text{ M}$$

For sulphuric acid, when

$$[H_2SO_4] = 1 \text{ M}$$

the correct way to write the concentrations is

$$[H^+] = 2 \text{ M and } [SO_4^{2-}] = 1 \text{ M}$$

When you want to react two solutions together you have to calculate the volumes to mix. Consider this problem: what volume of 2 M sulphuric acid do you need in order to neutralize 100 cm³ of 2 M sodium hydroxide?

Figure 3.23 Preparing a solution of known concentration

Molar masses:
Na = 22.99 g mol^{-1}
Cl = 35.45 g mol^{-1}

COMMENT

Notice that the **volume of solution** measures **1 dm³**; solutions are NOT prepared by weighing out the required amount of solid and adding 1 dm³ of liquid.

First you need to write a balanced equation for the reaction:

sodium hydroxide and **sulphuric acid** $\longrightarrow$ **sodium sulphate** and **water**

Write down the formulae:

$\quad$ **NaOH** $\quad$ and $\quad$ **H$_2$SO$_4$** $\quad$ $\longrightarrow$ $\quad$ **Na$_2$SO$_4$** $\quad$ and $\quad$ **H$_2$O**

Balance the equation:

$\quad$ **2NaOH(aq) + H$_2$SO$_4$(aq) $\longrightarrow$ Na$_2$SO$_4$(l) + 2H$_2$O(aq)**

So: **2 moles of sodium hydroxide react with 1 mole of sulphuric acid**.

Since a reaction always takes place in the ratios shown in its equation we can deduce a general expression. For a reaction of A with B:

$$\frac{\text{Amount of A from the equation (mol)}}{\text{Amount of B from the equation (mol)}} = \frac{\text{Amount of A in solution}}{\text{Amount of B in solution}}$$

$$= \frac{\text{Volume (cm}^3/1000) \times \text{Concentration of A (mol dm}^{-3})}{\text{Volume (cm}^3/1000) \times \text{Concentration of B (mol dm}^{-3})}$$

Use this relationship to confirm that the volume needed for the neutralization of 100 cm^3 2 M sodium hydroxide is 50 cm^3 of 2 M sulphuric acid.

> **COMMENT**
> You should not write 2[H$^+$] or [2H$^+$], and definitely not [H$_2^+$] = 2 M, because there is no such ion as H$_2^+$.
> [H$^+$] records the hydrogen ion concentration wherever they come from. We would still use [H$^+$] for a mixture of [HCl] and [H$_2$SO$_4$].

STUDY TASK

1. Write out the concentrations of the separate ions in a 1 M solution of phosphoric acid H$_3$PO$_4$ (for this calculation assume the acid is fully ionized). What are the concentrations in a 0.1 M solution?
2. How many moles of phosphoric acid must you measure out for
   a. 5 dm^3 of a 1 M solution
   b. 100 cm^3 of a 1 M solution
   c. 100 cm^3 of a 0.1 M solution?
3. What volume of 2 M sodium hydroxide should neutralize all the acidic hydrogen in 100 cm^3 of 0.1 M phosphoric acid?

COMMENT

In laboratories concentrations are usually written using the symbol M. Here are some typical reagent concentrations:

Reagent	Concentration
Concentrated sulphuric acid	[H$_2$SO$_4$] = 18 M
Concentrated nitric acid	[HNO$_3$] = 16 M
Concentrated hydrochloric acid	[HCl] = 10 M
Dilute acids, e.g. hydrochloric acid	[HCl] = 2 M usually
Dilute alkali, e.g. sodium hydroxide	[NaOH] = 2 M usually
Metal salts e.g. copper sulphate	[CuSO$_4$] = 0.1 M usually

3 Atoms, ions and acids

The solubility of calcium hydroxide

We can use an acid–base neutralization reaction to measure the solubility of calcium hydroxide.

EXPERIMENT 3.6

To find the solubility of calcium hydroxide in water by titration

Procedure

List the apparatus you expect to need for this experiment.

1 Put about 100 cm³ of pure water in a conical flask and add one spatula measure of solid calcium hydroxide. Fit the flask with a cork or rubber stopper and agitate the mixture thoroughly. Allow the mixture to stand for at least 24 hours so that the water becomes saturated with the calcium hydroxide.
2 Carefully decant the solution into a filter funnel and filter paper over a second conical flask, so as to collect the saturated solution of calcium hydroxide.
3 Titrate 10 cm³ portions of this solution with 0.050 M hydrochloric acid, using methyl orange or bromophenol blue as indicator. Repeat the titrations until two successive results agree to within 0.1 cm³. If you are not familiar with how to use the burette and pipette properly ask for help.

Write a short account of the procedure for this experiment in your notebook and record your results as follows.

Titration of 10.0 cm³ portions of saturated calcium hydroxide solution with 0.050 M hydrochloric acid solution

	1st titration	2nd titration, etc.
1st burette reading	...	...
2nd burette reading	...	...
Volume delivered	...	...
Volume of acid used	... cm³	...

Calculation

1 From the **average** volume of 0.050 M hydrochloric acid used, calculate the moles of hydrochloric acid used in the titration.
2 Write the equation for the reaction between calcium hydroxide and hydrochloric acid, checking the formulae and the balancing carefully.
3 Use this equation to work out how many moles of calcium hydroxide reacted with the hydrochloric acid in the titration.
4 This number of moles of calcium hydroxide must have been in 10 cm³, so how many moles of calcium hydroxide would have been in 1000 cm³ (1 dm³)?
5 From this, calculate the **mass** of calcium hydroxide in 1 dm³ of saturated solution. You will need to look up the molar mass of calcium hydroxide, or of the elements it contains.

COMMENT
The solubility of most substances varies with temperature, so the temperature of the saturated solution of calcium hydroxide should be recorded along with its solubility.

3.7 Study task: The role of calcium in agriculture

QUESTIONS
Read the passage below, then answer the questions based on it.
1. Why are calcium compounds added to agricultural land?
2. How does the soil pH affect the growth of plants?
3. Summarize the information about the effect of soil pH on the growth of particular plants.
4. Explain how the calcium content of soil can vary.

Calcium compounds are the principal factor in controlling the pH of the soil, and this affects the ability of plants to absorb nutrients through the roots.

The pH of the soil influences the concentration of plant nutrients in the soil solution and hence their availability. For example at a pH of about 5 the concentration of aluminium and manganese is higher than at a pH of 7. Some plants grow best at a low soil pH and are checked at higher values. Tea is a wellknown example of a crop which thrives in very acid soils and it contains far more aluminium than most plants.

Some species of forest trees do not thrive in soils of high pH. Sitka Spruce, for example, was found to make the best growth at pH 5 and failed to grow well on neutral and alkaline soils, but growth was depressed below pH 5 – a narrow range of optimum pH. On the other hand, sugar beet does not grow well under acid conditions and the optimum pH for this crop is around 6.5–7.0.

Crops are roughly graded in their tolerance for soil acidity: lucerne, sugar beet, and barley are only considered suitable for neutral or slightly acid soils (pH 7.0–6.5); wheat grows well on more acid soils (pH 6.5–6.0); and potatoes and rye on soils of pH 5.0 – too acid for sugar beet and barley.

Figure 3.24 Barley on an unlimed plot (right) the soil has a low calcium content; it has become acid in reaction and the barley crop has failed. The plot on the left has been well limed and has a good calcium content and a pH of about 6.5; the barley crop is good there.

Rain water, which contains carbonic acid, H_2CO_3, leads, as it percolates through the soil, to the replacement of cations such as Ca^{2+} by H^+. Fertilizers such as ammonium sulphate $(NH_4)_2SO_4$, are also involved in cation exchange.

$$Ca\text{-soil} + 2NH_4^+(aq) \longrightarrow (NH_4)_2\text{-soil} + Ca^{2+}(aq)$$

The NH_4^+ ions held in the soil are converted first to NO_2^- and then to NO_3^- as the result of bacterial action. Hydrogen ions are produced simultaneously to balance these anions and these replace ammonium ions in the exchange complex, making the soil more acid (the more exchangeable H^+, the more acid the soil).

Exchangeable calcium and other ions are removed from the soil by growing crops. Some examples of the quantities involved are given in the following table (1 hectare = $10\,000\,m^2$).

Crop	Yield /tonnes ha^{-1}	Calcium removed /kg ha^{-1}	Magnesium removed /kg ha^{-1}
Potatoes	30	6	4
Hay	3.8	26	10
Wheat (grain and straw)	3.8	13	9

Except in chalk soils the weathering of mineral fragments in the soil is generally not enough to replace the calcium that is lost in these various ways, and the deficiency has to be made good by the addition of either calcium hydroxide (slaked lime) or calcium carbonate (ground limestone).

Summary

At the end of this Topic you should be able to:
a interpret line emission spectra qualitatively as evidence for the existence of electron energy levels within atoms
b demonstrate understanding of the term ionization energy and interpret the sequence of ionization energies for a particular element in terms of the arrangement of electrons in quantum shells and sub-shells
c understand the pattern of electron distribution in quantum shells and sub-shells, and use the standard notation to describe the electronic arrangement of atoms and ions, using the Periodic Table as a guide
d interpret the formation of ionic compounds in terms of electron transfer and construct 'dot-and-cross' diagrams of simple ionic compounds
e interpret electron density maps of simple ionic compounds
f demonstrate understanding of the term ionic radius
g interpret the trends in atomic and ionic radii and ionization energies down a group and across a period, in terms of electronic structures
h recall the Brønsted–Lowry theory of acid–base behaviour and the meaning of strong and weak acids and bases
i calculate concentrations in mol dm^{-3} (M)
j perform calculations related to acid–base titrations

Review questions

*Indicates that the *Book of data* is needed.

3.1 On the graph the first ionization energy of some elements is plotted against the atomic number of the elements.

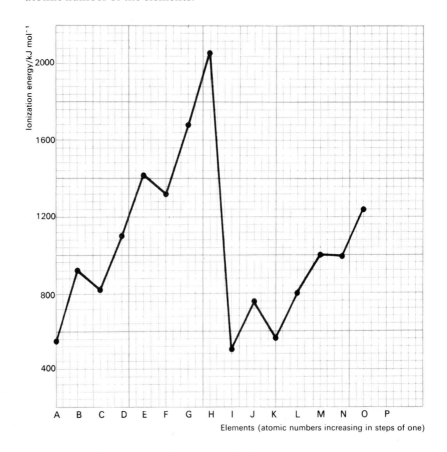

a State two of the elements likely to be alkali metals.
b State one of the elements likely to be a noble gas.
c Which one of the elements would you expect to be in the same group of the Periodic Table as the element **C**?
d How does the first ionization energy of the elements vary with the increasing atomic number?

3.2 The electron energy levels of a certain element can be represented by

$1s^2 2s^2 2p^6 3s^2 3p^1$

Sketch a graph showing the general form which you would expect for the first five ionization energies of the element.

3.3 The electron energy levels of a certain element can be represented by

$1s^2 2s^2 2p^6 3s^2 3p^6 3d^{10} 4s^2 4p^6 5s^2$

a What is the atomic number of the element?
b In which group of the Periodic Table will the element be found?
c The element forms an ionic bond when it reacts with oxygen. What will be the charge on the ion of the element?

3.4 The following table shows the first three ionization energies in kJ mol^{-1} of elements in the *same* group of the Periodic Table (the letters are **not** the symbols for the elements).

Element	1st ionization energy	2nd ionization energy	3rd ionization energy
A	376	2420	3300
B	403	2632	3900
C	419	3051	4412
D	496	4563	6913
E	520	7298	11815

a Which of these elements should have the largest atomic number?
b In which group of the Periodic Table should the elements be placed? Give reasons for your answer.

3.5 Refer to the following table of ionization energies in kJ mol^{-1} of five elements (the letters are **not** the symbols for the elements).

Elements	1st ionization energy	2nd ionization energy	3rd ionization energy	4th ionization energy
A	520	7298	11815	–
B	578	1817	2745	11578
C	1086	2353	4621	6223
D	496	4563	6913	9544
E	590	1145	4912	6474

a Which of the elements, when it reacts, is most likely to form a 3^+ ion?
b Which pair of the elements, **A** to **D**, is most likely to be in the same group of the Periodic Table?
c Which of the elements would require the most energy to convert one mole of atoms into ions carrying one positive charge?
d Which of the elements would require the most energy to convert one mole of atoms into ions carrying two positive charges?

*3.6 What mass of each of the following is dissolved in 250 cm^3 of a solution of concentration 0.100 mol dm^{-3}?

a Hydrochloric acid, HCl
b Sulphuric acid, H_2SO_4
c Sodium hydroxide, NaOH
d Potassium manganate(VII), $KMnO_4$
e Sodium thiosulphate, $Na_2S_2O_3.5H_2O$.

*3.7 How many moles of each solute are contained in the following solutions? (Express your answer as a decimal, when necessary.)

  a 25 cm^3 of sodium chloride, 0.100 mol dm^{-3}
  b 10 cm^3 of sodium chloride, 2.00 mol dm^{-3}
  c 12.2 cm^3 of nitric acid, 1.56 mol dm^{-3}
  d 12.2 cm^3 of sulphuric acid, 1.56 mol dm^{-3}.

*3.8 What is the concentration, in mol dm^{-3}, of each of the following solutions?

  a 5.85 g of sodium chloride, NaCl, in 1000 cm^3 of solution
  b 5.85 g of sodium chloride, NaCl, in 250 cm^3 of solution
  c 3.16 g of potassium manganate(VII), KMnO$_4$, in 2 dm^3 of solution
  d 6.20 g of sodium thiosulphate, Na$_2$S$_2$O$_3$.5H$_2$O, in 250 cm^3 of solution.

*3.9 Calculate the concentration, in mol dm^{-3}, of the following. (Assume that the compounds are fully ionized.)

  a Hydrogen ion in 1 dm^3 of solution containing 3.65 g of hydrogen chloride
  b Hydroxide ion in 1 dm^3 of solution containing 17.1 g of barium hydroxide, Ba(OH)$_2$
  c Sulphate ion in a solution of aluminium sulphate, Al$_2$(SO$_4$)$_3$.12H$_2$O, of concentration 0.1 mol dm^{-3}.
  d Aluminium ion in a solution of aluminium sulphate, Al$_2$(SO$_4$)$_3$.12H$_2$O, of concentration 0.1 mol dm^{-3}.

*3.10 10.0 cm^3 of a saturated solution of barium hydroxide were exactly neutralized by 24.0 cm^3 of 0.2 M hydrochloric acid.

  a What indicator would be suitable for the titration and what is its colour change?
  b Write the equation for the reaction.
  c Calculate the concentration of the saturated solution of barium hydroxide.
  d Calculate the solubility of barium hydroxide in g dm^{-3}.

Examination questions

3.11 Strontium oxide is made from the mineral celestine by strongly heating it with carbon to form a sulphide.

$$SrSO_4(s) + 4C(s) \longrightarrow SrS(s) + 4CO(g)$$

The sulphide formed is reacted with sodium hydroxide.

$$Sr(s) + 2NaOH(aq) \longrightarrow Sr(OH)_2(s) + Na_2S(aq)$$

Sodium sulphide is removed with water, and thermal decomposition of the strontium hydroxide is used to produce the oxide.

  a Suggest **one** safety precaution which should be used when heating celestine with carbon in the laboratory. Give a reason.
  b Outline the practical procedure you would use to remove sodium sulphide and obtain pure, dry strontium oxide.

c i Write a balanced equation for the thermal decomposition of strontium hydroxide.
ii What was the percentage yield if 1.3 g of strontium oxide were obtained from 4.6 g of strontium sulphate? (Molar masses Sr = 88, S = 32, O = 16 g mol^{-1})
d i State the electronic configuration of strontium (atomic number = 38).
ii Draw a diagram to show the electronic structure of strontium oxide, showing the outer shells of electrons only.
e i Using a sample of strontium oxide, what procedure would you use to obtain the flame colour of strontium?
ii What is the flame colour of strontium?
iii What is the origin of flame colours?

3.12 Beryllium (Be, atomic number 4) and barium (Ba, atomic number 56) are in the same Group of the Periodic Table.

a What is the electronic configuration, in terms of s, p and d electrons, of
i the beryllium atom
ii the barium atom?
b i To what Group of the Periodic Table do these two elements belong?
ii What is the oxidation state common to both elements in their compounds?
iii State in terms of ionization energies why there is little likelihood of a higher oxidation state for either element.
c i Which of the two elements has the higher first ionization energy? Give a reason for this.
ii Would their ions be smaller or larger than their atoms? Give a reason for this.

3.13 The first ionization energies of several elements of consecutively increasing atomic number are shown below. The letters are **not** the symbols for the elements.

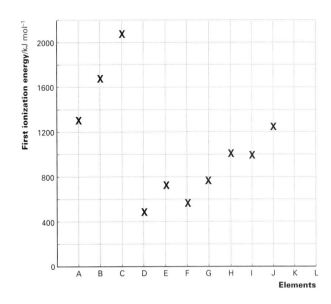

a What are your estimates of the first ionization energies of the elements **K** and **L**?
b i Which elements, from **A–L**, are noble gases?
ii Which elements are alkali metals?
iii Which of the elements **A** to **D** consist of small molecules?

iv Which element could have the electron configuration $1s^2 2s^2 2p^6 3s^2 3p^1$?
v **E** forms an ionic compound with **J**, formula **EJ$_2$**. Give the formulae of two further ionic compounds of **E** with other elements from **A** to **L**.
c Explain why there is a general rise in first ionization energy from element **D** to element **J**.

3.14 The element chlorine has atomic number 17.
a State in full the electronic configuration of the chlorine atom.
b i Draw a diagram to show the change in *outer-shell* electrons which occurs when chlorine reacts with sodium to form sodium chloride.
ii Name the element whose atoms have the same electronic configuration as chloride ions in sodium chloride.
c Draw a diagram to show the arrangement of outer-shell electrons in the molecule of tetrachloromethane.
d i Name the types of bond present in sodium chloride and tetrachloromethane.
ii Describe the difference between these two bonds in terms of electrons.

*3.15 Carry out appropriate calculations, using the *Book of data* for any information you need, in order to predict what will happen at each stage of the following experiment and describe what you expect to see.
i 1 g of calcium hydroxide was mixed with 100 cm^3 of water at room temperature.
ii 10 cm^3 of 2 mol dm^{-3} sulphuric acid was then added to the mixture.
iii 5 drops of Full-range Indicator were then added.

TOPIC 4

Energy and reactions

There is an energy change when nearly all chemical reactions take place, whether the reaction takes place in a car engine, our bodies, fireworks or a laboratory. The study of these energy changes is as much a part of chemistry as the study of changes in composition or structure as the result of reactions.

Figure 4.1 Fireworks in the City of London with the Old Bailey law court in the foreground

4.1 Energy from chemical reactions

Your teacher may show you some typical energy changes of reactions to start you thinking about energy in chemistry.

EXPERIMENT 4.1 Measuring some energy changes

We are going to start with two simple experiments to give you a first impression of how energy changes can be measured. To calculate the total energy change in a chemical reaction we need to know the amounts (in moles) that react, and we have to measure any temperature change that takes place.

The energy changes of reactions in solution are often measured using an electrical compensation calorimeter linked to a joulemeter. Otherwise we also need to know the heat capacity of the apparatus we are using. The heat capacity is the quantity of energy needed to raise the temperature of the apparatus by one degree Celsius.

4 Energy and reactions

You are going to look at an oxidation–reduction reaction and a neutralization. For each reaction you have to calculate the mass of solid reagent to use. The molar amounts are stated in the *Procedure*. Have your calculations checked before you start the experiment.

SAFETY Zinc dust is flammable.

Procedure

1 The reaction between copper(II) sulphate and zinc

$$Cu^{2+}(aq) + Zn(s) \longrightarrow Cu(s) + Zn^{2+}(aq)$$

Measure 25.0 cm³ of 0.2 M copper(II) sulphate solution into a well insulated container and then measure its temperature. Add 0.01 mol (an excess) of zinc powder, Zn. Stir gently and continuously and note the highest temperature reached. Work out the temperature change to the nearest 0.1 °C

2 The reaction between citric acid and sodium hydrogencarbonate

$$C_6H_8O_7(aq) + 3HCO_3^-(s) \longrightarrow C_6H_5O_7^{3-}(aq) + 3CO_2(g) + 3H_2O(l)$$

Repeat part **1** using 25 cm³ of 1.0 M citric acid and 0.1 mol (an excess) of sodium hydrogencarbonate, $NaHCO_3$. Record the temperature change as before.

- Record the equations of the reactions and whether they were exothermic or endothermic.
  From your results calculate the energy change in kilojoules per mole for 1.0 mol of copper(II) sulphate, $CuSO_4$, and 1.0 mol of citric acid, $C_6H_8O_7$, respectively.

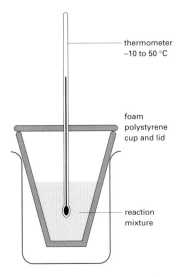
Figure 4.2 Measuring energy changes

How to do the calculation

We are going to assume that all the energy produced in the reactions is exchanged between the reactants and the water in the calorimeter, and no energy is transferred to the air, the glass of the thermometer, the material of the calorimeter, or even the products of the reaction. We are also going to assume 1 cm³ of water weighs 1 g and therefore needs 4.18 joules to change in temperature by 1 °C.

The relationship we are going to use is:

Energy exchanged between = Specific heat capacity × mass of the × temperature change
reactants & surroundings   of the solution   solution
(Q/joules)   (c/J g^{-1} K^{-1})   (m/g)   (ΔT/K)

$$Q = cm\Delta T$$

which with our assumptions becomes:

$$Q = 4.18 \times \text{volume of solution} \times \text{temperature change}$$

To complete the calculation you need to work out how much energy would have been exchanged if you had used 1 mol of copper sulphate, $CuSO_4$, and 1 mol of citric acid, $C_6H_8O_7$.

4 Energy and reactions

Using a joulemeter

The energy that has been transferred to the calorimeter can be measured directly, using a joulemeter. This electrical energy is equal to the energy change of the chemical reaction taking place in the calorimeter. The apparatus and how to use it is described in Topic 18.

If the number of moles taking part is known, the energy change per mole can be calculated.

Calculation

Suppose x joules are recorded by the joulemeter as supplied to the calorimeter. Now 100 cm³ of 0.2 M copper(II) sulphate solution contain

$$\frac{100}{1000} \times 0.2 = 0.02 \text{ mol of copper ions}$$

If 0.02 mol of copper ions in reaction with zinc gave x joules, then the energy change for the reaction is

$$\frac{x \text{ joules}}{0.02 \text{ mol}} \quad \text{or} \quad \frac{x}{0.02} \text{ joules per mol}$$

Work out the energy change of the reaction, in kilojoules per mole, giving it the correct sign.

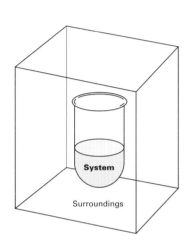

Figure 4.3 The system is the sample or reaction mixture. Outside the system are the surroundings, which include the apparatus.

Recording energy changes

When we investigate energy changes, we must state the conditions under which the changes are measured, so that the results of different experiments can be compared.

The energy change that we shall be measuring in this Topic is the **enthalpy change**. This can be considered as the energy that would be exchanged with the **surroundings** if the reaction occurred in such a way that the temperature of the **system** before and after the reaction were the same. The reaction mixture is referred to as the system.

The reaction must also take place at constant pressure to allow for changes in volume between the reactants and products, such as the carbon dioxide gas produced.

The enthalpy change of a reaction is the energy exchanged with the surroundings at constant pressure

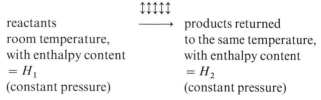

reactants   ⟶   products returned
room temperature,   to the same temperature,
with enthalpy content   with enthalpy content
$= H_1$ $= H_2$
(constant pressure)   (constant pressure)

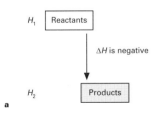

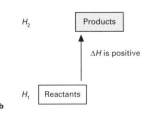

Figure 4.4 **a** An exothermic reaction, **b** an endothermic reaction

The symbol Δ (Greek capital delta, equivalent to our D) is used to denote the change in the value of a physical quantity. So the change in enthalpy in going from reactants to products is given by

$$\Delta H = H_2 - H_1$$

The standard conditions ($^\ominus$) for measuring enthalpy changes are:

temperature	298 K, equivalent to 25 °C
pressure	1 atmosphere
physical state	normal state at 298 K and 1 atmosphere
solutions	1.0 M

To measure an enthalpy change we insulate the system from its surroundings, and allow the energy of the reaction to change the temperature of the system. We then calculate how much energy would have to be put into or taken from the system to bring it back to its initial temperature. This amount of energy is the **enthalpy change**.

When energy is given out from the system to the surroundings during the reaction, which we observe as our apparatus getting hot, the reaction is said to be **exothermic** and the enthalpy change ΔH is **negative**.

Alternatively when energy is taken into the system from the surroundings during the reaction, observed as our apparatus getting cold, the reaction is said to be **endothermic** and ΔH is **positive**.

For the reaction

$$\text{Mg(s)} + \text{Cl}_2\text{(g)} \longrightarrow \text{MgCl}_2\text{(s)} \qquad \Delta H = -641.3 \text{ kJ mol}^{-1}$$

the value of ΔH is for the amounts shown in the equation, that is, for one mole of magnesium atoms, Mg, one mole of chlorine molecules, Cl_2, and one mole of magnesium chloride, MgCl_2. Normally, a **standard** enthalpy change is quoted and it is then given the symbol $\Delta H^\ominus$. This means that the enthalpy change was measured in conditions fixed by convention.

> **The standard enthalpy change for a reaction, symbol $\Delta H^\ominus$, refers to the amounts shown in the equation, at a pressure of 1 atmosphere, at a temperature of 298 K, with the substances in the physical states normal under these conditions. Solutions must have a concentration of 1 mol dm^{-3}.**

For example:

$$\text{Fe}_2\text{O}_3\text{(s)} + 2\text{Al(s)} \longrightarrow 2\text{Fe(s)} + \text{Al}_2\text{O}_3\text{(s)} \quad \Delta H^\ominus_{\text{reaction}}(298) = -851.5 \text{ kJ mol}^{-1}$$

$$\text{NH}_3\text{(g)} + \text{HCl(g)} \longrightarrow \text{NH}_4\text{Cl(s)} \quad \Delta H^\ominus_{\text{reaction}}(298) = -176 \text{ kJ mol}^{-1}$$

For important reactions we have a shorthand which saves us from having to write the full equation. The magnesium chloride reaction is an example; the enthalpy change is called the **standard enthalpy change of formation** of magnesium chloride. It is given the symbol $\Delta H^\ominus_f(298)$ [MgCl$_2$(s)].

$$\text{Mg(s)} + \text{Cl}_2\text{(g)} \longrightarrow \text{MgCl}_2\text{(s)} \qquad \Delta H^\ominus_f(298) = -641.3 \text{ kJ mol}^{-1}$$

> **The standard enthalpy change of formation of a compound, symbol $\Delta H^\ominus_f(298)$, is the enthalpy change that takes place when one mole of the compound is formed from its elements under the standard conditions.**

For example:

$$2\text{Fe(s)} + 1\tfrac{1}{2}\text{O}_2\text{(g)} \longrightarrow \text{Fe}_2\text{O}_3\text{(s)} \quad \Delta H^\ominus_f(298)[\text{Fe}_2\text{O}_3\text{(s)}] = -824.2 \text{ kJ mol}^{-1}$$

$$\tfrac{1}{2}\text{H}_2\text{(g)} + \tfrac{1}{2}\text{Cl}_2\text{(g)} \longrightarrow \text{HCl(g)} \quad \Delta H^\ominus_f(298)[\text{HCl(g)}] = -92.3 \text{ kJ mol}^{-1}$$

Figure 4.5 Lord Kelvin, who had the idea of an 'absolute zero' temperature

> **COMMENT**
> The standard temperature is stated in **kelvin** (symbol K). The lowest temperature that is possible is 0 K, which is −273 °C. As the kelvin was chosen to be the same size (or temperature interval) as the degree celsius (or centigrade), it follows that
> 0 °C is 273 K
> 25 °C is 298 K
> and in general t °C is $(273+t)$ K

4.2 Hess's Law

As the law of conservation of energy applies to chemical processes just as much as to any other process, then, when one set of substances is converted to another set, by whatever route, the total energy change must be the same.

The First Law of Thermodynamics states that energy is always conserved.

If this were not so, it would be possible to go from A + B to C + D by one route and then back again by a different route with an overall gain of energy. The law of conservation of energy tells us that this is impossible: the energy change must be the same by whatever route we travel from A + B to C + D.

In 1840 Germain Hess discovered this particular application of the law of conservation of energy experimentally, and it is generally referred to as **Hess's Law**.

Hess's Law states that the total enthalpy change accompanying a chemical change is independent of the route by which the chemical change takes place.

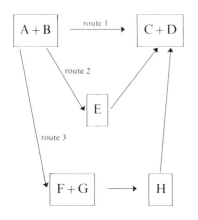

Figure 4.6

The correctness of Hess's law can be illustrated by the following example. You should have determined the enthalpy change ΔH for the displacement of copper by zinc:

$$Cu^{2+}(aq) + Zn(s) \longrightarrow Cu(s) + Zn^{2+}(aq) \qquad \Delta H = -216 \text{ kJ mol}^{-1}$$

If the copper obtained is used to displace silver from a solution of silver ions, we would find that ΔH is:

$$2Ag^{+}(aq) + Cu(s) \longrightarrow 2Ag(s) + Cu^{2+}(aq) \qquad \Delta H = -147 \text{ kJ mol}^{-1}$$

But if zinc is to displace silver directly, we find that ΔH is:

$$2Ag^{+}(aq) + Zn(s) \longrightarrow 2Ag(s) + Zn^{2+}(aq) \qquad \Delta H = -363 \text{ kJ mol}^{-1}$$

You can see that the overall enthalpy change in kilojoules per mole is the same whichever route you take.

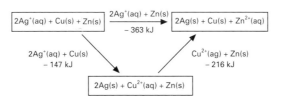

The great value of Hess's Law is that it can be used to calculate enthalpy changes that cannot be determined by experiment. For example $\Delta H_f^{\ominus}(298)$ values have been determined directly for many oxides but for some this is not possible. $\Delta H_f^{\ominus}(298)$ for magnetic iron oxide, Fe_3O_4, is an example of a standard enthalpy change of formation which cannot be measured directly. This is because when iron is burnt completely in oxygen it is impossible to prevent iron(III) oxide, Fe_2O_3, forming.

In these, and in many hundreds of other instances, values for the standard enthalpy changes of formation have been obtained in indirect ways, by means of calculations using Hess's Law.

Determining standard enthalpy changes of reaction

From now on **we will quote $\Delta H^\ominus(298)$ as $\Delta H^\ominus$**, as all our standard values will be for 298 K.

In Topic 1 we quoted an energy of reaction for a change that does not actually occur:

$$Fe_2O_3(s) + 3Cu(s) \longrightarrow 2Fe(s) + 3CuO(s)$$

How was the value found when no experimental measurement is possible? The answer is that a Hess cycle was constructed using standard enthalpy changes of formation:

$$2Fe(s) + 1\tfrac{1}{2}O_2(g) \longrightarrow Fe_2O_3(s) \qquad \Delta H_f^\ominus[Fe_2O_3(s)] = -824.2 \text{ kJ mol}^{-1}$$

$$Cu(s) + \tfrac{1}{2}O_2(g) \longrightarrow CuO(s) \qquad \Delta H_f^\ominus[CuO(s)] = -157.3 \text{ kJ mol}^{-1}$$

The balanced equations of the three reactions were combined in a Hess cycle:

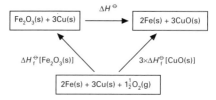

From this diagram we deduced the value of $\Delta H_{\text{reaction}}^\ominus$ as follows:

$$\Delta H_f^\ominus[Fe_2O_3(s)] + \Delta H_{\text{reaction}}^\ominus = 3\Delta H_f^\ominus[CuO(s)]$$

$$-824.2 \text{ kJ mol}^{-1} + \Delta H_{\text{reaction}}^\ominus = 3(-157.3 \text{ kJ mol}^{-1})$$

$$\Delta H_{\text{reaction}}^\ominus = +352.3 \text{ kJ mol}^{-1}$$

By using this procedure we can use Hess's Law to calculate most other enthalpy changes which cannot be determined directly.

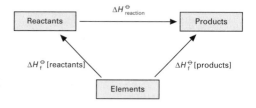

Figure 4.7 An outline Hess cycle for all reactions

$$\Delta H_{\text{reaction}}^\ominus = \Delta H_f^\ominus[\text{products}] - \Delta H_f^\ominus[\text{reactants}]$$

INVESTIGATION 4.2 Determining an enthalpy change that cannot be measured directly

The enthalpy change when potassium hydrogencarbonate decomposes on heating is not easily measured by any direct method:

$$2KHCO_3(s) \longrightarrow K_2CO_3(s) + CO_2(g) + H_2O(l)$$

Carry out an investigation to determine the enthalpy change for the decomposition of potassium hydrogencarbonate. Make a risk assessment before starting any experiments.

4.3 Uses of standard enthalpy changes of formation

STUDY TASK
Look up the following compounds in the *Book of data*, and write in your notes their names, formulae, and standard enthalpy changes of formation. This will give you an idea of the range and pattern of values that exist.
Lithium chloride
Sodium chloride
Potassium chloride
Sodium oxide
Magnesium oxide
Aluminium oxide
Hydrogen chloride
Hydrogen bromide
Hydrogen iodide
Carbon dioxide
Nitrogen dioxide
Water

Standard enthalpy changes of formation of inorganic compounds are given in table 5.3 in the *Book of data*.

We can use standard enthalpy changes of formation to find the enthalpy change that takes place in a reaction, without doing an experiment in every case.

As an example let us look at the reaction of ammonia gas with hydrogen chloride gas to give ammonium chloride. $\Delta H^{\ominus}_{\text{reaction}}$ can be calculated as follows.

First write down the equation for the reaction:

$$NH_3(g) + HCl(g) \longrightarrow NH_4Cl(s)$$

The standard enthalpy changes of formation that you need are:

$\Delta H^{\ominus}_f[NH_3(g)] \quad = -46 \text{ kJ mol}^{-1}$
$\Delta H^{\ominus}_f[HCl(g)] \quad = -92 \text{ kJ mol}^{-1}$
$\Delta H^{\ominus}_f[NH_4Cl(s)] = -314 \text{ kJ mol}^{-1}$

Then draw a Hess cycle showing the formation of the compounds on both sides of the equation from their elements:

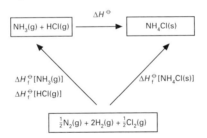

The total enthalpy change must be the same by whatever route the ammonium chloride is formed (whether it is formed direct from its elements, or through the intermediates of ammonia and hydrogen chloride).

Therefore

$$\Delta H^{\ominus}_f[NH_4Cl(s)] = \Delta H^{\ominus}_f[NH_3(g)] + \Delta H^{\ominus}_f[HCl(g)] + \Delta H^{\ominus}_{\text{reaction}}$$

That is

$$-314 = -46 - 92 + \Delta H^{\ominus}_{\text{reaction}}$$

So

$$\Delta H^{\ominus}_{\text{reaction}} = 46 + 92 - 314 = -176 \text{ kJ mol}^{-1}$$

This example shows that standard enthalpy changes of chemical reactions can be calculated from the standard enthalpy changes of formation of the reactants and products. This is the great value of standard enthalpy changes of formation; they make it possible to calculate enthalpy changes which cannot otherwise be found.

REVIEW TASK

As a further example let us look at the enthalpy changes involved in the thermal decomposition of iron(II) sulphate. It will be a little easier if we assume the salt is anhydrous.

1. You studied this reaction in Topic 1. To get the reaction to go you had to heat continuously. Does this suggest to you that the reaction is exothermic or endothermic?
2. What is the balanced equation of the decomposition of the anhydrous salt?
3. Construct a balanced Hess cycle for the reactions involved.
4. Look up the standard enthalpy changes of formation in the *Book of data*. Now use the Hess cycle to calculate $\Delta H^\ominus_{\text{reaction}}$.
5. Your answer should be +194.8 kJ mol^{-1}

More enthalpy changes

There are other energy changes that are specially defined. These are changes for which the data are regularly needed when constructing Hess cycles. The appropriate data are collected together in tables in the *Book of data*.

The enthalpy change of atomization of an element, $\Delta H^\ominus_{\text{at}}$ refers to the enthalpy change when one mole of gaseous atoms is formed from the element in its stable physical state at 298 K.

The standard enthalpy change of atomization of an element, symbol $\Delta H^\ominus_{\text{at}}(298)$, is the enthalpy change that takes place when one mole of gaseous ATOMS is made from the element in its standard physical state under standard conditions.

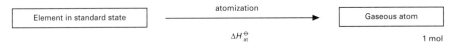

Equation	Enthalpy change of atomization /kJ mol^{-1}
Mg(s) ⟶ Mg(g)	$\Delta H^\ominus_{\text{at}}[\text{Mg(g)}] = +147.7$
Fe(s) ⟶ Fe(g)	$\Delta H^\ominus_{\text{at}}[\text{Fe(g)}] = +416.3$
$\frac{1}{2}$O$_2$(g) ⟶ O(g)	$\Delta H^\ominus_{\text{at}}[\text{O(g)}] = +249.2$
$\frac{1}{2}$H$_2$(g) ⟶ H(g)	$\Delta H^\ominus_{\text{at}}[\text{H(g)}] = +218$

More values are listed in table 5.2 in the *Book of data*.

The energy change which takes place when gaseous atoms of an element acquire electrons and form single negatively charged ions is known as the **electron affinity**, E_{aff}, of that element. Values are listed in table 5.10 in the *Book of data*. Some examples are given overleaf.

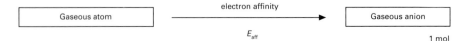

Equation	Electron affinity E_{aff} /kJ mol^{-1}
Cl(g) + e$^-$ ⟶ Cl$^-$(g)	−349
Br(g) + e$^-$ ⟶ Br$^-$(g)	−325
I(g) + e$^-$ ⟶ I$^-$(g)	−295

The opposite change is the loss of an electron from a gaseous atom to form a single positively charged ion. This enthalpy change was introduced in the previous Topic and is known as the **ionization energy**, E_{mj}, of an element. Values are listed in table 4.1 in the *Book of data*. Some examples are:

Equation	Ionization energy E_{mj} /kJ mol^{-1}
Li(g) − e$^-$ ⟶ Li$^+$(g)	+520
Na(g) − e$^-$ ⟶ Na$^+$(g)	+496
K(g) − e$^-$ ⟶ K$^+$(g)	+419

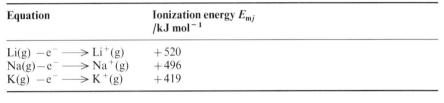

So far we have been concerned with enthalpy **changes**. Absolute enthalpies cannot be determined, only differences between them. For purposes of calculation, however, a value of zero is given to the enthalpy of elements in their standard state at 298 K.

Examples are:

$H^{\ominus}[O_2(g)] = 0$
$H^{\ominus}[Na(s)] = 0$

In the case of elements or compounds which can exist in different forms, the most stable form is chosen as the standard. In the case of carbon the most stable form is graphite.

$H^{\ominus}[C(graphite)] = 0$

It follows from this definition that the standard enthalpy change of formation of an element in its standard state is zero (as we will use in some Hess cycles):

C(graphite) ⟶ C(graphite) $\Delta H_f^{\ominus}[C(graphite)] = 0$

However, when the form of the element is changed there will be an enthalpy change:

C(graphite) ⟶ C(diamond) $\Delta H_f^{\ominus}[C(diamond)] = +1.90$ kJ mol^{-1}

4.4 The Born–Haber cycle: lattice energies

When ions form crystals the bonding which occurs is very strong, so ionic compounds are not easily decomposed to their elements. The energy given out when ions come together to form a crystalline solid is called the **lattice energy** of the compound. Lattice energies always have negative values.

> **The lattice energy of an ionic crystal is the standard enthalpy change of formation of the crystal lattice from its constituent ions in the gas phase.**

$$Na^+(g) + Cl^-(g) \longrightarrow Na^+Cl^-(s) \qquad \Delta H^\ominus_{lattice}(298) = \text{lattice energy}$$

The direct determination of lattice energies is not possible, but values can be calculated from an energy cycle, known as a **Born–Haber cycle**.

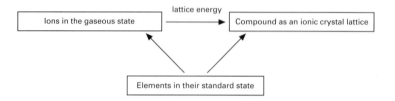

For sodium chloride the Born–Haber cycle is:

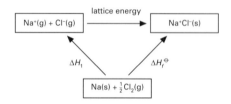

The standard enthalpy change of formation of sodium chloride can be measured directly, by the reaction of sodium with chlorine in a calorimeter. If the energy required to convert sodium metal into gaseous ions, and chlorine molecules into gaseous ions, can be obtained, ΔH_1 will be known, and it is then possible to obtain a value for the lattice energy.

ΔH_1 has to be obtained in stages. Taking the sodium first,

$$Na(s) \xrightarrow[\text{standard enthalpy change of atomization}]{} Na(g) \xrightarrow[\text{ionization energy}]{-e^-} Na^+(g)$$

The two energy values required are the **standard enthalpy change of atomization** of sodium, for the conversion of solid sodium into gaseous sodium consisting of separate atoms:

$$Na(s) \longrightarrow Na(g) \qquad \Delta H^\ominus_{at} = +107.3 \text{ kJ mol}^{-1}$$

and the **ionization energy**, for the conversion of gaseous atoms into gaseous ions:

$$Na(g) \longrightarrow Na^+(g) + e^- \qquad E_{m1} = +496 \text{ kJ mol}^{-1}$$

Taking the chlorine we have

$$\tfrac{1}{2}Cl_2(g) \xrightarrow{\text{standard enthalpy change of atomization}} Cl(g) \xrightarrow{+e^-\ \text{electron affinity}} Cl^-(g)$$

The two energy values required are the **standard enthalpy change of atomization** of chlorine, for the conversion of gaseous chlorine molecules into gaseous chlorine atoms,

$$\tfrac{1}{2}Cl_2(g) \longrightarrow Cl(g) \qquad \Delta H^\ominus_{at} = +121.7 \text{ kJ mol}^{-1}$$

and the **electron affinity**, which is the energy change occurring when a chlorine atom accepts an electron and becomes a chloride ion,

$$Cl(g) + e^- \longrightarrow Cl^-(g) \qquad E_{aff} = -348.8 \text{ kJ mol}^{-1}$$

(Each of these can be determined experimentally, although the determination of electron affinity is difficult.)

Finally we need the enthalpy change of formation of sodium chloride:

$$Na(s) + \tfrac{1}{2}Cl_2(g) \longrightarrow Na^+Cl^-(s) \qquad \Delta H^\ominus_f = -411.2 \text{ kJ mol}^{-1}$$

and then the lattice energy can be determined.

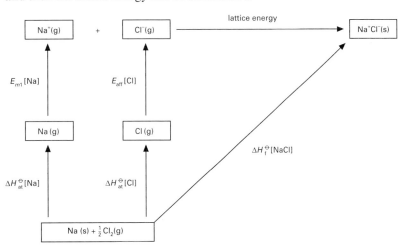

Figure 4.8 The Born–Haber cycle for sodium chloride

From figure 4.8 you can work out the lattice energy by substituting the appropriate values:

$$\Delta H^\ominus_{\text{lattice}} = \Delta H^\ominus_f[NaCl(s)] - \{\Delta H^\ominus_{at}[Na(s)] + \Delta H^\ominus_{at}[Cl(g)] + E_{m1}[Na(g)] + E_{aff}[Cl(g)]\}$$

$$= -411.2 - \{+107.3 + 121.7 + 496 - 348.8\} \text{ kJ mol}^{-1}$$

$$\Delta H^\ominus_{\text{lattice}} = -787.4 \text{ kJ mol}^{-1}$$

Lattice stability and ion formation

In order to form positive ions an element must be ionized; that is, electrons must be removed from its atoms. This requires energy. For the resulting compound to be stable the lattice energy must be large enough to compensate for the energy involved in ionization. This is commonly achieved for M^+ and M^{2+} compounds, but not for M^{3+} or M^{4+}:

Equation	Ionization energy E_{mj} /kJ mol^{-1}
$Mg(g) - e^- \longrightarrow Mg^+(g)$	+738
$Mg^+(g) - e^- \longrightarrow Mg^{2+}(g)$	+1451
$Mg^{2+}(g) - e^- \longrightarrow Mg^{3+}(g)$	+7733

Magnesium will form ionic compounds but magnesium(III) is unknown.

The successive ionization energies of carbon are 1086, 2353, 4621, and 6223 kJ mol^{-1}, and the very large amount of energy involved in ionizing carbon cannot be recovered in any lattice energy. As a result carbon does not form C^{4+} ions; it does not in fact form stable positive ions at all but forms bonds by another method.

In order to form negative ions, atoms must gain electrons. Some values of electron affinity are:

Equation	Electron affinity E_{aff} /kJ mol^{-1}
$Cl(g) + e^- \longrightarrow Cl^-(g)$	−349
$O(g) + e^- \longrightarrow O^-(g)$	−141
$O^-(g) + e^- \longrightarrow O^{2-}(g)$	+798

From these figures you will see that the formation of chloride ions from chlorine **atoms** is exothermic, but the formation of $O^{2-}(g)$ ions from oxygen **ions** is endothermic. The formation of all X^{2-} ions (X is any element) is endothermic because energy has to be supplied to overcome the repulsion between the X^- negative ion and the incoming electron.

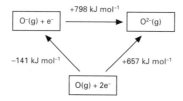

Broadly speaking, the successful formation of a compound depends on whether the energy needed for forming the cation and anion is recovered in the energy released as lattice energy. When the ionization energies cannot be recovered the compound is unlikely to be formed.

QUESTIONS

Use table 5.9 in the *Book of data* to answer the following questions about trends in lattice energies.

1. How do the lattice energies of the Group 1 fluorides vary down the group?
2. How do the lattice energies of the Group 1 chlorides differ from the lattice energies of the Group 2 chlorides?
3. How do the lattice energies of the Group 2 chlorides differ from the lattice energies of their oxides?
4. Try to suggest explanations for the trends you have found.

Theoretical values for lattice energies

Using the principles of electrostatics it is possible to calculate lattice energies. The oppositely charged ions are imagined being brought from infinity until they are in contact. It is assumed that the ions are spherical, separate entities, each with its charge distributed uniformly. Some of the theoretical values that have been calculated are given in the table.

Look at the values for the alkali metal halides, and compare the theoretical values with the Born–Haber experimental values. For one or two, use your calculator to work out the percentage difference between the theoretical and experimental values.

Compound	Theoretical value	Experimental value (Born–Haber cycle)	Difference
NaCl	−770	−780	10
NaBr	−735	−742	7
NaI	−687	−705	17
KCl	−702	−711	9
KBr	−674	−679	5
KI	−636	−651	15
AgF	−920	−958	38
AgCl	−833	−905	72
AgBr	−816	−891	75
AgI	−778	−889	111
ZnO	−4142	−3971	171

Lattice energies/kJ mol^{-1}

In the cases of the alkali metal halides, the excellent agreement between the theoretical and experimental values is strong evidence that the simple model of an ionic crystal is a good one.

Now look at the values for the silver halides. Calculate the percentage difference between the theoretical and the experimental values. Do you think that the ionic model accurately represents the bonding situation in the silver halides? If not, some other model is required.

More values are given in table 5.9 in the *Book of data*.

Polarization of ions

Spectroscopic studies of the vapours of the alkali metal halides show that these contain diatomic molecules, MX, and that the atoms are closer together than in the ionic solid. This is illustrated by lithium bromide and lithium iodide. Values for these compounds are given in the table.

For the atoms to move closer together a higher concentration of electrons is needed between the two nuclei. This will result in a distortion of the electron charge-cloud of one ion, or both, from a spherical distribution. Figure 4.9 illustrates the effect, which is known as the 'polarization of ions'.

Internuclear separation/nm		
Halide	Crystal	Vapour
LiBr	0.275	0.217
LiI	0.300	0.239

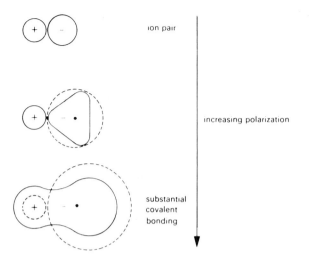

Figure 4.9 Increasing polarization of a negative ion by a positive ion

The polarization of ions represents the start of a transition from ionic bonding to covalent bonding.

STUDY TASK
What factors might affect the extent to which an electron charge-cloud around an ion is distorted?

What type of ion would be best at causing distortion? What type of ion would be most easily distorted? Consider the sizes of ions, and the number of charges on ions. Make a list of examples for both cations and anions.

Is there any evidence from lattice energies to support your ideas? Look at table 5.9 in the *Book of data*.

Summary

At the end of this Topic you should be able to:
a demonstrate understanding of the terms: standard conditions, standard enthalpy changes of formation, atomization and ionization; electron affinity; lattice energy; and polarization of ions
b plan investigations for determining enthalpy changes and justify the procedures involved
c calculate enthalpy changes from experimental data, with the energy changes given in joules
d select appropriate data to calculate, using Hess's law:
   i enthalpy changes of reaction
   ii lattice energies, using the Born–Haber cycle
e interpret differences between experimental (Born–Haber cycle) values and theoretical calculations of lattice energies in terms of structure and bonding.

Review questions

*Indicates that the *Book of data* is needed.

4.1 100 cm^3 of 0.02 M copper(II) sulphate solution were put in a calorimeter and an excess of magnesium powder was added. 1052 J were produced.
   a Calculate the moles of copper(II) ions used.
   b Calculate $\Delta H^\ominus$ for the reaction in kJ mol^{-1}.
   c Is energy evolved or absorbed during the reaction?
   d Write the equation for the reaction, together with the enthalpy change, giving it the correct sign.

4.2 50 cm^3 of 0.05 M silver nitrate solution were placed in an electrical compensation calorimeter and an excess of copper powder was added. 184 J had to be supplied to give the same rise in temperature as that which resulted from the reaction.
   a Calculate the moles of silver ions used.
   b Calculate $\Delta H^\ominus$ for the reaction in kJ mol^{-1} per mole of copper used.
   c Write the equation for the reaction, together with the enthalpy change, giving it the correct sign.

4.3 Using your results from questions **4.1** and **4.2**, calculate the enthalpy change per mole of magnesium atoms when magnesium reacts with silver nitrate solution. Write the equation for the reaction, together with the enthalpy change, giving it the correct sign. What assumptions have you made in your calculation?

*4.4 Calculate $\Delta H^\ominus(298)$ for the following reactions.
   a $SO_2(g) + 2H_2S(g) \longrightarrow 3S(s) + 2H_2O(l)$
   b $N_2O(g) + Cu(s) \longrightarrow CuO(s) + N_2(g)$
   c $NH_4Cl(s) \longrightarrow NH_3(g) + HCl(g)$
   d $Mg(s) + \frac{1}{2}O_2(g) \longrightarrow MgO(s)$
   e $H_2(g) + S(s) \longrightarrow H_2S(g)$
   f $CO_2(g) + 2Mg(s) \longrightarrow 2MgO(s) + C(s)$
   g $Na(s) + \frac{1}{2}Cl_2(g) \longrightarrow NaCl(s)$
Note: you will need the answers to these questions in Topic 10 question 2.

4.5 Suppose you were given the enthalpy changes for the following reactions.

$$2Fe(s) + 1\tfrac{1}{2}O_2(g) \longrightarrow Fe_2O_3(s)$$
$$Ca(s) + \tfrac{1}{2}O_2(g) \longrightarrow CaO(s)$$

What further information, if any, would you require in order to calculate the enthalpy changes of each of the following reactions?
   a $3Ca(s) + Fe_2O_3(s) \longrightarrow 3CaO(s) + 2Fe(s)$
   b $Ca(s) + CuO(s) \longrightarrow CaO(s) + Cu(s)$
   c $2Fe(s) + 3CuO(s) \longrightarrow Fe_2O_3(s) + 3Cu(s)$

*4.6 Draw a fully labelled Born–Haber cycle for the formation of calcium oxide.

*4.7 Use your *Book of data* to determine the lattice energy of sodium monoxide, Na_2O.

Examination questions

4.8 The enthalpy change of precipitation of magnesium carbonate was investigated by mixing magnesium nitrate solution with sodium carbonate solution.

In a first experiment 10.0 cm³ of 1.0 M magnesium nitrate solution at 20.0 °C was placed in a plastic beaker and an equal volume of 1.0 M sodium carbonate solution at 20.0 °C was added. The temperature rose to 22.1 °C. The experiment was performed a second time, using **50.0 cm³** of 1.0 M solution, and then performed a third time, using 50.0 cm³ of **0.20 M** solution.

- **a** What range and what sensitivity should you select for the thermometer to be used in this experiment?
- **b** Calculate the temperature rise you would expect in
  - **i** the second experiment
  - and **ii** the third experiment.
- **c** Which of the three experiments is likely to lead to the **most accurate** determination of a temperature rise? Justify your answer.
- **d** Which of the three experiments is likely to lead to the **least accurate** determination of a temperature rise? Justify your answer.

*4.9 How would you determine experimentally the enthalpy change for the reaction:

$$Ca(OH)_2(aq) + 2HCl(aq) \longrightarrow CaCl_2(aq) + 2H_2O(l)$$

State exactly what apparatus you would use and what measurements you would make.

How would the measurements be used to calculate the molar enthalpy change of the reaction?

Look up some enthalpy changes of neutralization in table 5.7 of the *Book of data* and comment on the values quoted.

4.10 The reaction between magnesium and water is extremely slow at room temperature, which is one important reason why it is not possible to measure the enthalpy change directly for the reaction:

$$Mg(s) + H_2O(l) \longrightarrow MgO(s) + H_2(g)$$

A student proposes to determine this enthalpy change indirectly by measuring the enthalpy changes for the reactions:

$$Mg(s) + 2H^+(aq) \longrightarrow Mg^{2+}(aq) + H_2(g)$$

$$MgO(s) + 2H^+(aq) \longrightarrow Mg^{2+}(aq) + H_2O(l)$$

Write a set of experimental instructions for the student, including details of apparatus and chemicals to be used, and show how the results would be used to calculate the enthalpy change for the reaction of magnesium with water.

The student suggests that it ought to be possible to measure the enthalpy change directly for the reaction:

$$Ca(s) + H_2O(l) \longrightarrow CaO(s) + H_2(g)$$

Is the student correct? Explain your answer.

4.11 This question uses the following data about boron and its compounds. On the graph the number of electrons removed is plotted against the logarithm of the appropriate ionization energy and in the table standard enthalpy changes are listed for various processes.

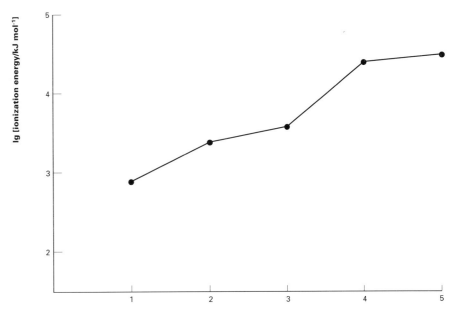

Number of electrons removed from boron

Process	$\Delta H^\ominus(298)/\text{kJ mol}^{-1}$
First electron affinity of oxygen	− 140
Second electron affinity of oxygen	+ 790
Standard enthalpy change of atomization of oxygen	+ 250
Standard enthalpy change of atomization of boron	+ 590
Standard enthalpy change of formation of the oxide of boron	− 1270
First ionization energy of boron	+ 800
Second ionization energy of boron	+ 2400
Third ionization energy of boron	+ 3700
Fourth ionization energy of boron	+25000
Fifth ionization energy of boron	+32800

a From the graph deduce the most likely formula for the oxide of boron. Justify your answer.
b i Draw a labelled Born–Haber cycle for the formation of the oxide of boron (assumed to be ionic).
 ii Calculate a lattice energy for the oxide of boron.
c i One mole of the oxide of boron reacts with three moles of water and the product dissolves in water to give a weakly acidic solution. Write an equation, with state symbols, for the reaction occurring in aqueous solution.
 ii Would you expect your value for the lattice energy of the oxide of boron to be in good or poor agreement with a theoretically derived value? Give your reasons.

4 Energy and reactions

4.12 This question is about the stability and structure of calcium(II) iodide. The following experimental data are provided.

	$\Delta H^\ominus(298)/\text{kJ mol}^{-1}$
First ionization energy of calcium	+ 590
Second ionization energy of calcium	+1145
Enthalpy change of atomization of calcium	+ 178
Enthalpy change of formation of calcium(II) iodide	− 534
Enthalpy change of atomization of iodine per mol of I	+ 107
Electron affinity of iodine atoms	− 295

Calcium(II) iodide crystals have a layer structure in which the calcium–iodine separation is 0.31 nm and the iodine–iodine separation is 0.43 nm. The usual radii are given below.

	$R(\text{cov})/\text{nm}$	$R(\text{ion})/\text{nm}$
Calcium	0.17	0.09
Iodine	0.13	0.22

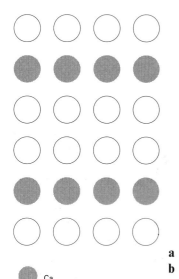

Ca
I

The layer structure can be represented as shown in the margin (not to scale).
- **a** Explain in full the meaning of $\Delta H^\ominus(298)$
- **b** The enthalpy change of formation of calcium(I) iodide is -58 kJ mol^{-1}. Which of the two compounds is likely to be the more stable one relative to the elements? Give your reason.
- **c** Construct a **labelled** Born–Haber cycle for the formation of calcium(II) iodide.
- **d** From your Born–Haber cycle, calculate the lattice energy of calcium(II) iodide.
- **e** In crystalline calcium(II) iodide, what bonding would you predict? Show how the data in the table of radii support your prediction.
- **f** By what other calculation could the bonding in calcium(II) iodide be confirmed?

*4.13 What do you understand by the term 'lattice energy'? What experimental data are needed in order to calculate the value for the lattice energy of an ionic compound using the Born–Haber cycle?

Using the *Book of data* to provide the necessary data, calculate a value for the lattice energy of cadmium oxide, CdO, by means of a Born–Haber cycle.

Compare the value you obtain with the theoretical value given in table 5.9 of the *Book of data*. Do you think the two values are significantly different? What do you deduce from their similarity or difference?

On what basis are theoretical values for lattice energies calculated?

TOPIC 5

The halogens and redox reactions

This topic is about the halogens, the elements of Group 7 of the Periodic Table. You will be considering the reactivity of both the elements and some of their compounds, and how that reactivity is related to their electronic structure and their position in the Periodic Table. You will also be using 'oxidation numbers', which are very useful when looking for patterns in redox reactions.

All the halogens are in use commercially, either as the free element or in compounds. For example, chlorine is used as a disinfectant, both as the element and in compounds; halogens and their compounds are important as anaesthetics and solvents. Halogen compounds have in the past been used as coolants in refrigerators and as propellants in aerosols but we are now aware of their danger to health and the environment. Knowledge of their chemistry enables us to find substitutes for them.

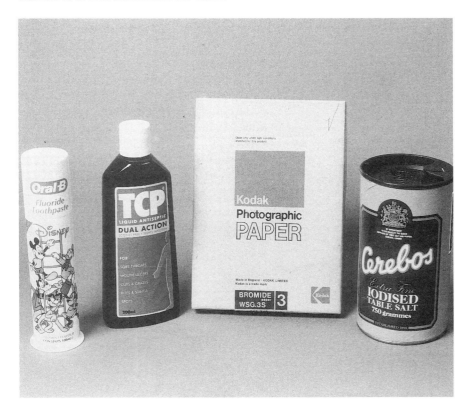

Figure 5.1 How the halogens are useful to us

5.1 Sources of the halogens

The halogens are obtained from a variety of sources and extracted by a number of different methods depending on convenience and cost. Some information is summarized in figure 5.2.

Halogen	Abundance (parts per million by mass) in rocks	in the sea	Source	Relative cost of sodium halide (laboratory grade)	Chemical process to obtain the free element
Fluorine	700	1.4	fluorite, CaF_2, e.g. Derbyshire 'Blue John'	5	Electrolysis of a solution of potassium fluoride in anhydrous hydrogen fluoride.
Chlorine	200	19 000	rock salt, NaCl, and sea water	1	Electrolysis of an aqueous saturated solution of sodium chloride.
Bromine	3	67	sea water	2	Oxidation of bromide solution by chlorine.
Iodine	0.3	0.05	caliche, $NaNO_3$, containing $NaIO_3$	7	Reduction of iodate solution by sodium hydrogensulphite.

Figure 5.2

Several fluorine compounds occur as minerals in rocks, but these are usually so widely dispersed that few deposits are large enough to be worked economically.

By contrast iodine, with much the lowest overall abundance of the halogens, occurs in extensive deposits in northern Chile between the Pacific ocean and the Andes mountain range. The rock is called **caliche** and the principal mineral is sodium nitrate, but it also contains iodine compounds in varying proportions, usually high enough to make the extraction of the element economically feasible. For example, one sample of caliche was analysed at 500 grams of sodium iodide per tonne and 1500 grams of sodium iodate(v), $NaIO_3$, per tonne.

Figure 5.3 The Atacama salt lake in Chile. It is so large you can find it on a map of South America quite easily.

Sea-water is a good source of chemicals, and chlorine and bromine are among the elements whose compounds are obtained from it. However by far the most important source of chlorine compounds is rock salt, which is mainly sodium chloride. Rock salt deposits are formed by the evaporation of sea-water, a process that has produced vast rock salt deposits in many parts of the world, including Britain, and is still doing so today in places like the Dead Sea.

The occurrence of hydrogen fluoride and hydrogen chloride gases in nature is surprising, as they are so reactive. They are usually associated with volcanic action; in the Valley of Ten Thousand Smokes in Alaska, for example, eruptions early this century produced over one million tonnes of hydrogen chloride and nearly a quarter of a million tonnes of hydrogen fluoride per year.

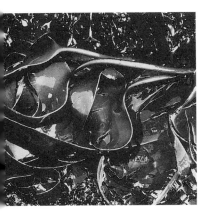

A number of plants and animals concentrate halogen compounds from their environment to a remarkable extent. The *Laminariaceae* seaweeds (see photograph alongside) can contain 800 parts per million of iodine in fresh wet weed, taken in as iodide ions from the sea where iodine is present at a concentration of only 0.05 part per million.

Certain species of marine snail, called *Murex brandaris*, found principally in the Eastern Mediterranean, take in bromide ions to produce compounds which are fine dyes, such as Tyrian purple (dibromoindigo). The tea plant, *Camellia sinensis*, takes in fluoride ions from the soil to the extent that dried tea leaves contain about 100 parts per million of fluorine, which results in a concentration of about 1 part per million in a cup of tea.

STUDY TASK

1. The passage gives details of the iodine compounds in one sample of caliche. How many tonnes of caliche of this composition would be needed for the extraction of one mole of iodine molecules, I_2?
   What assumptions have you made in your calculation?
2. Find out how the UK obtains its supplies of the halogens, including sources, the methods of extraction used and some idea of the annual quantities needed. This may be done as a group activity, with each group presenting findings about one of the halogens to the rest of the class as, for example, a poster or an OHP transparency.
3. Similarly, find out about the range of uses of and products made from each halogen.
4. Make notes summarizing the findings about each halogen.

EXPERIMENT 5.1

The extraction of iodine from seaweed

For this experiment you need a *Laminariaceae* seaweed that has been dried: other varieties of seaweed contain much less iodine. After releasing the iodine by oxidation you will concentrate it by a process called 'solvent extraction'.

Procedure

Burn about 2 g of dried seaweed to ash on a tin lid in a fume cupboard if possible, because of the smell. Transfer the ash to a small beaker and boil with 25 cm³ of water for about 3 minutes.

Filter into a boiling tube, and add 3 cm³ of dilute sulphuric acid and 10 cm³ of dilute (20 volume) hydrogen peroxide to the solution.

> **COMMENT**
> '20 volume' hydrogen peroxide when fresh will give off 20 cm³ of oxygen gas per 1 cm³ of solution.

Pour your mixture into a separating funnel and add 10 cm³ of hydrocarbon solvent. Stopper the funnel and mix the contents well by inverting the funnel several times – release any build up of pressure by opening the tap when the funnel is upside down. Run the upper layer into a small conical flask and label it as 'Iodine extract in hydrocarbon solvent'.

- How might you obtain iodine from your solution? Write a description of solvent extraction, using diagrams where you think they will be helpful.

5.2 Redox reactions and oxidation numbers

REVIEW TASK
Working in groups, list the main features of redox reactions.

In Topic 1 you were shown how restricting redox reactions to reactions involving oxygen was too limiting; in this section we will develop a more general definition and use it to introduce the idea of oxidation numbers.

EXPERIMENT 5.2

The reactions between halogens and halide ions

This experiment investigates the relative reactivity of the halogen elements towards the halide anions.

Use the halogen elements chlorine, bromine, and iodine in solution: chlorine and bromine are dissolved in water; iodine is dissolved in aqueous potassium iodide, as the solubility of iodine in water is low.

Procedure

1 Set up four test-tubes containing about 1 cm³ each of solutions of potassium chloride, potassium bromide, and potassium iodide, and water as a control.
a Add two or three drops of chlorine solution to each.

- Have reactions taken place? What are the products? Use the colour changes as a guide.
  Write equations for any reactions you see. Would the addition of a hydrocarbon solvent help you in reaching a decision?

SAFETY ⚠
Handle the solutions with care; chlorine is toxic, bromine and iodine are corrosive. Avoid inhaling any vapours, and do not allow the solutions to come into contact with your skin or clothing.
Fluorine is too hazardous for use under ordinary laboratory conditions.
Remember to wear eye protection.

b Add an equal volume of hydrocarbon solvent to each test-tube, stopper the test-tubes and shake.

- Why do you think the halogens are more soluble in hydrocarbon solvent than water? This will be considered in detail in Topic 9.

2 Now repeat the experiment, using in turn bromine solution and iodine solution. Is a definite trend in reactivity observable in this experiment?

- Write ionic equations for every reaction that took place.
  Draw up a table for recording your results. A suitable table is shown overleaf.

	Action on			
Solution added	water	potassium chloride solution	potassium bromide solution	potassium iodide solution
Chlorine solution				
etc.				

Oxidation and reduction by electron transfer

You should have realized from the equations you have just written that the reactions involved the transfer of an electron from one halogen to another.

When the reactions are analysed into component reactions such as

$$2Br^-(aq) \longrightarrow Br_2(aq) + 2e^-$$
$$\text{and } 2e^- + Cl_2(aq) \longrightarrow 2Cl^-(aq)$$

you can see how the reactions involve the transfer of electrons.

In the first half-reaction, each bromide ion loses an electron when it is oxidized; in the second half-reaction each chlorine atom gains an electron when it is reduced. In the complete reaction bromide ion is the reducing agent and chlorine is the oxidizing agent.

$$\underset{\text{reducing agent}}{2Br^-(aq)} + \underset{\text{oxidizing agent}}{Cl_2(aq)} \longrightarrow 2Cl^-(aq) + Br_2(aq)$$

COMMENT
These are known as **half-reactions**.

To summarize:

Loss of electrons is an *oxidation* process

And the opposite process:

Gain of electrons is a *reduction* process

It follows from these definitions that compounds that gain electrons in reactions are acting as oxidizing agents; those that lose electrons are acting as reducing agents.

COMMENT
The mnemonic OIL RIG 'Oxidation Is Loss, Reduction Is Gain' will help you remember this.

The chemists' toolkit: oxidation numbers

The halogens combine with almost all other elements, as well as each other, and so have a large number of compounds. It is not easy to name halogen compounds in a way that is both unambiguous and chemically helpful. One way in which they can be classified is according to the 'oxidation number' of the halogen atom in the compound.

When you look at formulae by Periodic Table groups you can see definite patterns. This can be seen, for example, when comparing the formulae of the chlorides and oxides of the elements of Period 3.

In some cases only one of several possible compounds has been selected.

| NaCl | $MgCl_2$ | $AlCl_3$ | $SiCl_4$ | PCl_3 | BrCl |
| Na_2O | MgO | Al_2O_3 | SiO_2 | P_2O_3 | Br_2O |

We will begin by considering ionic compounds. In ionic compounds such as those listed above, the charge on the ion of each element is taken as the oxidation number of that element. In NaCl, therefore, the oxidation number of sodium is $+1$ and that of chlorine is -1, and in sodium monoxide, Na_2O, sodium and oxygen have oxidation numbers of $+1$ and -2 respectively.

Just as the overall positive and negative charges of an ionic compound balance and their sum is zero, so the sum of the oxidation numbers in any compound is zero.

With the oxidation number of oxygen fixed as -2 in all its common compounds, the use of oxidation number can be extended to molecular compounds. For example, in the molecular compound CO_2 the oxidation number of carbon must be $+4$ for the total sum of the oxidation numbers to be zero.

Extensions of this sort enable one to assign an oxidation number to any element in any compound, once the empirical formula of that compound has been determined experimentally.

Rules for assigning oxidation numbers

1. The oxidation number of any uncombined element is zero.
2. The oxidation number of each of the atoms in a compound counts separately, and their algebraic sum is zero.
3. The oxidation number of an element existing as a monatomic ion is the charge on that ion.
4. In a polyatomic ion, the algebraic sum of the oxidation numbers of the atoms is the charge on the ion.
5. Many elements have invariable oxidation numbers in their common compounds, including

Group 1 metals	$+1$
Group 2 metals	$+2$
Al	$+3$
H	$+1$ except in metal hydrides
Group 7 halogens	-1 except in compounds with oxygen
O	-2 except in peroxides

COMMENT
Hydrogen is -1 in $LiAlH_4$
Chlorine is $+1$ in Cl_2O
Oxygen is -1 in H_2O_2 (the structure is H—O—O—H)
Transition metals each have a range of oxidation numbers.

The application of these rules is relatively straightforward. For example, in binary compounds of one metal with one non-metal, there is no difficulty in deciding which sign should be given to which element; the metal is given a positive sign and the non-metal a negative one. For many other compounds the signs can be decided by using the invariable oxidation numbers given in rule **5**. The signs are always relative to other elements. For example, the oxidation number of sulphur in sodium sulphide, Na_2S, is -2; its oxidation number in sulphur dioxide, SO_2, however, is $+4$.

As an example of the rule about polyatomic ions consider the SO_4^{2-} ion: the oxidation number of oxygen is fixed as -2, the total for oxygen is therefore -8, so the oxidation number of sulphur must be $+6$ for the algebraic sum to be the charge on the ion (-2).

QUESTIONS

1 Work out the oxidation numbers of each element in the compounds

$KBr \quad BaO \quad Al_2S_3 \quad NO_2 \quad NH_3 \quad SO_3 \quad NaNO_3 \quad Na_2CO_3 \quad NaClO_4$

2 Work out the oxidation numbers of vanadium in VO_2^+ and VO_3^-. Can you write an equation involving protons that shows that these ions are related by an acid–base reaction rather than a redox reaction?

Oxidation and reduction

A change in the oxidation number of an element in a reaction can be used to discover whether the element has been oxidized or reduced. In a particular reaction, a substance which increases the oxidation number of an element is called an **oxidizing agent**, whereas one which decreases the oxidation number of an element is called a **reducing agent**. The word 'increases' is taken to mean 'makes more positive or less negative'.

Oxidation numbers and the Stock notation

The Roman numerals used in the naming of compounds of metals are, in fact, the oxidation numbers of these elements. This system of naming is known as the **Stock notation**, after the chemist who devised it. It is usually only used when it is useful to have a simple way of distinguishing between similar compounds. For example, in the oxide of copper, CuO, the oxidation number of copper is $+2$ and the compound is known as copper(II) oxide. In the oxide Cu_2O the oxidation number of copper is $+1$ and the compound is known as copper(I) oxide.

Stock notation is used less widely for distinguishing between the compounds of non-metals. Compounds such as NO and NO_2 are referred to by the names nitrogen monoxide and nitrogen dioxide, rather than nitrogen(II) oxide and nitrogen(IV) oxide.

When naming oxoacids, the oxidation number of the central atom in the acid is written after the rest of the name, which always ends in '–ic'. For example, H_3PO_4 is phosphoric(V) acid.

The salts and ions of oxoacids are named by writing the oxidation number of the central atom after the rest of the name, which always ends in '–ate'. For example, in Na_2SO_3 (anion SO_3^{2-}), sulphur has an oxidation number of $+4$ and is called sodium sulphate(IV).

The salts of the common acids are usually named without including the appropriate oxidation number: Na_2SO_4 is sodium sulphate, not sodium sulphate(VI) and $NaNO_3$, sodium nitrate, is not sodium nitrate(V).

5.3 The range of halogen compounds

Oxidation number chart for chlorine

+7	Cl_2O_7	$HClO_4$	$K^+ClO_4^-$
+6	$Cl_2O_6 \rightleftharpoons 2ClO_3$		
+5		$HClO_3$	$K^+ClO_3^-$
+4	ClO_2		
+3		$HClO_2$	$K^+ClO_2^-$
+2			
+1	Cl_2O	$HClO$	K^+ClO^-
0	Cl_2		
−1		HCl	K^+Cl^-, PCl_3

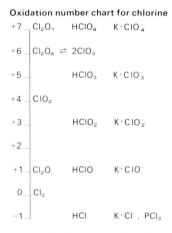

Figure 5.4 Oxidation number chart for chlorine

Oxidation numbers are invariable for some elements, but those of a number of elements, such as the halogens, may have different values in different compounds. The range of oxidation numbers of such elements is very well shown by constructing an oxidation number chart. The formulae of the various compounds can be written on the chart, as shown in figure 5.4.

In this book the Stock names for these compounds will be used, but as the old names may still be found, both names are included in figure 5.5. The potassium salts are given as examples.

Formula of compound	Oxidation no. of Cl	Stock name	Old name
$KClO_4$	+7	potassium chlorate(VII)	potassium perchlorate
$KClO_3$	+5	potassium chlorate(V)	potassium chlorate
$KClO_2$	+3	potassium chlorate(III)	potassium chlorite
$KClO$	+1	potassium chlorate(I)	potassium hypochlorite
KCl	−1	potassium chloride	potassium chloride

Figure 5.5

This section will be devoted to reactions of the halogens (oxidation number 0) with alkalis, and then later sections will deal with compounds with oxidation number −1, oxidation number +1, and oxidation number +5.

EXPERIMENT 5.3a

Oxidation number 0: the reactions of the halogens with alkalis

SAFETY ⚠

The alkalis used in this experiment are very corrosive, especially to eyes. Eye protection MUST be worn, and great care should be taken in doing the experiment. Protective gloves should be worn when clearing up any spillages.

Procedure

Take 2 cm^3 samples of solutions of each of the halogens in water and add a few drops at a time of 1 M sodium hydroxide solution. It should be easy to see what happens to the bromine and iodine because the solutions are coloured; the chlorine is less easy to observe. Record your observations as follows.

Halogen solution	Observations on adding alkali	Equation

An interpretation of the reactions of the halogens with alkalis

Halogens react with cold sodium hydroxide solution according to the pattern set by chlorine:

$$Cl_2(g) + 2NaOH(aq) \longrightarrow NaCl(aq) + NaClO(aq) + H_2O(l)$$

The compound with formula NaClO is called sodium chlorate(I) or sodium hypochlorite.

QUESTIONS
1 Turn the equation into an **ionic equation**, leaving out the sodium ions since these do not undergo chemical change.
2 What changes of oxidation number does the chlorine undergo?

A reaction in which the same element both increases and decreases in oxidation number is called a **disproportionation reaction** – we say that chlorine 'disproportionates' when it reacts with alkalis.

When the solution is hot, chlorate(I) ions themselves disproportionate so that the overall reaction between chlorine and hot sodium hydroxide is

$$3Cl_2(aq) + 6NaOH(aq) \longrightarrow 5NaCl(aq) + NaClO_3(aq) + 3H_2O(l)$$

or, ionically,

$$3Cl_2(aq) + 6OH^-(aq) \longrightarrow 5Cl^-(aq) + ClO_3^-(aq) + 3H_2O(l)$$

QUESTIONS
1 What changes of oxidation number does the chlorine now undergo?
2 What is the equation for the reaction between iodine and hot potassium hydroxide solution?

EXPERIMENT 5.3b The preparation of potassium iodate(v)

The equation you wrote in answer to question **2** above represents a reaction which can be used to prepare samples of potassium iodide and potassium iodate(v) from iodine.

Procedure
1 Take about 10 cm³ of 4 M potassium hydroxide solution (TAKE CARE) in a boiling-tube and heat it in a beaker of boiling water. Cautiously add solid iodine (TAKE CARE), a little at a time, until there is a very slight excess, that is, until the iodine colour is only just visible. Then add the minimum quantity of 4 M potassium hydroxide solution (a drop or two) to react with the excess of iodine to give a solution which is a very pale yellow colour.
2 The hot solution now contains both potassium iodate(v) and potassium iodide dissolved in water. The solubility of these two substances varies with temperature as shown in figure 5.6. Use the graph to predict the identity of the crystals formed by cooling the hot solution.

Allow the solution to cool to room temperature and filter off the crystals, using suction filtration with a Buchner funnel (see figure 5.7). Wash the crystals with about 5 cm³ of water and transfer them to fresh filter paper to dry them.
3 Transfer the solution in the Buchner flask to an evaporating basin and cautiously evaporate the solution to dryness, using a water bath for heating if necessary.

> **SAFETY** ⚠
> The alkali used in this experiment is very corrosive, especially to eyes. You **MUST** wear safety glasses and take great care. Wear protective gloves when clearing up any spillages. Iodine is corrosive so do not touch any.

> **SAFETY** ⚠
> The mixture has a tendency to 'spit' towards the end of the evaporation.

5 The halogens and redox reactions 112

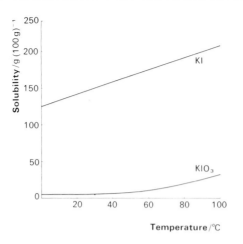

Figure 5.6 The change in solubility of potassium iodate(V) and potassium iodide with temperature

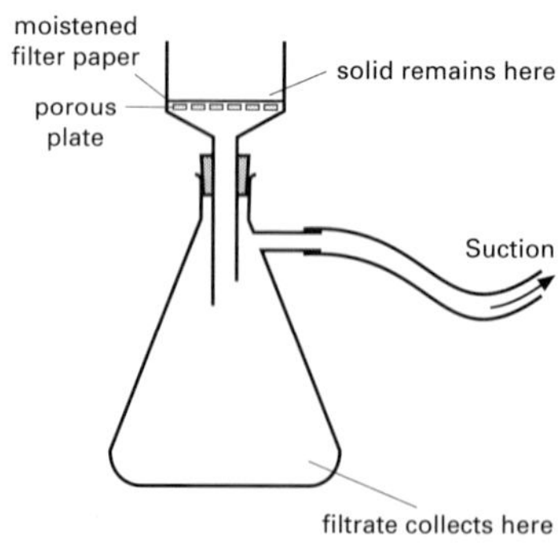

Figure 5.7 A Buchner flask prepared for use

- Write up this experiment explaining how you used the graph to identify your products. Label your products and keep them. When you have completed the next sections you should be able to devise tests to confirm your deductions from the graph. Consider whether your tests will work if your products are impure.

EXPERIMENT 5.3c

The reaction between iodine and sodium thiosulphate

The reaction to be investigated in this section has a practical value in the quantitative analysis of oxidizing agents. You will first be investigating the reaction itself and then using it to analyse the samples you obtained in experiment 5.3b.

Procedure

Titrate 10 cm^3 samples of 0.010 M iodine, I_2, solution with 0.010 M sodium thiosulphate, $Na_2S_2O_3$, solution. You can measure the iodine solution using a burette or a pipette; if the latter, you must use a pipette filler to fill it. The sodium thiosulphate solution must be delivered from a burette.

You will probably be able to do these titrations without using an indicator because the iodine solution is yellow–red in colour and the products of the reaction are colourless. Nevertheless the endpoint can be 'sharpened' considerably by adding a few drops of 1% starch when the iodine colour has become very pale. A very dark blue colour is produced which suddenly disappears at the endpoint of the titration.

Record the details of the experiment in your notes and give your titration results in the form of a table as in experiment 3.6.

Show that your titration results are consistent with the equation for the reaction, which is

$$2Na_2S_2O_3(aq) + I_2(aq) \longrightarrow 2NaI(aq) + Na_2S_4O_6(aq)$$

or ionically,

$$2S_2O_3^{2-}(aq) + I_2(aq) \longrightarrow 2I^-(aq) + S_4O_6^{2-}(aq)$$

Record these equations in your notes.

Work out the oxidation number of sulphur in sodium thiosulphate, $Na_2S_2O_3$, and in sodium tetrathionate, $Na_2S_4O_6$. It is interesting that the oxidation number of sulphur in sodium tetrathionate contains a fraction. This does not invalidate the use of the oxidation number and the situation is not unusual, particularly in organic chemistry.

This reaction may be used to estimate the concentrations of oxidizing agents which will oxidize iodide ions to iodine. Either or both of the following experiments may now be done.

EXPERIMENT 5.3d To determine the purity of samples of potassium iodate(v)

Part 1

You are going to find the percentage purity of the potassium iodate(v) from experiment 5.3b. It is quite possible that this contains small amounts of other substances. This experiment is intended to find out how much of a weighed sample of your product is actually potassium iodate(v) and to express this as a percentage.

Procedure

1 Weigh out accurately about 0.05 to 0.1 g of your potassium iodate(v), dissolve it in pure water in a beaker, and transfer the solution through a funnel to a 100 cm³ volumetric flask. Rinse out the beaker several times with water and add the rinsings to the flask. Then make up the volume of the solution to the mark on the neck of the flask with pure water. Mix the contents of the flask well.

2 To 10.0 cm³ portions of this potassium iodate(v) solution, taken with a pipette and pipette filler or with a burette, add about 10 cm³ of approximately 0.1 M potassium iodide and about 10 cm³ of 1 M sulphuric acid. The effect of this is to liberate iodine according to the equation:

$$IO_3^-(aq) + 5I^-(aq) + 6H^+(aq) \longrightarrow 3I_2(aq) + 3H_2O(l)$$

3 Titrate each sample with 0.010 M sodium thiosulphate, using 1% starch as indicator. Record your results in a table.

4 Use your results to calculate the percentage purity of your potassium iodate(v) as follows.

Calculation

1 How many moles of sodium thiosulphate, $Na_2S_2O_3$, were used in an average titration?
2 How many moles of iodine molecules did these react with? (See the equation in the previous experiment.)
3 How many moles of iodate(v) ions are involved in producing this iodine? (See equation above.)
4 What mass of potassium iodate(v) is this?
5 The mass of pure potassium iodate(v) in 100 cm³ of solution is 10 times this.

6 Calculate the percentage purity of the potassium iodate(v) according to the relationship:

$$\text{percentage purity} = \frac{\text{mass of KIO}_3 \text{ as calculated in } \mathbf{5}}{\text{mass of crude KIO}_3} \times 100$$

Part 2

You are going to find the percentage of potassium iodate(v) remaining in the potassium iodide from experiment 5.3b. The principal impurity in the potassium iodide is likely to be potassium iodate(v). When this mixture is acidified, iodine will be liberated according to the equation:

$$IO_3^-(aq) + 5I^-(aq) + 6H^+(aq) \longrightarrow 3I_2(aq) + 3H_2O(l)$$

Procedure

Weigh out accurately about 0.5 g of crude potassium iodide and dissolve it in water, making up the solution to 100 cm³ as in the previous experiment. Acidify 10 cm³ portions of the solution with 10 cm³ of 1 M sulphuric acid, titrate with 0.010 M sodium thiosulphate using starch as indicator, and record your results in a table.

Calculate the percentage of potassium iodate(v) in the sample in the same way as in the previous experiment.

5.4 Oxidation number −1: the properties of the halides

The halides of many elements are toxic or corrosive. There are, however, some properties of halides that can be investigated with relative safety using compounds of the alkali metals, and they are included in the next experiment.

EXPERIMENT 5.4 Some reactions of the halides

Procedure

1 The silver halides

Use 0.1 M solutions of potassium (or sodium) chloride, bromide and iodide. Where you can, attempt to estimate roughly the proportions of the solutions needed for complete reaction.

a To 1 cm³ portions of each of the halide solutions, add a few drops of 0.1 M silver nitrate solution.
b To the precipitates obtained in part **a** add ammonia solution.
c Obtain a second set of silver halide precipitates and leave them exposed to the light for an hour.

- Record your results in a table, noting similarities and differences between the reactions.

> **COMMENT**
> The photochemical change which occurs when silver bromide is exposed to sunlight is used in black-and-white photography. The silver ions are converted to silver metal which remains as an opaque image on the photographic film.

> **SAFETY** ⚠
> Concentrated sulphuric acid and solid phosphoric acid are corrosive. Use **solid potassium (or sodium) chloride, bromide, and iodide. You MUST wear eye protection.**

2 The action of concentrated sulphuric acid on the potassium salts

a Put about 0.1 g of the solid salt into a test-tube (about enough to fill the rounded end of the tube if it is 100 × 16 mm) and add about 10 drops of concentrated sulphuric acid (TAKE CARE). Warm the reaction mixture gently if necessary.

- Identify as many products as you can (test with strips of filter paper moistened with lead(II) ethanoate, and potassium dichromate(VI) solutions). Note the similarities and differences between the reactions. Record and explain your observations as fully as you can.

b Repeat part **a**, using phosphoric acid (TAKE CARE) in place of sulphuric acid. Note any difference.

3 The properties of the hydrogen halides

Use the reactions in **2** to prepare and collect samples of hydrogen chloride, hydrogen bromide, and hydrogen iodide. The apparatus shown in figure 5.8 is convenient for this purpose. A good yield of gas is obtained if solid 100 per cent phosphoric acid is used, but you must use **dry** test-tubes to collect your samples of gas.

Mix about 2 g halide with an equal quantity of solid phosphoric acid (TAKE CARE) in the side-arm test-tube. Cork it securely. Put a **dry** test-tube round the delivery tube and warm the mixture gently until gas is evolved. Collect at least three tubes of gas, corking them with a dry cork when apparently full.

Use the tubes of gas to investigate:

> **SAFETY** ⚠
> Do this reaction in a fume cupboard as toxic gases are formed.

a *The solubility of the gas in water.* Invert a tube of gas in a beaker of water and remove the cork.

- If the water rises rapidly the gas is readily soluble. Is there a residue of undissolved gas, and if so, what do you suppose it is?

b *The reaction of the gas with ammonia gas.* Hold a drop of fairly concentrated ammonia solution in the mouth of an open test-tube, using a glass tube or rod.

- What do you observe, and what do you suppose is formed?

c *The stability of the gas towards heat.* Heat the end of a length of nichrome wire or a glass rod to dull red heat, and plunge it into a tube of gas; if no change occurs in the gas, try again with the wire hotter.

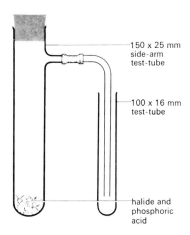

Figure 5.8 Apparatus for making the hydrogen halides

- What do you observe?
  Record the properties of these hydrogen halides in a table in your notes, and write equations for the reactions you have seen.

5.5 Oxidation number +1: the reactions of sodium chlorate(I)

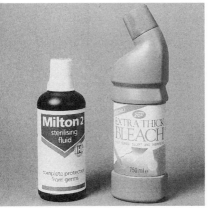

Figure 5.9 Products based on sodium chlorate(I)

Sodium chlorate(I) is stable only in solution, and is made by reaction between chlorine and cold sodium hydroxide solution. It is alternatively known as sodium hypochlorite, and is sold in solution as 'bleach', or as a very dilute solution used as a sterilizing agent for babies' feeding bottles.

$$Cl_2(aq) + 2NaOH(aq) \longrightarrow NaCl(aq) + NaClO(aq) + H_2O(l)$$

On standing, particularly in sunlight, the solution evolves oxygen. Because of this, bleach solutions are normally supplied in opaque plastic containers or in brown bottles.

$$2NaClO(aq) \longrightarrow 2NaCl(aq) + O_2(g)$$

What type of reagent would you expect sodium chlorate(I) to be in redox reactions?

EXPERIMENT 5.5a

Some reactions of sodium chlorate(I)

> **SAFETY** ⚠
> Handle sodium chlorate(I) solution with care. You MUST wear eye protection.

1 Reaction with iron(II) ions, Fe^{2+}(aq)

Add some sodium chlorate(I) solution to a solution of iron(II) sulphate. Test to find out the oxidation number of the iron in the product.

- Record your observations and try to write a balanced equation for the reaction.

2 Reaction with iodide ions, I^-(aq)

Add some sodium chlorate(I) solution to a solution of potassium iodide.

- Record your observations and try to write a balanced equation for the reaction.

QUESTIONS
1. What changes of oxidation number occur during these reactions?
2. Read the passage overleaf on balancing redox equations, then use the rules to help you balance the equations of the reactions you have just carried out.
3. In the second experiment you should have seen the colour diminish in intensity a moment or two after it appeared: try to work out why.

The chemists' toolkit: balancing redox equations

It is sometimes helpful to use oxidation numbers in balancing the equation for a redox reaction.

In a simple example, the oxidation number method is not really necessary:

$$2Br^-(aq) + Cl_2(aq) \longrightarrow 2Cl^-(aq) + Br_2(aq)$$
$$2 \times (-1) \quad (0) \quad\quad 2 \times (-1) \quad (0)$$

But if you needed to balance

$$ClO_3^- + H^+ + Fe^{2+} \longrightarrow Cl^- + Fe^{3+} + H_2O$$

the process can be made much easier by using oxidation numbers.

First you have to identify the elements which actually change in oxidation number. The chlorine changes from $+5$ in ClO_3^-(aq) to -1 in Cl^-(aq), a change of 6 units. The iron changes from $+2$ in Fe^{2+}(aq) to $+3$ in Fe^{3+}(aq), a change of 1 unit.

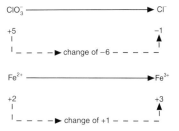

The total change in oxidation number must be the same in both directions, so there must be six Fe^{2+}(aq) ions reacting with every ClO_3^-(aq) ion. The first stage of balancing thus gives:

```
            ┌ - - - - - - ► change of +6 - - - - - - - ┐
            │                                           ▼
1 ClO₃⁻  +  H⁺  +  6 Fe²⁺ ─────────────► 1 Cl⁻  +  6 Fe³⁺  +  H₂O
 +5                 +2                    -1         +3
  │                                        ▲
  └ - - - - - - - ► change of -6 - - - - - ┘
```

The remaining balancing numbers may now be inserted, so that the balanced equation is

$$ClO_3^-(aq) + 6H^+(aq) + 6Fe^{2+}(aq) \longrightarrow Cl^-(aq) + 6Fe^{3+}(aq) + 3H_2O(l)$$

To balance a redox equation:
1 **Write down oxidation numbers.**
2 **Calculate changes in oxidation numbers which occur.**
3 **Balance to give a total oxidation number change of zero.**
4 **Balance for oxygen, hydrogen and water.**
5 **Check that the $+$ and $-$ charges balance.**

INVESTIGATION 5.5b

Make a risk assessment before starting. Bromine is a dangerous chemical.

The reaction of bromine with sodium thiosulphate

Carry out an investigation to determine the equation of the reaction between bromine and sodium thiosulphate in dilute solutions.

5.6 Oxidation number +5: the reactions of the potassium halates(V)

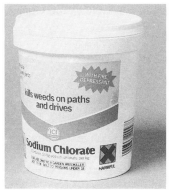

The halates(V) are vigorous oxidizing agents which must be handled with great care. Potassium and sodium chlorate(V) are very dangerous indeed, and mixtures of the solids with many other substances explode in a violent and unpredictable manner. Because of this it is essential to check the names of the various compounds of the halogens with great care before carrying out any experiments.

Remember that when the words chlorate, bromate, or iodate in a name are **not** followed by a Roman numeral, the names are probably the old names for chlorate(V), bromate(V), and iodate(V).

Figure 5.10 Some weed killers are based on sodium chlorate(V), $NaClO_3$

EXPERIMENT 5.6 Some reactions of the potassium halates(V)

The three potassium halates(V) which are used in this set of experiments have the formulae $KClO_3$, $KBrO_3$, and KIO_3.

Procedure

1 Action of heat

Using dry, hard-glass test-tubes, investigate the action of heat on each of the solid potassium halates(V).

SAFETY
Wear eye protection.

- What gas do you expect to be given off? Test for the possible gases which might be given off.
  Compare the three halates(V) in their ease of decomposition.

2 Reaction with iron(II) ions, Fe^{2+} (aq)

Make a solution of each of the halates(V) in turn and acidify each with 1 M sulphuric acid. Add the acidified solutions to samples of iron(II) sulphate solution.

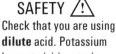
SAFETY
Check that you are using **dilute** acid. Potassium bromate (V) is a toxic substance.

- Describe and try to explain what you see.

3 Reaction with iodide ions, I^-(aq)

Add acidified samples of each of the halates(V) to portions of potassium iodide solution.

- Which species have been oxidized and which reduced in the reactions that occurred? You have encountered one of these reactions previously in this Topic.

5.7 The properties of the halogens

Write a general account of halogen chemistry in your notes, using the outline below as a guide. You should begin by explaining what a halogen is and then list the characteristic properties with carefully selected examples in each case. Use your experimental results from the previous sections together with the *Book of data*.

Consult reference books and textbooks as additional sources for examples or to clarify your experimental results. Use a spreadsheet programme if one is available to display your data as bar charts. This should make trends and patterns in the data easier to recognize.

1 Trends in physical properties

The physical properties of the halogens show considerable variation down their Periodic Table group.

Include in your survey of their properties a description of their appearance and their hazardous properties, together with data on molar masses, melting and boiling points, and electronic configurations. Report any trends that you can identify in the data. The data you need are in tables 5.2 and 4.2 of the *Book of data*.

We suggest you use a table to record your findings, similar to this one.

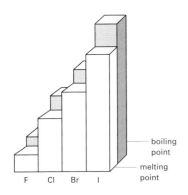

Figure 5.11 Trends in melting and boiling points

COMMENT
You may prefer to record temperatures in °C in which case convert the temperatures in kelvins in the *Book of data* into °C by subtracting 273.

Name	Formula, molar mass	Appearance of element	Hazardous properties	Melting point	Boiling point	Electronic configuration
Fluorine	F_2	yellow gas				
Chlorine	Cl_2					
Bromine	Br_2					
Iodine	I_2					

The appearance and physical properties of the halogens

2 Electronic structure of the halide ion

Look up the first ionization energies and the electron affinities in tables 4.1 and 5.10 in the *Book of data*. Do the values support the formation of anions rather than cations in reactions? Write out the electronic configuration of the anions.

Also record the sizes of the atoms and anions as their radii; to get a feel for the relative sizes you should make scale drawings. The data you need are in table 4.4 (use van der Waals radii for the atomic radii: the meaning of van der Waals radius will be explained in Topic 9).

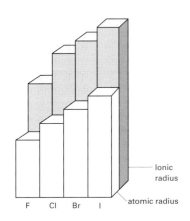

Figure 5.12 Trends in size

COMMENT
Atomic radii are quoted in nanometres, 1 nm = 10^{-9} metre.

Name	Symbol	First ionization energy /kJ mol^{-1}	Electron affinity /kJ mol^{-1}	Atomic radius /nm	Ionic radius /nm
Fluorine	F				
Chlorine	Cl				
Bromine	Br				
Iodine	I				

3 Variable oxidation numbers

Find out the oxidation numbers of a variety of types of halogen compounds and draw up a table similar to the one in figure 5.5, but do not bother to list the old names.

4 Relative reactivity of the compounds

For each oxidation number from -1 to $+5$ compare the relative reactivity of the halogen compounds to thermal decomposition and redox reagents. You will need your laboratory results for this task.

5 Reactivity and structure across a period of the Periodic Table

Good examples of periodicity amongst the elements can be obtained by studying information on various physical properties of their chlorides as these change with atomic number. For example, there is a transition from ionic bonds on the lefthand side of the Periodic Table to covalent bonds on the righthand side of the Table.

A convenient period to choose is Period 3, which starts with sodium and ends with argon, as samples of the elements and chlorides are likely to be available for you to look at. For convenience, some of the information is summarized in figure 5.13.

COMMENT
By **periodicity** chemists mean properties which change in regular patterns that match the sequencing of the elements in the Periodic Table.

Principal chloride	Formula and state at room temperature	Structure	Bonding	Standard enthalpy change of formation at 289 K/kJ mol^{-1}
Sodium chloride	NaCl(s)	ionic lattice	ionic	-411
Magnesium chloride	MgCl$_2$(s)	layer lattice	ionic	-641
Aluminium chloride	AlCl$_3$(s)	layer lattice	covalent	-704
Silicon tetrachloride	SiCl$_4$(l)	SiCl$_4$ molecules	covalent	-687
Phosphorus trichloride	PCl$_3$(l)	PCl$_3$ molecules	covalent	-320
Disulphur dichloride	S$_2$Cl$_2$(l)	S$_2$Cl$_2$ molecules	covalent	-59
Chlorine	Cl$_2$(g)	Cl$_2$ molecules	covalent	0
Argon	no chloride	–	–	–

Formula	Effect of adding the chloride to water	Equation	Other known chlorides
NaCl(s)	Dissolves readily	NaCl(s) + aq ⟶ Na$^+$(aq) + Cl$^-$(aq)	–
MgCl$_2$(s)	Dissolves readily with very slight hydrolysis	MgCl$_2$(s) + aq ⟶ Mg^{2+}(aq) + 2Cl$^-$(aq)	–
AlCl$_3$(s)	Hydrolyses	AlCl$_3$(s) + 3H$_2$O(l) ⟶ Al(OH)$_3$(s) + 3H$^+$(aq) + 3Cl$^-$(aq)	–
SiCl$_4$(l)	Hydrolyses	SiCl$_4$(aq) + 4H$_2$O(l) ⟶ SiO$_2$(s) + 4H$^+$(aq) + 4Cl$^-$(aq)	many
PCl$_3$(l)	Hydrolyses	PCl$_3$(l) + 3H$_2$O(l) ⟶ H$_3$PO$_3$(aq) + 3H$^+$(aq) + 3Cl$^-$(aq)	PCl$_5$, P$_2$Cl$_4$
S$_2$Cl$_2$(l)	Hydrolyses	S(s), H$^+$(aq), and Cl$^-$(aq) are amongst the products	SCl$_2$, SCl$_4$
Cl$_2$(g)	Hydrolyses	Cl$_2$(g) + H$_2$O(l) ⇌ HClO(aq) + H$^+$(aq) + Cl$^-$(aq)	–

Figure 5.13 Reactivity and structure of the chlorides across a period of the Periodic Table

> **COMMENT**
> When you have finished it will help you to remember the most important points if you summarize your work in the form of a poster, at A3 size or even larger. Use plenty of illustrations.

6 Sources and uses

Record at least one source from which each halogen can be manufactured, one industrial use and one example of a use made by plants or animals.

5.8 Study task: The halogens in human metabolism

QUESTIONS

Read the passage, and answer the questions based on it.

1. Explain why the halogens are toxic to life.
2. In which parts of the body is fluoride ion found at relatively high concentrations?
3. Give two reasons why the concentration of fluoride ion must be carefully controlled when it is added to drinking water.
4. What is an 'antibacterial solution'?
5. Calculate the concentration of chloride ion (in mol dm^{-3}) in blood plasma.
6. Which halide ion can be oxidized in the body to halogen? Suggest why only this ion is oxidized.
7. List the halogens in order of increasing importance in our diet. Justify the order in terms of their functions.
8. Summarize in about 100 words the role of the halogens in human metabolism.

The halogens are powerful oxidizing agents. In all living cells there are a large number of complex organic molecules which are extremely sensitive to even mild oxidizing systems, and for this reason the halogens in general are toxic to life, except at very low concentrations. However, their reduction products, that is the halides, occur in many living systems at varying concentrations.

The element fluorine is too powerful an oxidizing agent to occur free in nature and it is extremely toxic to all living systems. On the other hand, the fluorides are widely distributed in the plant and animal kingdom. They are present in some of the food that we eat, and the fluoride ion is easily absorbed and slowly excreted. It becomes concentrated in the supporting structures such as the bones. In humans, there is a high concentration in teeth, especially in the enamel. At the present time there appears to be no known specific role for the fluoride ion in metabolism.

A large number of dental studies have shown that fluoride ions in low concentrations in drinking water can effectively arrest the development of tooth decay in children. As a result of these studies, fluoride ions are now added to the drinking water supply in certain areas in Great Britain. However, the concentration of the fluoride ion must be critically controlled because at slightly

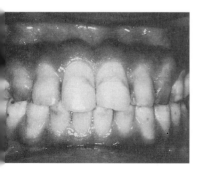

Figure 5.14 Mottled teeth due to excess fluoride

higher concentrations the ion causes mottling of the dental enamel, and at much higher concentrations the ion is toxic to life. But at the correct concentration in drinking water it is beneficial to the healthy development of teeth.

Chlorine, like fluorine, is very toxic to living systems and was used as a chemical weapon in the First World War. This halogen is sufficiently soluble in water to be useful as an antibacterial solution and for this reason it is often added to the drinking water at concentrations of 0.1 to 0.5 part per million.

The chloride ion is the principal anion found in the fluid which bathes our body cells (the extracellular fluid); blood plasma forms a significant proportion of this fluid. In this extracellular fluid the chloride ion plays an important role in the maintenance of the water potential between the intracellular fluid and the extracellular fluid. The concentration of chloride ions in the blood plasma is about 365 mg per 100 cm^3 which closely resembles the concentration of chloride in sea water. From this fact attempts have been made to draw conclusions that life originated in the sea.

Apart from the abundance of the chloride ion in blood plasma, it is also present in sweat and saliva. Consequently, during bouts of hard physical activity, or if we have to live in a hot climate, situations in which we would sweat more than usual, it is essential that we increase the intake of salt to compensate for the increased losses of sodium and chloride ions. Muscular cramp is one of the first symptoms of this salt deficiency.

Bromine is also too powerful an oxidizing agent to be encountered in living systems but the bromide anion occurs in small amounts. This ion is readily absorbed from the diet and, unlike the fluoride or the chloride ion, the bromide ion exhibits a highly specific effect on the central nervous system. Bromides depress the higher centres of the brain so that at the correct dosage the effect is one of sedation, but at higher dosage the effect is drowsiness and sleep. Bromides, unlike the fluorides, are not concentrated in any one tissue in the body and are eliminated in much the same way as the chlorides, namely by urinary excretion.

The element iodine is essential for humans and all other mammals, and we derive much of our daily requirement from small amounts of iodide ions which are present in common salt as a trace contamination. In mammals the iodide is concentrated by a small endocrine gland which is located in the neck and called the thyroid. The iodide-trapping mechanism in the thyroid is not fully understood but after the anion is concentrated it is then subjected to an oxidation–reduction reaction and converted from iodide to iodine. The iodine is then involved in a series of reactions which eventually yield the hormone thyroxine. This hormone is then secreted by the thyroid gland and circulates via the blood; it is picked up by almost all cells and tissues. Thyroxine influences the rate of metabolism in body tissues, in particular the rate of oxygen uptake. In certain communities where the drinking water is low in iodide and the diet does not contain any other sources of iodide, there is the possibility of iodine deficiency disease developing. This can be prevented by supplying such communities with common salt to which iodide has been added.

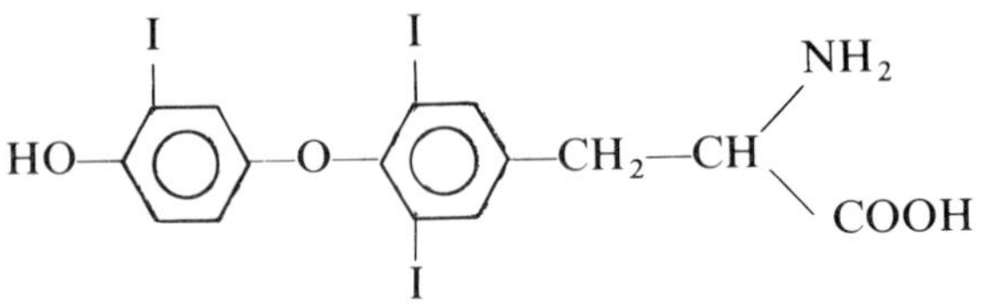

Figure 5.15 The structure of thyroxine

Summary

At the end of this Topic you should be able to:
a demonstrate understanding of:
  i oxidation number, and oxidation and reduction in terms of oxidation number changes
  ii the nomenclature and formulae of inorganic compounds, including Stock notation
b recall the characteristic physical and chemical properties of the elements of Group 7 and their compounds
c identify, and make predictions from, the trends in the physical and chemical properties of the halogens and their compounds
d demonstrate understanding of iodine–thiosulphate titrations for the quantitative analysis of oxidizing agents, and plan investigations using the technique
e evaluate information by extraction from text and the *Book of Data* about the sources of the halogens, and their role in human metabolism.

Review questions

5.1 Consider the **first** element in each of the following reactions and state whether its oxidation number goes: **A** up, **B** down, **C** remains the same.

a $Ag^+(aq) + e^- \longrightarrow Ag(s)$
b $Zn(s) + 2H^+(aq) \longrightarrow Zn^{2+}(aq) + H_2(g)$
c $2Sr(s) + O_2(g) \longrightarrow 2SrO(s)$
d $2Na(s) + Cl_2(g) \longrightarrow 2NaCl(s)$
e $Cl_2(g) + 2Na(g) \longrightarrow 2NaCl(s)$
f $Co^{2+}(aq) + \frac{1}{2}Cl_2(g) \longrightarrow Co^{3+}(aq) + Cl^-(aq)$
g $H_2(g) + Cl_2(g) + aq \longrightarrow 2HCl(aq)$
h $I^-(aq) + \frac{1}{2}Br_2(aq) \longrightarrow \frac{1}{2}I_2(aq) + Br^-(aq)$
i $\frac{1}{2}O_2(g) + H_2(g) \longrightarrow H_2O(g)$
j $BaCl_2(s) + aq \longrightarrow Ba^{2+}(aq) + 2Cl^-(aq)$
k $Cu^{2+}(aq) + Cu(s) \longrightarrow 2Cu^+(aq)$
l $Zn(s) + Pb^{2+}(aq) \longrightarrow Zn^{2+}(aq) + Pb(s)$
m $Hg(l) \longrightarrow Hg(g)$

5.2 Name the compounds whose formulae are given below, showing the oxidation number of the metal.

a $CuSO_4.5H_2O$ b Cu_2O c $Fe(OH)_3$ d FeS
e $PbCO_3$ f $PbCl_4$ g $CrBr_2$ h Co_2O_3
i Mn_2O_3 j MnO_3 k Mn_2O_7 l UF_6
m TlF_3 n $TlClO_3$ o TiI_4 p $Sr(NO_3)_2$

5.3 Use the oxidation number method to balance the following equations:

a $Zn(s) + Fe^{3+}(aq) \longrightarrow Zn^{2+}(aq) + Fe^{2+}(aq)$
b $Al(s) + H^+(aq) \longrightarrow Al^{3+}(aq) + H_2(g)$
c $Cu^{2+}(aq) + I^-(aq) \longrightarrow CuI(s) + I_2(aq)$
d $Fe(s) + Fe^{3+}(aq) \longrightarrow Fe^{2+}(aq)$
e $Sn(s) + HNO_3(l) \longrightarrow SnO_2(s) + NO_2(g) + H_2O(l)$
f $Cl_2(aq) + OH^-(aq) \longrightarrow Cl^-(aq) + ClO^-(aq) + H_2O(l)$
g $SO_2(aq) + Br_2(aq) + H_2O(l) \longrightarrow H^+(aq) + SO_4^{2-}(aq) + Br^-(aq)$
h $As_2O_3(s) + I_2(aq) + H_2O(l) \longrightarrow As_2O_5(aq) + H^+(aq) + I^-(aq)$
i $MnO_4^-(aq) + H^+(aq) + Fe^{2+}(aq) \longrightarrow Mn^{2+}(aq) + Fe^{3+}(aq) + H_2O(l)$
j $[Fe(CN)_6]^{4-}(aq) + Cl_2(aq) \longrightarrow [Fe(CN)_6]^{3-}(aq) + Cl^-(aq)$

5.4 A solution containing 24.8 g of sodium thiosulphate, $Na_2S_2O_3.5H_2O$, in 1 dm^3 was prepared. 23.6 cm^3 of this solution reacted exactly with 25.0 cm^3 of an aqueous solution of iodine.

a What is the concentration, in mol dm^{-3}, of the thiosulphate ions, $S_2O_3^{2-}$(aq)?
b What is the concentration, in mol dm^{-3}, of iodine, I_2(aq)?
c What is the concentration of the iodine in g dm^{-3}?

5.5 20 dm^3 of air, contaminated with chlorine, were bubbled through an excess of aqueous potassium iodide. The iodine so formed reacted exactly with 45.0 cm^3 of 0.100 M sodium thiosulphate, $Na_2S_2O_3$. Calculate the mass of chlorine in the sample of air.

5.6 When acidified sodium chlorate(v) is added to an acidified solution of tin(II) chloride, the chlorine is reduced to oxidation number -1 and the tin is oxidized to the ions Sn^{4+}(aq).

a What do you understand by the term 'reducing agent'?
b What is the reducing agent in the reaction described?
c i By how many units does each atom of chlorine go down in oxidation number?
  ii By how many units does each atom of tin go up in oxidation number?
  iii How many tin(II) ions react with one chlorate(v) ion?
  iv How many H^+(aq) ions are required to react with one chlorate(v) ion, assuming that all the oxygen atoms in the chlorate ion eventually become part of water molecules?
  v Hence write the ionic equation for the reaction.

Examination questions

5.7 This question is about the reaction of iodine with potassium hydroxide, which forms potassium iodate(v) and potassium iodide.

$$3I_2(s) + 6KOH(aq) \longrightarrow KIO_3(aq) + 5KI(aq) + 3H_2O(l)$$

The stoichiometric amount of iodine was added to 100 cm³ of 4.0 M potassium hydroxide solution. The reaction mixture was warmed until reaction was complete and was then cooled to 20 °C. As the reaction mixture cooled, white crystals were precipitated.

The solubility curves of potassium iodate(v) and potassium iodide are given in the diagram (molar masses H = 1, O = 16, K = 39, I = 127 g mol^{-1}).

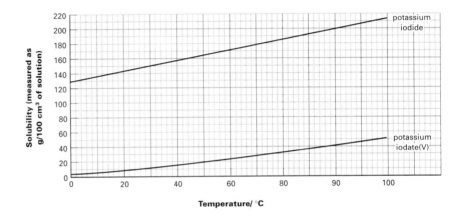

- **a** What mass of solid iodine was added to 100 cm³ of 4.0 M potassium hydroxide?
- **b** How could you tell when the reaction was complete?
- **c** What mass of potassium iodate(v) and potassium iodide would be formed by the reaction?
- **d** At what temperature would the white crystals start to appear as the reaction mixture cooled? Assume no water is lost during the reaction.
- **e** What would be the composition of the white crystals when the reaction mixture had been cooled to 20 °C?

5.8 Domestic bleach solutions sold in shops usually contain 15% sodium hypochlorite, NaClO.

- **a i** Give the Stock (systematic) name for sodium hypochlorite.
- **ii** Outline a method for making a solution of sodium hypochlorite from chlorine and a common laboratory reagent.
- **iii** Suggest a use for sodium hypochlorite solution other than bleaching.
- **b i** Calculate the number of moles of sodium hypochlorite in 100 cm³ of 15% solution. (15% means 15 g of NaClO are present in 100 cm³ of solution; molar mass of NaClO = 74.4 g mol^{-1})
- **ii** Hence calculate the concentration of the bleach solution in mol dm^{-3} of NaClO.

c The label on the bleach container includes a hazard symbol and safety warnings.

> ✗ **WARNING – IRRITANT**
>
> Contains sodium hypochlorite. Contact with acids liberates toxic gas. Irritating to eyes and skin.
> FIRST AID
> If splashed on skin or clothes wash off immediately with water.
> If splashed in eyes wash thoroughly with water and obtain medical advice.
> If swallowed, wash out mouth immediately then drink plenty of water or milk and obtain medical attention. Show the container to the doctor.

    i What is the significance of the symbol?
    ii Explain why there is a warning not to mix the bleach solution with acid.
    iii What other important safety warning would you expect to find on a bleach container?

d On standing, particularly in sunlight, bleach solution evolves oxygen.

$$2NaClO(aq) \longrightarrow 2NaCl(aq) + O_2(g)$$

    i Which element is oxidized in this reaction? State the oxidation number of the element before and after the reaction.
    ii How should bleach solution be stored to minimize its decomposition?

5.9 An aqueous solution of chlorine can be used as a disinfectant, for example in swimming pools.

a The amount of chlorine in pool water can be determined by adding excess potassium iodide solution which reacts to form iodine:

$$Cl_2(aq) + 2I^-(aq) \longrightarrow 2Cl^-(aq) + I_2(aq)$$

The amount of iodine formed is found by titration with sodium thiosulphate solution of known concentration.

    A student carried out the determination of chlorine in a sample of pool water. A complete record of all the measurements obtained is given below:

> Volume of water sample tested = 1000 cm³
>
> Initial reading of burette   = 7 cm³
> Final reading of burette   = 16.3 cm³
> Volume added from burette = 9.3 cm³
>
> Concentration of sodium thiosulphate
> = 0.00500 mol dm⁻³

    i Write a balanced equation, including state symbols, for the reaction between iodine and sodium thiosulphate in aqueous solution.
    ii The record of measurements reveals faults in both the **procedure** and the **recording** of measurements. State one fault in each.
    iii Calculate the concentration, in mol dm^{-3}, of chlorine molecules, Cl_2, in the sample of pool water using the results obtained by the student.
    iv At these low concentrations, the endpoint of an iodine–sodium thiosulphate titration, at which the last sign of the yellow-brown colour of iodine solution disappears, is very difficult to see.
        What indicator would you use to make this endpoint more visible?
        What would be the colour change at the endpoint?

b The disinfecting action of chlorine in swimming pools is due to the presence of chloric(I) acid, HClO, formed by the reaction of chlorine with water.

    In many swimming pools, chemicals other than chlorine are used to form the chloric(I) acid. This is partly because the use of chlorine gas causes much more corrosion of metal parts in the pool than does chloric(I) acid.

    Compounds used to chlorinate pool water in this way include calcium chlorate(I) and chlorine dioxide, ClO_2.
    i Suggest one other reason why the use of chlorine itself is undesirable.
    ii Write down the formula for calcium chlorate(I).
    iii Chlorine dioxide undergoes a disproportionation reaction when it reacts with water, one of the products being chloric(I) acid. By use of oxidation numbers or otherwise, predict a balanced equation for this reaction.

5.10 There are definite trends in the properties and reactions of compounds containing elements from Group 7 of the Periodic Table.
    Describe some of these trends, giving equations for the types of reaction you mention, and offer explanations where possible.

5.11 Choose any one of the halogens.
    Describe its distribution in the world and how it can be manufactured, supporting your answer with appropriate chemical data.
    Describe some of the uses of the halogen chosen, both as an element and in its compounds, and discuss its importance in animal metabolism.

5.12 Write an account of the redox chemistry of chlorine and its compounds.
    Indicate how at least one compound of each of the principal oxidation states (-1, $+1$ and $+5$) may be obtained from chlorine. Illustrate your answer with appropriate equations.
    1.00 g of a tin alloy was dissolved in hydrochloric acid to give exactly 250 cm^3 of a solution containing Sn^{2+} ions.
    25.0 cm^3 of this solution was titrated with an aqueous iodine solution of concentration 0.050 M of I_2, and required 10.0 cm^3 for complete oxidation to Sn^{4+} ions. Calculate the percentage of tin in the alloy.
    (You may assume that the other metals in the alloy are not oxidized by the iodine solution after dissolving in acid.)
    Why was the alloy not dissolved in nitric acid?

READING TASK 2

ORIGIN OF THE CHEMICAL ELEMENTS

The elements on the Earth vary enormously in their abundance. Oxygen, silicon and iron are common; many elements, such as gold, are millions of times as rare. Most of the elements are important to us in various ways. Nearly thirty elements are essential to life. These include carbon, oxygen, hydrogen and nitrogen, but also some quite rare ones such as selenium. In this century, industry has come to make use of nearly all the elements. The manufacture of a modern touch-dialling telephone, for example, will involve no less than 42 of them!

Many features of the Earth's chemical composition, however, are not at all typical of the Universe as a whole. We know that the two lightest elements, hydrogen and helium, make up more than 99% of the visible Universe, with the others being present in very small proportions. On Earth, hydrogen, and especially helium, are much rarer because they are gases except at very low temperatures, and they largely escaped into space when the Earth formed.

The abundance of elements in space is very important, not only because it influenced the ultimate composition of the Earth, but also because it can provide many clues to how the elements were originally formed. Clearly, a satisfactory theory about the origin of the elements should be able to account for the abundances that we observe. How do we know these abundances? If you measure the spectrum of sunlight, by splitting the light into different colours, or wavelengths, with a prism or diffraction grating, you see many dark lines running across. When chemists put different elements in a flame, they see the same lines in spectra produced in the laboratory. The lines are there because different elements absorb or emit light corresponding to characteristic, and extremely precise, wavelenths. The strengths of these lines show us how much light is absorbed or emitted, and, therefore, the amount of each element that is present.

Astronomers have carried out this kind of analysis on the Sun and on many stars and galaxies, building up a picture of the abundance of elements in space. The information is supplemented by analysing meteorites.

The combination of data from the solar spectrum and the chemical analysis of meteorites allows us to know fairly precisely what the overall composition of the Solar System is. The abundances found are shown in figure R2.1. Notice the scale and the enormous range of abundances: each scale mark in the abundance scale differs from the neighbouring one by a factor of 10. For every 10^{12} atoms of hydrogen there are about 10^{11} of helium, fewer than 10^9 of the next commonest elements, carbon and oxygen, and fewer than 10 each of some rare elements such as uranium.

The first few minutes: hydrogen and helium

According to the big bang theory, the Universe began as a fireball of extraordinarily dense and hot matter. In the early stages, it was so hot that not even atomic nuclei – let alone the molecules and solids familiar in everyday life – could exist. The chemical elements could not have been present in the beginning, but must have been made subsequently.

A few seconds from the beginning, the temperature was around $10^{10}\,°C$. This is the maximum temperature at which atomic nuclei can exist. Protons were constantly changing into neutrons and vice versa, giving a ratio of about 1 neutron to every 7 protons. Free neutrons are

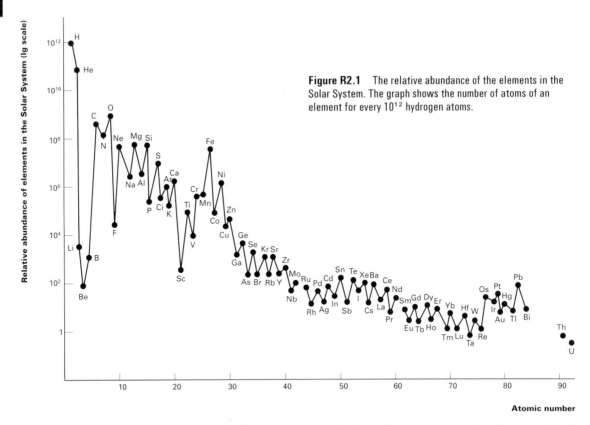

Figure R2.1 The relative abundance of the elements in the Solar System. The graph shows the number of atoms of an element for every 10^{12} hydrogen atoms.

unstable and decay to protons and electrons. Before this decay process could occur, there was time for neutrons and protons to combine, forming deuterium, a heavy form of hydrogen. At the high temperature, the deuterium nuclei reacted rapidly with more protons, and the ultimate product was the stable nucleus of helium (with two protons and two neutrons). Any further nuclear fusion reactions, making heavier elements from helium, could not happen to an appreciable extent because the temperature was too low by the time helium was made.

It seems, therefore, that 99% of material in the Universe today, the hydrogen and helium, owes its origin to the early stages. The agreement between theory and observation is impressive, and is one of the strongest pieces of evidence that ideas about the big bang are correct: no other theory of the origin of the Universe can explain the existence of hydrogen and helium in their observed proportions.

The heavier elements

Although elements heavier than helium make up only one per cent of the Universe, they are essential to us in many ways. A universe made of hydrogen and helium would be a very dull place in chemical terms! To make the heavier elements requires high temperatures sustained over a much longer period of time than just after the big bang.

But, such conditions do exist now – at the centre of stars. It is here that most of the remaining elements are made. A star begins when a large mass of gas contracts under its own gravity. Compression raises the temperature in the centre, to the point at which nuclei can start to fuse to form heavier nuclei. The output of energy from the nuclear fusion keeps stars hot, and prevents any further contraction, at least until the nuclear fuel has been used up.

The first reaction to begin, at a temperature of about 10^7 °C, is the fusion of protons to form helium; this reaction occurs in a number of steps, in some of which protons are converted into neutrons. This is the so-called **hydrogen burning** phase of stars. It is not burning in the everyday sense of the word. Apart from helium, hydrogen burning does not produce new elements, but it is important because the energy produced keeps stars going for much of their lives. The hydrogen burning phase leads to the build-up of a core of helium in the centre of the star. When the hydrogen is exhausted, and the output of energy from the reaction declines, the centre of the star starts to contract again and becomes even hotter. As the core shrinks, the outer parts of the star expand. The star grows into a **red giant**.

What happens next depends on the star's mass. In the case of stars that have a relatively low mass, the core of helium simply becomes a compact object no larger than the Earth, known as a **white dwarf**, in which the helium nuclei are closely packed. The outer layers escape into space. If a star is more massive than 0.4 times the Sun, the core becomes so hot, around 10^8 °C, that the helium nuclei can react to form heavier nuclei. These fusion reactions require higher temperatures because the nuclei are more highly charged, and so need more energy to overcome their mutual electrostatic repulsion before they can fuse.

Two helium nuclei form beryllium (with four protons) but this nucleus is quite unstable and reacts quickly with further helium nuclei, to form first carbon and then oxygen. These two elements are the commonest in the Universe, after hydrogen and helium. The relative amounts made depend on the temperature of the star, which in turn is controlled by its mass. As helium is consumed, a core of carbon and oxygen builds up. For a star with a mass between 0.4 and 8 times that of the Sun, this is the end of fusion reactions. The core becomes a white dwarf that is composed of carbon and oxygen.

In the most massive stars, the core gets so very hot that carbon and oxygen can in turn fuse together, forming elements as heavy as sulphur. Further reactions happen in stages, eventually producing iron (which has 26 protons) and a number of elements with similar masses. The reactions stop here, because iron has the most stable nucleus of all elements, and cannot fuse under these conditions.

Around the iron core there are various layers in the star where the other reactions are still going on, so in cross-section the star tends to resemble an onion. As well as the reactions that build hydrogen up to iron, other fusion processes are going on in these layers. These minor reactions can build up nuclei that are heavier than iron, in what astronomers call the s-process (meaning slow). The s-process occurs when some reactions produce neutrons, which are captured by other nuclei, so increasing their mass. Once a neutron has been captured, it may change into a proton. In this way, the s-process can produce elements up to bismuth (which has 83 protons).

Elements of a supernova

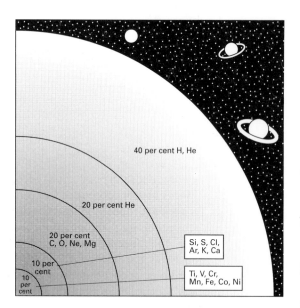

Figure R2.2 The shell structure of a heavy star just before its explosion in a supernova, shown by mass. The inner shells occupy very much less space than shown in the diagram.

The life of a star reaches its final stage when the core of iron builds up in the centre. The iron nuclei cannot produce energy by fusion, but the gravitational force is remorseless: it continues to compress the core, raising the temperature to billions of degrees. Some of the elements formed in the core begin to disintegrate in this inferno, and the very centre of the star collapses suddenly into a dense mass of solid neutrons. The outer layers fall in, then bounce back, spewing the contents of the star out into space in a supernova explosion.

The explosion itself creates more heavy elements, by producing a flood of neutrons that are absorbed by existing nuclei. Unlike the s-process, where neutrons add on to nuclei one by one, there are now so many neutrons that several attach to a nucleus at once. The r-process (rapid) can make elements as heavy as uranium.

In a supernova explosion, the star becomes very much brighter, sometimes as brilliant as a billion Suns. Over the past 50 years, astronomers have found hundreds of supernovae in distant galaxies. When a supernova occurs in our galaxy or a near neighbour galaxy, it is sometimes bright enough to be easily visible with the naked eye. We can find several supernovae in historical reports, including an observation by Chinese astronomers in AD 1054. The remains of this supernova now form the Crab Nebula, a cloud of hot gas still expanding outwards from the explosion. Products from supernovae spread out, and eventually mix up with more gas. They then become incorporated into later generations of stars formed from the gas, eventually forming planets as well. The most recent supernova visible to the unaided eye was seen in 1987. Its gas spectrum shows that many elements were made in the explosion, including some that are radioactive and have slowly decayed since 1987.

Apart from direct observations on the remnants of old supernovae, the best evidence for the theory that the elements are produced in stars is that calculations confirm the observed abundances of elements. Such calculations are difficult and require the power of modern supercomputers. But the agreement is good. It appears from such calculations that almost all the material of the Solar System, apart from the hydrogen and helium remaining from the big bang, was produced by supernovae during the first few billion years of our galaxy's existence.

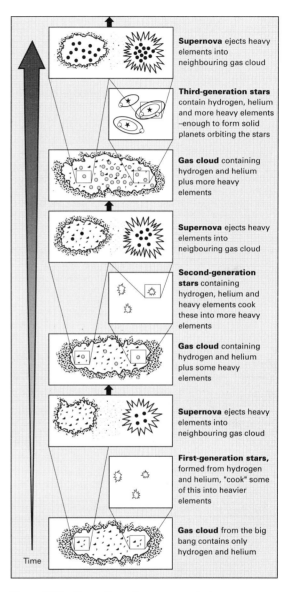

Figure R2.3 Generations of supernova explosions are needed to produce the heavy elements that form planets

Questions

Reread the passage and answer these questions.

1 What does the zig–zag pattern of relative abundance suggest about the relative stability of different nuclei?

2 Draw a diagram to illustrate the build-up of nuclei from hydrogen as far as sulphur.

3 Tell the story of the creation of a single atom of uranium starting from the 'big bang'. Try to use no more than 150 words.

TOPIC 6

Covalent bonding

Any model of the bonding in molecules must be able to account for the formula of the molecule, its structure, and the forces which hold the atoms together.

The electron density for the simplest possible covalent structure, the H_2^+ molecule ion, was calculated from theory by C. A. Coulson (figure 6.1). This consists of two hydrogen nuclei but only one electron; it thus has a net positive charge. The molecule H_2^+ was chosen for the calculations because of its simplicity.

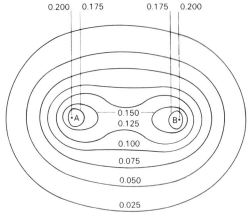

Figure 6.1 Electron density map for the H_2^+ molecule ion. Contours are in electrons per 10^{-30} m³.

- What do you notice about the contours that is different from the contours in ionic compounds?
  What does this tell us about the electron density between the nuclei of the atoms in the molecule?

The electron density map, determined by X-ray diffraction, for crystals of 4-methoxybenzoic acid is shown in figure 6.2.

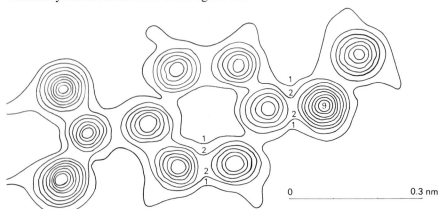

Figure 6.2 Electron density map of 4-methoxybenzoic acid. Contours are in electrons per 10^{-30} m³.

- What can you say about the electron density between adjacent atoms in this molecule?

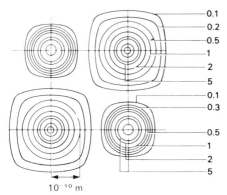

Figure 6.3 Electron density map for sodium chloride

The sets of maps in figures 6.1, 6.2 and 6.3 show that in structures consisting of ions the electron density drops to zero between the ions, and the ions are discrete entities; but in molecules there is a substantial electron density between the two bonded atoms. Thus it seems that in bonds in molecules the electrons are *shared*: bonds formed by electron sharing are known as **covalent bonds**.

6.1 Electron sharing in covalent molecules

Covalent bonding exists between atoms when electrons are shared, usually in pairs. In many cases, the number of atoms involved is such as to enable a noble gas electron structure to be built up around each atom. Figure 6.5 shows how this is done in the case of methane, CH_4.

The hydrogen atoms have a share in the electrons from the carbon atom, thus acquiring helium structures; and the carbon atom has acquired the neon structure by sharing electrons from the hydrogen atoms. When atoms of non-metals are joined together it is in general by covalent bonds.

- Now try to draw dot-and-cross diagrams showing the electronic configurations in the molecules of hydrogen, H_2; chlorine, Cl_2; hydrogen chloride, HCl; chloromethane, CH_3Cl; methanol, CH_3OH; ethane, CH_3CH_3; ethanol, CH_3CH_2OH; ethene, $CH_2\!=\!CH_2$; and oxygen, O_2.

Single bonds

Refer to the shape of the molecules shown in figure 6.4 and to their electronic configurations. Remember that a covalent bond consists of a negative charge-cloud. Why do you think that the molecules have the shapes that they do?

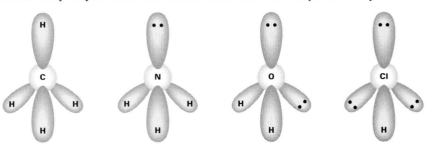

Figure 6.4 Electron-cloud models of CH_4, NH_3, H_2O and HCl

Figure 6.5 The formation of covalent bonds

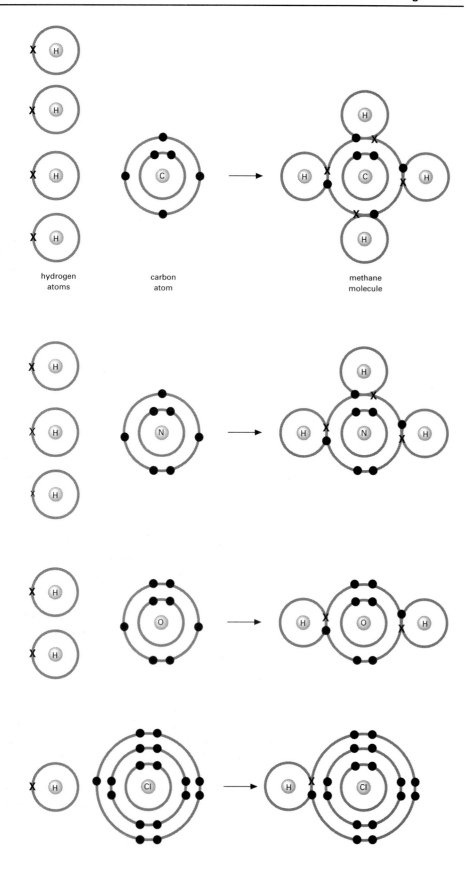

methane

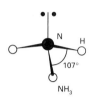
NH₃

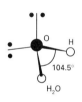
H₂O

Figure 6.6 Bond angles in CH$_4$, NH$_3$ and H$_2$O

- What do you notice about the spatial arrangements of the bonds in the molecules of CH$_4$, NH$_3$, H$_2$O and HCl? What does this suggest about the interaction between electron charge-clouds?
  Why are the molecules of ammonia and water not planar and linear respectively?

The bond angles in methane, ammonia, and water molecules are given in figure 6.6.

- What do these figures suggest?

If the electron density of the hydrogen molecule ion had been displayed in three dimensions, instead of the two-dimensional representation given in figure 6.1, it would be seen that the electron density is symmetrical about the axis joining the hydrogen atoms. Bonds of this type are known as σ-bonds (sigma bonds). An important example of a σ-bond is the single bond between carbon atoms, as in the ethane molecule, CH$_3$—CH$_3$ (figure 6.7a and 6.8a). We shall be considering the carbon–carbon σ-bond in Topic 7.

Multiple bonds

From your consideration of the shapes of molecules containing single bonds, and the changes in the bond angles from one hydride to another as shown in figure 6.6, you should have come to the following conclusions:

1 Pairs of electrons try to get as far away from each other as they can. This results in a tetrahedral distribution when there are four electron pairs.
2 'Lone' pairs of electrons, that is, pairs of electrons not shared between two atoms, repel one another more strongly than shared pairs do. This causes some distortion of the tetrahedral arrangement.

If you did not come to these conclusions, you may like to go back over the evidence to see how well they account for the observations.

- On the basis of these ideas, which were developed for single bonds, can you suggest a shape for the molecule of ethene, CH$_2$=CH$_2$? Look again at the dot-and-cross diagrams you drew, based on figure 6.5.

A double bond, as in ethene, consists of two pairs of shared electrons. However, it does not follow that a double bond can be thought of as two single bonds, although it might seem so from our usual method of representing double bonds (figure 6.7).

Evidence from both chemical properties and structural studies suggests that the electron density in a double bond is **asymmetric** about the axis joining the two nuclei. In the case of the carbon-carbon double bond the electron density distribution corresponds to a σ-bond (sigma bond) plus what is described as a π-bond (pi bond). The π-bond consists of two electron clouds, as shown in figure 6.8b, which are not symmetrical about the axis joining the carbon nuclei. A σ-bond is shown in figure 6.8a.

The ideas developed above enable us to predict correctly the shapes of a surprisingly large number of molecules and other structures; but they are subject to some limitations. Some of these will be discussed later.

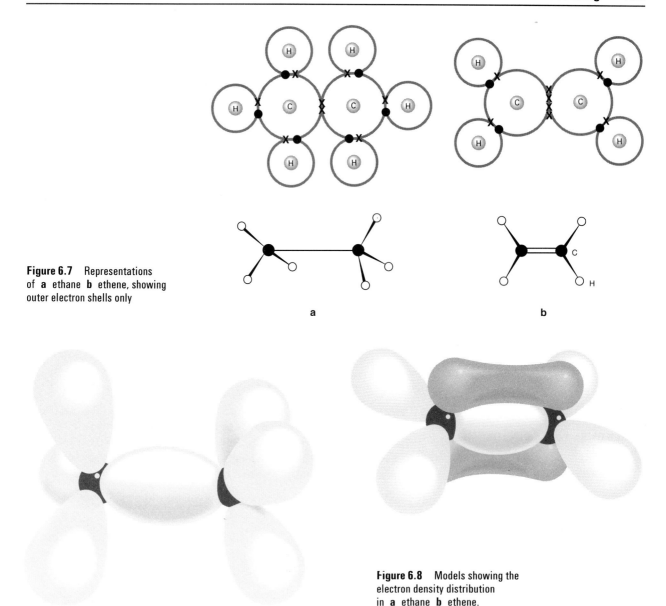

Figure 6.7 Representations of **a** ethane **b** ethene, showing outer electron shells only

Figure 6.8 Models showing the electron density distribution in **a** ethane **b** ethene.

6.2 Bond energies

The energy involved in the making and breaking of a covalent bond is measured by a different approach to the determination of lattice energies that is such a useful method for ionic compounds.

A first step is to determine the enthalpy changes of combustion of the compound, and of the constituent elements; and then to use Hess's Law to calculate the enthalpy change of formation of the compound. Some examples will illustrate the method.

> **The standard enthalpy change of combustion of a substance, symbol $\Delta H_c^\ominus(298)$, is defined as the enthalpy change that occurs when one mole of the substance undergoes complete combustion under standard conditions.**

For a compound containing carbon, for example, complete combustion means the conversion of the whole of the carbon to carbon dioxide, as shown in the following equation.

$$C_{12}H_{22}O_{11}(s) + 12O_2(g) \longrightarrow 12CO_2(g) + 11H_2O(l)$$

sucrose $\quad \Delta H_c^\ominus(298) = -5639.7 \text{ kJ mol}^{-1}$

The enthalpy change of formation of sucrose can then be calculated from the enthalpy change of combustion by the use of a Hess cycle of the form:

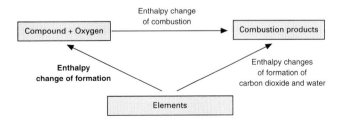

COMMENT

The standard enthalpy changes of formation of water and of carbon dioxide can be measured directly by calorimetry experiments, and these two values are known to a very high degree of accuracy.

$C(graphite) + O_2(g) \longrightarrow CO_2(g) \quad \Delta H_f^\ominus(298) = -393.5 \text{ kJ mol}^{-1}$

$H_2(g) + \tfrac{1}{2}O_2(g) \longrightarrow H_2O(l) \quad \Delta H_f^\ominus(298) = -285.8 \text{ kJ mol}^{-1}$

Let us take methane as an example. First, write down the equation for the standard enthalpy change of formation (that is, from the elements in their standard states).

$$C(graphite) + 2H_2(g) \xrightarrow{\Delta H_f^\ominus} CH_4(g)$$

Now write in the equations for burning both the reactants and the methane in oxygen, as follows.

$$C(graphite) + O_2(g) \xrightarrow{\Delta H_f^\ominus} CO_2(g)$$

$$2H_2(g) + O_2(g) \xrightarrow{\Delta H_f^\ominus} 2H_2O(l)$$

$$CH_4(g) + 2O_2(g) \xrightarrow{\Delta H_c^\ominus} CO_2(g) + 2H_2O(l)$$

and the data for these reactions are

$\Delta H_f^\ominus[CO_2(g)] \quad = -393.5 \text{ kJ mol}^{-1}$

$2 \times \Delta H_f^\ominus[H_2O(l)] = -571.6 \text{ kJ mol}^{-1}$

$\Delta H_c^\ominus[CH_4(g)] \quad = -890.3 \text{ kJ mol}^{-1}$

We may now substitute the methane example in the general Hess cycle, given earlier:

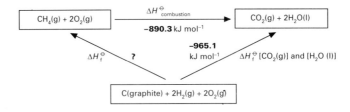

The overall energy change in going from elements to combustion products must be the same whatever the route. Equating the two routes,

$$-965.1 = \Delta H_f^\ominus[CH_4(g)] - 890.3$$

$$\Delta H_f^\ominus[CH_4(g)] = -74.8 \text{ kJ mol}^{-1}$$

STUDY TASK

The compound methanol, CH_3OH, cannot be prepared directly from its elements. Given that the standard enthalpy change of combustion of methanol, $\Delta H_c^\ominus$, is -726.0 kJ mol^{-1}, calculate the standard enthalpy change of formation of the compound.

$$C(\text{graphite}) + 2H_2(g) + \tfrac{1}{2}O_2(g) \longrightarrow CH_3OH(l)$$

The other standard enthalpy changes of combustion that you need have been given earlier in this Topic.

How can enthalpy changes of combustion give information about the energy required to break individual bonds?

One way of attempting to answer this question would be to find a series of substances which are closely related to each other, and which differ from each other by some fixed unit of structure. Then by studying such substances it might be possible to see whether that fixed unit of structure makes any consistent contribution to the overall energy situation.

An example would be the series of alcohols:

$CH_3CH_2CH_2OH$	Propan-l-ol
$CH_3CH_2CH_2CH_2OH$	Butan-l-ol
$CH_3CH_2CH_2CH_2CH_2OH$	Pentan-l-ol
$CH_3CH_2CH_2CH_2CH_2CH_2OH$	Hexan-l-ol
$CH_3CH_2CH_2CH_2CH_2CH_2CH_2OH$	Heptan-l-ol
$CH_3CH_2CH_2CH_2CH_2CH_2CH_2CH_2OH$	Octan-l-ol

Each compound differs from the next by one —CH_2— unit.

In this series, the question can be posed 'Does the —CH_2— group make a specific contribution to the enthalpy change of combustion of alcohols?' One method of finding out would be to burn the alcohols and measure the enthalpy change per mole of each.

EXPERIMENT 6.2 To find the enthalpy changes of combustion of some alcohols

In this experiment, you will find the enthalpy change of combustion of one of the series of alcohols and compare your results with those obtained by other members of the class.

Procedure

Use the combustion calorimeter with which you have been provided.

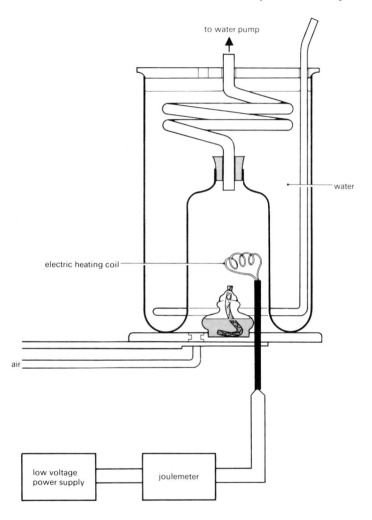

Figure 6.9 Combustion calorimeter

1 Put water in the calorimeter up to the level shown and mount it on its stand. Attach a water pump on the top of the copper spiral outlet, and adjust the pump so as to draw a moderately rapid stream of air through the spiral.
2 Almost fill the small spirit lamp with the alcohol to be used, then light the wick, and adjust the length so as to give a flame about 1.5–2 cm high. Put the lamp under the calorimeter and watch its behaviour.

If the flame remains reasonably steady but slowly goes out this probably indicates that not enough air is being supplied; adjust the water pump so that more air is drawn through.

If the flame is very unsteady and goes out, it may indicate that too great a rush of air is being drawn in; adjust the water pump accordingly.

3 When you have adjusted the height of the wick and the flow of air so that the lamp remains alight with a good flame, extinguish the lamp and put the cap over the wick. Weigh it on a balance reading to 0.001 g.

4 Stir the water in the calorimeter and record the temperature, using a thermometer reading to 0.1 °C.

5 Remove the cap from the spirit lamp, light the wick, and without delay put the lamp under the calorimeter. Stir the calorimeter periodically. When a rise in temperature of between 10 °C and 11 °C has been obtained blow out the flame, remove the lamp from the stand, and replace the cap. Stir the water thoroughly and note the maximum temperature that is reached.

6 Reweigh the spirit lamp as soon as possible after extinguishing it.

7 Using a joulemeter, supply electrical energy to the calorimeter via the heater until a similar temperature rise to that produced by the burning alcohol is obtained.

Calculation

Calculate the moles of alcohol molecules burned and hence ΔH_c, the enthalpy change of combustion, in kilojoules per mole.

QUESTIONS

1 Heat losses Do you think the method has taken heat losses from the calorimeter into account satisfactorily?

2 Combustion products The term 'standard enthalpy change of combustion' implies complete combustion, in this instance to carbon dioxide and water. Can we be sure that this has taken place? What might have been formed instead, to some extent?

3 Do you think that the various sources of error will tend to make your value higher or lower than the true one?

4 Compare your results with those of others in the class, and see whether they provide any answer to the question with which this section began.

Using a spreadsheet plot a graph of ΔH_c for the alcohols from methanol to octan-1-ol against the number of carbon atoms. Use your own values and those of other members of your group or take values from the *Book of data*.

You will find that the difference in value between successive alcohols in the series is about the same; and, of course, the structural difference between successive alcohols is the CH_2 group of atoms.

When one extra CH_2 group burns, extra energy must be supplied to break one **extra C—C** bond and **two extra C—H** bonds; and extra energy is released by the formation of **four extra C—O** bonds and **two extra O—H** bonds.

If a fixed amount of energy is associated with this number of bonds broken and bonds formed, as is found experimentally, it seems likely that each individual bond has its own energy that must be supplied to break it, or that will be released when it is formed.

Bond energies in other compounds

Consider now the alkane series of hydrocarbons and begin with the first member, methane, CH_4. It seems reasonable to assume that the energy associated with the C—H bonds must be reflected in the total amount of energy required to break the molecule into its constituent atoms.

This total amount of energy can be found by using Hess's Law. The equation is

$$CH_4(g) \longrightarrow C(g) + 4H(g)$$

The enthalpy changes are shown on the diagram:

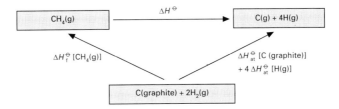

$$\Delta H^\ominus + \Delta H_f^\ominus[CH_4(g)] = \Delta H_{at}^\ominus[C(graphite)] + 4\Delta H_{at}^\ominus[H(g)]$$

Putting in the values:

$$\Delta H^\ominus = -(-74.8) + 716.7 + (4 \times 218)$$
$$= +1663.5 \text{ kJ mol}^{-1}$$

So, for the reaction,

$$H-\underset{\underset{H}{|}}{\overset{\overset{H}{|}}{C}}-H(g) \longrightarrow C(g) + 4H(g) \quad \Delta H^\ominus = +1663.5 \text{ kJ mol}^{-1}$$

If the bonds are equal in strength, then the bond energy of one C—H bond should be

$$+\frac{1663.5}{4} = +415.9 \text{ kJ mol}^{-1}$$

Denoting the bond energy of the C—H bond by E(C—H) we have

$$E(\text{C—H}) = +415.9 \text{ kJ mol}^{-1}$$

Now consider ethane, C_2H_6. A similar calculation to the one above shows that for the reaction

$$C_2H_6(g) \longrightarrow 2C(g) + 6H(g) \quad \Delta H^\ominus = +2826.1 \text{ kJ mol}^{-1}$$

This reaction involves the breaking of six C—H bonds and one C—C bond. Denoting the bond energy of the C—C bond by E(C—C), we have

$$+2826.1 = E(\text{C—C}) + 6E(\text{C—H})$$

Substituting the value 415.9 for E(C—H)

$$+2826.1 = E(\text{C—C}) + (6 \times 415.9)$$

So $\quad E(\text{C—C}) = +330.7 \text{ kJ mol}^{-1}$

The bond energy E(C—Cl) has been determined, using several compounds. Below are shown the compounds and the values obtained for them:

Compound		E(C—Cl) /kJ mol^{-1}
Cl—C(Cl)(Cl)—Cl	tetrachloromethane	+327
H—C(H)(H)—Cl	chloromethane	+335
H—C(H)(H)—C(H)(H)—Cl	chloroethane	+342

From these examples it can be seen that the bond energy value is approximately the same in each case, though it depends upon the compound from which it was determined, to some extent; that is, the environment of the bond affects the value. The X—Y bond energy will vary somewhat, depending upon the nature of the other atoms or groups of atoms which are attached to X and Y.

But if an average bond energy is taken this can often be very useful. Tables have therefore been prepared giving average bond energies. Average bond energies per mole of bonds are denoted by the symbol E.

An approximate value for the enthalpy change involved in the atomization of a compound from the gaseous state can be obtained by adding up the average bond energies for all the bonds in the molecule of that compound.

Bond	E/kJ mol^{-1}
C—H	413
C—C	347
C—O	358
C—Cl	346
O—H	464

A fuller table is given in the *Book of data*, table 4.6.

QUESTIONS

1 Using table 4.6 of bond energies given in the *Book of data*, work out an approximate value for the energy needed to atomize one mole of the alcohol propan-l-ol, $CH_3CH_2CH_2OH$.

2 Make a table showing the bond energies for the hydrides across the Periodic Table, C—H, N—H, O—H, and F—H, and then insert the vertical series F—H, Cl—H, Br—H, and I—H. What are the trends in the ease of breaking the bonds, and what information can you deduce from them?

Bond lengths and bond energies

The greater electron density between nuclei joined by multiple bonds causes a greater force of attraction between the nuclei and is reflected in shorter bond lengths and greater bond energies. Examine the figures given in the table overleaf to confirm this. Are the bond energies of multiple bonds simple multiples of the bond energies of single bonds?

Bond	Compound(s)	Bond length/nm	Bond energy/kJ mol^{-1}
C—C	alkanes	0.154	347
C=C	alkenes	0.134	612
C≡C	alkynes	0.121	838
C—N	amines	0.147	286
C=N	(average)	0.132	615
C≡N	nitriles	0.116	887
C—O	alcohols	0.143	336
C=O	ketones	0.122	749
C≡O	carbon monoxide	0.113	1077

Bond lengths and bond energies

6.3 Dative covalency

The two electrons which form a covalent bond between two atoms do not necessarily have to come one from each atom; both may come from one of the atoms.

Look back at the electronic structure of water; of the eight electrons around the oxygen atom, four are not shared with any other atom. Water molecules readily react with protons to form hydroxonium ions, as we described in Topic 3 as part of the Brønsted–Lowry theory of acid–base behaviour.

$$H_2O + H^+ \longrightarrow H_3O^+$$

This can be interpreted in terms of electron sharing as follows:

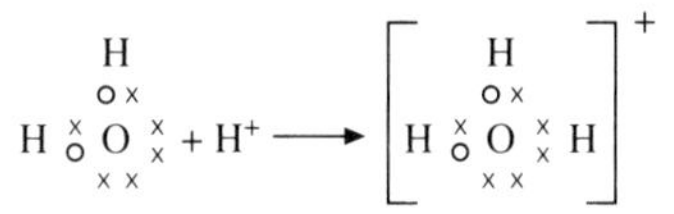

Since one pair of the shared electrons has come from one atom the bonding is sometimes known as **dative covalency**, and the bond is indicated by an arrow.

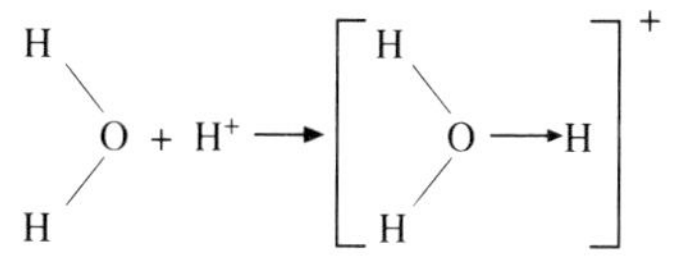

Another example of dative covalency occurs in the ammonium ion, NH_4^+. As the next diagram shows, in the ammonium ion the hydrogen atoms each have a share in 2 electrons, giving a helium structure; and the nitrogen atom has a share in 8 electrons, giving a neon structure. The ion has an overall +1 charge:

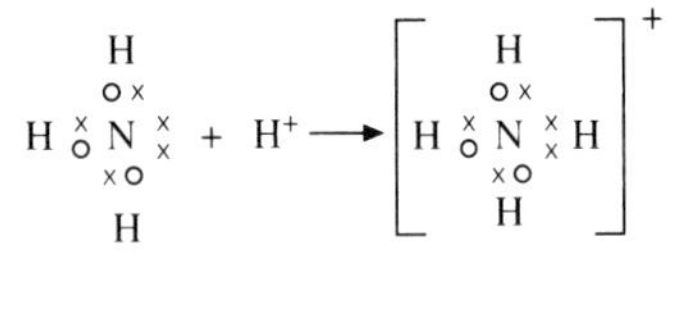

$$H-\underset{\underset{H}{|}}{\overset{\overset{H}{|}}{N}} + H^+ \longrightarrow \left[H-\underset{\underset{H}{|}}{\overset{\overset{H}{|}}{N}} \rightarrow H \right]^+$$

The charge came originally from the proton; now it is distributed all over the ion, and is not located on any particular atom. As all the N—H bonds have the same length, and the hydrogen atoms are indistinguishable, this suggests that the ammonium ion should be represented as:

$$\left[H-\underset{\underset{H}{|}}{\overset{\overset{H}{|}}{N}}-H \right]^+ \quad \text{or} \quad NH_4^+$$

in the same way as the hydroxonium ion:

$$\left[\underset{H}{\overset{H}{\diagdown}} O-H \right]^+ \quad \text{or} \quad H_3O^+$$

The carbon monoxide molecule and the nitric acid molecule also have dative covalent bonds.

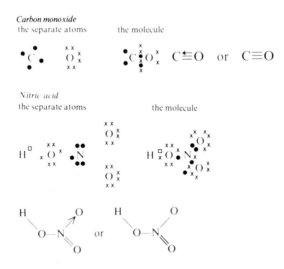

The bonding in the two NO bonds is not, however, to be taken as different; this point is discussed later.

- Would it be possible, in terms of electrons, to form an ion H_4O^{2+}? Try to draw such a structure.
  Have you ever heard of such an ion? Suggest a reason for your answer.

6.4 The chemists' toolkit: electronegativity

Two electrons shared between two atoms constitute a covalent bond between these atoms. It is reasonable to ask whether the electrons are always shared equally between the two atoms, or whether some elements are more 'electron-attractive' than others. It is found that elements do differ considerably in their electron-attractiveness. The term used for this is **electronegativity**.

The electronegativity of an atom represents the power of an atom in a molecule to attract electrons to itself.

Many attempts have been made to allot numerical values for the electronegativities of the elements, but so far no wholly satisfactory method has been devised; the most widely used values are based on the method of the American chemist Linus Pauling. But whatever numerical scale is used, the trends in the values of the electronegativities of the elements in the Periodic Table are clear.

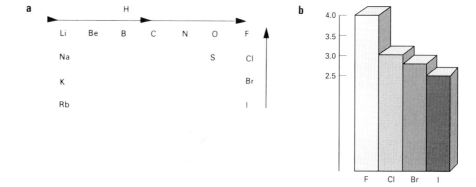

Figure 6.10 **a** Trends in electronegativity in the Periodic Table **b** The electronegativities of the halogens decrease steadily down the group

It can be seen, from the trends, that the most electronegative element is fluorine; chlorine, oxygen, and nitrogen are also very electronegative.

When two atoms bonded covalently are atoms of the same element, then the attractions of their nuclei for the bonding electrons are the same, and the bonding electrons will be shared equally between them. But if the atoms are not of the same element, the two nuclei exert different degrees of attractive force on the bonding electrons, and these electrons are displaced towards one atom. The next diagram illustrates this.

This unequal sharing of electrons is known as **bond polarization**. It represents the departure of the bond from being purely covalent, and it introduces some ionic character into the bond.

Thus the polarization of ions (see Topic 4.4) represents the existence of some covalent character in the ionic bonding; and the polarization of a covalent bond represents the existence of some ionic character in the covalent bond.

One important conclusion from this section so far is that wholly ionic and wholly covalent bonds are extreme types, and examples occur over the whole range of intermediate types: bonds can be partially ionic and partially covalent in character.

Electronegativity and polar molecules

Using the dot-and-cross diagrams above for the HCl molecule and the CH_3CH_2Cl molecule as a starting-point, draw representations of the electron charge-cloud distributions in the two molecules. Do this both for the polarized bonds and the lone pairs of electrons. Superimpose on these drawings a series of + signs to indicate the positions of the atomic nuclei. Now consider whether you think the 'centre of gravity' of the positive charges, and the 'centre of gravity' of the negative charges coincide.

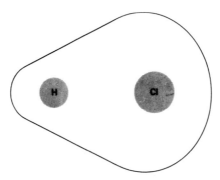

Figure 6.11 Shape of the HCl molecule

Figure 6.11 represents the shape of the HCl molecule. Copy the figure into your notebook, and draw in by means of a + and a − the relative positions of the centre of positive charge and the centre of negative charge.

- What implications do you think this has for the properties of the molecule?
  Suggest relative positions for the centre of positive charge and the centre of negative charge in the molecules shown in figure 6.12.
  What elements in addition to oxygen and chlorine would be likely to produce these effects?

In asymmetric molecules such as HCl and CH_3CH_2Cl the centre of positive charge does not coincide with the centre of negative charge, and a permanent dipole results. Such molecules are said to be **polar**. Highly electronegative elements such as F, Cl, O, and N cause polarity in molecules. They do so partly by virtue of the bond polarization which they produce, and partly by virtue of their lone pairs of electrons.

Polarity in molecules has important effects on the physical and chemical properties of the substances, and on the mechanisms by which they undergo reaction.

Look back to the charge-cloud diagrams for the molecules shown in figure 6.4. Decide which of these molecules are polar and which are non-polar. Draw them and label them as such.

Figure 6.12 Are these molecules polar?

6.5 Delocalization of electrons

Do single and double covalent bonds always represent the electron distribution adequately when electrons are shared?

We shall examine structural evidence for some compounds, with a view to obtaining an answer to this question.

Nitric acid, and the nitrate ion

Refer back to the electron diagram for the structure of nitric acid, and answer the following questions in your notes.

- Would any other arrangements of the bonds have done equally well? If so, what are they?
  In a range of other compounds the average N=O bond length is 0.114 nm and the average N—O bond length is 0.136 nm. On the basis of your diagram what would you expect the bond lengths to be in nitric acid?

Electron diffraction and microwave spectroscopic studies of nitric acid vapour show the molecule to have the structure shown below.

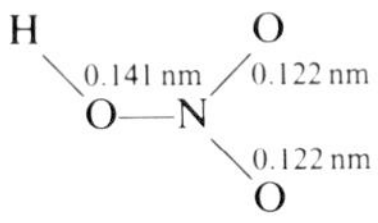

bond lengths 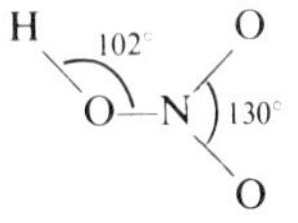   bond angles

- Are the bond lengths what you expected?

Draw an electron diagram for the nitrate ion, NO_3^-; and then draw a bond diagram using a line — to represent a single covalent bond and an arrow → to represent a dative covalency. The nitrate ion has one electron which has been transferred to it from an atom which is now a positive ion.

- Is there any other way in which the bonds could be arranged? What would you expect the bond lengths in the nitrate ion to be?

The structure of the nitrate ion has been found to be as shown below.

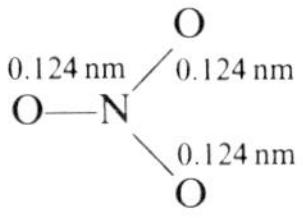

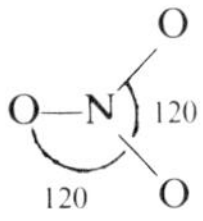

- What can be concluded about the nature of the bonds in the nitrate ion?

Methanoic acid, and the methanoate ion

The structural formula of methanoic acid is

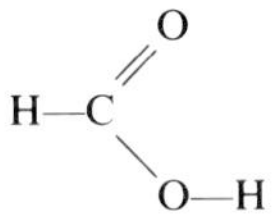

- On the basis of this diagram, what would you expect the bond lengths to be in methanoic acid?

An electron diffraction study of methanoic acid vapour shows it to have the structure shown here.

```
         O                          O
0.109 nm╱ 0.123 nm              H—C╲ 122°
H—C                                 
0.136 nm╲                          OH
         O—H
         0.097 nm

    bond lengths              bond angles
```

- Are the bond lengths what you expected?

Earlier in this Topic it was seen that the shapes of molecules could be explained by supposing that pairs of electrons repel one another; thus bonds, and lone pairs of electrons, tend to get as far away from one another as possible.

- Are the bond angles in methanoic acid what you would expect from this theory?
  Would a small departure from the simple predicted angle seem a likely situation?

Sodium methanoate has the formula $HCO_2^- Na^+$, and that of the methanoate ion is HCO_2^-. Draw an electron structure for the methanoate ion; then draw a diagram using — for a single covalent bond and = for a double covalent bond.

- Are alternative diagrammatic arrangements of the bonds possible?
  What sort of length would you expect the C—O distance to be?

X-ray diffraction studies of sodium methanoate show the methanoate ion to have the structure given below.

```
         O                          O
       ╱ 0.127 nm                  ╱
H—C                            H—C╲ 124°
       ╲ 0.127 nm                  
         O                          O

    bond lengths              bond angles
```

- Are the bond lengths within the limits which you expected?
  What can you say about the nature of the C—O bonds in the methanoate ion?
  Use the evidence from nitric acid, the nitrate ion, methanoic acid, and the methanoate ion. What can be said about the likely bond lengths in bond systems where single and double bonds may be represented diagrammatically by two or more interchangeable arrangements?
  Can the structures of such compounds be described adequately in terms of the types of bonds which you have so far studied?

Where equivalent alternative bond structures can be drawn for a compound the actual situation is neither of these; the bond lengths in these situations often prove to be equal, and the available electrons must therefore be distributed equally among the atoms concerned. It is believed that each atom is bonded to the next by two electrons between the nuclei, forming a single

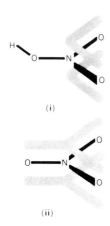

Figure 6.13 Delocalization of electrons in (i) nitric acid, (ii) the nitrate ion

covalent bond; and that the remaining available electrons are distributed as charge-clouds above and below all the atoms concerned. These electron charge-clouds are not associated with any particular atom but are mobile over the whole atomic system; they are thus known as **delocalized electrons**. Representations of the situation in nitric acid and the nitrate ion are given in figure 6.13.

Benzene

The organic compound benzene, C_6H_6, has a molecular structure in which delocalization of electrons occurs. The evidence for this is discussed in Topic 7.7, and the chemical properties of benzene are considered at the same time. You will find that these properties are very different from those that might be expected for a compound whose molecules were not delocalized in structure.

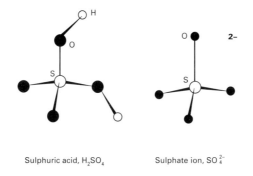

Sulphuric acid, H_2SO_4 Sulphate ion, SO_4^{2-}

QUESTION
Work out an electron structure for sulphuric acid and the sulphate ion. Do your proposals match the shapes shown in the diagrams?

6.6 Bonding, structure and chemical behaviour

This is a good moment to review what you have learnt about bonding and apply your knowledge to looking for patterns in reactions.

STUDY TASK
1 Look again at figure 5.13 about the properties of the chlorides of the elements of Period 3 in the Periodic Table. What pattern is there in the structure and bonding of the chlorides and their behaviour with water?
2 For the elements of Period 3 and Group 1 collect information about the structure and bonding of their oxides and their acid–base behaviour with water. Is there a pattern in your information?
3 For the same elements collect information about their behaviour with chlorine, oxygen and water. Is there a pattern in the behaviour of the elements with chlorine, oxygen and water and their position in the Periodic Table?

Summary

At the end of this Topic you should be able to:
a demonstrate understanding of the covalent bond in terms of electron sharing
b interpret and construct 'dot-and-cross' diagrams of simple covalent compounds
c demonstrate understanding of the terms bond length and bond angle
d predict and interpret the shapes and bond angles in simple covalent molecules using electron-pair repulsion theory
e demonstrate understanding of the terms σ-bond and π-bond, including the electron density in each type
f select data in order to predict the nature of the bonding in a given substance, including dative covalency, bonding of intermediate type, bond polarity and delocalization
g apply your knowledge of delocalization effects to predict or explain the shapes and bond angles in given molecules
h demonstrate understanding of the terms enthalpy change of combustion and bond energy
i select appropriate data to calculate bond energies, using Hess's law.
j interpret the behaviour of chlorides and oxides towards water in terms of their structure and bonding
k identify patterns in chemical behaviour of the elements relative to their position in the Periodic Table.

Review questions

* Indicates that the *Book of data* is needed.

*6.1 Ethanol burns in an excess of air according to the equation

$$C_2H_5OH(l) + 3O_2(g) \longrightarrow 2CO_2(g) + 3H_2O(l)$$

Calculate the value for the enthalpy change for this reaction in the following way.

a Write down the equation for the reaction.
b Draw a Hess cycle showing the formation of the compounds on both sides of the equation, from the same elements.
c Using the *Book of data*, find the standard enthalpy changes of formation for the products, and write them in the correct place on your Hess cycle.
d Do the same for the reactants.
e Calculate the required enthalpy change, giving it the correct sign.

6 Covalent bonding

6.2 1.000 g of each of the following alcohols was burned in a combustion calorimeter. In each case the quantity of energy required to give the same temperature rise as in the reaction was determined; the energy was supplied electrically.

Alcohol	Energy required
Methanol CH_3OH	22.34 kJ
Ethanol C_2H_5OH	29.80 kJ
Propanol C_3H_7OH	33.50 kJ
Butanol C_4H_9OH	36.12 kJ

Calculate the standard enthalpy change of combustion of each alcohol. Comment on the results you obtain and, using a graphical method, make a prediction of the enthalpy change of combustion for pentanol, $C_5H_{11}OH$.

*6.3 Calculate $\Delta H^\ominus(298)$ for the following reactions.

a $CH_3OH(l) + 1\tfrac{1}{2}O_2(g) \longrightarrow CO_2(g) + 2H_2O(g)$
b $CH_4(g) + 2O_2(g) \longrightarrow CO_2(g) + 2H_2O(l)$
c $CH_4(g) \longrightarrow C(g) + 4H(g)$

*6.4 Using the standard enthalpy changes of formation of the compounds, and those of atomization of the elements, calculate the enthalpy changes for the following reactions.

a $HF(g) \longrightarrow H(g) + F(g)$
b $HCl(g) \longrightarrow H(g) + Cl(g)$
c $HBr(g) \longrightarrow H(g) + Br(g)$
d $HI(g) \longrightarrow H(g) + I(g)$

What generalizations do your answers indicate about the energies of the bonds in the hydrides of Group 7 elements?

*6.5 Using the standard enthalpy changes of formation of the compounds, and those of atomization of the elements, calculate the enthalpy changes for the following reactions.

a $NH_3(g) \longrightarrow N(g) + 3H(g)$
b $PH_3(g) \longrightarrow P(g) + 3H(g)$
c $AsH_3(g) \longrightarrow As(g) + 3H(g)$
d $SbH_3(g) \longrightarrow Sb(g) + 3H(g)$

What generalizations do your answers indicate about the energies of the bonds in the hydrides of Group 5 elements?

6.6 Given that:

$$P(g) + 3Cl(g) \longrightarrow PCl_3(g) \quad \Delta H = -983 \text{ kJ}$$
$$P(s) + 1\tfrac{1}{2}Cl_2(g) \longrightarrow PCl_3(g) \quad \Delta H = -305 \text{ kJ}$$
$$P(g) + 3H(g) \longrightarrow PH_3(g) \quad \Delta H = -958 \text{ kJ}$$
$$P(s) + 1\tfrac{1}{2}H_2(g) \longrightarrow PH_3(g) \quad \Delta H = -8.4 \text{ kJ}$$
$$P(s) \longrightarrow P(g) \quad \Delta H = +314 \text{ kJ}$$

calculate the bond energies of the following bonds.

a P—Cl in PCl_3
b P—H in PH_3
c Cl—Cl in Cl_2
d H—H in H_2.

6.7 Sulphur forms a chloride, SCl_2, in the gas phase. Draw a diagram of the molecule showing the shape you would expect it to have, and indicating the approximate value you would expect for the bond angle.

6.8 The carbonate ion, CO_3^{2-}, has a planar structure as shown in the diagram. The diagram shows only internuclear separations and shape, not bonding.

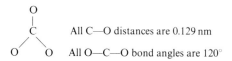

All C—O distances are 0.129 nm
All O—C—O bond angles are 120°

a What structures can you draw for this ion using the dot-and-cross method?
*b Select any *one* of the structures you have drawn. What bond lengths does it suggest the ion should have? Do these agree with the observed internuclear separations? If not, how do they differ?
c Suggest a way of explaining the nature of the bonds in the carbonate ion.
d It has been suggested that molecules of carbonic acid, H_2CO_3, are present in very low concentration in aqueous solutions of carbon dioxide. How, if at all, would you expect the bond lengths and angles between carbon and oxygen atoms in the carbonic acid molecule to differ from those in the CO_3^{2-} ion?

*6.9 Use the general trends in electronegativity to explain the following:

a Sodium hydride has a structure which contains the ions Na^+ and H^-.
b Carbon hydride (methane) has a covalent molecular structure and the electrons are evenly shared in the bonds between the carbon and hydrogen atoms in the methane molecules.
c Chlorine hydride (hydrogen chloride) gas has a covalent molecular structure but the molecule has a dipole

$$\underset{\delta+ \quad \delta-}{\text{H—Cl}}$$

d Lithium forms a crystalline fluoride $Li^+ F^-$ whereas oxygen forms a gaseous fluoride OF_2.

Examination questions

6.10 Carbon and silicon are both in Group 4 of the Periodic Table. Although there are similarities between them, there are also differences, such as the molecular nature of carbon dioxide contrasting with the giant lattice of silicon(IV) oxide. This question is concerned with both similarities and differences.

a State the ground state electronic configurations of the gaseous atoms of carbon and silicon.

b Draw a 'dot-and-cross' diagram to illustrate the electronic structure of a molecule of carbon dioxide, showing outer shells only.

c A series of hydrides of silicon, the silanes (for example, SiH_4, Si_2H_6, Si_3H_8), which are analogous to the alkanes, can be prepared.

i Write an equation for the combustion of SiH_4 in air.

ii Comment on the fact that silane, SiH_4, is spontaneously flammable in air, while methane requires ignition in order to burn.

d The standard molar enthalpy changes of combustion at 298 K for methane and silane are:

$$\Delta H_c^\ominus[CH_4(g)] = -890 \text{ kJ mol}^{-1}$$
$$\Delta H_c^\ominus[SiH_4(g)] = -1517 \text{ kJ mol}^{-1}$$

The major part of this large difference can be attributed to bond energy differences.

Explain this statement, showing how bond energy values are involved in determining the enthalpy changes of combustion.

The following bond energy data may be helpful:

$E(C-H) = 413$ kJ mol^{-1}
$E(Si-H) = 318$ kJ mol^{-1}
$E(C=O) = 805$ kJ mol^{-1} in $CO_2(g)$
$E(Si-O) = 466$ kJ mol^{-1} in $SiO_2(s)$

A calculation is NOT required.

e By analogy with the chemistry of appropriate carbon compounds:

i Predict the formula for a simple silicate ion, containing only one silicon atom, and give its charge.

ii Predict what you would observe when dilute solutions of hydrochloric acid and sodium silicate are mixed. Name and give a formula for the product which contains silicon.

6.11 The element tin has the electronic configuration:

$1s^2 2s^2 2p^6 3s^2 3p^6 3d^{10} 4s^2 4p^6 4d^{10} 5s^2 5p^2$

Tin forms a series of halides of general formula SnX_2. It also forms another series of halides of general formula SnX_4, which are prepared by direct combination of tin with the appropriate halogen. The preparation of SnI_4 is typical of such preparations:

I Add a solution of iodine (6.35 g) in tetrachloromethane (25 cm^3) to 2.00 g of granulated tin contained in a 100 cm^3 round-bottomed flask.

II Heat under reflux until the reaction just starts. Further heating is unnecessary as the reaction is exothermic.

III Filter the hot solution through a pre-heated funnel.

        IV Cool the filtrate in an ice-bath, when orange crystals of the product are formed. The crystals have a melting point of 144 °C.

a In which Group of the Periodic Table is tin?

b What is the atomic number of tin?

c i Give the systematic chemical name for $SnCl_2$.

  ii Write down the electronic configuration of the tin ion in SnF_2.

  iii Write an equation (with state symbols) for a reaction by which $SnCl_2$ might be prepared from a sample of tin.

d In the preparation of SnI_4 described above:

  i Show by means of a calculation which element (tin or iodine) was present in excess. (Molar masses: Sn = 119, I = 127 g mol^{-1}).

  ii What technique would be most appropriate for purifying the product?

  iii The mass of pure product obtained was 5.6 g. What was the percentage yield?

e i What **two** properties of SnI_4 described in this question suggest that it is composed of covalently bonded molecules?

  ii What **two** reasons can you give for SnI_4 having a covalent structure, while SnF_2 has an ionic structure?

  iii What shape would you predict for the SnI_4 molecule?

6.12 This question is about interhalogen compounds which are compounds of two or more halogen elements only. Examples are ClF, ClF_3, ICl, BrF, BrCl, BrF_5 and IF_7. Several others are known. Like the halogens themselves, interhalogen compounds are usually quite reactive.

a Draw diagrams to show the electronic structure ('dot-and-cross' diagrams) of the following interhalogens (only the outer-shell electrons need be shown):

  i Iodine monochloride, ICl

  ii Bromine pentafluoride, BrF_5.

b Draw a diagram to show the shape you would expect for a molecule of bromine pentafluoride, BrF_5.

c Most of the interhalogen compounds known are fluorides. Suggest a reason for this.

*6.13 List the formulae of the simplest hydrides of the elements lithium to fluorine using tables 5.3 and 5.5 in the *Book of data*.

Classify the hydrides as ionic or covalent, justifying your classification by reference to appropriate data, and draw dot-and-cross diagrams to illustrate their electronic structures.

Are the physical properties of the hydrides of lithium to fluorine consistent with their positions in the Periodic Table? Explain any pattern of inconsistency that you notice.

TOPIC 7

Hydrocarbons and halogenoalkanes

In this second Topic on organic chemistry, we shall begin by considering some of the reasons for the great diversity of carbon compounds, and more of the rules necessary for naming them. Next, we shall consider the results of experiments on four different types of carbon compounds. This will help us to examine an interpretation of the fundamental ways in which carbon compounds react. We shall also meet the important reactions of these carbon compounds and learn something of their social and industrial importance.

This Topic contains a number of ideas that will be new to you. However, the remaining Topics on organic chemistry contain a further exploration of these ideas, rather than a large number of additional ones. It is important, therefore, that you learn the essential ideas about the types of organic reactions before you proceed to Topic 12. The last section is a summary which is designed to help you learn the essential framework of ideas and reactions that have been included in the Topic.

7.1 The variety of molecular structure in organic compounds

REVIEW TASK

Working in groups, list from your work on Topic 2:
1 The names of the first five alkanes.
2 The molecular, structural and displayed formulae of propane, and work out the molar mass.
3 The structural formulae of the isomers having molecular formular C_4H_{10}.

Although the chemistry of carbon can be studied as a part of Group 4 of the Periodic Table, carbon also has a special chemistry of its own, a chemistry that has been able to flourish in the conditions on this planet. If we compare carbon compounds with those of silicon, as silicon is the nearest neighbour to carbon in Group 4 of the Periodic Table, two major differences can be identified:

1 Carbon compounds frequently contain long chains of carbon atoms,

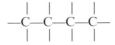

whereas silicon compounds commonly contain chains made up of silicon and oxygen,

$$-O-\underset{|}{\overset{|}{Si}}-O-\underset{|}{\overset{|}{Si}}-$$

2 Carbon atoms never bond to more than four other atoms.

At a simple level the first of these differences between carbon and silicon chemistry can be interpreted as due to differences in bond energies (look again at Topic 6 if necessary).

E(C—C)   347 kJ mol^{-1}
E(C—H)   413 kJ mol^{-1}
E(C—O)   358 kJ mol^{-1}

For single bonds these are all high values of similar magnitude, which means that the bonds can be classified as strong bonds of about the same strength. Thus there will be no strong tendency for a C—C bond or a C—H bond to be replaced by a C—O bond (under standard conditions).

Now look at the corresponding values for silicon:

E(Si—Si)   226 kJ mol^{-1}
E(Si—H)   318 kJ mol^{-1}
E(Si—O)   466 kJ mol^{-1}

The Si—Si bond can be classified as a weak bond with a strong tendency to be replaced by Si—O bonds. So, on the Earth, silicon is found in silicate rocks while carbon can be found in a rich variety of carbon–hydrogen compounds as well as in rocks containing metal carbonates.

The second difference can be described as a difference in possible electron arrangements in Group 4 elements. When carbon atoms have formed four bonds, they have eight electrons in the second electron shell, and this shell is incapable of further expansion. This restricts the possibilities of further chemical attack.

Silicon atoms, however, can form an outermost electron shell with more than eight electrons because of the availability of an empty 3d energy level. So compounds such as the silicon hydrides, for example, SiH_4, are not resistant to chemical attack.

For example, some silicon hydrides ignite at once on mixing with air

$$Si_2H_6(g) \xrightarrow{air} SiO_2(s)$$

but alkanes do not ignite unless heated.

However, bond strengths in carbon compounds, and the limitation of carbon to form compounds with no more than eight electrons in the outermost shell around the carbon atom, are only part of the reason for the diversity of carbon compounds.

Let us look at the tetrahedral arrangement of groups around the carbon atom more fully. If you make a model of the straight-chain C_4H_{10} molecule,

7 Hydrocarbons and halogenoalkanes

> **QUESTION**
> Make models of the molecules and write down the structural formulae of all the isomers with the formula C_6H_{14}. You should finish with five isomers. Clearly we need some additional rules in order to name all these isomers

you will find that the model can be twisted into a number of different shapes. Is this property of the model shared by the molecule and, if so, can any of the different shapes be described as different compounds? The different shapes cannot correspond to different compounds because chemists have only found properties corresponding to one straight-chain compound. It follows, therefore, that the different shapes are all shapes of one compound which must twist just as our model can twist. A glance at a model of the ethane molecule, C_2H_6, shown in figure 7.1, should make this clear. The rotation about the C—C single bond which makes possible the movement seen in the figure is a property of the molecule. So A and B are merely ethane molecules at different stages of a continuous rotation, not isomers of different structures.

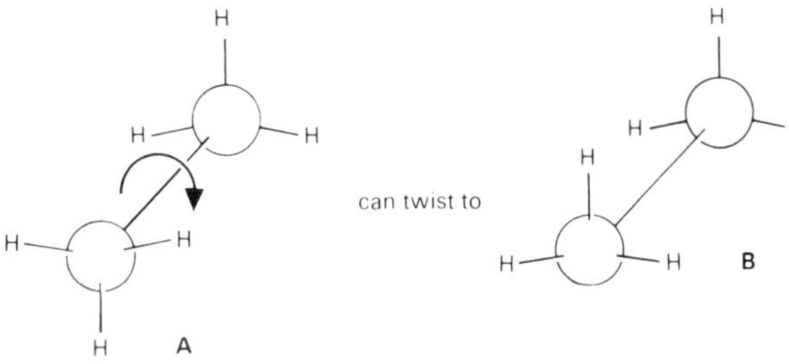

Figure 7.1 Free rotation about a C—C bond

The chemists' toolkit: naming organic compounds

The rules for naming compounds are settled by international agreement through the International Union of Pure and Applied Chemistry. They are usually known as IUPAC rules. The complete set of rules is published in a very large book, and there are regular additions to keep up with the synthesis of new compounds.

You should remember that compounds containing carbon and hydrogen with only single bonds between the carbon atoms are known as **saturated hydrocarbons**. They occur as three main types.

1 Alkanes containing unbranched carbon atom chains

These are hydrocarbons in which the molecules are made up of straight chains of carbon atoms. The general name for these is **alkanes**. Names for individual compounds all have the ending '-ane'. For example,

$$CH_3-CH_2-CH_2-CH_2-CH_3 \quad \text{pentane}$$

For convenience we will repeat the names of the alkanes given in Topic 2 in the table alongside.

Number of carbon atoms in chain	Molecular formula	Name
1	CH_4	methane
2	C_2H_6	ethane
3	C_3H_8	propane
4	C_4H_{10}	butane
5	C_5H_{12}	pentane
6	C_6H_{14}	hexane
7	C_7H_{16}	heptane
8	C_8H_{18}	octane
9	C_9H_{20}	nonane
10	$C_{10}H_{22}$	decane
11	$C_{11}H_{24}$	undecane
12	$C_{12}H_{26}$	dodecane
20	$C_{20}H_{42}$	eicosane

2 Alkanes containing branched chains

Hydrocarbons having branched chains of carbon atoms are still known as alkanes, but the rules for the individual names are more complicated. An example is

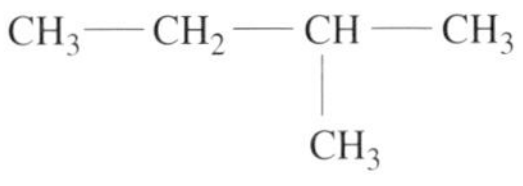

$$CH_3-CH_2-CH-CH_3$$
$$\quad\quad\quad\quad\quad\quad | $$
$$\quad\quad\quad\quad\quad CH_3$$

2-methylbutane

In order to name these compounds, groups of atoms known as **alkyl** groups are used. These are derived from hydrocarbons with unbranched carbon chains by removing one hydrogen atom from the end carbon atom of the chain. For example CH_3—CH_2—CH_3 (propane) becomes CH_3—CH_2—CH_2— with one bond unoccupied. Alkyl groups are named from the parent hydrocarbon by substituting the ending '-yl' for the ending '-ane'. Thus, $CH_3CH_2CH_2$— is the **propyl** group. A list of alkyl groups is given below.

Hydrocarbon	Alkyl group	Formula for alkyl group
Methane	Methyl	CH_3—
Ethane	Ethyl	C_2H_5—
Propane	Propyl	C_3H_7—
Butane	Butyl	C_4H_9—
Pentane	Pentyl	C_5H_{11}—
Hexane	Hexyl	C_6H_{13}—
and so on		

Branched chain hydrocarbons are named by combining names of alkyl groups with the name of an unbranched chain hydrocarbon. The simplest is

$$CH_3-\underset{\underset{CH_3}{|}}{CH}-CH_3$$

which is called methylpropane. The hydrocarbon name is always derived from the longest continuous chain of carbon atoms in the molecule. The position of the alkyl group forming the side chain is obtained by numbering the carbon atoms in the chain. The numbering is done so that the lowest number(s) possible are used to indicate the side chain (or chains). Thus

$$CH_3-\underset{\underset{CH_3}{|}}{CH}-CH_2-CH_2-CH_3$$

is named 2-methylpentane, not 4-methylpentane which would be obtained by numbering from the other end of the chain. When there is more than one substituent alkyl group of the same kind, the figures indicating the positions of the groups are separated by commas, for example

$$CH_3-\underset{\underset{CH_3}{|}}{CH}-\underset{\underset{CH_3}{|}}{CH}-CH_3 \quad \text{is 2,3-dimethylbutane, and}$$

$$CH_3-CH_2-\underset{\underset{CH_3}{|}}{\overset{\overset{CH_3}{|}}{C}}-CH_3 \quad \text{is 2,2-dimethylbutane}$$

Different alkyl groups are placed in alphabetical order in the name for a branched chain hydrocarbon, for example,

$$CH_3-CH_2-\underset{\underset{\underset{CH_3}{|}}{\underset{CH_2}{|}}}{CH}-CH_2-\underset{\underset{CH_3}{|}}{CH}-CH_3 \quad \text{is named 3-ethyl-5-methylhexane}$$

3 Alkanes containing a ring of carbon atoms

Hydrocarbons having one or more rings of carbon atoms (to which side chains may be attached) are called **cycloalkanes**. For example,

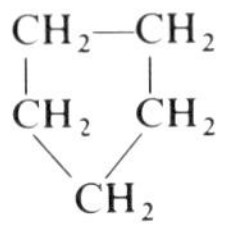

$$\begin{array}{c} CH_2-CH_2 \\ |\quad\quad | \\ CH_2\quad CH_2 \\ \backslash\quad / \\ CH_2 \end{array} \quad \text{cyclopentane}$$

They are named from the corresponding unbranched chain hydrocarbon by adding the prefix '**cyclo–**'.

All the carbon atoms in an unsubstituted cycloalkane ring are equivalent, so far as substitution is concerned, so that if only one alkyl group is added as a substituent there is no need to number the carbon atoms. Thus methylcyclohexane is

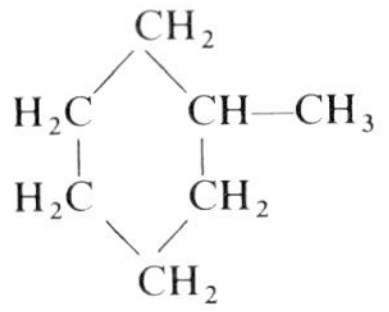

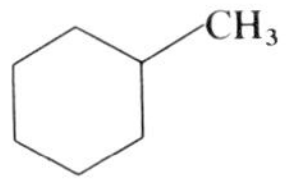

Names and structures of some functional groups

In this table the structures of the functional groups are printed out in the second column so as to show the atomic linkages. When these structures are repeated in the examples given in the fourth column they are printed on one line only, so as to show this abbreviated method of writing them.

Figure 7.2 The main functional groups

Class of compound	Structure of the functional group	Example of a compound Name	Formula
Alkene	$\mathrm{C}=\mathrm{C}$	propene	$CH_2=CH-CH_3$
Arene	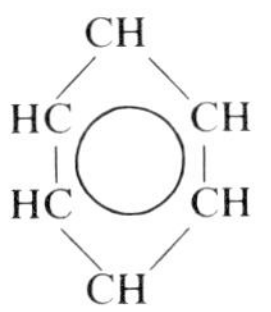 (benzene ring)	benzene	C_6H_6 or 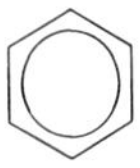 ⌬
Alcohol	—OH	propanol	$CH_3-CH_2-CH_2-OH$
Amine	—NH_2	propylamine	$CH_3-CH_2-CH_2-NH_2$
Nitrile	—C≡N	propanenitrile	CH_3-CH_2-CN
Halogeno	—Cl etc.	1-chloropropane	$CH_3-CH_2-CH_2-Cl$
Aldehyde	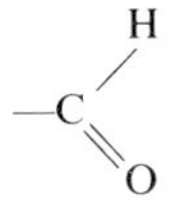 —CHO	propanal	CH_3-CH_2-CHO
Ketone	$\mathrm{C}=\mathrm{O}$	propanone	$CH_3-CO-CH_3$

Class of compound	Structure of the functional group	Example of a compound Name	Formula
Carboxylic acid	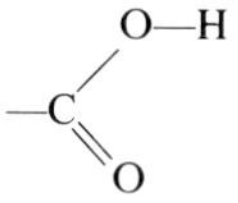 −C(=O)−O−H	propanoic acid	$CH_3-CH_2-CO_2H$
Carboxylate ion	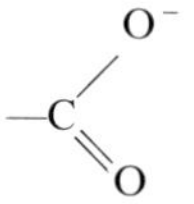 −C(=O)−O$^-$	sodium propanoate	$CH_3-CH_2-CO_2^-\ Na^+$
Acyl chloride	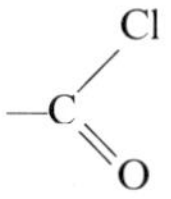 −C(=O)−Cl	propanoyl chloride	CH_3-CH_2-COCl
Acid anhydride	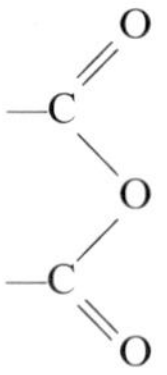 −C(=O)−O−C(=O)−	propanoic anhydride	$(CH_3-CH_2-CO)_2O$
Amide	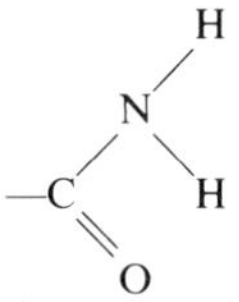 −C(=O)−NH$_2$	propanamide	$CH_3-CH_2-CONH_2$
Nitro compound	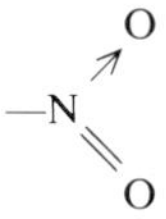 −NO$_2$	nitrobenzene	$C_6H_5NO_2$ or NO_2-phenyl 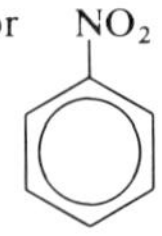
Sulphonic acid	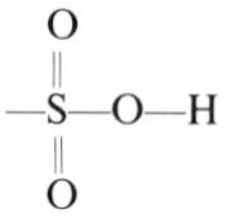 −S(=O)$_2$−O−H	benzenesulphonic acid	$C_6H_5SO_2OH$ or SO_2OH-phenyl 
Ether	−C−O−C−	ethoxyethane	$CH_3-CH_2-O-CH_2-CH_3$

7.2 The alkanes

The alkanes are saturated hydrocarbons; they have only single bonds in their structure (they are saturated) and are composed only of hydrogen and carbon (hydrocarbons).

Alkanes are the major components of crude oil so it is reasonable to claim that they are **the** compounds of the twentieth century: the largest international companies make their living from alkanes and the richest men made their fortunes from alkanes. Countries with crude oil have been able to transform their way of life from that of subsistence farming to one with an increasing contribution from modern technology. The money derived from crude oil has enabled governments to build schools, roads, and hospitals, but it has also

Figure 7.3 Going to work in China and the USA

disrupted traditional ways of life and been used to buy modern war weapons.

The price of crude oil, and ensuring a supply of crude oil, have come to dominate modern economic thinking and modern industrial planning. Consumption of crude oil varies from year to year but the USA takes about 25% of the total while China uses less than 5%. Total world consumption per year has risen by 50% since the first edition of this book in 1970, yet over the same period known reserves have risen from 30 years consumption to around 40 years. So how long will the world's oil reserves last?

Exploration for new oil fields is usually planned to ensure that newly discovered fields can be brought into production by the time they are needed. And that can be a long time: in the North Sea the period from exploration to bringing the first oil ashore in the UK was about 15 years.

Nevertheless, we can be sure that the supply of crude oil is finite so that shortages will develop as oil fields run dry, or as countries restrict output to conserve their source of wealth. It remains to be seen if the world will manage an orderly change to new technologies or whether the rich will outbid the poor in a desperate scramble for oil, and whether cars, aircraft and other transport will come to a halt before adequate alternatives are ready. Without care, the future of oil may prove as dark as its past has proved bright.

But how suitable is crude oil for all the potential uses? The composition of a typical barrel of crude oil and the corresponding demand in the market place are illustrated in figure 7.4. Matching supply to demand is the job of the oil refinery. This is a complex task because not only must the demand for alkanes be satisfied but also most of the demand for unsaturated hydrocarbons and arene hydrocarbons. We shall return to this question later in the Topic.

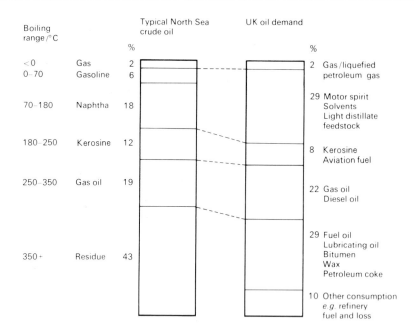

Figure 7.4 Matching supply to demand

One alkane behaves much like another, so we can describe most of the properties we are concerned with by describing ethane, CH_3—CH_3. The structure of ethane is illustrated in figure 7.5.

Figure 7.5 The structure of ethane. **a** A model of the molecule showing the distribution of the electrons **b** A diagram showing the bond lengths and bond angles

Notice that the single bond between the carbon atoms (the σ or sigma bond) is symmetrical, which is consistent with the free rotation about the C—C bond discussed in the previous section.

The average bond energies in alkanes are:

E(C—C)   347 kJ mol^{-1}
E(C—H)   413 kJ mol^{-1}

If you compare these values with other bond energies in the *Book of data*, they will be seen to be towards the top of the range. What does this suggest about the likely reactivity of the alkanes? Are the molecules likely to be polar?

A typical infra-red spectrum of an alkane is shown in figure 7.6. See Topic 18 for an account of how infra-red spectra are obtained.

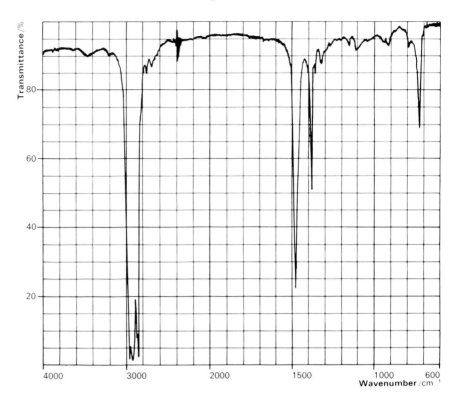

Figure 7.6 The infra-red spectrum of an alkane (decane)

Use the *Book of data*, table 3.3, to check the assignment of the peaks to the appropriate bonds.

The change in boiling points with increase in number of carbon atoms for the straight-chain alkanes can be seen in figure 7.7. You will consider the possible reasons for this trend in Topic 9.

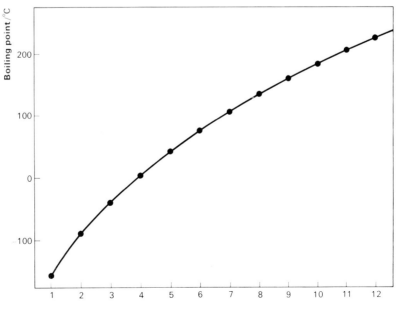

Figure 7.7 The change in boiling points of straight-chain alkanes with increase in number of carbon atoms

7 Hydrocarbons and halogenoalkanes 164

Experiments with alkanes

We shall begin our experimental investigations with a study of some alkanes: hexane, C_6H_{14}, and poly(ethene), $\mathrm{+CH_2-CH_2+}_n$. Poly(eth**ene**) is an alk**ane** in spite of its name.

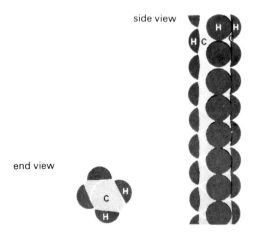

Figure 7.8 The molecular chain in poly(ethene)

EXPERIMENT 7.2 The properties of some alkanes

Carry out the following tests on a sample of each of your alkanes. Note your observations in tabular form carefully because you will be doing similar tests on other compounds later in this Topic and you will be expected to compare the behaviour of the different types of compound.

Figure 7.9 Cyclohexane caught fire and destroyed this factory in 1974

> **SAFETY** ⚠
> Be careful when using hexane as it is volatile and highly flammable. Concentrated sulphuric acid is corrosive and should be washed off **at once** if you get a drop on your skin, and be careful when picking up bottles of the acid as it can dribble down the outside; avoid inhaling bromine fumes and keep the solution off your skin; 20% potassium hydroxide is corrosive. Remember to wear eye protection.

Procedure

1 Combustion

If possible, do this experiment in a fume cupboard. Keep a sample of the liquid alkane in a stoppered boiling-tube well away from any flame. Dip a combustion spoon into the sample, so that it is no more than wet with alkane. Set fire to the alkane on the combustion spoon.

- Is the flame clean or sooty? A clean flame is necessary for a substance to be used as a fuel.

2 Oxidation

To 0.5 cm³ of a mixture of equal volumes of 0.01 M potassium manganate(VII) and 1 M sulphuric acid in a test-tube add two or three drops of the liquid alkane or two granules of poly(ethene). Shake the contents and try to tell from any colour change if manganate(VII) oxidizes the alkane.

- Is there any sign of reaction?

3 Action of bromine

To 0.5 cm³ of bromine water (TAKE CARE) in a test-tube add a few drops of the liquid alkane or two granules of poly(ethene).

- What, if anything, happens to the colour of the bromine?

4 Action of bromine in sunlight

> **SAFETY** ⚠
> Protect your eyes from the light.

To 10 cm³ of hexane in a test-tube add several drops of 2% bromine in an inert solvent. Loosely cork the tube and irradiate with sunlight or a photoflood light.
Prepare a second similar tube and leave it in a dark place. Compare the intensity of the bromine colour at intervals.

- Is there any evidence of reaction? Can you identify any fumes evolved? If no fumes are apparent try tipping the contents of the test-tube into a beaker.

5 Action of sulphuric acid

Put 1–2 cm³ of concentrated sulphuric acid in a test-tube held in a rack (TAKE CARE). Add 0.5 cm³ of the liquid alkane or two or three granules of poly(ethene).

- Do the substances mix or are there two separate layers in the test-tube?

6 Action of alkali

To 0.5 cm³ of the liquid alkane or two or three granules of poly(ethene) in a test-tube add 1–2 cm³ of 20% potassium hydroxide (TAKE CARE) dissolved in ethanol. Mix the liquids by shaking the tube gently.

- Is there any sign of reaction?

7 Catalytic cracking

Put some 'light paraffin' in a test-tube to the depth of 1–2 cm. Light paraffin is a mixture of alkanes, containing molecules with about 12 carbon atoms.
Push in some loosely packed ceramic fibre until all the paraffin has been soaked up. Now add 2–3 cm depth of aluminium oxide granules and clamp the test-tube horizontally so that the granules form a layer in the test-tube. Connect the test-tube for collection of gas over water as shown in figure 7.10.

Figure 7.10 Apparatus for the catalytic cracking of an alkane

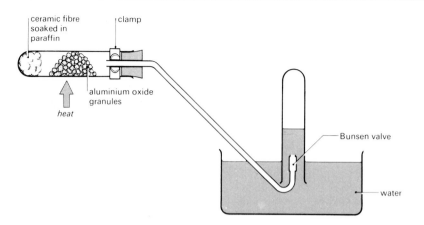

Heat the aluminium oxide strongly and continuously but be careful not to melt the rubber stopper, nor to allow the delivery tube to become blocked. The paraffin should get hot enough to evaporate without needing direct heat.

Collect three or four tubes of gas (discard the first one: why?), and when the delivery of gas slows down, lift the apparatus clear of the water to avoid it being sucked up into the hot test-tube.

Carry out the following two tests on the gas collected.

a *Test for flammability*. Any flame will be more visible if the test-tube is held upside down.

b *Test for reaction with bromine*. Add a little bromine water (TAKE CARE).

■ Is the gaseous product an alkane or can different properties be observed?

An interpretation of the photochemical experiment with alkanes

The lack of positive results when one is doing experiments with alkanes in the laboratory probably seems disappointing but, for most of their uses, the chemical inertness of the alkanes is their greatest asset. Compounds that are non-corrosive to metals (lubricating oils), harmless to our skin (petroleum jelly), and safe in contact with foods (poly(ethene)), are enormously useful to us.

The reaction of the alkanes that we need to consider here is that of hexane with bromine in sunlight. Since the reaction needs light in order to take place (unless the reactants are heated to over 300 °C), it is known as a **photochemical reaction**. It is a general reaction between alkanes and bromine or chlorine. The process by which reaction takes place must depend on the absorption of the energy of the photons that make up the radiation. Other experiments on the reaction of methane with chlorine have shown that the process does not need many photons: many thousands of product molecules are produced for each photon absorbed. So by what process does this reaction occur? To simplify the equations involved we will consider the methane-chlorine reaction as a typical example of what is involved.

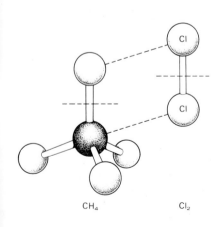

Figure 7.11 A ball-and-link representation of possible attack by chlorine on methane

Look at a ball-and-link model, such as the one shown in figure 7.11. You may think that the reaction could begin with the simultaneous breaking of the C—H and Cl—Cl bonds, followed by the making of C—Cl and H—Cl bonds. Calculations using the appropriate bond energies show that this would be an exothermic reaction, with an enthalpy change of about 100 kJ mol^{-1}.

$$CH_4 + Cl_2 \longrightarrow CH_3Cl + HCl \qquad \Delta H^\ominus = -100 \text{ kJ mol}^{-1}$$

But there are two problems about this proposal. Firstly, the molecules would need to come together with an exact orientation for reaction to happen, and that would be a very rare collision amongst all the collisions occurring. Secondly, the distance between the atoms would have to be reduced to about a bond length and the force of repulsion between the electron clouds would have to be overcome.

How much energy can be provided by the absorption of radiation?

When ultraviolet light is absorbed the energy provided is

$$E(\text{ultraviolet photons}) \approx 400 \text{ kJ mol}^{-1}$$

This is enough energy to break up the chlorine molecule into uncharged chlorine atoms with unpaired electrons, which are given the symbol Cl·

$$Cl_2 \longrightarrow Cl\cdot + Cl\cdot \qquad \Delta H^\ominus = +242 \text{ kJ mol}^{-1}$$

However, it is scarcely enough to break a methane molecule

$$CH_4 \longrightarrow CH_3\cdot + H\cdot \qquad \Delta H^\ominus = +435 \text{ kJ mol}^{-1}$$

and certainly not enough to produce ions.

$$Cl_2 \longrightarrow Cl^+ + Cl^- \qquad \Delta H^\ominus \approx +1130 \text{ kJ mol}^{-1}$$
$$CH_4 \longrightarrow H^+ + CH_3^- \qquad \Delta H^\ominus \approx +1700 \text{ kJ mol}^{-1}$$

We can therefore conclude that the first step in this photochemical reaction is probably the absorption of an ultraviolet photon by a chlorine molecule, resulting in the formation of two chlorine atoms. These chlorine atoms, each with an odd number of electrons, are known as **free radicals**.

Note carefully the difference between breaking a Cl—Cl bond to form free radicals, and the different pattern of breaking that produces ions.

To form a free radical the process is

$$:\!\overset{\cdot\cdot}{\underset{\cdot\cdot}{Cl}}\!:\!\overset{\times\times}{\underset{\times\times}{Cl}}\!\overset{\times}{} \longrightarrow :\!\overset{\cdot\cdot}{\underset{\cdot\cdot}{Cl}}\!\cdot + \overset{\times\times}{\underset{\times\times}{\times Cl}}\!\overset{\times}{}$$

This is known as **homolytic fission**; 'homo-' is from ancient Greek, meaning 'same'.

To form ions the process is

$$:\!\overset{\cdot\cdot}{\underset{\cdot\cdot}{Cl}}\!:\!\overset{\times\times}{\underset{\times\times}{Cl}}\!\overset{\times}{} \longrightarrow \left[:\!\overset{\cdot\cdot}{\underset{\cdot\cdot}{Cl}}\!:\right]^+ + \left[\overset{\times\times}{\underset{\times\times}{\times Cl}}\!\overset{\times}{}\right]^-$$

This is known as **heterolytic fission**, 'hetero-' meaning 'different'.

How does the reaction proceed? When the chlorine radical attacks a methane molecule there seem to be two possibilities for products. Either a hydrogen radical (we have to have a radical product if we start with an odd number of electrons in the reactants) and chloromethane are produced

$$Cl\cdot + CH_4 \longrightarrow H\cdot + CH_3Cl \qquad \Delta H^\ominus = +108 \text{ kJ mol}^{-1}$$

or a methyl radical and hydrogen chloride are produced.

$$Cl\cdot + CH_4 \longrightarrow CH_3\cdot + HCl \qquad \Delta H^\ominus = +4 \text{ kJ mol}^{-1}$$

COMMENT

To calculate the energy of radiation, E_{mol} in kJ mol^{-1}, for 1 mole of photons the relationship used is

$$E_{mol} = Lh\nu$$

where

$L = 6.02 \times 10^{23}$ mol^{-1} (the Avogadro constant)
$h = 6.62 \times 10^{-37}$ kJ s (the Planck constant)
ν = frequency of the radiation in hertz, s^{-1}

According to the *Book of data*, table 3.1, an approximate value for the frequency of infra-red radiation is 10^{13} s^{-1}, for visible light is 10^{14} s^{-1}, and for ultraviolet light is 10^{15} s^{-1}. If you carry out this calculation you will find that only ultraviolet photons have sufficient energy to be of interest to us.

Thus, on the basis of the energy involved, we can conclude that the second step in the reaction is likely to be the formation of a **methyl free radical**. Similar considerations will lead us to the next step.

$$CH_3\cdot + Cl_2 \longrightarrow CH_3Cl + Cl\cdot \qquad \Delta H^\ominus = -97 \text{ kJ mol}^{-1}$$

These two last steps taken together produce the correct end-products, hydrogen chloride and chloromethane, but they also link together in an apparently endless chain for as long as methane and chlorine molecules are available.

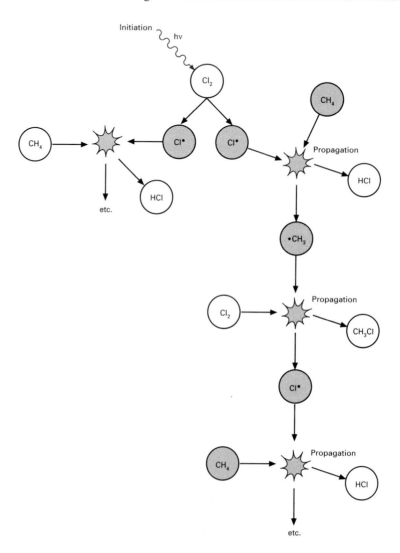

Figure 7.12 A chain reaction

Is this chain likely to go on indefinitely? Well, not really, because to keep going it depends on free radicals colliding only with ordinary molecules. In practice, free radicals are bound to collide with each other, and if you think about the possibilities, you should realize there are three:

$$Cl\cdot + Cl\cdot \longrightarrow Cl_2$$
$$CH_3\cdot + CH_3\cdot \longrightarrow CH_3-CH_3$$
$$\text{and } Cl\cdot + CH_3\cdot \longrightarrow CH_3Cl$$

These collisions will all terminate the chain (figure 7.13). Experiments have shown that for each original photon absorbed, on average 10 000 molecules of chloromethane are produced. So how many links are there on average in a chain before the chain is terminated?

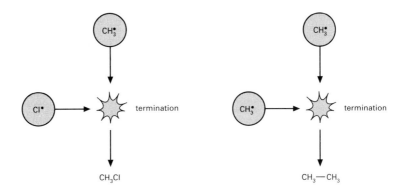

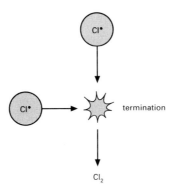

Figure 7.13 Chain termination

Free radical chain reactions are also important in the formation of polymers, such as poly(ethene), and in the combustion of hydrocarbons, especially petrol. We shall return to these topics later.

Reactions of the alkanes

1 Combustion

It is the combustion of the alkanes that provides their most important uses:

$$CH_4 + 2O_2 \longrightarrow CO_2 + 2H_2O \qquad \Delta H_c^\ominus = -890 \text{ kJ mol}^{-1}$$
methane

$$C_4H_{10} + 6\tfrac{1}{2}O_2 \longrightarrow 4CO_2 + 5H_2O \qquad \Delta H_c^\ominus = -2877 \text{ kJ mol}^{-1}$$
butane

These equations represent very familiar processes because the oxygen for combustion normally comes from the air, while methane is the main component of domestic gas and propane and butane are components of bottled gas.

7 Hydrocarbons and halogenoalkanes

Other industrial products, including petrol, jet fuel (kerosine), diesel oil, paraffin, heating oil, and candlewax are mixtures of saturated and unsaturated hydrocarbons, but the main components are alkanes with appropriate boiling points. In normal use, the alkane fuels have the great advantage of burning with a relatively clean flame and producing non-toxic products. However, if the air supply is restricted, carbon monoxide can be produced:

$$CH_4 + 1\tfrac{1}{2}O_2 \longrightarrow CO + 2H_2O$$

Carbon monoxide is dangerously toxic.

The particular case of the combustion of petrol will be dealt with in more detail next in this Topic.

2 Photochemical reactions with halogens

The photochemical reaction of chlorine and bromine with alkanes

$$\underset{\text{hexane}}{C_6H_{14}} + Br_2 \longrightarrow \underset{\text{bromohexane}}{C_6H_{13}Br} + HBr$$

is not a useful method of preparing halogenoalkanes. This is because further reaction takes place, forming a mixture of products. Photochemical and free radical organic reactions are important, however, in other contexts.

This photochemical process is a chain reaction involving free radicals in the following sequence (as explained earlier in this section).

Figure 7.14 Butane gas for heating and cooking

$$Cl_2 \xrightarrow{h\nu} 2Cl\cdot \qquad \text{chain \textbf{initiation}}$$

$$\left.\begin{array}{l} CH_4 + Cl\cdot \longrightarrow HCl + CH_3\cdot \\ Cl_2 + CH_3\cdot \longrightarrow CH_3Cl + Cl\cdot \end{array}\right\} \text{chain \textbf{propagation}}$$

$$\left.\begin{array}{l} Cl\cdot + Cl\cdot \longrightarrow Cl_2 \\ CH_3\cdot + CH_3\cdot \longrightarrow CH_3\!-\!CH_3 \\ Cl\cdot + CH_3\cdot \longrightarrow CH_3Cl \end{array}\right\} \text{chain \textbf{termination}}$$

The overall reaction is

$$CH_4 + Cl_2 \longrightarrow CH_3Cl + HCl$$

This is followed by reaction with more chlorine in further photochemical chain reactions producing

$$CH_3Cl + Cl_2 \longrightarrow CH_2Cl_2 + HCl$$
$$CH_2Cl_2 + Cl_2 \longrightarrow CHCl_3 + HCl$$
$$CHCl_3 + Cl_2 \longrightarrow CCl_4 + HCl$$

3 Catalytic cracking

This is an important process in the petrochemical industry where much of the fraction from the distillation of crude oil with a boiling range 200–300 °C (C_{10} to C_{20} alkanes) is heated to 500 °C in the presence of a silica–alumina catalyst to produce unsaturated hydrocarbons, alkenes, and short-chain alkanes useful for petrol. The high boiling residues are used as fuel oil.

7 Hydrocarbons and halogenoalkanes

Figure 7.15 Catalytic cracking of the naphtha fraction from crude oil to produce ethene. (Shell, Singapore)

Summary: the alkanes

Draw up charts similar to figure 7.16. Add colour and illustrations that you feel might help you learn the properties of the alkanes.

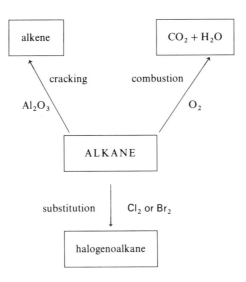

Figure 7.16 Summary of the reactions of the alkanes

7.3 Study task: Octane number of petrol hydrocarbons

QUESTIONS
Read the passage below and answer these questions based on it.
1. What is meant by pre-ignition?
2. How is octane number related to pre-ignition?
3. Explain the difference between catalytic cracking and catalytic reforming.
4. Suggest an equation for the isomerization of an unbranched alkane to give a branched chain alkane.
5. Suggest a balanced equation for the cracking of eicosane, $CH_3(CH_2)_{18}CH_3$, with ethene as one of the products.
6. Acids catalyse the process of alkylation. Use the Brønsted-Lowry definition of an acid to suggest how this happens.
7. Write the structural formula of one isomer of C_8H_{18} other than iso-octane (2,2,4-trimethylpentane). Do you think the octane number for your isomer will be about the same, higher, or lower than 100?
8. What is a free radical?
9. What is the effect of chain branching on the reactivity of free radicals?

When it is first obtained from the Earth, crude oil is a complex mixture of hydrocarbons with sulphur compounds and inorganic impurities. These hydrocarbons may contain one to more than fifty carbon atoms, and are mostly alkanes (with straight or branched chains), together with naphthenes and arenes. Petroleum is separated into fractions by distillation, for example, gasoline, naphtha, kerosine, gas oil, and diesel oil.

In general, the percentage of motor gasoline or petrol in crude oils is not enough to meet the heavy demands for motor car use. So it is necessary to devise ways whereby a larger proportion of the hydrocarbons in crude oil can be made use of as petrol. The value of hydrocarbons for use in petrol can be judged from their 'octane number'. Heptane is given an octane number of 0 and 2,2,4–trimethylpentane (iso–octane) is given an octane number of 100. The higher the number, the less the tendency to pre-ignite in a car engine – that is, the less the tendency to explode under compression before the spark is passed. A second explosion when the spark is passed results in the two shock waves producing a characteristic 'knocking' in the engine.

Four processes of importance for producing petrol-grade hydrocarbons are catalytic cracking, catalytic reforming, alkylation, and isomerization.

Catalytic cracking
One method of obtaining more petrol is to heat the larger hydrocarbon molecules so that they break down. In early years the process of thermal cracking was used, although much of the petroleum was broken down too extensively. In the 1930s, the higher compression in petrol engines called for fuels with a higher octane rating. The value for the products of thermal cracking was only 70–80. Fortunately, it had been discovered that the cracking of hydrocarbons in the presence of a catalyst (catalytic cracking) gave a petrol

containing more branched hydrocarbons and an octane rating of 90–95.

The first catalytic cracking unit was built in 1936 in the USA, at New Jersey. It contained a fixed bed of catalyst pellets composed of acid-treated clays. From a knowledge of the chemical composition of clays, various synthetic silica–alumina catalysts were also developed and these are still widely used. More recently, crystalline aluminosilicates, known as zeolites or molecular sieves, have also come into use as cracking catalysts.

Catalytic reforming

This is now one of the most important processes for the production of motor gasolines. Adding a metallic component to a cracking catalyst gives petrol with an even higher octane number. Platinum is used exclusively as this component and highly purified alumina is used in place of silica–alumina. The process is known as catalytic reforming, but 'platforming' and other commercial names are often used. The improvement in octane number is due largely to the higher percentage of arenes in the product. The process is therefore also a source of arenes for the chemical industry. Some of the chemical reactions which are carried out at the same time by reforming catalysts are:
- Dehydrogenation of cyclohexanes to arenes.
- Dehydrocyclization of alkanes and alkenes to arenes.
- Isomerization of unbranched chain to branched chain alkanes.
- Hydrocracking to hydrocarbons of lower molar mass.

In a reforming catalyst, the platinum is highly dispersed over the alumina, perhaps as platinum atoms or small groups of atoms. Both the platinum and the alumina play a catalytic role.

Alkylation

Another means of obtaining high octane blending stocks is to join some of the smaller molecules in the right way, that is, using C_3–C_4 hydrocarbons. The process of alkylation involves the reaction of a branched chain alkane (for example, 2-methylpropane) and an alkene (for example, propene or butene). The catalysts are, or contain, acids; sulphuric acid, hydrofluoric acid, and phosphoric acid are used.

Isomerization

As alkanes with branched chains have a higher octane number than those with straight chains, a process for converting straight C_5 or C_6 chains to branched chains has been developed. The catalyst used is a specially prepared platinum material kept in an active state by adding an activator to the reactants.

Octane number and molecular structure

The relationship of octane number to molecular structure can be seen in the tables of C_7 alkanes, the cyclic compounds, and the C_7 alkenes.

C_7 alkanes	Octane number	C_7 alkanes	Octane number
Heptane	0	2,4-dimethylpentane	77
2-methylhexane	41	3,3-dimethylpentane	95
3-methylhexane	56	3-ethylpentane	64
2,2-dimethylpentane	89	2,2,3-trimethylbutane	113
2,3-dimethylpentane	87		

7 Hydrocarbons and halogenoalkanes

Cyclic compounds	Octane number
(Hexane)	(26)
Cyclohexane	77
Methylcyclohexane	104
Benzene	108
Methylbenzene	124

C_7 alkenes	Octane number
Hept-1-ene	68
5-methylhex-1-ene	96
2-methylhex-2-ene	129
2,4-dimethylpent-1-ene	142
4,4-dimethylpent-1-ene	144
2,3-dimethylpent-2-ene	165
2,4-dimethylpent-2-ene	135
2,2,3-trimethylbut-1-ene	145

You can see that the more branches to the carbon chain the higher the octane number and hence the value of isomerization:

$$CH_3-CH_2-CH_2-CH_2-CH_2-CH_2-CH_3 \xrightarrow{isomerization} CH_3-CH_2-CH_2-C(CH_3)_2-CH_3$$
octane number 0 → octane number 89

The conversion of alkanes to cycloalkanes and the dehydrogenation of cycloalkanes by the process of catalytic reforming also enhance the octane number of the petrol fraction from crude oil:

$$CH_3-CH_2-CH_2-CH_2-CH_2-CH_3 \xrightarrow{catalytic\ reforming} \text{cyclohexane} + H_2$$
octane number 26 → octane number 77

$$\text{cyclohexane} \xrightarrow{catalytic\ reforming} \text{benzene} + 3H_2$$
octane number 108

If you compare the table of C_7 alkenes with the previous one of C_7 alkanes, you can see that the formation of an unsaturated compound enhances the octane number of a hydrocarbon.

$$CH_3-CH_2-CH_2-C(CH_3)_2-CH_3 \xrightarrow{catalytic\ reforming} CH_2=CH-CH_2-C(CH_3)_2-CH_3 + H_2$$
octane number 89 → octane number 144

Why do these changes in molecular structure enhance octane numbers? The answer lies in the process of combustion. The conditions of temperature and pressure in a car engine result in the production of free radicals. The more reactive the free radicals, the greater the chance of an uncontrolled chain reaction such as pre-ignition explosion or knocking in the engine.

$$CH_3-CH_2-CH_2-CH_2-CH_2-CH_3 \longrightarrow 2CH_3-CH_2-CH_2\cdot$$
octane number 26 → reactive radicals

$$CH_3-C(CH_3)_2-CH_2-CH(CH_3)-CH_3 \longrightarrow CH_3-C(CH_3)_2-CH_2\cdot + \cdot CH-CH_3$$
octane number 100 → less reactive radicals

The function of the petrol additive, tetraethyl lead, was to help to control the free radical chain reaction. When the free radicals reacted with tetraethyl lead the chain was terminated because the final product is an unreactive lead atom:

$$(CH_3CH_2)_4Pb + 4\cdot CH(CH_3)-CH_3 \longrightarrow 4CH_3-CH_2-CH(CH_3)-CH_3 + \cdot Pb\cdot$$

By removing the hydrocarbons of low octane number which cause knocking the need for the addition of tetraethyl lead has been eliminated.

7.4 The halogenoalkanes

In the whole field of naturally occurring materials, there are practically no organic halogen compounds. Thus, almost all of them must be produced synthetically. Those few which do occur in nature are found in rather obscure situations. Examples include the iodine compound thyroxine, a hormone produced by the thyroid gland, a shortage of which causes goitre and cretinism; the bromine compound Tyrian purple (mentioned in Topic 5), present in the sea-snail, *Murex brandaris*, which was extracted and used by the Romans for dyeing their statesmen's robes; and the chlorine compound chloromethane, produced by some marine algae.

Owing to the considerable reactivity of the halogen atoms in organic compounds many of these compounds are manufactured as 'intermediates', that is, for conversion into other substances. There is also a range of organic halogen compounds which are important products of industry, for example

$\pmb{+}CH_2\pmb{-}CHCl\pmb{+}_n$ the polymer PVC, poly(chloroethene)
$CHClBr\pmb{-}CF_3$ an anaesthetic, Fluothane

Figure 7.17 Manufacturing 1,2-dichloroethane

When hydrogen atoms in alkanes are replaced by halogen atoms, we have compounds of the type known as **halogenoalkanes**. These are named by using the name for the alkane from which they are derived and adding 'chloro', 'bromo', or 'iodo'. For example,

CH_3Cl **chloro**methane
$CH_3\pmb{-}CH_2\pmb{Br}$ **bromo**ethane

Two halogenoalkanes can be derived from propane. They are distinguished by numbering the carbon atoms as for the branched chain alkanes:

CH_3—CH_2—CH_2Cl 1-**chloro**propane
CH_3—$CHCl$—CH_3 2-**chloro**propane

Halogenoalkanes with side chains are named in the same way as the corresponding alkanes:

CH_3—CH—CH_2Cl 1-**chloro**-2-**methyl**propane
      |
     CH_3

CH_3—CCl—CH_3 2-**chloro**-2-**methyl**propane
      |
     CH_3

For halogenoalkanes with more than one halogen atom the full name of the alkane is used, preceded by the number of the carbon atom to which the halogen atoms are attached, with 'di', 'tri', etc., to indicate the total number of halogen atoms. For example,

CH_2Br—CH_2Br 1,2-**dibromo**ethane

In considering the possible reactivity of the halogenoalkanes, we should first look at the strength of the bonds. We shall use the bond energies:

$CH_3Cl \longrightarrow CH_3\cdot + Cl\cdot$ $E(C\text{—}Cl) = 351 \text{ kJ mol}^{-1}$
$CH_3Br \longrightarrow CH_3\cdot + Br\cdot$ $E(C\text{—}Br) = 293 \text{ kJ mol}^{-1}$
$CH_3I \longrightarrow CH_3\cdot + I\cdot$ $E(C\text{—}I) = 234 \text{ kJ mol}^{-1}$

so

$E(C\text{—}Cl) > E(C\text{—}Br) > E(C\text{—}I)$

If we change the molecular structure of the alkyl group, we find that the bond energy also changes:

         CH_3
         |
CH_3CH_2—C—Br $E(C\text{—}Br) = 284 \text{ kJ mol}^{-1}$
         |
         H

      CH_3
      |
CH_3—C—Br $E(C\text{—}Br) = 263 \text{ kJ mol}^{-1}$
      |
      CH_3

Secondly, we can look at the polarity of some molecules, which is measured by a quantity called the **dipole moment** (in units called debyes, symbol D).

The existence of dipole moments in molecules is due to bond polarization (see Topic 6.4). They can however only be a guide to the polarity of individual bonds, since a dipole moment must be the sum of all the bond polarities, but we can suggest that

$C^{\delta+}\text{—}Cl^{\delta-} > C^{\delta+}\text{—}Br^{\delta-} > C^{\delta+}\text{—}I^{\delta-}$

	Dipole moment/D
1-chlorobutane	2.16
1-bromobutane	1.93
1-iodobutane	1.88

On the basis of these data and your knowledge of halogen chemistry, what reagents might attack halogenoalkanes?

The characteristic infra-red stretching wavenumbers (see Topic 18) are affected by the molar masses of the halogens.

The C—Br and C—I wavenumbers lie outside the usual range of infra-red spectrometers and this makes infra-red spectra less useful in the case of organic halogen compounds. The infra-red spectrum of 1-chlorobutane is shown in figure 7.18. The mass spectra of organic halogen compounds are interesting because they show the isotopic composition of chlorine and bromine. The ratio of ^{35}Cl to ^{37}Cl is about 3 to 1, while that of ^{79}Br to ^{81}Br is about 1 to 1.

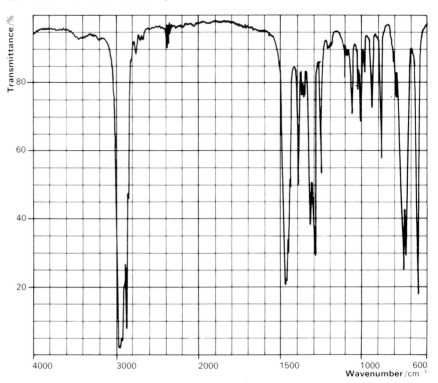

Figure 7.18 The infra-red spectrum of 1-chlorobutane

The mass spectrum of 1-chlorobutane is shown in figure 7.19. See Topic 18 for an account of how mass spectra are obtained.

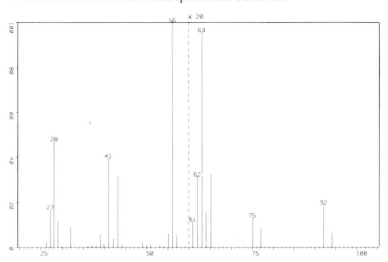

Figure 7.19 A reading from a mass spectrometer giving the mass spectrum of 1-chlorobutane. Relative abundance is shown as a percentage on the vertical axis. It is multiplied by 20 for masses (shown on the horizontal axis) above 60.

Figures for the abundances of the principal fragments, relative to the abundance of the fragment of mass 56.1, are given in the following table.

Mass	Relative abundance %	Mass	Relative abundance %
26.3	2.41	56.1	100.00
27.2	16.90	57.1	5.58
28.1	47.25	62.0	1.58
29.0	11.70	63.0	4.79
32.0	9.20	64.1	0.79
39.0	5.83	65.1	1.62
40.9	39.80	75.1	0.67
42.0	3.96	77.0	0.42
43.1	31.43	92.0	0.96
48.9	2.37	94.0	0.33
55.1	6.20		

Experiments with halogenoalkanes

In these experiments, we shall concentrate on reactions that are likely to occur with the halogen atom, remembering that ionic reagents are likely to be favoured. We shall compare the three halogenoalkanes

CH_3—CH_2—CH_2—CH_2—Cl CH_3—CH_2—CH_2—CH_2—Br
1-chlorobutane   1-bromobutane

CH_3—CH_2—CH_2—CH_2—I
1-iodobutane

and also look at the influence of the structures of the alkyl group.

The structures are known as **primary**, **secondary**, or **tertiary** on the basis of the number of alkyl groups joined to the carbon atom which is bonded to the halogen atom in the compound:

CH_3—CH_2—CH_2—CH_2Cl CH_3—CH_2—$\underset{\underset{Cl}{|}}{CH}$—$CH_3$ CH_3—$\underset{\underset{Cl}{|}}{\overset{\overset{CH_3}{|}}{C}}$—$CH_3$
1-chlorobutane (primary)   2-chlorobutane (secondary)   2-chloro-2-methylpropane (tertiary)

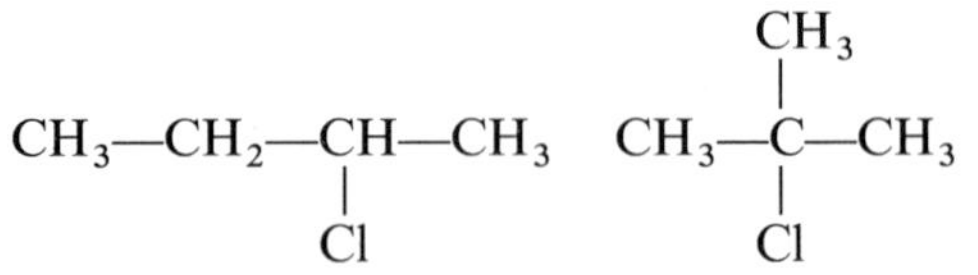

EXPERIMENT 7.4a Preparation of halogenoalkanes

Replacement of the hydroxyl group by halogen is fairly easy with tertiary alcohols, but more difficult with secondary and primary alcohols. Using a tertiary alcohol, the hydroxyl group can be replaced using aqueous hydrochloric acid, but for other alcohols, phosphorus halides or strong dehydrating agents are required.

Depending on the availability of materials you should be able to carry out one of these preparations.

SAFETY ⚠️

Concentrated sulphuric acid is dangerous if spilt on your skin; avoid inhaling the vapour from concentrated hydrochloric acid; halogenoalkanes are flammable and many have toxic vapours.

Procedure

1 A halogenoalkane from a primary alcohol

Place about 5 cm³ of ethanol in a small pear-shaped flask and cautiously, in amounts of 0.5 cm³ at a time, add about 5 cm³ of concentrated sulphuric acid (TAKE CARE). Ensure that the contents of the flask are well mixed, and cool them by holding the flask under a running cold water tap.

Quickly add 6 g of powdered potassium bromide and arrange the apparatus as shown in figure 7.20.

Heat the flask, with a low Bunsen burner flame, to distil the halogenoalkane which forms. Collect the product under water to ensure complete condensation. Do not remove the flame without first removing the delivery tube, or water may be sucked up into the hot flask.

Next, take the conical flask, tip off the bulk of the water, and remove as much as possible of what remains, using a dropping pipette.

- Write down what you see and smell. What is the evidence that you have changed ethanol into something new? Your product is bromoethane. Write an equation for the reaction and work out which bond was broken in the ethanol.

2 A halogenoalkane from a tertiary alcohol

You can use 2-methylpropan-2-ol (tertiary butanol) to prepare 2-chloro-2-methylpropane. The product is a volatile liquid with a distinctive odour, and a boiling point T_b of 50 °C. A yield of 85% can be achieved.

$$(CH_3)_3COH \xrightarrow{\text{concentrated HCl}} (CH_3)_3CCl$$

Place 20 cm³ (0.21 mole) of 2-methylpropan-2-ol and 70 cm³ of concentrated hydrochloric acid in a large conical flask. Stopper and shake at intervals, releasing any pressure after each shaking, until a top layer of 2-chloro-2-methylpropane is fully formed. This usually takes about twenty minutes.

Add about 6 g of powdered anhydrous calcium chloride (TAKE CARE, irritant), shake until dissolved and transfer to a separating funnel (see figure 2.16). Allow to separate and discard the lower aqueous layer.

Add about 20 cm³ of 0.1 M sodium hydrogencarbonate to the tertiary chloride in the funnel. Shake the funnel carefully, releasing the pressure of carbon dioxide frequently. Allow the layers to separate and discard the lower aqueous layer again. Repeat the washing with sodium hydrogencarbonate solution until no more carbon dioxide is formed.

Transfer the tertiary chloride to a small conical flask, add a little anhydrous sodium sulphate and cork securely. Shake occasionally for about five minutes to dry the chloride.

Meanwhile set up the apparatus for distillation (see figure 2.15). Filter the product directly into the distillation flask through a small funnel fitted with a plug of cottonwool. Distil the liquid and collect the fraction boiling between 50 °C and 52 °C.

Weigh your product and calculate the yield based on the moles of alcohol used.

Using the same molar proportions, 2-methylbutan-2-ol gives 2-chloro-2-methylbutane, T_b 86 °C.

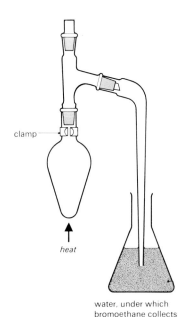

Figure 7.20 Apparatus for the preparation of a halogenoalkane

7 Hydrocarbons and halogenoalkanes

EXPERIMENT 7.4b

The reactions of the halogenoalkanes

Except where otherwise stated, use 2-chloro-2-methylpropane for the experiments, as it is much the cheapest of the halogenoalkanes you will be using.

> **SAFETY**
> Halogenoalkanes are flammable and many have toxic vapours; 20% potassium hydroxide is corrosive.

Figure 7.21 Hazard warning on a solvent storage tank

Procedure

1 Combustion

Keep the halogenoalkane in a stoppered boiling-tube well away from any flame. Dip a combustion spoon into it. Set fire to the halogenoalkane on the combustion spoon and note how readily it burns.

2 Reaction with aqueous alkali

To 1 cm^3 of 20% potassium hydroxide in ethanol (TAKE CARE) add an equal volume of water followed by 0.5 cm^3 of 2-chloro-2-methylpropane. Shake the tube from side to side for a minute. To test for chloride ions add an equal volume of 2 M nitric acid to neutralize the potassium hydroxide (test with indicator paper to ensure that your solution is acidic) and then add a few drops of 0.02 M silver nitrate. If chloride **ions** are present a white precipitate of silver chloride will appear.

- What new organic compound has been formed?

3 A comparison of halogenoalkanes

a Arrange three test-tubes in a row and add three drops of halogenoalkane in the sequence 1-chlorobutane, 1-bromobutane, 1-iodobutane. Now add 2 cm^3 of ethanol to each test-tube to act as a solvent.

In each of three different test-tubes, heat 2 cm^3 of 0.02 M silver nitrate to near boiling and then, as quickly as possible, add it to each halogenoalkane, in sequence, starting with 1-chlorobutane. Note the order in which precipitates appear and try to relate the reactivity of the halogenoalkanes to their dipole moments and to the halogen–carbon bond energies.

- Which factor appears to be the more important when considering the rates of reaction?

b Repeat the experiment in a further three test-tubes using the halogenoalkanes 1-chlorobutane (primary), 2-chlorobutane (secondary), and 2-chloro-2-methylpropane (tertiary).

- Comment on your results as you did for part **a**.

4 Reaction with alcoholic alkali

To 2 cm³ of 20% potassium hydroxide in ethanol (TAKE CARE) add 0.5 cm³ of 2-chloro-2-methylpropane. Push a loose plug of ceramic fibres into the mixture and arrange the test-tube for collection of gas (figure 7.22). Heat gently and collect two to three test-tubes of gas. Test the gas for flammability; also test it with bromine water.

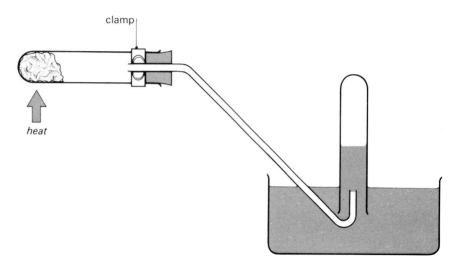

Figure 7.22

- What new organic compound has been formed? What type of reaction has taken place?

An interpretation of the halogenoalkane experiments

We can write an equation for the reaction in part **2** of experiment 7.4b, between 2-chloro-2-methylpropane and potassium hydroxide in aqueous solution.

$$(CH_3)_3C-Cl + K^+OH^- \longrightarrow (CH_3)_3C-OH + K^+Cl^-$$

This reaction tells us a great deal about organic halogen reactions in general. Firstly, we can see that a hydroxide ion has been exchanged for a chloride ion. Because of this, the reaction is known as a **substitution reaction**. Secondly, we need to consider the process by which the reaction occurred. We need to consider what species leaves the organic molecule, the **leaving group**, and what species attacks, the **attacking group**.

1 During the reaction a chloride ion has left the molecule. Being a chloride **ion** it will have taken with it the electron pair that formed the covalent bond:

$$(CH_3)_3C\!:\!\ddot{\underset{\cdot\cdot}{Cl}}\!: \longrightarrow (CH_3)_3C^+ + \left[:\ddot{\underset{\cdot\cdot}{Cl}}:\right]^- \text{(charged)}$$

You should contrast this with the process that would produce a free radical:

$$(CH_3)_3C\!:\!\ddot{\underset{\cdot\cdot}{Cl}}\!: \longrightarrow (CH_3)_3C\cdot + \cdot\ddot{\underset{\cdot\cdot}{Cl}}: \text{(uncharged)}$$

2 Since the chlorine atom takes away the bonding electrons, the group that attacks the tertiary chlorobutane molecule will need to have an unshared pair of electrons available for bonding to the carbon atoms, for example:

:O—H$^-$

3 Because of the polarity of the carbon atom, $C^{\delta+}$, it will be an advantage for the attacking group to be negatively charged.

Attacking groups with an unshared pair of electrons available for forming a new covalent bond are known as nucleophiles.

In addition, they are often negatively charged. 'Nucleus' is Latin meaning 'little nut' and the suffix '–phile' is derived from ancient Greek and means 'loving'; so 'nucleophile' means 'nucleus–loving'.

So far we have used the idea of bond polarity to interpret the process by which a halogenoalkane might react. Is polarity also a guide to the relative ease of reaction? Look at your results for part **3** of experiment 7.4b, on the reaction between silver nitrate and five different halogenoalkanes. What are the leaving groups? What are the possible attacking groups? Can this be described as a 'nucleophilic substitution reaction'? If so, it follows that the discussion about the reaction between potassium hydroxide and 2-chloro-2-methylpropane is also applicable to these reactions.

Now compare bond polarities (page 176) with the relative ease of formation of the silver halide precipitates. Finally, compare bond energies (page 176) with the relative ease of formation of the precipitates.

■ Which factor, bond polarity or bond strength, is the best guide to ease of reaction in this particular case?

The reaction in part **4** of experiment 7.4b is of a different type. The gaseous product was an unsaturated hydrocarbon, which means that both halogen and hydrogen have been lost from the 2-chloro-2-methylpropane molecule.

$$\underset{\underset{Cl\ H}{|\ \ |}}{\overset{\overset{CH_3}{|}}{CH_3\!-\!C\!-\!CH_2}} \longrightarrow \underset{}{\overset{\overset{CH_3}{|}}{CH_3\!-\!C\!=\!CH_2}} + HCl$$

This is known as an **elimination reaction**. Chemists think that the hydroxide ion, a powerful base, extracts a proton from the halogenoalkane and a chloride ion separates from the molecule. Notice that elimination does not occur by the departure of hydrogen and chlorine as hydrogen chloride in the same step in the reaction process. The overall reaction can be written as:

$$CH_3-\underset{\underset{Cl}{|}}{\overset{\overset{CH_3}{|}}{C}}-CH_3 + K^+OH^- \longrightarrow CH_3-\overset{\overset{CH_3}{|}}{C}=CH_2 + K^+Cl^- + H_2O$$

Reactions of the halogenoalkanes

1 Substitution by a hydroxyl group

$$CH_3-CH_2-CH_2-CH_2\textbf{Br} + K^+OH^- \longrightarrow$$
$$CH_3-CH_2-CH_2-CH_2\textbf{OH} + K^+Br^-$$

This equation can be written more briefly as

$$CH_3(CH_2)_3Br + OH^- \longrightarrow CH_3(CH_2)_3OH + Br^-$$

The hydroxide ion is a strong **nucleophile**, attacking the terminal carbon atom and substituting for the bromine atom. The reaction is, therefore, described as a **nucleophilic substitution**.

The reaction is not used to prepare the common alcohols because they are more readily and cheaply prepared by other reactions (see, for example, experiment 7.5, part **4**) but the reaction may be useful when chemists want to substitute a hydroxyl group into a complex compound.

2 Substitution by an amine group

$$CH_3(CH_2)_3Br + 2:NH_3 \longrightarrow \underset{\text{butylamine}}{CH_3(CH_2)_3NH_2} + NH_4^+Br^-$$

In this reaction ammonia uses an unshared pair of electrons and therefore functions as a **nucleophile**. An alcoholic solution of ammonia is needed, and heating is carried out under pressure to give an adequate concentration of ammonia. Yields are not good and the reaction is complicated by the formation of secondary and tertiary alkyl amines, R_2NH and R_3N (see Topic 12), since the amines themselves are nucleophiles.

3 Elimination in the synthesis of alkenes

$$\underset{\substack{\text{2-bromo-2-}\\\text{methylpropane}}}{(CH_3)_3CBr} + K^+OH^- \longrightarrow \underset{\text{methylpropene}}{CH_3\overset{\overset{CH_3}{|}}{C}=CH_2} + K^+Br^- + H_2O$$

This is a good method of introducing double bonds into complex molecules. The reagent, concentrated potassium hydroxide (a strong base), is the same as the one used to substitute for halogen (in reaction **1**). But because the solvent is changed from ethanol and water to ethanol alone, and because the reaction is carried out at a higher temperature, **elimination** becomes the more favoured reaction. The hydroxide ion here acts as a strong **base**, in contrast to its behaviour as a nucleophile in reaction **1**.

Preparation of halogenoalkanes from alcohols

Heating an alcohol with a mixture of sodium bromide and concentrated sulphuric acid replaces the —OH group with a bromine atom. This is an example of a **nucleophilic substitution reaction**. A mixture of sodium bromide and concentrated sulphuric acid produces hydrogen bromide. (Bromine forms too so the mixture turns orange-brown; see Topic 5.4.

$$CH_3CH_2OH + HBr \longrightarrow \underset{\text{bromoethane}}{CH_3CH_2Br} + H_2O$$

The reaction is rather easier with tertiary alcohols and concentrated hydrochloric acid can sometimes be used. The alcohol molecule is first protonated (see below) and this activated form will then react with nucleophiles to form the substitution product.

$$CH_3CH_2\text{—O—H} \xrightarrow[\text{protonation}]{H^+} CH_3CH_2\text{—}\overset{+}{\underset{H}{\overset{H}{O}}} \xrightarrow[\text{nucleophile}]{Br^-} CH_3CH_2\text{—Br} + H_2O$$

The reactions are essentially Brønsted–Lowry acid–base reactions, as were described in Topic 3, in which a proton is transferred to the alcohol and then water is displaced by the nucleophilic base.

Summary: the halogenoalkanes

Draw up a chart similar to figure 7.23 in which to record the properties of the halogenoalkanes. Consider including any personal touches that might help you recall the important ideas.

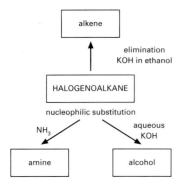

Figure 7.23 Summary of the reactions of the halogenoalkanes

7.5 The alkenes

If the supply of crude oil ever became seriously restricted, the most obvious effect would be the lack of petrol for cars, of diesel for buses and lorries, and aviation fuel for aircraft. But another effect would be the lack of alkenes for the synthesis of polymers, detergents, solvents, and many other chemicals.

In experiment 7.2 (part 7) you did an experiment on the cracking of paraffin to obtain short chain alkanes and alkenes from long chain alkanes (gases from liquids). In the petroleum industry much of the fraction from crude oil boiling in the range 150–300 °C (C_{10} to C_{20} alkanes) is cracked over a catalyst of aluminium and silicon oxides at 500 °C. The products include branched chain alkanes and alkenes which are valuable components of petrol, and the gaseous alkenes which are vital to the chemical industry.

		$T_b/°C$
$CH_2\text{=}CH_2$	Ethene	−104
$CH_3CH\text{=}CH_2$	Propene	−47
$CH_3\text{—}CH_2\text{—}CH\text{=}CH_2$	But-1-ene	−6
$CH_3\text{—}CH\text{=}CH\text{—}CH_3$	But-2-ene	+ 4 (*cis*: see below)
		+ 1 (*trans*: see below)

It is possible to separate but-2-ene into two components with significant differences in boiling point and other physical properties but with no significant differences in chemical reactivity. The explanation is that but-2-ene exists as two **isomers**. Chemists have suggested that there cannot be free rotation about the double bond so that

$$\begin{array}{c} CH_3 \quad H \\ \diagdown \diagup \\ C \\ \| \\ C \\ \diagup \diagdown \\ CH_3 \quad H \end{array} \quad \text{cannot change to} \quad \begin{array}{c} CH_3 \quad H \\ \diagdown \diagup \\ C \\ \| \\ C \\ \diagup \diagdown \\ H \quad CH_3 \end{array}$$

cis-but-2-ene *trans*-but-2-ene

The term *cis* means 'on this side of' (from the Latin) and *trans* means 'across'. These isomers are known as **geometric isomers**. (See figure 7.25.)

This is quite different from the C—C single bond where free rotation is possible. The difference is due to the difference in the electron clouds making up the single and double bonds (figure 7.24).

Figure 7.24 The structure of ethene
a A model of the structure showing the distribution of the electrons
b A diagram showing the bond lengths and bond angles

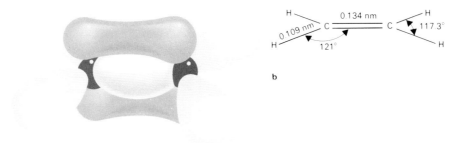

In a C—C single bond (σ bond) the electron cloud is symmetrical about the central axis but in the C=C double bond the geometry is different. The CH_2=CH_2 molecule is flat and the electron clouds which make up the second bond (π bond) are above and below the plane of the molecule. Note that a double bond consists of a σ bond plus a π bond.

Breaking a double bond requires more energy than breaking a single bond.

$$CH_3CH_2—CH_2CH_3 \longrightarrow CH_3CH_2\cdot + \cdot CH_2CH_3$$
$$E(C—C) = +346 \text{ kJ mol}^{-1}$$
$$CH_3CH=CHCH_3 \longrightarrow CH_3CH\colon + \colon CHCH_3$$
$$E(C=C) = +610 \text{ kJ mol}^{-1}$$

But if we consider just the π bond we find it is weaker than a σ bond. By calculation from the data above, you should be able to confirm that the difference between the two bond energies is only 264 kJ mol^{-1}.

The infra-red spectra of alkenes are significantly different from those of alkanes, as might be expected from the differences in structure (figure 7.26). In the infra-red spectrum of oct-1-ene there is an absorption peak due to C=C stretch at 1650 cm^{-1} and another peak at 1825 cm^{-1}. There is also a peak due to C—H stretch in =CH_2 at 3100 cm^{-1}. The other peaks are mainly due to CH_3 and CH_2 groups.

7 Hydrocarbons and halogenoalkanes

COMMENT

Two readily available geometric isomers are rubber and gutta-percha. Compare the properties of a rubber band (*cis*) with the strips of gutta-percha (*trans*) (sometimes used by florists to tie up bunches of flowers).

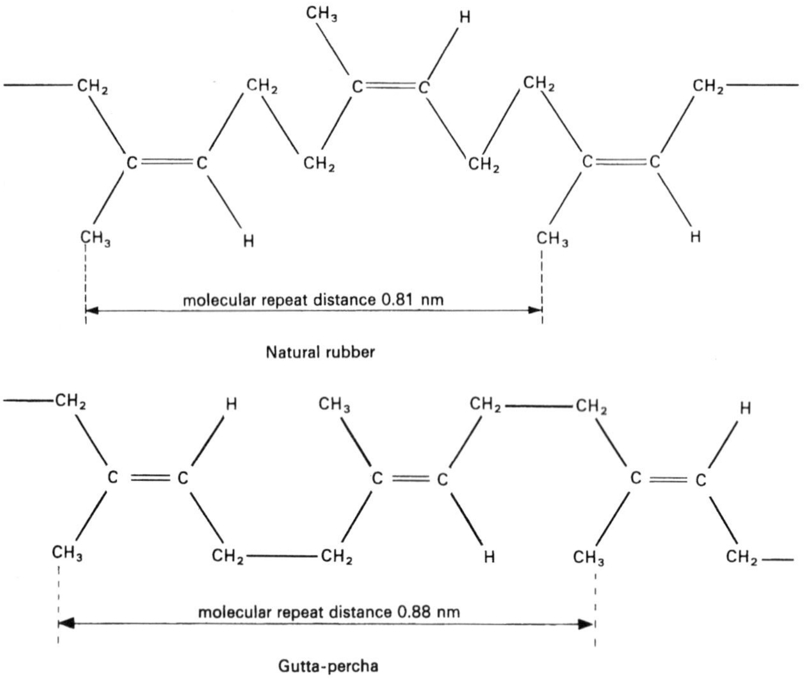

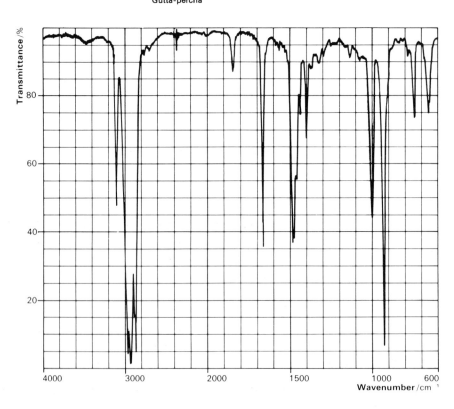

Figure 7.25 The effect of *cis* and *trans* structures

Figure 7.26 The infra-red spectrum of oct-1-ene

The rules for naming unsaturated hydrocarbons are similar to those used for the corresponding saturated compounds, explained in section 7.1. The only additional problem which arises is that of locating the double bond. This is done by using the carbon atom of lower number, of the pair of carbon atoms connected by the double bond.

$$CH_2{=}CH{-}CH_2{-}CH_3 \quad \text{but-1-ene (not but-2-ene or but-3-ene)}$$

There is no need to do this for the first two members of the series, $CH_2{=}CH_2$ ethene (also called ethylene), and $CH_3{-}CH{=}CH_2$ propene (also called propylene).

With four carbon atoms and one double bond structural isomers are possible.

$$CH_2{=}CH{-}CH_2{-}CH_3 \quad \text{but-1-ene}$$
$$CH_3{-}CH{=}CH{-}CH_3 \quad \text{but-2-ene}$$

The same rule is used for compounds containing more than one double bond, for example

$$CH_2{=}CH{-}CH{=}CH{-}CH_3 \quad \text{penta-1,3-diene}$$

(The 'di' indicates two double bonds; note also that 'a' is added to the hydrocarbon root when more than one double bond is present.)

Branched chain alkenes are dealt with as for alkanes, for example

$$\underset{\underset{CH_3}{|}}{CH_2{=}C}{-}CH_2{-}CH_3 \quad \text{2-methylbut-1-ene}$$

Cycloalkenes follow similar rules to cycloalkanes.

Experiments with alkenes

For these experiments we shall repeat most of the reactions carried out with alkanes in section 7.2, but we shall be using unsaturated instead of saturated hydrocarbons.

EXPERIMENT 7.5 The reactions of the alkenes

For this experiment you should use the alkene cyclohexene. Handle cyclohexene with care: it is highly flammable.

cyclohexene
obtained from petroleum

Record your results in the form of a table, so that they can be compared with the results of the similar experiments with the alkanes.

Procedure
1 Combustion

Keep the liquid alkene in a stoppered boiling-tube well away from any flame. Dip a combustion spoon into the sample. Set fire to the alkene on the combustion spoon.

> **SAFETY**
> You are using the same reagents as in 7.2, so you must take the same care. The polymerization reaction should be carried out in a fume cupboard, as irritant fumes are evolved; di(dodecanoyl) peroxide is an irritant.

- What are the products of combustion? How do they account for the sooty and luminous flame?

2 Oxidation

To 0.5 cm³ of a mixture of equal volumes of 0.01 M potassium manganate(VII) solution and 1 M sulphuric acid in a test-tube add a few drops of the alkene. Shake the contents and try to tell from any colour change of the manganate(VII) if it oxidizes the alkene.

3 Action of bromine

To 0.5 cm³ of bromine water (TAKE CARE) in a test-tube add a few drops of the alkene.

- What happens to the colour of the bromine?

4 Action of sulphuric acid

Put 1–2 cm³ of concentrated sulphuric acid (TAKE CARE) in a test-tube held in a test-tube rack and add 0.5 cm³ of the alkene. Shake the test-tube **gently**.

- Do the substances mix or are there two separate layers in the test-tube?

5 Polymerization

In this experiment you will prepare a sample of poly(methyl 2-methylpropenoate), Perspex.

In a test-tube, mix 5 cm³ of methyl 2-methylpropenoate (methyl methacrylate: T_b 100 °C) (TAKE CARE) and 0.1 g di(dodecanoyl) peroxide to start the reaction. Stand a wooden splint in the reaction mixture as a means of testing the viscosity of the liquid.

Heat a water bath to boiling, remove the Bunsen burner, and place the test-tube in the water bath. Allow the test-tube to stand in the slowly cooling water bath and try stirring the mixture with the wooden splint every 5 minutes.

methyl 2-methylpropenoate

- Try to write an equation for the polymerization of methyl 2-methylpropenoate. Does the product have the same empirical formula as the starting material?

An interpretation of the experiments with alkenes

You should have seen that the alkenes are readily reactive to substances such as concentrated sulphuric acid and bromine, whereas the alkanes are unreactive. Furthermore, it can be shown that the reactions are giving only a single product. Since the reagents are adding to the C=C double bond and not removing any atoms from the alkene, the reactions are known as **addition reactions**:

By what process do these reactions occur? From the study of a large number of alkene reactions chemists have proposed a mechanism that is consistent with all the evidence available. The π bond contributes an electron pair to the formation of a new bond with the positively polarized attacking group.

In this and similar reactions the attacking group is called an **electrophile (electron-seeking)**. Electrophiles are commonly acidic compounds, as in the above reaction in which sulphuric acid is reacting as an electrophile. The sulphuric acid donates a proton, H^+, to the cyclohexene molecule, which provides an electron pair to form the new bond to the proton.

This is the common pattern of electrophilic reactions.

The electrophile is an electron-deficient compound that can form a new covalent bond, using an electron pair provided by the carbon compound. The commonest electrophilic reagent is the proton, H^+.

The reaction ends with the addition of a hydrogensulphate ion to the cyclohexyl ion, because a positively charged carbon atom in the cyclohexyl ring (called a **carbocation**) is still a reactive species.

cyclohexyl hydrogensulphate

This type of reaction is described as an **electrophilic addition**.

The reaction with bromine may not follow the same course because the bromine molecule is not polar and does not have a proton to donate to the double bond. Read the passage quoted below and try to decide whether the results can be explained in terms of the course described for the electrophilic addition of sulphuric acid or whether we shall have to propose a different process for the addition of bromine.

> 'A 1 dm³ pressure bottle was filled two-thirds full with a saturated salt solution, and sufficient halogen was added to saturate it. The bottle was closed with a cap containing a bicycle valve, and a moderate pressure (4 or 5 atmospheres) of ethene was added from a cylinder. The bottle was well shaken for one minute or until the pressure had become practically atmospheric, and more ethene was added. When the solution had become colourless, more halogen and ethene were introduced. The process was continued until a sufficient amount of oil had been accumulated. This was separated from the aqueous solution, washed with water, dried with calcium chloride, and examined by determination of density, refractive index, or boiling point. In each case a mixture was obtained, and a partial separation was made by fractional distillation.

From ethene, bromine, and sodium chloride solution a mixture of 1,2-dibromoethane and 1-bromo-2-chloroethane was obtained. The product mixture, an oil, contained about 46% of C_2H_4ClBr as estimated from the refractive index, 1.51 (the value for $C_2H_4Br_2$ in the literature is 1.53; for $C_2H_4Cl_2$, 1.44).

1-bromo-2-nitratoethane was obtained from ethene, bromine, and sodium nitrate solution. Before distillation the product mixture was washed with sodium hydrogen carbonate solution to remove any trace of nitric acid. The oil began to boil at 132 °C ($C_2H_4Br_2$) but a portion boiled at 163–5 °C (the boiling point of $C_2H_4BrNO_3$ given in the literature is 164 °C). In the distillation the last trace exploded with evolution of brown fumes of nitrogen dioxide, recognized also by their odour.'

These results were obtained by A. W. Francis, and published in the *Journal of the American Chemical Society* in 1925.

QUESTIONS

An equation for one reaction carried out by Francis could be written as:

$$CH_2{=}CH_2 + Br_2 + NO_3^- \longrightarrow CH_2Br{-}CH_2ONO_2 + Br^-$$

1 Can you see how this product might have come about?

Consider first the nitrato (—ONO_2) group, which is derived from the nitrate (NO_3^-) ion.

2 Would you describe the nitrate ion as a free radical, an electron-deficient or an electron-rich species, electrophilic or nucleophilic?

3 So would the nitrate ion be most likely to attack a carbon grouping of the free radical type, a π bond, or a carbocation?

Now consider the bromo (—Br) group, which is derived from the bromine (Br_2) molecule.

4 Can a bromine–bromine bond break to give free radicals, electron-deficient or electron-rich species?

5 Which of these would react with a π bond to give a molecule with a charge that the nitrate ion might attack?

6 Finally, does the process you propose give a bromide (Br^-) ion as one of the products?

Reactions of the alkenes

1 Addition of halogens

$$CH_3{-}CH{=}CH_2 + Br_2 \longrightarrow \underset{\text{1,2-dibromopropane}}{CH_3{-}CHBr{-}CH_2Br}$$

This is an electrophilic addition with the Br—Br molecule being polarized in part by the π bond. The reaction needs to be carried out without heat and in the absence of sunlight, to avoid free radical substitution (see section 7.2). It is an important reaction for the preparation of dihalogenoalkanes.

Figure 7.27 Alcohols are manufactured from alkenes by absorption in sulphuric acid (BP Chemicals, Scotland)

2 Addition of hydrogen halides

In normal laboratory conditions the hydrogen halide acts as an electrophile, forming the intermediate $CH_3—\overset{+}{C}H—CH_3$ with the positive charge on the —CH— group.

$$CH_3—CH=CH_2 + HBr \longrightarrow \underset{\text{2-bromopropane}}{CH_3—CHBr—CH_3}$$

In sunlight or with a suitable catalyst the alternative product, 1-bromopropane, is obtained.

3 Addition of sulphuric acid

This is an important reaction because the product, an alkyl hydrogensulphate, will take part in further reactions producing alcohols as shown in the following reaction scheme. These reactions are carried out in industry.

$$CH_2=CH_2 \xrightarrow{H_2SO_4} CH_3—CH_2—OSO_3H \xrightarrow{H_2O} \underset{\text{ethanol}}{CH_3—CH_2—OH}$$

The addition of sulphuric acid follows the same pattern as the electrophilic addition of hydrogen halides.

4 Formation of diols by potassium manganate(VII)

Reaction with cold acidified potassium manganate(VII) produces compounds known as **diols**. The complete balanced equation for this reaction is complicated. As the interest is chiefly centred on the organic compounds, we can write a simplified version in this way.

$$CH_2=CH_2 \xrightarrow{KMnO_4} \underset{\text{ethane-1,2-diol}}{CH_2OH—CH_2OH}$$

An alternative name for the product is **glycol**. It is commonly used in antifreeze for car radiators and is manufactured by a more efficient method than manganate(VII) oxidation.

5 Reduction with hydrogen

Reduction by hydrogen requires the use of a metal catalyst such as nickel. The metal has to be finely divided, and in the case of nickel the catalyst is made by treating a special nickel–aluminium alloy with sodium hydroxide. This dissolves away the aluminium and leaves the nickel (known as Raney nickel) in a very finely divided state.

$$CH_2=CH_2 + H_2 \xrightarrow{\text{Ni catalyst}} CH_3—CH_3$$

The reaction is useful for preparing alkanes or for saturating some of the double bonds in natural oils. In this way, liquid fats can be converted to solids for use in margarine (although saturated fats are considered to be a contributor to heart complaints and unsaturated fats are thought to be safer to eat).

6 Polymerization

Ethene molecules will join together in suitable conditions to form very long chain alkane molecules. The reaction involves free radicals:

$$n\text{CH}_2\!\!=\!\!\text{CH}_2 \xrightarrow{\text{free radicals}} -\!\!\left[\text{CH}_2\!\!-\!\!\text{CH}_2\right]\!\!-_n$$

A variety of substances such as peroxides which yield **free radicals** are used to initiate the reaction. This reaction allows the formation of side chains and the product has a low density. This is known as an **addition polymerization** because the unsaturated monomer, ethene, reacts to give the polymer as the only product.

Summary: the alkenes

Draw up a chart similar to figure 7.28. Add any personal touches that you feel will help you remember and understand the reactions of the alkenes.

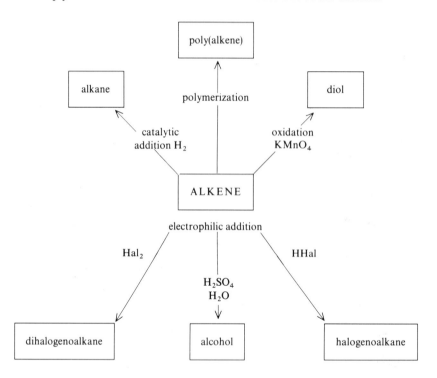

Figure 7.28 Reactions of alkenes

7.6 Study task: Polymerization

QUESTIONS
Read the passage below, then answer these questions based on it.
1 How many ethene monomer molecules have to polymerize to form a polymer of molar mass 3000 g mol^{-1}?
2 Sketch a short length of poly(ethene) chain to show the difference between branched and unbranched chains.
3 Suggest why branching occurs in free radical addition polymerization but not in catalytic polymerization.
4 What do you think were the significant differences between the experimental conditions used in the research and the original manufacturing process?

Poly(ethene) is manufactured by two distinct processes. In the original process, ethene is polymerized in the liquid state at pressures up to 3000 atmospheres and temperatures up to 300 °C. This reaction allows the formation of side chains and the product has a low density.

$$n\text{CH}_2{=}\text{CH}_2 \xrightarrow{\text{free radicals}} {+}\text{CH}_2{-}\text{CH}_2{+}_n$$

In the 1950s Karl Ziegler developed a process in which ethene could be polymerized at a relatively low pressure and temperature: less than 50 atmospheres and below 100 °C. In the Ziegler process the polymerization takes place on the surface of catalyst particles to give straight unbranched chains and the product has a higher density. The catalysts used are organometallic complexes, often based on chromium or vanadium.

Poly(chloroethene), PVC, is usually prepared as a suspension in water. Typically, equal quantities of water and chloroethene and a small quantity of a surface active agent are stirred together in an autoclave to produce a suspension of chloroethene in water. Potassium peroxodisulphate(VI) is added and the temperature raised to about 60 °C. The droplets polymerize to form solid particles of poly(chloroethene) which can be recovered by filtration. The water dilutes the reaction mixture and helps to keep it cool.

$$n\text{CH}_2{=}\text{CHCl} \xrightarrow{\text{free radicals}} {+}\text{CH}_2{-}\text{CHCl}{+}_n$$

The potassium peroxodisulphate(VI) is called an **initiator** rather than a catalyst. It increases the rate of reaction by producing the free radicals which initiate the reaction, but is decomposed in the process and not reformed.

The discovery of poly(ethene)

High pressure studies had been started by ICI in connection with the Haber process but in 1931 attention was directed at the possibility of bringing about organic chemical reactions, particularly those of the condensation type, which at lower pressures require the aid of catalysts. Two research chemists, Gibson and Fawcett, reported in April 1933:

'Ethene and benzaldehyde
Attempts have been made to react ethene and benzaldehyde. At 170 °C and 2000 atm pressure, reaction occurred slowly to yield a hard waxy solid

containing no oxygen, melting at 113 °C and analysing to $(CH_2)_n$, apparently an ethene polymer. In some cases during the investigation of this reaction, a violent reaction has occurred with considerable rise in pressure and on opening up the bomb it has been found that complete carbonization of the charge has taken place. In these experiments (that is, in which carbonization has occurred) simple decomposition of ethene has apparently taken place.'

By July about half a gram of polymer had been obtained in all, but the explosive carbonization could not be controlled and work ceased.

To pause on the brink of an important discovery may seem strange but it should be remembered that in 1933 the common plastics, such as Bakelite, were rigid materials while the poly(ethene) obtained was soft and the quantity available was insufficient for technical evaluation. Nor was polymerization the object of the research so the high pressure studies were continued using carbon monoxide instead of the explosive ethene. However in 1935 another chemist, Perrin, looked again at the polymerization of ethene and by good fortune obtained sufficient yields for the importance of the discovery to be realized:

> *December 1935*
> In one experiment at 2000 atm and 170 °C where the gas was compressed directly into a steel bomb, the reaction proceeded steadily to give about 8 g in four hours of a white waxy solid polymer of ethene. The molar mass of the polymer is about 3000. The properties of this substance are being studied.
> *January 1936*
> Further quantities of the polymer have been made. A preliminary figure shows that the electrical volume resistivity of the moulded polymer is greater than 5×10^{14} ohm cm^{-3} showing that it falls into the class of good dielectrics. It has been possible to make thin transparent films of the polymer which have considerable strength and toughness.'

7.7 Benzene and some substituted benzene compounds

The structure of benzene, C_6H_6, provided chemists with a major problem. The principal difficulty was the absence of isomers of monosubstituted derivatives of benzene, such as chlorobenzene, C_6H_5Cl. An acceptable structure must therefore be one in which all six hydrogen atoms would occupy equivalent positions.

A major step towards the solution to the problem was taken by Kekulé, then Professor of Chemistry at Ghent in Belgium, in 1865. He later described how he came to propose the structure illustrated below.

Translation from
FINLAY, ALEXANDER (1937)
100 years of chemistry.
Duckworth.

'I turned my chair to the fire and dozed. Again the atoms were gambolling before my eyes. This time the smaller groups kept modestly in the background. My mental eye, rendered more acute by repeated visions of this kind, could now distinguish larger structures, of manifold conformation; long rows, sometimes more closely fitted together; all twining and twisting in snake-like motion. But look! What was that? One of the snakes had seized hold of its own tail, and the form whirled mockingly before my eyes. As if by a flash of lightning I awoke.'

7 Hydrocarbons and halogenoalkanes

Arthur Koestler, in his book *The act of creation* (Hutchinson, 1964), describes this as probably the most important dream in history since the interpretation by Joseph of Pharaoh's dream of seven fat and seven lean cows (Genesis, chapter 40). 'The serpent biting its own tail', he writes, 'gave Kekulé the clue to discovery which has been called "the most brilliant piece of prediction to be found in the whole range of organic chemistry", and which, in fact, is one of the cornerstones of modern science.'

Kekulé was not actually the first scientist to propose ring structures for organic compounds. An Austrian schoolteacher, Josef Loschmidt, had published a booklet in 1861 but the ideas were not clearly explained; we know that Kekulé had read the booklet but dismissed it as confused. Perhaps that is the real origin of Kekulé's dream in 1865.

 Loschmidt structure

The modern evidence for the symmetry of the benzene ring is based on X-ray diffraction studies. The unusual nature of the bonding is seen from a comparison of the bond lengths of benzene with those of cyclohexene.

Carbon–carbon single bond in cyclohexane	0.15 nm
Carbon–carbon double bond in cyclohexene	0.13 nm
Carbon–carbon bonds of benzene	0.14 nm

The bonding in benzene cannot therefore be described as three double bonds plus three single bonds, but must be considered as a delocalized electron cloud spread out over the whole ring, as in figure 7.30.

Figure 7.29 These stamps were issued to commemorate the centenary of Kekulé's proposal of a ring structure for benzene

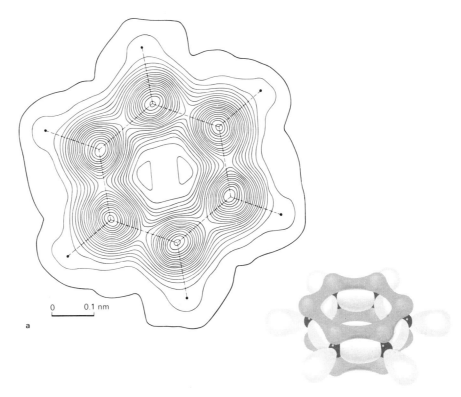

Figure 7.30 a Electron density map of benzene at −3 °C. Contours are at 0.25 electron per 10^{-30} m^3.
b A model showing the delocalized electron cloud in benzene.

When drawing a structure to indicate the molecule of benzene certain difficulties arise; a single line is normally used to represent two electrons, and two lines to represent four electrons. As neither of these is appropriate for the carbon–carbon bonds in benzene, this representation is often used:

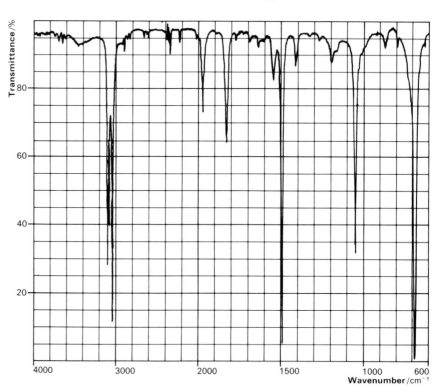

The infra-red spectrum of benzene shows that the carbon–carbon double bonds are not the normal C=C or C—C bond types.

QUESTION
Use table 3.3 in the *Book of data* to identify the origin of the main absorptions in the infra-red spectrum of benzene.

Figure 7.31 The infra-red spectrum of benzene.

Thermochemical data

The influence of the structure of benzene on its reactions can be looked at by considering the enthalpy change which takes place when hydrogen is added.

You have already seen that cyclohexene reacts with hydrogen to form cyclohexane.

$$\Delta H^\ominus = -120 \text{ kJ mol}^{-1}$$

Use these data to calculate the enthalpy change of hydrogenation for a molecule with the Kekulé structure:

⬡ + 3H₂ ⟶ ⬡ $\Delta H^\ominus = ?$

You can compare your results with the known value for benzene.

⬡ + 3H₂ ⟶ ⬡ $\Delta H^\ominus = -208 \text{ kJ mol}^{-1}$

On the basis of this result, it is reasonable to deduce that the benzene ring is less likely to take part in addition reactions than other unsaturated compounds would be.

The naming of arenes

Arenes were originally called the **aromatic hydrocarbons**. Two examples are:

benzene (C_6H_6)   naphthalene ($C_{10}H_8$)

The group C_6H_5—, derived from benzene, is known as the phenyl group. Many substitution products, when one substituent only is involved, are commonly known by non-systematic names, for example

methylbenzene (not phenylmethane); also known as toluene

When drawing such structures the convention is that where a group is attached to the benzene ring a hydrogen atom has been removed.
   In the next section you will be doing some experiments to compare the reactions of some arenes.

Experiments with arenes

Benzene has been shown to be toxic and mildly carcinogenic and its use for teaching is illegal. We shall therefore need to do our experiments with various derivatives of benzene such as methylbenzene and methoxybenzene.

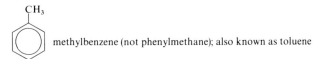

methylbenzene (toluene)   methoxybenzene (anisole)

7 Hydrocarbons and halogenoalkanes

The hazards involved in experimenting with benzene have not been recognized for long and in older books you may find suggestions for experiments involving benzene itself. Such experiments should be avoided.

Although safer than benzene itself, the substitutes used in this experiment are flammable and have a harmful vapour. Take care when handling them.

Record your results in such a way that you can compare them with those of the experiments you did with alkanes (7.2) and with alkenes (7.5). The experiments with methylbenzene will enable you to compare the reactivity of arenes with that of alkanes and alkenes.

In the experiments with methoxybenzene you will be able to consider the nature of the reactions of arenes. The methyl group and the methoxy group are unreactive in the conditions of the experiments, so any reactions you observe are likely to be reactions of the benzene ring.

EXPERIMENT 7.7

The reactions of the arenes

This experiment is in two parts: first the reactions of methylbenzene, then of methoxybenzene.

Procedure for methylbenzene

1 Combustion

Keep a sample of liquid methylbenzene in a stoppered boiling-tube well away from any flame. Dip a combustion spoon into the sample. Set fire to the methylbenzene on the combustion spoon.

- Note the luminosity and sootiness of the flame, which is characteristic of compounds with a low hydrogen to carbon ratio.

2 Oxidation

To 0.5 cm^3 of a mixture of equal volumes of 0.01 M potassium manganate(VII) solution and 1 M sulphuric acid in a test-tube add a few drops of methylbenzene. Shake the contents of the tube and try to tell from any colour changes of the manganate(VII) whether it oxidizes the methylbenzene.

3 Action of bromine

To 1–2 cm^3 of 2% bromine (TAKE CARE) dissolved in an inert solvent in a test-tube add a few drops of methylbenzene.

- What happens to the colour of the bromine?

4 Action of sulphuric acid

Place 1–2 cm^3 of concentrated sulphuric acid (TAKE CARE) in a test-tube held in a test-tube rack and add 0.5 cm^3 of the methylbenzene. Shake the test-tube **gently**.

- Do the substances mix or are there two separate layers in the test-tube?

Procedure for methoxybenzene

Methoxybenzene has a benzene ring that is fairly reactive, so by using appropriate reagents you should be able to observe some typical reactions of the benzene ring.

SAFETY ⚠
Methylbenzene and methoxybenzene are flammable; bromine is very dangerous to your skin and the vapour is particularly dangerous if inhaled; anhydrous aluminium chloride dust is corrosive and causes severe burns in contact with moisture; concentrated sulphuric and nitric acids are corrosive to the skin.

STUDY TASK
Compare your results with those obtained with alkanes and with alkenes. You should be able to see that the benzene ring is comparable to the alkanes in stability. Also, it is remarkably resistant to the reagents that readily took part in addition reactions with alkenes.

1 Bromination

To 0.5 cm³ of 2% bromine (TAKE CARE) in an inert solvent in a test-tube add a few drops of methoxybenzene.

- What happens to the colour of the bromine? What are the fumes that are evolved (test them with ammonia)? Did alkenes give off fumes in this reaction? Has an addition reaction occurred?

2 Sulphonation

To 0.5 cm³ of methoxybenzene in a test-tube add 1 cm³ concentrated sulphuric acid (TAKE CARE). Shake the test-tube **gently** to mix the contents (does the tube get hot?), then **cautiously** add 4 cm³ water.

- Is there a product which is soluble in water?

3 Friedel–Crafts reaction

To 1 cm³ of methoxybenzene in a test-tube add a small spatula measure of **anhydrous** aluminium chloride (TAKE CARE) followed by 1 cm³ of 2-chloro-2-methylpropane. If necessary, warm the mixture in a beaker of hot water.

- What are the fumes that are evolved? Test with moist indicator paper.

4 Nitration

To 1 cm³ of water add 1 cm³ of concentrated nitric acid (TAKE CARE) followed by a few drops of methoxybenzene. Warm in a water bath and observe the formation of coloured products.

An interpretation of the substitution reaction of the benzene ring

The difference between alkene reactions and benzene ring reactions can be seen most clearly in the bromination reaction. An alkene such as cyclohexene undergoes an addition reaction with bromine:

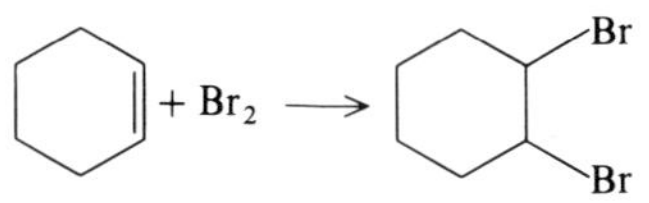

But when a benzene ring reacts, hydrogen bromide is produced and this means that a hydrogen atom has been displaced from the benzene ring:

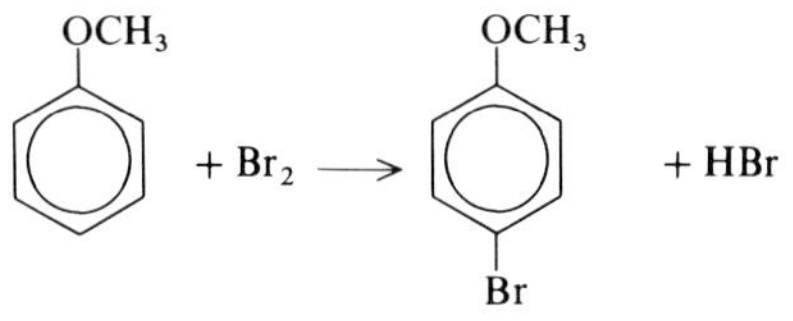

4-methoxy-bromobenzene

The product will react with more bromine to give

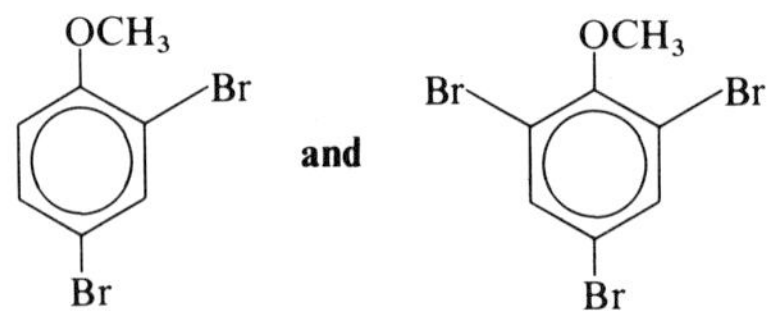

[structures: 2-bromo-4-bromo-methoxybenzene and 2,4,6-tribromo-methoxybenzene] and

with more hydrogen bromide being produced.

These reactions of the benzene ring are known as **substitution reactions**.

We can now consider how the bromine substitution reaction of the benzene ring takes place.

We have already said that this type of reaction, in which a hydrogen atom of a benzene ring is replaced by another atom, is known as a substitution reaction. You have seen that it takes place quite easily with methoxybenzene but not with methylbenzene. Methylbenzene **will** undergo such a reaction but a catalyst (iron is suitable) is needed.

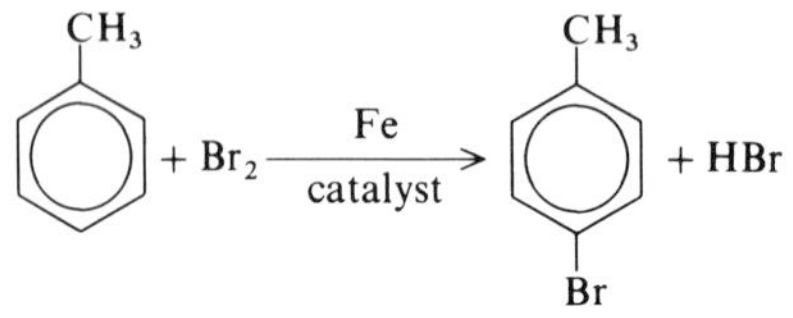

[reaction: methylbenzene + Br₂ →(Fe catalyst) 4-bromomethylbenzene + HBr]

The major product is the monobromo-compound, although the yields of dibromo- and tribromo-methylbenzene can be increased by heating.

QUESTIONS

1 What evidence is there of the nature of the attacking group? It has been found that the reaction of iodine monochloride, I—Cl, with methoxybenzene produces only iodine substitution products.

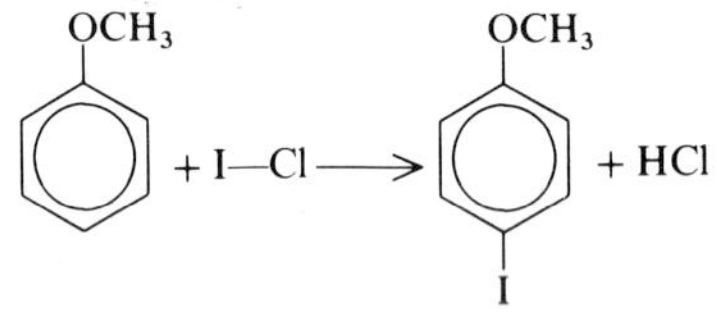

[reaction: methoxybenzene + I—Cl → 4-iodomethoxybenzene + HCl]

2 What is the attacking atom in this reaction? What polarization would you expect in I—Cl? So what is the charge on the attacking atom in the reaction?
3 Now consider the leaving group. What atom is lost from the benzene ring in the reaction? Will this atom more easily carry a positive or a negative charge when it leaves the benzene ring? Is this consistent with the charge which, you have suggested, the attacking atom will bring to the benzene ring?

You should now have a hypothesis about the charge on the attacking agent in a benzene ring substitution and also about the nature and charge of the leaving group. We can see if this hypothesis is consistent with the relative ease of attack on methylbenzene and methoxybenzene.

QUESTIONS

1 What polarization of the benzene ring is required to facilitate attack by the iodine atom of iodine monochloride? The polarization of the benzene ring caused by substituents will be indicated by the dipole moment of the molecules.

Molecule	Direction of dipole	Dipole moment/D
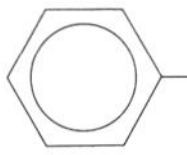 C$_6$H$_5$—OCH$_3$	←—∣ $\delta+$	1.38
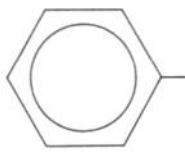 C$_6$H$_5$—CH$_3$	←—∣ $\delta+$	0.36
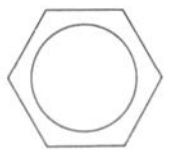 C$_6$H$_6$		0.0

2 Do the dipole moments change in parallel with the reactivity of the benzene ring?

If you examine an electron cloud model of methoxybenzene you will see that the p-electrons on the oxygen are available to interact with the delocalized π-electrons in the benzene ring. This is considered to be the source of the greater reactivity of methoxybenzene. Check that this theory is consistent with your hypothesis about the nature of the attacking group.

Finally, let us look at the function of the iron catalyst. Iron reacts with bromine to form iron(III) bromide:

$$2Fe + 3Br_2 \longrightarrow 2FeBr_3$$

This in turn induces polarization in other bromine molecules:

$$FeBr_3 + Br_2 \longrightarrow Br^{\delta+}\text{—}Br^{\delta-}.FeBr_3$$

Reaction of this last compound with methylbenzene regenerates the iron(III) bromide, and the catalyst is therefore iron(III) bromide and not iron.

Thus, the function of the catalyst is to provide a bromine atom carrying the correct charge for attack on the benzene ring.

Reactions of the benzene ring

We can now see that the special reaction of the benzene ring can be described as an **electrophilic substitution**, with electron-deficient reagents attacking the benzene ring. The benzene ring can donate a pair of electrons to the attacking group. This theory can be enlarged to interpret the positions on the benzene ring that are attacked, but we shall not be following the theory as far as that.

1 Halogenation

Bromine, usually in the presence of a catalyst, such as iron(III) bromide to make the bromine molecules more electrophilic, substitutes a bromine atom for a hydrogen atom.

$$C_6H_6 + Br_2 \longrightarrow C_6H_5Br \text{ (bromobenzene)} + HBr$$

2 Sulphonation

Fuming sulphuric acid is used for sulphonation, giving products which are often water-soluble. Refluxing for several hours is often necessary. The electrophile is considered to be sulphur trioxide, SO_3.

$$C_6H_6 + SO_3 \longrightarrow C_6H_5SO_3H \text{ (benzenesulphonic acid)}$$

In this reaction, benzene gives benzenesulphonic acid. This compound, like sulphuric acid, ionizes in water.

$$C_6H_5SO_3H + H_2O \longrightarrow C_6H_5SO_3^- + H_3O^+$$

Sulphonation is used in the manufacture of a wide range of substances including sulphonamide drugs, detergents, and dyestuffs.

3 Alkylation by the Friedel–Crafts reaction

Chloroalkanes, RCl, in the presence of aluminium chloride, will form a complex, $R^+AlCl_4^-$, in which R^+ acts as an electrophile (R represents any alkyl group). For example, if R is CH_3CH_2, the equation is

$$CH_3CH_2Cl + AlCl_3 \longrightarrow CH_3CH_2^+ AlCl_4^-$$

$$CH_3CH_2^+ AlCl_4^- + C_6H_6 \longrightarrow C_6H_5CH_2CH_3 \text{ (ethylbenzene)} + AlCl_3 + HCl$$

The aluminium chloride is a catalyst so only small quantities are needed to carry out the reaction.

This reaction has important industrial applications because it links alkyl chains to arene rings (see Topic 16.1).

4 Nitration

Nitric acid in the presence of concentrated sulphuric acid produces the electrophile NO_2^+. The reaction of benzene with the electrophile NO_2^+ substitutes a nitro group, NO_2, for a hydrogen atom.

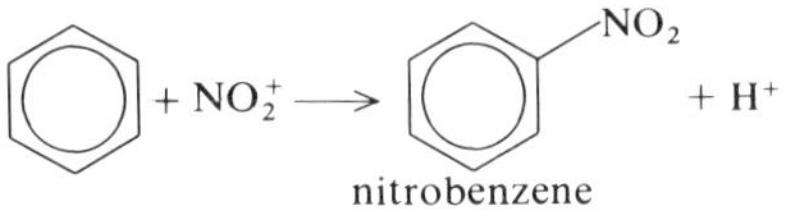

$$C_6H_6 + NO_2^+ \longrightarrow C_6H_5NO_2 + H^+$$
nitrobenzene

Reactions of this type are known as **nitrations**. They are used in the manufacture of explosives (such as TNT, trinitrotoluene) and dyestuffs.

5 Addition reactions of benzene

In severe conditions benzene will undergo some addition reactions. Thus hydrogen in the presence of a nickel catalyst will react with benzene to form cyclohexane. A temperature of 200 °C is necessary, and a pressure of 30 atmospheres is used to keep the reaction in the liquid phase. This reaction is the main source of the high purity cyclohexane needed for the manufacture of nylon.

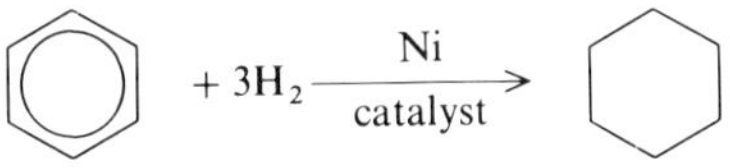

$$C_6H_6 + 3H_2 \xrightarrow[\text{catalyst}]{\text{Ni}} C_6H_{12}$$
cyclohexane

Chlorine will also add to benzene when irradiated with ultra-violet light, and this is another example of a free radical reaction.

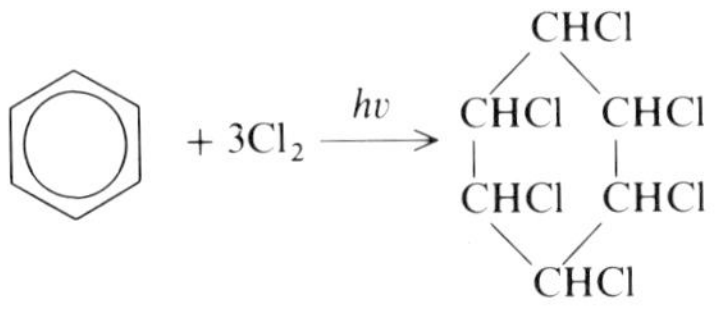

$$C_6H_6 + 3Cl_2 \xrightarrow{h\nu} C_6H_6Cl_6$$
1,2,3,4,5,6-hexachlorocyclohexane

Summary: benzene and some compounds

Draw up your own chart for the reactions of benzene using the earlier charts as a guide. You can include more (or less) information, depending on how helpful you find charts as a method of learning.

7.8 Survey of reactions and reagents in Topic 7

This Topic has contained a number of important ideas that you need to understand before you proceed to the next organic chemistry Topic. You will also need to learn the particular reagents, and conditions, for each reaction. When this basic information has been learned you should make sure you understand the interpretations of the reactions.

These lists should be treated as statements to be understood, not definitions to be memorized. You should find an example of your own choice for each item, to help you to understand the ideas involved.

Bond breaking

- **Homolytic bond breaking** involves the breaking of a bond so that the electrons are equally shared between the atoms, or free radicals.

$$Cl-Cl \longrightarrow Cl\cdot + Cl\cdot$$

- **Heterolytic bond breaking** involves the breaking of a bond so that the electrons are unequally distributed between the two atoms, forming ions.

$$H-Cl \longrightarrow H^+ + :Cl^-$$

Types of reaction

- **A substitution reaction** is a reaction in which one group replaces another in a molecule.

$$CH_3CH_2Br + NaOH \longrightarrow CH_3CH_2OH + NaBr$$

- **An addition reaction** is one in which one or more groups are added onto a molecule, to give a single product.

$$CH_2=CH_2 + Br_2 \longrightarrow CH_2Br-CH_2Br$$

- **An elimination reaction** is one in which one or more groups are removed from a molecule. (Notice that this is the reverse of an addition reaction.)

$$CH_3CH_2Br \longrightarrow CH_2=CH_2 + HBr$$

- **A polymerization reaction** is one in which molecules with a small molar mass join up to become molecules with a large molar mass.

$$nCH_2=CH_2 \longrightarrow -[CH_2-CH_2]_n-$$

- **A chain reaction** is one in which molecules of product are produced at each cycle of a process that usually repeats itself a large number of times.

$$CH_4 \xrightarrow{\cdot Cl} \cdot CH_3 + HCl \xrightarrow{Cl_2} CH_3Cl + \cdot Cl$$

This process includes the stages of **initiation**, **propagation**, and **termination**.

Types of reagent

- **Nucleophiles** are attacking groups with a pair of electrons available for forming a new covalent bond. They are often negatively charged.

$$:OH^- \quad :NH_3$$

- **Electrophiles** are attacking groups with a vacancy for a pair of electrons; a new covalent bond results when the vacancy is filled. Electrophiles are often positively charged.

$$H^+ \quad NO_2^+ \quad SO_3 \quad R^+(AlCl_4^-)$$

- **Free radicals** are uncharged attacking groups with an odd number of electrons, so they possess only one of the electron pair needed for the formation of a new covalent bond.

$$\cdot Cl \quad \cdot CH_3$$

Summary

At the end of this Topic you should be able to:

a demonstrate understanding of

    i the nomenclature and corresponding displayed and structural formulae for: alkanes, alkenes, arenes, cycloalkanes and cycloalkenes, and halogenoalkanes

    ii the following terms as associated with the structure of organic molecules: empirical and molecular formulae; isomerism, structural and geometric isomers

    iii the following terms as associated with organic reactions: homolytic and heterolytic fission; free radical, photochemical reaction; chain reaction, initiation, propagation, termination; nucleophile, electrophile; addition, substitution and elimination reactions

b deduce the groups in organic compounds from infra-red spectra using simple correlation tables

c recall the typical behaviour of the alkanes and alkenes, limited to:
combustion and cracking
treatment with

    i acidified potassium manganate(VII)

    ii halogens

    iii concentrated sulphuric acid

polymerization of alkenes

d recall the typical behaviour of halogenoalkanes, limited to:
combustion
treatment with

    i aqueous alkali

    ii alcoholic alkali

    iii aqueous silver nitrate

    iv alcoholic ammonia

e recall the characteristic behaviour of benzene, as typical of arenes, limited to :
combustion
treatment with

    i bromine

    ii concentrated nitric and sulphuric acid

    iii concentrated sulphuric acid

alkylation by the Friedel–Crafts reaction
addition reactions with hydrogen and with chlorine

f interpret the reactions of alkanes, alkenes, halogenoalkanes and arenes in terms of the processes of bond-breaking and bond-making by nucleophilic or electrophilic attack and by reference as appropriate to electron pair availability, bond polarization and bond energy

g evaluate information by extraction from text and the *Book of data* about:

    i the value to society of organic materials

    ii the importance of the work of the organic chemist in the development and manufacture of these by safe, economic and environmentally acceptable routes.

Review questions

* Indicates that the *Book of data* is needed.

Alkanes

7.1 Draw the structural formula of each of the following:

   a 2-methylbutane
   b 2,2-dimethylpropane
   c 2,3,3-trimethylpentane
   d 3-ethyl-4,4-dimethylheptane.

*7.2 a Use the *Book of data* to find the enthalpy change of formation of methane and the enthalpy changes of atomization of carbon and hydrogen. Determine the enthalpy change of formation of methane from free gaseous atoms. From this, calculate the average bond energy of the carbon–hydrogen bond.

   b Use the result you obtained in **a**, together with the appropriate data for ethane, to calculate the bond energy of the carbon–carbon bond in ethane.

Halogenoalkanes

7.3 Write the structural formulae of:

   a 2-bromo-2-methylpropane
   b 4,4-dichloro-3-ethylhexane.

7.4 Write down the structures and name all the compounds with the molecular formula C_4H_9Br.

7.5 Explain how, starting with the appropriate mono- or dihalogenated alkane, you might synthesize the following:

   a CH_3—$CHOH$—CH_3 (propan-2-ol)
   b $H_2N(CH_2)_6NH_2$ (1,6-diaminohexane)
   c CH_2=CH—CH_2—CH_3 (but-1-ene).

7.6 a Draw the electronic structure (outer shell only) of the water molecule, H_2O.

   b Draw the electronic structures of the *atom* and *free radical* that would be obtained if one of the O—H bonds in a water molecule underwent *homolytic fission*.
   c Draw the electronic structure of the *ions* that would be obtained if one of the O—H bonds in a water molecule underwent *heterolytic fission*.
   d Explain why the ion OH^- is said to be nucleophilic.
   e Can you suggest a similar name which might be appropriate to describe the ion H^+?

7.7 Choose from the following list those reagents which you think might possess nucleophilic properties.

H^+ CN^- $CH_3\cdot$ OH^- Cu^{2+} H_2O NH_3 $Br\cdot$ Br^- CH_3NH_2 Na^+

Alkenes

7.8 Write structural formulae and give the names for as many compounds as possible with the molecular formula C_4H_8.

7.9 Which of the following substances can exist as geometric isomers (*cis*- and *trans*- forms)?

$(CH_3)_2C=CH_2$ $CH_3CH=CHCH_3$ $ClCH=CCl_2$ $CH_3CH=CHCH_2CH_3$
　　A　　　　　　　　　B　　　　　　　　　C　　　　　　　　　D

7.10 Which of the following reagents form stable addition products with ethene, with or without the use of catalysts?

$Br_2(l)$ $NaOH(aq)$ $H_2(g)$ $H_2SO_4(aq)$ $NH_3(aq)$ $HBr(g)$ $CuSO_4(aq)$ $H_2O(l)$

7.11 Give equations, reagents, and reaction conditions to show how you would perform the following syntheses:

a　CH_3CH_2Br　to　CH_2BrCH_2Br　(2 steps)
b　$CH_2=CH_2$　to　CH_2OHCH_2OH　(2 steps).

7.12 Classify the following species as free radicals, electrophiles, or nucleophiles, giving reasons for your choice by indicating the number of electrons in the outer shell of the significant atom.

Br^+ OH^- $CH_3\cdot$ H_2O CH_3^+ $Cl\cdot$ I^- H^+ CH_3NH_2

Why are there no metal cations such as Na^+ or Cu^{2+} in the list, although anions of non-metals are represented?

Arenes

7.13 Copy and complete the following table, using structural formulae instead of names for the stated substances.

Starting material	Benzene	Cyclohexene	Cyclohexane
Product	Bromobenzene	1,2-dibromo-cyclohexane	1-bromo-cyclohexane
Reagent used	Bromine	Bromine	Bromine
Essential condition for reaction to take place			
Type of reaction (addition, substitution, elimination)			
Type of reagent (electrophile, nucleophile, free radical)			

*7.14 This diagram shows some reactions of methylbenzene, A:

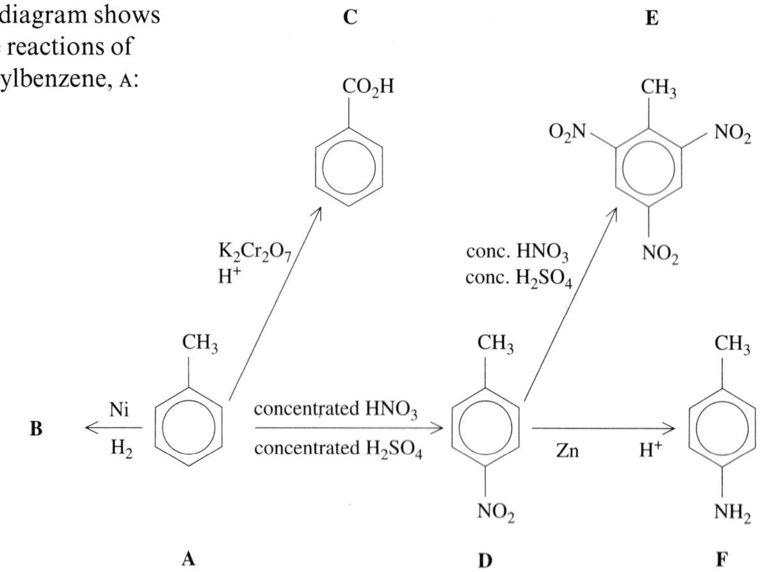

[Reaction scheme: A (methylbenzene) → B via Ni/H₂; A → C (benzoic acid, CO₂H) via K₂Cr₂O₇/H⁺; A → D (4-nitromethylbenzene) via concentrated HNO₃/concentrated H₂SO₄; D → E (2,4,6-trinitromethylbenzene, TNT) via conc. HNO₃/conc. H₂SO₄; D → F (4-aminomethylbenzene, NH₂) via Zn/H⁺]

a Give the name and structure of product B.
b Classify the following reactions, I–V, as oxidation, reduction, substitution, addition, or elimination.

   A ⟶ B,   A ⟶ C,   A ⟶ D,   D ⟶ E,   D ⟶ F
     I   II   III   IV   V

c What characteristics of product E lead to its use as an important explosive, TNT?
d Use the *Book of data* to suggest how substance C, benzoic acid, could be identified, using infra-red spectroscopy.

7.15 Chlorine reacts with boiling benzene in sunlight to give a mixture of several isomers, all with the molecular formula $C_6H_6Cl_6$.

a What type of reaction has taken place and what is the attacking species?
b Why is this type of reaction unusual for benzene?
c Suggest how the isomers differ from one another in structural terms.

*7.16 Starting from the standard enthalpy change of formation of benzene, calculate its enthalpy change of atomization, for the process:

$$C_6H_6(g) \longrightarrow 6C(g) + 6H(g)$$

Also, use the average bond energies in the *Book of data* to calculate this same value for the gaseous molecule

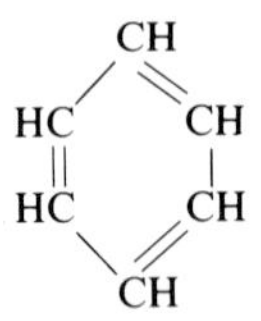

[Structure of cyclohexatriene with alternating single and double bonds shown]

(The enthalpy change of vaporization of benzene is: $\Delta h_{vap} = +39.4 \text{ kJ mol}^{-1}$)
Comment on the values you obtain.

Examination questions

7.17 This question is about the primary alcohol pentan-1-ol, $C_5H_{11}OH$, and some of its related compounds.

  a Give the structural formula and systematic name of another **primary** alcohol which is an isomer of pentan-1-ol.

  b One method of preparing 1-bromopentane is to put pentan-1-ol, water and sodium bromide into a flask and add concentrated sulphuric acid slowly from a tap funnel. The resulting mixture then needs to be heated for some time in order to obtain a reasonable yield.

  i Draw a labelled diagram of the apparatus you would use for carrying out this process in the laboratory.

  ii The reaction mixture often goes yellow or orange at this stage in the preparation. Name the substance likely to be responsible for this coloration and explain how it forms.

  iii This preparation normally gives a yield of 60%. Calculate the minimum mass of pentan-1-ol you would need to produce 15 g of 1-bromopentane by this method. (Molar masses: $H = 1, C = 12, O = 16, Br = 80 \, g\,mol^{-1}$)

  c Dehydration of pentan-1-ol yields pent-1-ene. Copy the diagram in the margin and add a sketch showing the electron density distribution in the carbon-carbon double bond.

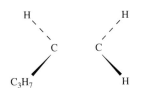

  d The compound shown below is among those secreted by insects to attract other members of the same species.

  $$CH_3CH_2CH_2CH_2CO_2H$$

  Such compounds are used to control insects, but to do so they need to be made synthetically. Give the reagents and conditions needed to synthesize these two compound using pentan-1-ol as one of the starting materials.

7.18 The reaction sequence opposite is used to make one of the chemicals needed to manufacture epoxy resins. Epoxy resins are polymers which are used in heavy-duty adhesives.

  a i Name the **type of reaction** and name the **mechanism** that takes place in Stage I.

  ii What is the reason for using ultra-violet light?

  iii Under other conditions a different reaction can occur between propene and chlorine. Write down the name and structural formula of the product of this reaction.

  b In Stage II the organic molecule, 3-chloropropene, undergoes **electrophilic** addition with chloric(I) acid.
  Explain why the 3-chloropropene molecule can undergo electrophilic attack.

  c In Stage III the reaction taking place is called an **elimination** reaction. Explain why this name can be applied to this reaction.

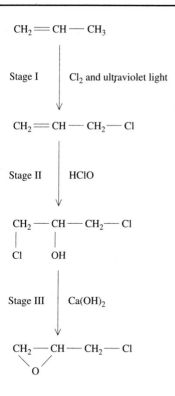

7.19 This question concerns compounds derived from ethane, C_2H_6.

a Parts of the infra-red spectrum of ethanol, C_2H_5OH, and ethane are shown below:

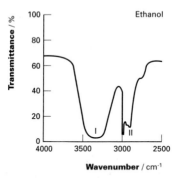
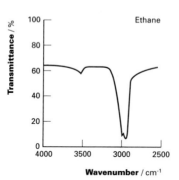

What chemical bonds do you think are responsible for the absorption of radiation at I and II?

b Starting from bromoethane, C_2H_5Br, what reactants and solvents would be required to produce each of the following compounds?
 i Ethanol
 ii Ethene
 iii Ethylamine.

c Reactants may be described as electrophilic, nucleophilic or free radical. Which of these descriptions fits the reactant described in **bi** and **biii**?

(Continued)

d Large quantities of ethene are converted industrially into chloroethene, C_2H_3Cl.
  i Draw a 'dot-and-cross' diagram to illustrate the electronic structure of a molecule of chloroethene; show outer shells only.
  ii Draw a displayed formula for ethene, and indicate on it your estimates for the values of the bond angles.
  iii Devise a two-stage reaction scheme to convert ethene into chloroethene.
  iv For what specific purpose are large quantities of chloroethene manufactured industrially?

7.20

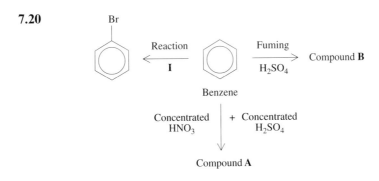

a State the reagent and conditions required for reaction I.
b Give the name and structural formula for compound A.
c Give the name and structural formula for compound B.

7.21 Give an account of the reactions of arenes.
How would you distinguish between an arene and an alkene?

7.22 Write an account of the chemical reactions of the three hydrocarbons: *cyclohexane, cyclohexene* and *benzene*. Give equations wherever possible, and point out important similarities and differences in their reactions.
How may the differences be accounted for in terms of the bonding in these molecules?

*7.23 The bond energy for the carbon–hydrogen bond is 413 kJ mol^{-1}. Explain what is meant by this statement.
Show how a value for the C—C bond energy may be obtained by the application of Hess's Law to the enthalpy change data of two closely related hydrocarbons such as methane and ethane.
Use Tables 4.6 and 5.2 in the *Book of data* to calculate a value of $\Delta H_f^\ominus(298)$ for the compound disilane:

From your answer predict whether disilane is more or less likely than ethane to decompose into its elements on heating.

TOPIC 8

How fast? Rates of reaction

Figure 8.1 How does a pressure cooker reduce cooking time?

What actually takes place during a chemical reaction? Why do some reactions take longer than others to go to completion? What factors influence the rate of a reaction? What principles govern the interactions that take place on the molecular level during a chemical reaction? Chemical kinetics, the study of the rates of chemical reactions, is concerned with all of these questions.

From your previous work in chemistry you may remember that the rate of a chemical reaction can depend on any or all of the following general factors:
- The concentration of the reactants
- The temperature
- The presence of catalysts.

In particular cases the rate may also depend on the state of subdivision of a solid reactant or the influence of electromagnetic radiation such as visible or ultra-violet light. This Topic is concerned with the study of the three general factors mentioned above. We shall not be concerned with the other variables, not because they are unimportant in specific instances (see, for example, the account of the effect of ultraviolet radiation on the rate of halogenation of alkanes in section 7.2), but because there are definite reasons for studying the three general factors. We shall start by seeing what the reasons are, and what we hope to gain by involving ourselves in the detail which follows.

8.1 Why we study rates of reactions

Rates of reaction are studied for three main reasons. The first reason is curiosity: observing that reactions can take place at very different rates is enough to challenge the curiosity of chemists and encourage them to investigate rates. Secondly, we would like to understand how to change the rate of reaction if we want to. In industry knowledge of rates of reaction is essential when considering the economics of a manufacturing process.

A third reason why a study of reaction kinetics is worthwhile is that a knowledge of the effect of concentration and temperature on the rate of a reaction provides important evidence about the mechanisms of the reaction – the individual steps by which a reaction takes place. The mechanisms of some organic reactions have been described in Topic 7.

Figure 8.2 A reaction vessel for the controlled polymerization of chloroethene to PVC

The effect of concentration on the rate of a reaction

Firstly we must decide what we mean by the **rate of a reaction**. In a reaction which takes place between two substances A and B, we might follow the reaction by observing how quickly substance A is used up. For a reaction in solution, we would probably measure the concentration of A (denoted by [A]), at various times, to see how it changes. In such a case, our measure of the rate of the reaction would be **the rate of change of concentration of A, symbol r_A.**

We might have chosen to do our measurements on substance B, and obtained a value of r_B, the rate of change of concentration of B. This might well have a different numerical value, so it is obviously essential to specify the substance to which we are referring by writing r_A or r_B.

In a reaction of the type

$$xA + yB \longrightarrow \text{products}$$

the rate of change of concentration of substance A will be found to follow a mathematical expression of the form

$$r_A = k[A]^a[B]^b[C]^c$$

An expression of this kind is called a **rate equation**. Substance C is included in this general expression as well as A and B, because rate equations sometimes include concentrations of substances that do not appear as reactants in the equation for the reaction. An example of this appears in experiment 8.2b.

It is important to understand the significance of each part of such a rate equation, so let us consider each part in turn.

1 Rate of reaction, r_A

The meaning of this has already been explained. The normal units would be mol dm^{-3} s^{-1}, but some other units may be more convenient.

Minutes might be used instead of seconds, or some property which is proportional to the concentration might be used instead of the concentration itself. We shall meet an example of this in experiment 8.2a, the first experiment.

2 Concentrations [A], [B], [C]

As in previous Topics, the use of square brackets in this context denotes 'concentration in moles per decimetre cubed'.

Again, in particular cases it may be more convenient to use a quantity which is proportional to the concentration instead of the concentration itself.

3 The indices *a*, *b*, *c*.

The index '*a*' is called the order of the reaction with respect to the reactant A, '*b*' is the order of the reaction with respect to the reactant B, and so on. The sum of all the indices is called the overall **order** of the reaction.

So overall order $= a+b+c$. In our work we will only meet reactions in which orders of reaction are whole numbers having the values 0, 1, or 2.

When we want to speak of orders of reaction we say that a reaction is 'first order with respect to A' or 'zero order with respect to B' or perhaps 'second order overall'.

Note that the orders of a reaction are experimental quantities; they cannot be deduced from the chemical equation for the reaction.

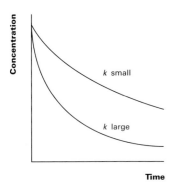

Figure 8.3 How concentration varies according to the value of k

4 The constant k

k is a constant of proportionality called the **rate constant**. The units of k depend on the order of the reaction, and can be worked out from the rest of the rate equation. For example, if the rate equation for the change of concentration of a substance A in a particular reaction is

$$r_A = k[A]^2[B]^1$$

then, substituting the units of the various quantities;

$$\text{mol dm}^{-3}\,\text{s}^{-1} = k\,(\text{mol dm}^{-3})^2(\text{mol dm}^{-3})$$

from which the units of k are $\text{dm}^6\,\text{mol}^{-2}\,\text{s}^{-1}$.

A knowledge of the rate constant has a practical use. It is often used to compare rates of reaction at different temperatures or for comparing the rates of different reactions with similar rate equations.

When studying the dependence of rate on reactant concentration you are usually asked to identify the orders of reaction and perhaps the rate constant.

8.2 Measuring rates of reaction

In designing an experiment or series of experiments to investigate the rate of a reaction there are several problems which have to be overcome.

One problem is that usually all the reactant concentrations vary at the same time and it is difficult to discover the effect of a particular reactant. This problem can be overcome by making the concentrations of all the reactants which we do not wish to study much larger than the one concentration we are interested in. The effect of this is that only this one concentration varies significantly and we can then see what the effect of this variation is on the rate of the reaction. In experiment 8.2b, method 1, for example, propanone will be reacting with iodine in the presence of an acid. There are three reactants but we shall want to find the order of the reaction with respect to only one of them, the iodine. It is arranged, therefore, that the propanone and acid concentrations are much higher than the iodine concentration so that effectively the iodine concentration is the only one of the three concentrations which varies. The other two concentrations do change, of course, but only very slightly.

A second problem with experimental design is that there is usually no easy way of finding directly the rate of a reaction at a particular time from the start of the reaction, since the rate itself may be changing as reactant concentration changes. In practice what we often do is to measure reactant concentration at various times from the start of the reaction, and from the gradient of a graph of reactant concentration against time it is usually possible to deduce what the order is. The order may then be checked by graphical or mathematical means.

A third problem is the very wide variation in the rates themselves. The time needed for a simple electron transfer between atoms to take place is about 5×10^{-16} second, whereas the time needed to bring about the change of half of a sample of one isomer of aspartic acid to another in fossil bones at ordinary temperatures is about 3×10^{12} seconds (which is about 100 000 years), probably the slowest known reaction. Such very fast or very slow reactions can only be studied by using special techniques. The reactions that you will study have been chosen with care, so that they have rates that are measurable by techniques available in a school laboratory.

> **COMMENT**
> The difference between the two isomers of aspartic acid is explained in Topic 14, on natural products.

EXPERIMENT 8.2a The kinetics of the reaction between calcium carbonate and hydrochloric acid

$$CaCO_3(s) + 2HCl(aq) \longrightarrow CaCl_2(aq) + CO_2(g) + H_2O(l)$$

In this experiment the problem is to find the order of the reaction with respect to hydrochloric acid. The calcium carbonate is in the form of marble. Fairly large pieces are used so that the surface area does not change appreciably during the reaction. On the other hand the hydrochloric acid is arranged to be in such quantity and concentration that it is almost all used up during the reaction.

Two methods are described; which one you use will depend on the apparatus available.

Procedure: Method 1

Put all the following items on the pan of a direct-reading, top-loading balance:
- A small conical flask containing about 10 g of marble in six or seven lumps
- A measuring cylinder containing 20 cm³ of 1 M hydrochloric acid
- A plug of cottonwool for the top of the conical flask.

Adjust the balance so that it is ready to weigh all these items. Pour the acid into the conical flask, plug the top with the cottonwool, and replace the measuring cylinder on the balance pan. Allow a few seconds to pass so that the solution is saturated with carbon dioxide; then start timing and taking mass readings. Record the mass of the whole reaction mixture and apparatus at intervals of 30 seconds until the reaction is over and the mass no longer changes. Record your results in the form of a table and work out the total loss of mass at each time interval. This loss of mass is due to the carbon dioxide escaping into the atmosphere.

Time t/s	Total mass m/g	Mass of CO_2 m_t/g	$m_{final} - m_t$/g

The fourth column in the table needs a little thought. When the reaction is over, the total mass of carbon dioxide evolved (m_{final}) is proportional to the concentration of the hydrochloric acid at the moment when timing started. Thus ($m_{final} - m_t$) is proportional to the concentration of hydrochloric acid at each time t.

■ Plot a graph of ($m_{final} - m_t$) against t, putting t on the horizontal axis.

Procedure: Method 2

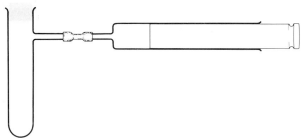

Figure 8.4

Set up the apparatus as in figure 8.4.

Place about 10 g of marble in six or seven lumps in the test-tube and have ready 10 cm³ of 1 M hydrochloric acid. Put the acid into the test-tube and allow a few seconds for the solution to become saturated with carbon dioxide. Put the stopper in place and start timing. Take readings of volume every 30 seconds until the reaction is over and the volume no longer changes. Record your results in the form of a table:

Time t/s	Volume of CO_2 V_t/cm³	$V_{final} - V_t$ /cm³

The third column in the table needs a little thought. When the reaction is over the total volume of carbon dioxide collected (V_{final}) is proportional to the concentration of hydrochloric acid at the moment when timing started, so ($V_{final} - V_t$) is proportional to the concentration of hydrochloric acid at each time, t.

- Plot a graph of ($V_{final} - V_t$) against t, putting t on the horizontal axis.

Discussion of results

Consider the rate equation in the form

$$r_{HCl} = k[HCl]^a$$

where r_{HCl} is the rate of change of concentration of hydrochloric acid. Note that $[CaCO_3]$ is not needed in the rate equation: because calcium carbonate is a solid its **concentration** does not vary even though its mass is decreasing.

The problem is: 'What is the order of reaction, a?'

Let us consider the various possibilities:

a If $a = 0$ (zero order) the graph will be a straight line (see figure 8.5a) since

$$r_{HCl} = k[HCl]^0 \text{ is the same as } r_{HCl} = k$$

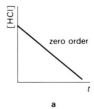  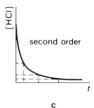

Figure 8.5   a   b   c

b If $a = 1$ (first order) the graph will be a curve such that the time it takes for the concentration of the reactant to be halved (the 'half life') is constant whatever value of the concentration you start from (see figure 8.5b).

c If $a = 2$ (second order) the graph will again be a curve but as a second-order curve is much 'deeper' than a first-order one, the half life is not constant but will increase dramatically as the reaction proceeds (see figure 8.5c).

Compare your results with the descriptions given for the various orders and identify the value of a.

Notice that this method assumes that the order is one of the three possibilities 0, 1, or 2.

Now that you have used one method of 'following' a reaction so as to investigate its order you are in a better position to appreciate other possible methods. The account which follows mentions several methods and you will be using one of these in the next experiment.

Methods of 'following' a reaction

1 By titration

A reaction mixture is made up and samples are withdrawn from it, using a pipette. Some means is then found of 'quenching' the reaction – slowing it abruptly at a measured time from the start of the reaction, perhaps by rapid cooling in ice or by removing the catalyst. The samples can then be titrated in some way which depends on what is in the reaction mixture. This is the principle of the method which is going to be used in experiment 8.2b (method 1).

2 By colorimetry

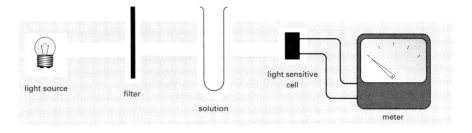

Figure 8.6 A colorimeter

If one of the reacting substances or products has a colour, the intensity of this colour will change during the reaction. The intensity can be followed using a colorimeter. See figure 8.6, and Topic 18 for a fuller description of the use of a colorimeter.

3 By dilatometry

In some reactions, in the liquid phase the volume of the whole mixture changes slightly during the reaction. If it does the change of volume can be followed by using an enclosed apparatus fitted with a capillary tube. This type of apparatus is called a dilatometer. See figure 8.7.

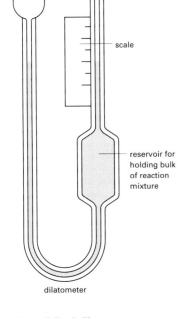

Figure 8.7 A dilatometer

4 By measurements of electrical conductivity

If the total number of ions in solution changes during a reaction it may be possible to follow the reaction by measuring the changes in electrical conductivity of the solution, using a conductivity meter. This uses alternating current so that actual electrolysis of the solution is avoided. See figure 8.8, and Topic 18 for a fuller description of the use of a conductivity meter.

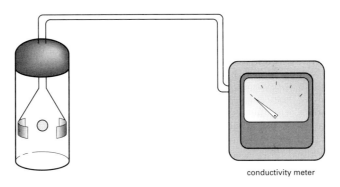

Figure 8.8 Measurement of the conductivity of a solution

5 By measuring a gaseous product

If a gas is given off it can be collected and its volume or mass measured. This was the principle behind the method used in experiment 8.2a.

6 By measurements of any other property which shows significant change

Possible properties not already discussed include pH and chirality (see Topic 14).

Now you can try another determination of reaction order by experiment.

EXPERIMENT 8.2b The kinetics of the reaction between iodine and propanone in acid solution

The reaction between iodine and propanone follows the overall equation

$$CH_3COCH_3(aq) + I_2(aq) \longrightarrow CH_3COCH_2I(aq) + H^+(aq) + I^-(aq)$$

The reaction is acid catalysed and is first order with respect to propanone and also first order with respect to hydrogen ions.

This reaction can be studied by measuring the rate of fading of the colour of iodine using a colorimeter, but the methods described below use different approaches.

Procedure: Method 1

This experiment is to investigate the order of the reaction with respect to iodine.
For convenience the solutions required are given identifying letters:
A is a 0.02 M solution of iodine, I_2, in potassium iodide
B is an aqueous solution of propanone (1.0 M)
C is sulphuric acid (1.0 M)
D is a solution of sodium hydrogencarbonate (0.5 M)
E is a solution of sodium thiosulphate (0.010 M)

1 In a titration flask labelled number *1* place 50 cm³ of A and in another, labelled number *2*, place 25 cm³ of B and 25 cm³ of C. Use measuring cylinders for these solutions.
2 Into each of several titration flasks labelled numbers *3*, *4*, etc. put 10 cm³ portions of D, again using a measuring cylinder.
3 Noting the time, pour the contents of flask *2* into flask *1* and shake well for about 1 minute. Then, noting the time, withdraw 10.0 cm³ of the reaction mixture, using a pipette and safety filler, and run this liquid into flask number *3*. Shake until bubbling ceases. The sodium hydrogencarbonate solution neutralizes the acid catalyst and quenches the reaction.
4 Titrate the contents of flask number *3* with E.
5 At measured and noted intervals from 5 to 10 minutes, repeat the extraction of a 10.0 cm³ sample from flask number *1* and the quenching and titration process.

Tabulate the results and draw a graph of titre of E against time from the start of the reaction. From your graph deduce the order of the reaction with respect to iodine.

Procedure: Method 2

In this method an attempt is made to estimate the rate of change of concentration of iodine, r_1, at the start, the initial rate. Just after the start of the reaction, the concentration of a reactant decreases almost linearly with time, whatever the order of the reaction; it is only later that the differences begin to show significantly. In this method, therefore, the concentrations of each of the three reactants are varied systematically in the four 'runs' and a direct estimate of r_1 made from:

$$r_1 \propto \frac{\text{volume of iodine solution used}}{\text{time for iodine colour to disappear}}$$

The procedure is as follows. Make up mixtures of hydrochloric acid, propanone solution, and water according to the table, using burettes. Start the reaction by adding the appropriate volume of iodine solution, measured into test-tubes from a burette, and measure the time in seconds for the colour of the iodine to disappear.

	Run 1	Run 2	Run 3	Run 4
Volume of 2 M HCl/cm^3	20	10	20	20
Volume of 2 M propanone/cm^3	8	8	4	8
Volume of water/cm^3	0	10	4	2
Volume of 0.01 M iodine/cm^3	4	4	4	2
Time for colour to disappear/s				
Rate (as indicated above)				

QUESTIONS

1 Why was water added to some of the reaction mixtures?
2 If you compare the mixtures for runs 1 and 2 you will see that the concentrations of propanone and of iodine are the same in both but the concentration of acid in run 2 is half of what it is in run 1.
What is the effect on the rate of change of concentration of iodine of halving the concentration of acid?
3 What, then, is the order of the reaction with respect to hydrogen ions?
4 Using similar arguments, what are the orders of the reaction with respect to propanone and to iodine?

8.3 Kinetics and reaction mechanism

You will have seen that in experiment 8.2b the order of the reaction with respect to iodine is zero or, to put it plainly, the rate of change of concentration of iodine, r_1, does not depend on the iodine concentration at all. Putting together all the kinetic evidence we have about the reaction gives the rate equation:

$$r_1 = k[\text{propanone}]^1[\text{hydrogen ion}]^1[\text{iodine}]^0$$

Or, since any quantity raised to the power 0 is 1, we can omit the iodine from the rate equation altogether:

$r_I = k[\text{propanone}]^1[\text{hydrogen ion}]^1$

At first sight it seems strange that a substance which appears in the chemical equation for the reaction does not affect its rate, whereas a substance which does not appear in the chemical equation does appear in the rate equation.

In organic chemistry, however, you have already met the idea that an organic reaction can occur in a number of successive steps. These steps are known as the **mechanism of the reaction** and there is no reason to suppose that all the steps take place at the same rate; in fact, it would be rather surprising if they did. When we realize this, it is clear that the whole reaction goes at the rate of the slowest of the steps in the mechanism. This slowest step is called the **rate-determining step**.

Quite a useful way of visualizing the idea of a rate-determining step is to imagine that a teacher has prepared some pages of notes and wants to collect them into sets with the help of some students. The notes are arranged in ten piles and one student collects a page from each of the piles (*step 1*). A second student takes the set of ten pages and tidies them ready for stapling (*step 2*). A third student staples the set of notes together (*step 3*).

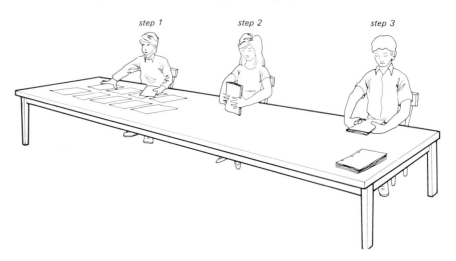

Figure 8.9 Which is the rate-determining step?

It is not hard to see that in this situation the overall rate of the process (the rate at which the final sets of notes are prepared) depends on the rate of *step 1*, the collecting of the sheets of notes, since this is by far the slowest step. It does not matter, within reason, how quickly the tidying or stapling is done; for the most part, the second and third students will be doing nothing while they wait for the first student to collect pages. The mechanism of the 'reaction' may thus be refined to

step 1   Student 1 collects pages   (slow)
step 2   Student 2 tidies set of pages   (fast)
step 3   Student 3 staples pages   (fast)

In order to speed up the process the teacher could offer to help. It would be of no value at all if this additional help were given to either student 2 or student 3 but more people at a time collecting up the pages in *step 1* would clearly make the whole process faster. *Step 1* is the rate-determining step.

We return now to the reaction which was investigated in experiment 8.2b. Various mechanisms might be proposed for the reaction, including this one:

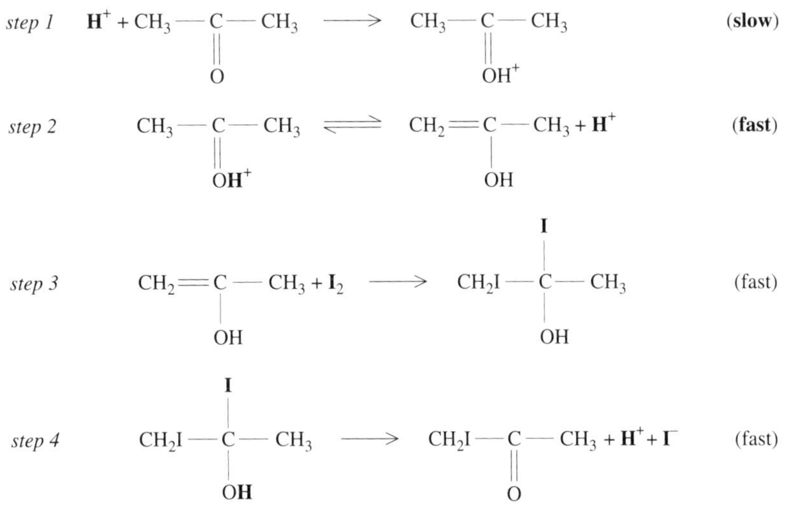

step 1 $H^+ + CH_3-\underset{\underset{O}{\|}}{C}-CH_3 \longrightarrow CH_3-\underset{\underset{OH^+}{\|}}{C}-CH_3$ (**slow**)

step 2 $CH_3-\underset{\underset{OH^+}{\|}}{C}-CH_3 \rightleftharpoons CH_2=\underset{\underset{OH}{|}}{C}-CH_3 + H^+$ (**fast**)

step 3 $CH_2=\underset{\underset{OH}{|}}{C}-CH_3 + I_2 \longrightarrow CH_2I-\underset{\underset{OH}{|}}{\overset{\overset{I}{|}}{C}}-CH_3$ (fast)

step 4 $CH_2I-\underset{\underset{OH}{|}}{\overset{\overset{I}{|}}{C}}-CH_3 \longrightarrow CH_2I-\underset{\underset{O}{\|}}{C}-CH_3 + H^+ + I^-$ (fast)

As the iodine molecules are not involved in the mechanism until after the rate-determining step, the overall rate of the reaction would not depend on the concentration of iodine. This is consistent with the outcome of experiment 8.2b.

It is important to realize that the mechanism suggested was not deduced from the kinetic evidence; the evidence was merely shown to be consistent with the suggestion.

Kinetic data is also used to decide between possible mechanisms for the hydrolysis of various bromoalkanes, using hydroxide ions. The general pattern of these reactions is

$$C_4H_9Br + OH^- \longrightarrow C_4H_9OH + Br^-$$

Two mechanisms have been proposed for this type of reaction, often referred to as the S_N1 and S_N2 mechanisms.

$C_4H_9Br \rightleftharpoons C_4H_9^+ + Br^-$ (slow: 1 molecule in the rate-determining step)

$C_4H_9^+ + OH^- \longrightarrow C_4H_9OH$ (fast: nucleophilic substitution overall)

or

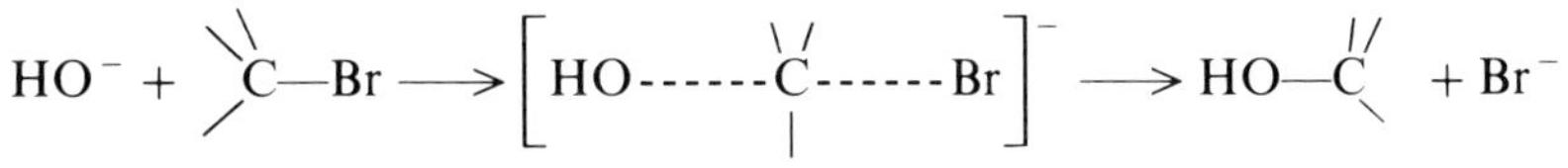

$HO^- + \overset{\diagdown}{\underset{\diagup}{C}}-Br \longrightarrow \left[HO\text{-----}\overset{\diagdown|}{\underset{\diagup}{C}}\text{------}Br \right]^- \longrightarrow HO-\overset{\diagup}{\underset{\diagdown}{C}} + Br^-$

The S_N2 mechanism is effectively a one-step process in which, for an instant of time, both the incoming hydroxide ion and the outgoing bromide ion are equally associated with the hydrocarbon group. In either of the two mechanisms, water molecules can replace hydroxide ions as the nucleophiles.

Here are some kinetics data about the hydrolysis of some bromoalkanes. Determine which of the two mechanisms is the more appropriate in each case, and answer the questions, recording the answers in your notes in such a way as to make it clear what each question was.

COMMENT
S_N1 means Substitution$_{Nucleophilic}$1 with one molecule in the rate-determining step.
S_N2 means Substitution$_{Nucleophilic}$2 with two molecules in the rate-determining step.

CASE A: The hydrolysis of 1-bromobutane

Equimolar quantities of 1-bromobutane and sodium hydroxide were mixed at 51 °C and the concentration of hydroxide ions was determined at various times, with the following results:

Time/hours	$[OH^-]$/mol dm^{-3}	Time/hours	$[OH^-]$/mol dm^{-3}
0.04	0.241	12.0	0.084
0.5	0.225	14.0	0.077
1.5	0.195	22.0	0.058
2.5	0.172	27.0	0.050
3.5	0.155	33.0	0.044
4.5	0.140	38.0	0.040
6.5	0.118	47.0	0.035
9.0	0.099	59.0	0.028

QUESTIONS

1. How might the results have been obtained practically?
2. From a suitable graph of the results, what is the order of the reaction?
3. Is this an overall order or an order with respect to a particular reactant?
4. Which mechanism is operating? Give reasons for your answer.

CASE B: The hydrolysis of 1-bromobutane

In this case a method was found of estimating the rate of change of concentration of 1-bromobutane, r_B, directly and the initial rate was found at various initial concentrations of hydroxide ions, and of 1-bromobutane.

Remember that the rate equation is of the general form

$$r_B = k[C_4H_9Br]^a[OH^-]^b \quad \text{where } a \text{ and } b \text{ are orders of reaction.}$$

	Initial concentrations/mol cm^{-3}		Initial rate/mol dm^{-3} s^{-1}
	$[OH^-]$	$[C_4H_9Br]$	
A	0.10	0.25	3.2×10^{-6}
B	0.10	0.50	6.6×10^{-6}
C	0.50	0.50	3.3×10^{-5}

QUESTIONS

1a. What is the effect on r_B when the concentration of 1-bromobutane is doubled?
 b. What is the order of the reaction with respect to 1-bromobutane?
2a. What is the effect on r_B when the concentration of hydroxide ions is increased five times?
 b. What is the order of the reaction with respect to hydroxide ions?
3. Which mechanism is operating?

CASE C: The hydrolysis of tertiary halogenoalkanes

The hydrolysis of 2-chloro-2-methylpropane using water

In this example the nucleophile is water so that the products of the reaction include a solution of hydrogen ions and chloride ions. Since the original mixture contains almost no ions, the conductivity of the solution increases as the reaction proceeds.

The results shown on the graphs (overleaf) were obtained by using a conductivity meter to measure the conductivity of the solution, with the data being logged and displayed by computer.

The solvent used was 15 cm^3 of ethanol/water in equal volumes. The upper graph was obtained by using 1.0 cm^3 of 2-**chloro**-2-methylpropane; the lower graph was obtained by using 0.50 cm^3 of the compound in the same conditions.

QUESTIONS

1 Use the graphs to estimate the rate of the reaction corresponding to each of the concentrations.
2 What is the effect on the rate of the reaction of doubling the concentration of the halogenoalkane?
3 What is the order of the reaction?
4 Is this an overall order of reaction or an order with respect to one reactant?
5 Do you have enough information to say which mechanism is the more appropriate? Explain how you arrived at your answer.

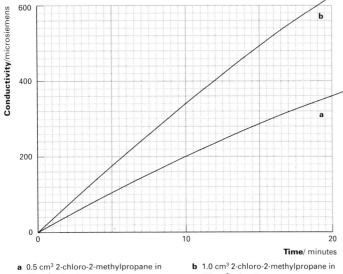

a 0.5 cm³ 2-chloro-2-methylpropane in 15 cm³ solvent
b 1.0 cm³ 2-chloro-2-methylpropane in 15 cm³ solvent

Figure 8.10 Variation in conductivity during the hydrolysis of a tertiary halogenoalkane, 2-chloro-2-methylpropane

QUESTIONS

1 What effect does doubling the concentration of hydroxide ions have on the time taken for the indicator to change colour?
2 What is the effect of an increase of hydroxide ion concentration on the rate of the reaction?
3 What is the order of the reaction with respect to hydroxide ions?
4 Which mechanism is the more appropriate in this case?

The hydrolysis of 2-bromo-2-methylpropane using sodium hydroxide solution

Here are the results from another experiment with 2-**bromo**-2-methylpropane which provide some further kinetics data, this time using hydroxide ions as the nucleophile.

In this experiment the concentration of 2-bromo-2-methylpropane remains the same but in successive runs of the experiment, the concentration of the hydroxide ions is increased. A few drops of an acid-alkali indicator are added to each mixture and a record is kept of the time taken for the indicator to change colour when the hydroxide ions are neutralized. The results are as follows:

Experiment number	$[C_4H_9Br] \times 10^2$/mol dm^{-3}	$[OH^-] \times 10^3$/mol dm^{-3}	Time/s
1a	2.5	1.25	9
1b	2.5	1.25	8
2a	2.5	2.50	17
2b	2.5	2.50	18
3a	2.5	3.75	26
3b	2.5	3.75	26
4a	2.5	5.00	37
4b	2.5	5.00	39

Overall we can say that there is a tendency for straight chain halogenoalkanes to hydrolyse by the S_N2 mechanism, whereas tertiary halogenoalkanes tend to hydrolyse by the S_N1 mechanism. However, some halogenoalkanes hydrolyse by a mechanism which has some characteristics of both of those described; and there can be competition from elimination reactions of the type

$$RCH_2CH_2Br + OH^- \longrightarrow RCH=CH_2 + Br^- + H_2O$$

8.4 The effect of temperature on the rate of reaction

Another factor which affects the rate of reactions is temperature. We shall begin this section by obtaining some experimental data for this effect.

EXPERIMENT 8.4

The effect of temperature on the rate of a reaction

The reaction we are going to investigate is between sodium thiosulphate and hydrochloric acid. The reaction between thiosulphate ions and hydrogen ions follows the equation

$$S_2O_3^{2-}(aq) + 2H^+(aq) \longrightarrow SO_2(aq) + S(s) + H_2O(l)$$

Procedure

1 Attach a piece of paper marked with a dark ink spot to the outside of a 400 cm³ beaker. Fill the beaker with water and clamp two boiling-tubes, A and B, vertically in the beaker so that they are about half immersed.

Position the beaker and the tube A so that you can see the ink spot through both the beaker and the tube A.

2 Add 10 cm³ of 0.10 M sodium thiosulphate solution to tube A and 10 cm³ of 0.50 M hydrochloric acid to tube B. Use labelled measuring cylinders to measure these liquids, and make quite certain that the two are not confused.

Allow time for both boiling-tubes to come to the same temperature as the water and then quickly add the acid from B to the thiosulphate in A.

Stir well with a thermometer and record the steady temperature.

Measure the time from the mixing of the solutions to the point when the dark spot just becomes completely obscured by the formation of a precipitate of sulphur.

3 Repeat the experiment at several different temperatures, say at 20, 25, 30, 40, 50, and 60 °C, using the same concentrations of solution and the same technique of observation.

In these particular circumstances it is possible to get a satisfactory measure of the initial rate of the reaction directly from the experimental results. In each experiment the initial rate of precipitation is given by

$$\text{rate} = \frac{\text{amount of sulphur}}{\text{time to obscure spot}}$$

SAFETY ⚠

Sulphur dioxide is formed in this reaction; do not inhale the gas, and dispose of your solutions as soon as possible.

If you muddle up the two measuring cylinders, sulphur precipitates will start to form in them, making the experimental results invalid.

The quantity 'amount of sulphur' is impossible to evaluate but as it is the same in each experiment we can say that the rate of the reaction is proportional to the reciprocal of the time taken for the spot to be obscured

$$\text{rate} \propto \frac{1}{\text{time to obscure spot}}$$

You will need the following columns in your results table:

Temperature /°C	Temperature T /K	$1/T$ /K^{-1}	Time t /s	$1/t$ (i.e. rate) /s^{-1}	ln (rate)

Plot a graph of ln (rate) on the vertical axis against $1/T$ on the horizontal axis. If the result is a downwards sloping straight line that tells us that the relationship between rate and temperature is of the form

$$\ln(\text{rate}) = \text{constant}_1 - \text{constant}_2(1/T)$$

COMMENT
The standard form for a straight line graph is $y = C + mx$, y and x are the variables while C and m are constants (m is the slope of the straight line).

The collision theory of reaction kinetics

One model which has been used to account for the dependence of rate of reaction on temperature is the collision theory.

In the collision theory we can picture the reactant particles as moving towards each other and colliding in such a way that bonds are broken and new bonds formed. The product particles may then move away from the site of the reaction.

In many reactions between gases it is the actual collision which controls the rate of the reaction. For a collision which actually results in a reaction the kinetic energy possessed by the colliding particles must be more than a certain minimum energy, E.

It can be shown that for a mole of colliding particles at a temperature T the rate of reaction can be found from

$$\ln(\text{rate}) = \ln(\text{collision rate}) - \frac{E_A}{R}(1/T)$$

where R is the gas constant, 8.314 J K^{-1} mol^{-1}, and E_A is called the activation energy. In a very simple gas reaction E_A would have about the same value as the bond energy per mole of the bonds that are being broken in the reaction.

Compare this equation with that given at the end of experiment 8.4:

$$\ln(\text{rate}) = \text{constant}_1 - \text{constant}_2(1/T)$$

and you will see that they are of the same form.

When we compare both the equations with the equation for a straight line

$$y = C + mx$$

we find that the slope m is equal to $-E_A/R$.

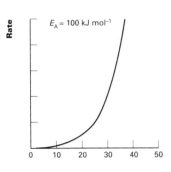
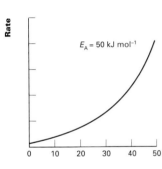
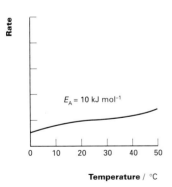

Figure 8.11 How the rate of reaction varies with temperature and activation energy

QUESTION

Measure the slope of your graph and calculate a value for the activation energy E_A for the reaction in experiment 8.4.

The generally accepted value is 46.6 kJ mol^{-1}.

COMMENT

The rate of the reaction depends on the collision rate and on the fraction of the collisions which have an energy greater than E_A per mole. The fraction of the particles having an energy greater than the value E_A is given by $e^{-E_A/RT}$.

Hence rate = collision rate × $e^{-E_A/RT}$

or taking logarithms to base e

$$\ln(\text{rate}) = \ln(\text{collision rate}) - \frac{E_A}{R}(1/T)$$

The Arrhenius equation

The relationship we have been using is called the **Arrhenius equation**:

$$\ln k = C - \frac{E_A}{R}(1/T)$$

where k is the rate constant
R is the gas constant

or, in its exponential form:

$$k = Ae^{-E_A/RT}$$

$\ln A$ = the constant C above

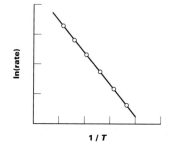

Figure 8.12 A graph plotted using the Arrhenius equation

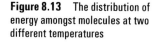

Figure 8.13 The distribution of energy amongst molecules at two different temperatures

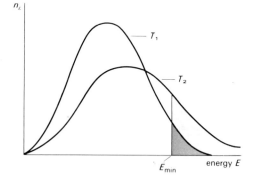

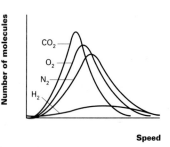

Figure 8.14 The range of molecular speeds for different gases at the same temperatures

In a very simple gas reaction, the quantity A is the molecular collision rate. Generally, however, A includes other factors, the most important of which is that successful reactions can only occur between molecules which are oriented correctly at the time of collision. Sometimes this is referred to as the **steric factor**.

The Arrhenius equation shows that when the temperature rises there is a large increase in the value of k, the rate constant. This corresponds to a large increase in the number of collisions occurring with the necessary minimum energy, the activation energy.

When you look at the distribution of energies amongst molecules at different temperatures T, as shown in figure 8.13, you should notice that a much higher proportion of the molecules have the necessary minimum energy for reaction, E_{min}, at the higher temperature, T_2, than at the lower temperature, T_1.

8.5 Catalysis

The essential feature of a catalyst is that it increases the rate of a chemical reaction without itself becoming permanently involved in the reaction. It does, however, become temporarily involved, by providing a route from reactants to products which has a lower activation energy.

This can be illustrated by an energy diagram (figure 8.15). Before the usual chemical reaction can take place the reactant molecules must be raised to a state of higher energy. They are said to be **activated** or to form an **activated complex**. Reactants and products are both at stable minima, while the activated complex is the state at the top of the energy barrier. A catalyst provides a new path in which the energy barrier is lower.

We can distinguish two types of catalysis, homogeneous and heterogeneous.

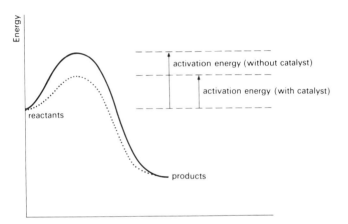

Figure 8.15 The effect of a catalyst on the activation energy of a reaction

When the catalyst and the reactants are all in one phase (for example all dissolved in the same solvent) the catalysis is described as **homogeneous**.

When the catalyst is in a different phase from the reactants (for example two reacting gases in contact with a solid catalyst) the catalysis is described as **heterogeneous**.

COMMENT

The acidic hydrolysis of esters is an example of **homogeneous** catalysis, with all the reactants and the catalyst of hydrogen ions dissolved in water

$$CH_3CO_2CH_3(aq) + H_2O(l) \xrightarrow{H^+(aq)} CH_3CO_2H(aq) + CH_3OH(aq)$$

The synthesis of ammonia is an example of **heterogeneous** catalysis because it involves the reaction of the gases hydrogen and nitrogen on a solid catalyst of iron

$$3H_2(g) + N_2(g) \xrightarrow{Fe(s)} 2NH_3(g)$$

Record some other examples that you have heard of.

From the equation

$$\ln k = C - \frac{E_A}{R}(1/T)$$

you should see that the rate constant k, and hence the rate of the reaction, will be greater when the activation energy is lower; this assumes, of course, that the term C remains constant in both the catalysed and non–catalysed reactions.

An example of this which occurs in homogeneous catalysis is the decomposition of ethanal catalysed by iodine. Ethanal decomposes without a catalyst at about 700 K to produce methane and carbon monoxide

$$CH_3CHO(g) \longrightarrow CH_4(g) + CO(g) \quad E_A \approx 200 \text{ kJ mol}^{-1}$$

The addition of a small proportion of iodine increases the rate of this reaction by several thousand times, with the activation energy falling to 134 kJ mol^{-1}. A series of gas reactions has occurred, involving free radicals thus

$$I_2 \rightleftharpoons I\cdot + I\cdot$$
$$CH_3CHO + I\cdot \longrightarrow CH_3CO\cdot + HI$$
$$CH_3CO\cdot \longrightarrow CH_3\cdot + CO$$
$$CH_3\cdot + I_2 \longrightarrow CH_3I + I\cdot$$
$$CH_3\cdot + HI \longrightarrow CH_4 + I\cdot \quad \text{continuing reaction}$$
$$CH_3I + HI \longrightarrow CH_4 + I_2 \quad \text{iodine regeneration}$$

$$CH_3CHO \longrightarrow CH_4 + CO \quad \text{overall reaction}$$

As a catalyst the iodine has formed some relatively stable intermediates and a reaction path or mechanism with a lower activation energy is created before the iodine is regenerated. In many homogeneous reactions, however, the intermediates are very reactive species, often present in very low concentrations, and difficult to identify.

The significance of catalysis is that, without the use of a catalyst, many major industrial processes would be uneconomic or impossible to realize while, in life-processes, the catalytic action of enzymes is crucial.

EXPERIMENT 8.5a A study of some catalysts

There are many examples of reactions which only give products slowly because the reaction route has a high activation energy. Catalysts are substances which participate in reactions opening new pathways of lower energy; the products are obtained faster and very little of the catalyst is used up in the reaction.

Procedure

1 Attempt to decompose 2 M hydrogen peroxide by the addition of small portions of metal oxides (MnO$_2$, Cu$_2$O, ZnO, MgO). Repeat the experiment with a small piece of liver.

- Do the oxides with catalytic activity have any common features? Consult a textbook about the biochemical significance of the result with liver.

2 Mix 10 cm^3 of 0.5 M potassium sodium tartrate (Rochelle salt) with 10 cm^3 of 2 M hydrogen peroxide in a 250 cm^3 beaker. Bring to the boil carefully and note any signs of a reaction. Stop heating and add 1 cm^3 of any 0.1 M cobalt(II) salt.

- What happens? Confirm that cobalt(II) ions do not change colour with either of the reagents separately.

> **SAFETY** ⚠
> Hydrogen peroxide and sodium hydroxide are corrosive; manganese(IV) oxide and copper(I) oxide are harmful; barium chloride is poisonous; urease-active meal is an irritant.

3 Sulphamic acid, NH_2SO_2OH, reacts with water to form ammonium sulphate. This experiment looks at the rate of the reaction. Dissolve a small portion of sulphamic acid, NH_2SO_2OH, in about 20 cm³ of cold water and divide into three equal portions in boiling tubes. Add 0.1 M barium chloride to each portion; there should be no precipitate at this stage. Add equal volumes of water, 2 M ethanoic acid and 2 M hydrochloric acid to separate portions and heat in a hot water-bath.

- Why does the rate of hydrolysis of sulphamic acid vary?

4 Mix equal volumes of 2 M urea, $CO(NH_2)_2$, and 2 M sodium hydroxide and divide into two portions. Add a small portion of urease-active meal to one of the portions. Wipe the test-tube mouths free of sodium hydroxide and place across each mouth a piece of moist red litmus paper. Heat the tubes in a hot water-bath, with the mouths well separated.

- Write the equation for the hydrolysis of urea. The activation energy has been measured and is 37 kJ mol⁻¹ for the catalysed reaction, and 137 kJ mol⁻¹ for the uncatalysed reaction.

Heterogeneous catalysis

STUDY TASK
Read the passage below, and answer the questions based on it.
1 Explain the difference between chemisorption and physical adsorption.
2 What type of bonding would you expect to be involved in physical adsorption?
3 Why is the strength of the chemisorption bonds formed in the ammonia synthesis critical to the success of the process?

Important events take place at the surface of a solid catalyst. When gas molecules approach solid surfaces there is a tendency for them to interact with the surface atoms. This interaction is called **adsorption**. It is the means by which new reaction paths are formed in heterogeneous catalysis.

Adsorption

To understand how heterogeneous catalysis occurs we must learn more about the process of adsorption. When new chemical bonds are definitely formed between a reactant molecule and the surface atoms of the catalyst we use the term **chemisorption**. Transition metals, for example nickel, are particularly able to chemisorb hydrogen. We may represent this process in the following simple way:

$$H_2 + \;—Ni—Ni— \;\longrightarrow\; \begin{array}{c} H\cdots H \\ \vdots \;\; \vdots \\ —Ni—Ni— \end{array} \;\longrightarrow\; \begin{array}{c} H \;\; H \\ | \;\; | \\ —Ni—Ni— \end{array}$$

gas   nickel surface atoms   surface complex   chemisorbed hydrogen atoms

A weaker form of adsorption may also occur, called **physical adsorption**. Here the forces which attract the molecule to the surface are like those which hold non-polar molecules together in a liquid (van der Waals forces; see Topic 9).

Ammonia synthesis catalysts

When ammonia is synthesized from nitrogen and hydrogen gases, by means of an iron catalyst, the following sequence of events occurs in the reactor:

1 Nitrogen gas and hydrogen gas diffuse to the surface of the catalyst.
2 The reactant gases are then adsorbed on the catalyst surface.

$$N_2(g) \longrightarrow 2N_{adsorbed} \quad \text{and} \quad H_2(g) \longrightarrow 2H_{adsorbed}$$

3 Nitrogen and hydrogen react on the catalyst surface, forming ammonia.

$$N_{adsorbed} + H_{adsorbed} \longrightarrow NH_{adsorbed} \xrightarrow{+H_{adsorbed}} NH_{2\,adsorbed} \xrightarrow{+H_{adsorbed}} NH_{3\,adsorbed}$$

4 Ammonia **desorbs** from the catalyst surface.

$$NH_{3\,adsorbed} \longrightarrow NH_3(g)$$

5 Ammonia diffuses away from the catalyst surface.

If any one of these events is much slower than all the others, it will be the **rate-determining step**. That is to say, the rate at which this event occurs will determine how rapidly ammonia can be synthesized. Problems arising in connection with **1** and **5** are very familiar to the chemical engineer but can often be successfully overcome so that both events are made sufficiently rapid.

Iron, ruthenium, osmium, molybdenum, and tungsten can be used as catalysts for the synthesis of ammonia. Therefore, on these metals, steps **2**, **3**, and **4** must also take place at a reasonable rate. Nevertheless, one of these steps must be rate-determining. After many experiments and much lively discussion between American, Japanese, Dutch, and British scientists, there is a majority view that the rate-determining step in ammonia synthesis is the adsorption of nitrogen (stage **2**).

$$N_2(g) \longrightarrow \begin{array}{c} N \quad N \\ \mathrel{\|}\mathrel{\|} \quad \mathrel{\|}\mathrel{\|} \\ -M-M- \end{array} \quad \text{(where M is a surface metal atom)}$$

Of course, we expect catalysts for ammonia synthesis to have the capacity to chemisorb nitrogen. This is not the whole story, but it is a help to know which metals can chemisorb nitrogen and which cannot. Figure 8.16 gives this information for nitrogen and for some other gases.

This information is not complete but we see that three of the metals, iron, molybdenum, and tungsten, which early workers found to be active ammonia synthesis catalysts, chemisorb nitrogen (+).

If a metal is going to be a good catalyst for the synthesis of ammonia, it must not chemisorb nitrogen too strongly, otherwise the nitrogen atoms are unreactive for the combination with hydrogen. On the other hand, if nitrogen is chemisorbed too weakly, only a few atoms will hold on to the surface at any one moment and the rate of ammonia synthesis will again be slow. The strength of chemisorption must be some suitable intermediate value if the synthesis rate is to be a maximum.

Metals			Gases O_2	CO	C_2H_4	H_2	N_2
Ti	Zr	Hf	+	+	+	+	+
V	Nb	Ta					
Cr	Mo	W					
Fe							
Co	Ni		+	+	+	+	−
Rh	Pd						
Ir	Pt						
Al	Cu		+	+	+	−	−
Ag*							
Zn	Cd		+	−	−	−	−
In							
Sn	Pb						

*CO is known to adsorb on silver

Figure 8.16 The ability of metals to chemisorb gases. + means that strong chemisorption is possible. − means that chemisorption is weak or not observed.

Now let us put these ideas together to find a good ammonia synthesis catalyst, that is, a metal which can chemisorb nitrogen but not too strongly. Figure 8.17 shows the approximate strength of nitrogen chemisorption and the rate of ammonia synthesis, plotted as a function of the position of a metal in the transition series. The ability to chemisorb nitrogen as atoms ceases to all intents and purposes to the right of iron, ruthenium, and osmium, so ruling out the metals to the right. To the left of these same metals, the strength of adsorption increases (there is a greater tendency to form bulk nitrides). Therefore we expect that iron, ruthenium, and osmium should have the greatest activity in ammonia synthesis. This is indeed the experimental result already found. It is fortunate that one of these metals, iron, is also very abundant, and therefore economically attractive as the basis of a catalyst for large scale ammonia synthesis.

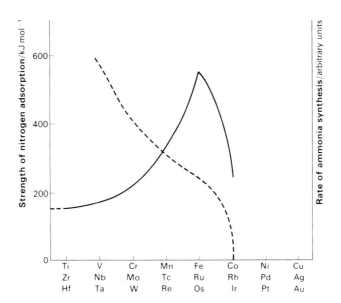

Figure 8.17 The rate of ammonia synthesis (continuous line) compared to the strength of nitrogen absorption on transition metals (broken line)

There is more information about the synthesis of ammonia in Topic 17.3.

INVESTIGATION 8.5b

The rate of reaction of magnesium with acid

Carry out an investigation into the rate of reaction of clean magnesium ribbon with acidic solutions.

Make a risk assessment before starting any experiments.

Summary

At the end of this Topic you should be able to:

a demonstrate understanding of the terms: rate of reaction, rate equation, order of reaction, rate constant, half life, rate-determining step, activation energy, catalyst, homogeneous and heterogeneous catalysis, physical adsorption and chemisorption

b plan an investigation for measuring extent of a reaction with time, and justify the procedures involved

c deduce from experimental data for simple zero, first- and second-order reactions only
    i half life
    ii order of reaction
    iii rate equation
    iv rate-determining step, related to possible reaction mechanisms
    v activation energy (by graphical methods only)

d predict the effects of concentration, temperature and catalysts on the rate of a reaction, and interpret them in terms of collision theory, the distribution of molecular kinetic energies, and alternative reaction pathways of lower activation energy (qualitatively only).

Review questions

8.1 Devise an appropriate procedure to monitor the kinetics of each of the following reactions:

a The decomposition of 4-hydroxy-4-methylpentan-2-one in the presence of hydroxide ions in aqueous solution

$$CH_3COCH_2C(CH_3)_2OH \longrightarrow 2CH_3COCH_3$$

b The effect of heat on a solution of 2,4,6-trinitrobenzoic acid

[structure: 2,4,6-trinitrobenzoic acid (aq)] $\longrightarrow$ [structure: 1,3,5-trinitrobenzene (aq)] $+ CO_2(g)$

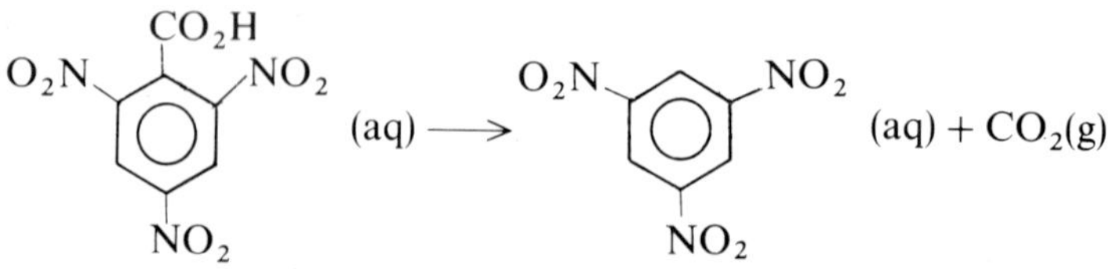

c The hydrolysis of 2-bromo-2-methylpropane in 80% aqueous ethanol

$$(CH_3)_3CBr(aq) + H_2O(l) \longrightarrow (CH_3)_3COH(aq) + H^+(aq) + Br^-(aq)$$

d The hydrolysis of ethyl ethanoate under acid conditions

$$CH_3CO_2CH_2CH_3 + H_2O \longrightarrow CH_3CO_2H + CH_3CH_2OH$$

e The reaction of hydrogen peroxide with potassium iodide in aqueous acid solution

$$H_2O_2(aq) + 2I^-(aq) + 2H^+(aq) \longrightarrow 2H_2O(l) + I_2(aq)$$

f The reaction of potassium bromate(v) with potassium bromide in aqueous acid solution

$$BrO_3^-(aq) + 5Br^-(aq) + 6H^+(aq) \longrightarrow 3Br_2(aq) + 3H_2O(l)$$

8.2 When hydrogen peroxide reacts with iodide ions in aqueous acid, iodine is liberated.

$$H_2O_2(aq) + 2H^+(aq) + 2I^-(aq) \longrightarrow 2H_2O(l) + I_2(aq)$$

The following table gives some experimental results for the reaction:

Initial reactant concentration/mol dm^{-3}			Initial rate of formation of I_2/mol dm^{-3} s^{-1}
[H_2O_2]	[I^-]	[H^+]	
0.010	0.010	0.10	1.75×10^{-6}
0.030	0.010	0.10	5.25×10^{-6}
0.030	0.020	0.10	1.05×10^{-5}
0.030	0.020	0.20	1.05×10^{-5}

a i What is the rate equation for this reaction?
  ii What is the value of the rate constant?
  iii What are the units of the rate constant?
b A proposed mechanism for this reaction is

$$\begin{aligned} H_2O_2 + I^- &\longrightarrow H_2O + IO^- \quad \text{(slow)} \\ H^+ + IO^- &\longrightarrow HIO \quad \text{(fast)} \\ HIO + H^+ + I^- &\longrightarrow I_2 + H_2O \quad \text{(fast)} \end{aligned}$$

Is this mechanism consistent with the rate equation for the reaction?
Give your reasons, discussing each of the three reactants in turn.

8.3 In acid solution, bromate ions slowly oxidize bromide ions to bromine.

$$BrO_3^-(aq) + 5Br^-(aq) + 6H^+(aq) \longrightarrow 3Br_2(aq) + 3H_2O(l)$$

The following experimental data have been determined using 1.0 M solutions.

Mixture	Volume of bromate /cm^3	Volume of bromide /cm^3	Volume of H^+(aq) /cm^3	Volume of water /cm^3	Relative rate of formation of bromine
I	50	250	300	400	1
II	50	250	600	100	4
III	100	250	600	50	8
IV	50	125	600	225	2

a i What is the rate equation for this reaction?
  ii What are the units of the rate constant?

b A proposed mechanism for this reaction is

$$H^+ + Br^- \longrightarrow HBr \quad \text{(fast)}$$
$$H^+ + BrO_3^- \longrightarrow HBrO_3 \quad \text{(fast)}$$
$$HBr + HBrO_3 \longrightarrow HBrO + HBrO_2 \quad \text{(slow)}$$
$$HBrO_2 + HBr \longrightarrow 2HBrO \quad \text{(fast)}$$
$$HBrO + HBr \longrightarrow H_2O + Br_2 \quad \text{(fast)}$$

Is this mechanism consistent with the rate equation for the reaction? Give your reasons, discussing each of the three reactants in turn.

8.4 The rate constant given by

rate of change of concentration of hydrogen iodide = $k[HI]^2$

for the reaction $2HI(g) \longrightarrow H_2(g) + I_2(g)$ varies with the temperature as in the following table.

Temperature /K	Rate constant/10^{-5} dm^3 mol^{-1} s^{-1}
556	0.0352
647	8.58
700	116
781	3960

Plot a graph of $\ln k$ (on the vertical axis) against $\dfrac{1}{T}$ (on the horizontal axis) and from the gradient of this graph find the activation energy of the reaction.

8.5 The rate constant given by

rate of formation of nitrogen = $k[C_6H_5N_2Cl]$

for the reaction

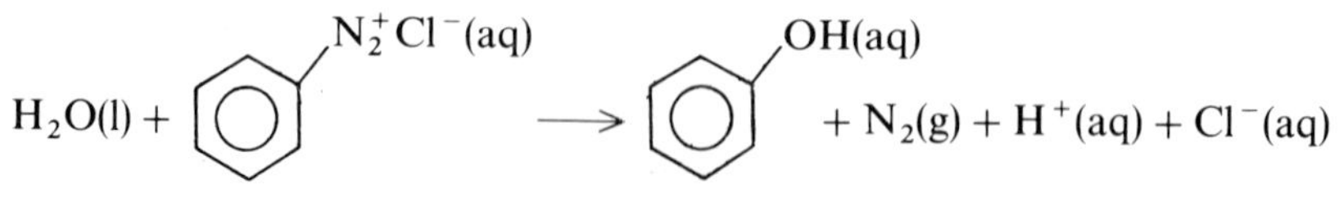

$$H_2O(l) + C_6H_5N_2^+Cl^-(aq) \longrightarrow C_6H_5OH(aq) + N_2(g) + H^+(aq) + Cl^-(aq)$$

varies with temperature as in the following table.

Temperature/K	Rate constant/10^{-5} s^{-1}
278.0	0.15
298.0	4.1
308.2	20
323.0	140

Find the activation energy of the reaction.

Examination questions

8.6 Hydrogen peroxide, H_2O_2, can decompose to form oxygen and water.

$$2H_2O_2(aq) \longrightarrow 2H_2O(l) + O_2(g)$$

a Describe how you would attempt to measure the rate of decomposition of hydrogen peroxide. You should include in your answer a description of what you would do, what measurements you would take and how you would process your results.

b The reaction is first order with respect to hydrogen peroxide.

i Write down the rate equation for the reaction.

ii What are the units of the rate constant?

8.7 When a lemonade bottle is left open, the lemonade goes 'flat' as carbon dioxide escapes from the carbonated water. Laboratory tests suggest that this is a first-order process. Describe how you would attempt to verify this by an experiment. You should include in your answer
- a description of the apparatus
- details of any chemicals used
- the measurements you would make
- an outline of how you would process the results
- the way you would deduce the order of the reaction.

8.8 The hydrolysis of methyl methanoate proceeds at a measurable rate in the presence of aqueous acid:

$$HCO_2CH_3 + H_2O \longrightarrow HCO_2H + CH_3OH$$

a What method would you use to investigate how the rate of this reaction varies with concentration of the reactants? Your answer should include an outline of the experimental procedure and the measurements you would make.

b In an investigation of this reaction the following results were obtained for the rate of hydrolysis of methyl methanoate at 298 K:

$[HCO_2CH_3]$/mol dm^{-3}	$[H^+]$/mol dm^{-3}	Initial rate $\times 10^3$/mol dm^{-3} s^{-1}
0.50	1.00	0.56
1.00	1.00	1.11
2.00	1.00	2.24
2.00	0.50	1.13
2.00	2.00	4.49

Deduce the rate equation for this reaction, and calculate the rate constant at 298 K. Explain how you arrive at your answers.

Using the value of your rate constant at 298 K and the data below, plot a suitable graph and hence find a value for the activation energy of the reaction.

Temperature/K	Rate constant/dm^3 mol^{-1} s^{-1}
313	3.95×10^{-3}
323	8.63×10^{-3}

8.9 When strips of cobalt foil are rotated rapidly at a constant rate in sodium peroxodisulphate(VI) solution, there is a slow reaction and the cobalt dissolves. The reaction can be followed by removing and weighing the foil at intervals.

Experiment I at 0.5 °C		Experiment II at 13.5 °C		Experiment III at 25 °C	
Time /min	Mass /mg	Time /min	Mass /mg	Time /min	Mass /mg
0	130	0	130	0	130
20	120	10	115	4	116
60	98	15	106	8	103
80	86	30	86	12	87

Determine a rate constant k at each temperature as a loss in mass per minute (mg min^{-1}), and then determine the activation energy E of the reaction using the relationship

$$k \propto 10^{-E/2.3RT}$$

What differences in the results would you predict if the cobalt foil were not rotated?

8.10 The reaction:

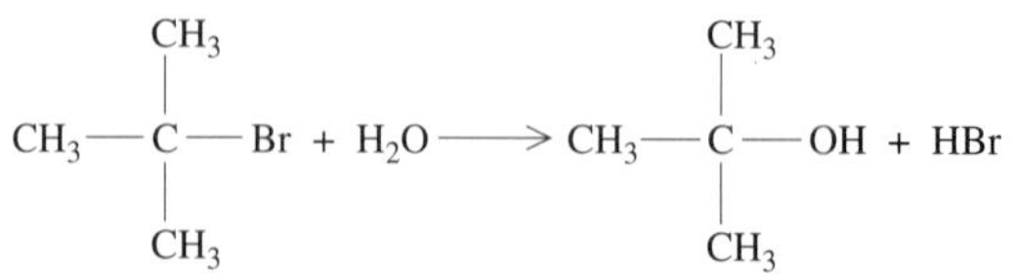

$$\underset{\text{Compound A}}{CH_3-\underset{\underset{CH_3}{|}}{\overset{\overset{CH_3}{|}}{C}}-Br} + H_2O \longrightarrow \underset{\text{Compound B}}{CH_3-\underset{\underset{CH_3}{|}}{\overset{\overset{CH_3}{|}}{C}}-OH} + HBr$$

takes place in 80% aqueous ethanol quite readily. It can be followed by measuring the electrical conductivity of the solution using a conductivity meter.

a i State the systematic chemical name for compound **A**.
ii To which class of compound does compound **B** belong?
iii Explain why the progress of this reaction may be followed by measuring changes in electrical conductivity.
iv Outline another method which could be used to follow *this* reaction.
b The graph overleaf shows some results derived from a series of experiments (I to IV) performed using the conductivity method. In each experiment a measured volume of compound **A** was added to a mixture of 40 cm³ of ethanol and 10 cm³ of water.

Experiment	Volume of compound A/cm³
I	0.5
II	0.8
III	1.2
IV	1.5

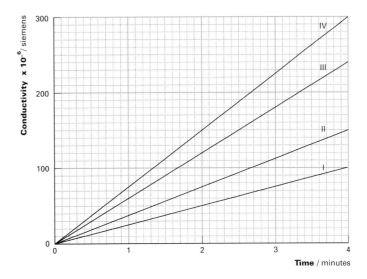

i   Plot a graph of initial rate, measured as the initial rate of change of conductivity, against initial concentration of compound **A**, measured as the volume of compound **A** taken. (The total volume in each of I to IV is assumed to be 50 cm^3)

ii This reaction is first order with respect to compound **A**. Explain how this may be deduced from the graph plotted above.

iii This reaction is zero order with respect to the concentration of water. Write a rate equation for the reaction.

8.11 The rate of nitration of the arenes by concentrated nitric acid can be studied using a dilatometer (see figure 8.7).

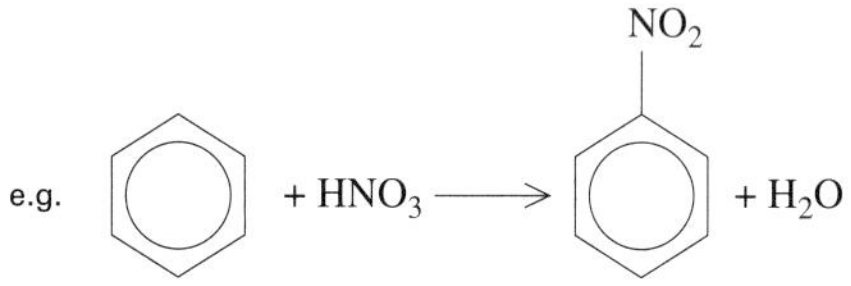

e.g. benzene + HNO$_3$ → nitrobenzene + H$_2$O

The reaction is accompanied by a small **decrease** in volume which can be observed by the movements of the liquid level in the capillary tube. The change in concentration of the arene may be calculated from the change in height of the liquid level.

The experiment was carried out first with benzene and then repeated with methylbenzene. The calculated concentrations of the arenes are plotted against time in the graph opposite.

a i   What important factor must be kept constant if accurate results are to be obtained?
  ii How could this be achieved?
b i   How do the rates of nitration for benzene and methylbenzene vary with time?
  ii What is the order of the reaction with respect to the arene implied by these results?
  iii The reaction is known to be first order with respect to nitric acid. Using this information and your answer to **bii**, give the rate equation for the reaction.
  iv What is the **minimum** number of steps in the reaction mechanism that would be consistent with this rate equation? Explain your answer.

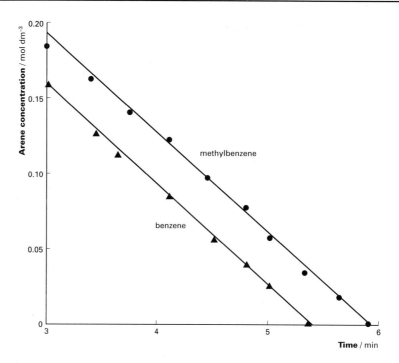

c The large scale nitration of arenes is usually achieved with a nitrating mixture of concentrated nitric and sulphuric acids which provides the attacking species NO_2^+ according to the equation:

$$HNO_3 + 2H_2SO_4 \longrightarrow NO_2^+ + H_3O^+ + 2HSO_4^-$$

Classify the nitration of arenes in terms of the nature of the reactant (nucleophile, electrophile etc) and the reaction type (addition, elimination etc).

8.12 In colour photography, the film consists of three layers of 'emulsion', each of which is a suspension containing silver halides in gelatin. Each layer is sensitive to one of the three primary colours in light: red, green and blue.

a The first stage in processing the film is development. Control of development is critical for the correct colours. Any variation in temperature must be allowed for by altering the time allowed for development. The instructions for one make of colour developer include a temperature/time table:

Temperature/°C	Development time
38	3 min
37	3 min 15 sec
36	3 min 40 sec
35	4 min 10 sec
34	4 min 35 sec
33	5 min 15 sec
32	5 min 45 sec

i Suggest a reason why control of the development is likely to be more accurate at a lower temperature.

ii At a given temperature:

$$\text{rate of development} \propto \frac{1}{\text{time required for normal development}}$$

Hence 1/(development time in seconds) may be used as a measure of the reaction rate at that temperature.

In the table below, most of the data in the temperature–time table above has been converted into values of 1/(temperature/K) and ln[1/(development time)].

Temperature/°C	1/temperature/K^{-1}	ln[1/(development time)]
38	3.215×10^{-3}	-5.19
37	3.226×10^{-3}	-5.27
36	3.236×10^{-3}	-5.39
35	3.247×10^{-3}	-5.52
34	3.257×10^{-3}	-5.62
33	3.268×10^{-3}	-5.75
32	?	?

What are the missing values in the table?

iii Plot a graph of ln[1/(development time)] on the y-axis against 1/(temperature).

iv The relationship between reaction rate and temperature is given by the equation:

$$\text{rate} \propto A e^{-E/RT}$$
$$or \quad \ln(\text{rate}) = \text{constant} - E/RT$$

where T is the temperature in kelvin, E is the activation energy, and R is the gas constant. Use your graph from **iii** to calculate the activation energy, E, for the development process. ($R = 8.31 \text{ J K}^{-1} \text{ mol}^{-1}$)

b The second stage of processing the film is called **bleaching**. This involves the conversion of the silver deposit back to a silver halide. An oxidizing agent is used, such as potassium hexacyanoferrate(III), $K_3\text{Fe(CN)}_6$, which contains the complex ion Fe(CN)_6^{3-}.

i The hexacyanoferrate(III) ion is reduced to the ion Fe(CN)_6^{4-}. Write down the name of this ion.

ii Write a balanced equation for the reaction.

READING TASK 3

PHOTOCHEMISTRY

Chemistry is all about energy and molecules: energy can help to make molecules, change them and break them. Photochemistry is all about exciting the electrons of molecules with discreet packages of light energy called **photons**, thereby making them jump from a lower to a higher energy state. Photons of ultraviolet and visible light can do this. Our eyes rely on such photochemical changes in the molecules of the retina to enable us to see. General interest in photochemistry has been awakened by the realization that we are being exposed to higher levels of ultraviolet radiation due to the depletion of ozone in the upper atmosphere and the discovery in 1985 by the British Antarctic Survey of the 'ozone hole'.

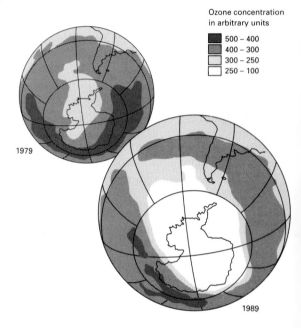

Figure R3.2 The Antarctic ozone hole, mapped by satellite in October 1989

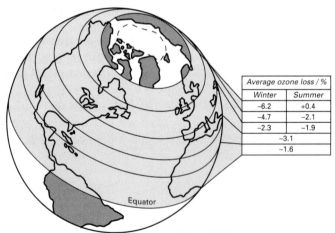

Figure R3.1 Ozone trends for the Northern Hemisphere in the last two decades

Average ozone loss / %	
Winter	Summer
−6.2	+0.4
−4.7	−2.1
−2.3	−1.9
−3.1	
−1.6	

Shorter wavelengths of light have more energy and cause bigger electron jumps. Light may even have enough energy to strip out an electron completely and turn a molecule into an ion. However, photochemists are more interested in light which excites electrons rather than light which ejects them. They study the remarkable behaviour of molecules once they have captured a photon of light. What happens next depends on the molecule: it may break apart, possibly to form highly reactive free radicals; or react chemically with another molecule; or rearrange itself into another isomer; or emit light.

Nearly all molecules respond to ultraviolet and visible light but some molecules are particularly sensitive. The best known naturally photosensitive substance is chlorophyll which enables plants to make carbohydrates with energy trapped from sunlight. Perhaps the best-known artificial photosensitive material is photographic film. Other examples include thin films, liquid crystals, and microelectronic devices that we now use to detect light, to count photons, and even to change the wavelength of light.

Light which produces photochemical changes generally has wavelengths of 200 to 740 nm (a nanometre is a billionth of a metre). This range includes both visible and ultraviolet light.

When a molecule absorbs a photon of radiation and is excited, how can we study the excited state and follow what happens next? The excited state may last a few seconds but more likely it will last for only a microsecond (a millionth of a second), or just a few femtoseconds (a million billionth of a second). We can carry out such studies thanks to the pioneering work of George Porter, the British chemist and Nobel prize winner, who developed a technique known as **flash photolysis**.

Chemists have used femtosecond flash photolysis to study photosynthesis. Chloroplasts, which occur in some cells of green plants, have a reaction centre with two chlorophyll molecules. The chloroplast absorbs a photon of light and in about one picosecond (a thousand billionth of a second) this causes an electron to be transferred out of the reaction centre leaving the system with enough energy to oxidize water in the cell to oxygen, which is released into the air:

$$2 H_2O(l) \longrightarrow O_2(g) + 4H^+(aq) + 4e^-$$

The hydrogen ions and the electrons released in the cell then bind to carbon dioxide to form carbohydrate molecules such as starch or cellulose. The process is complex but we know the oxidation takes place at a cluster of four magnesium atoms situated on one of the chlorophyll molecules. Photosynthesis enables an energetically unfavourable change to take place

$$6CO_2(aq) + 12H^+(aq) + 12e^- \longrightarrow C_6H_{12}O_6(aq)$$

which is the basis of plant life on Earth and its atmosphere of oxygen.

Shielding life on Earth

In the Cambrian period of geological time, about 600 million years ago, only one per cent of the atmosphere was oxygen. Life-threatening ultraviolet radiation from the Sun could reach the surface of the planet, so life only survived where it was shielded, in stagnant water for example. But as marine plants released more and more oxygen to the atmosphere, life on the surface became less hazardous, thanks primarily to the formation of a protective shield of ozone in the upper atmosphere, or stratosphere.

When molecular oxygen, O_2, is exposed to short wavelength ultraviolet, it is converted into ozone by a photochemical reaction. Light splits the oxygen molecule into two atoms of oxygen, $O\cdot$, each of which can then combine with another molecule of oxygen to form ozone, O_3. This process occurs in the stratosphere where there is now an estimated 3.5 billion tonnes of ozone. But the photochemistry does not stop there, because the ozone is also capable of absorbing longer wavelength ultraviolet. When it does, it throws off one oxygen atom and reverts to O_2. The result is that only a little ultraviolet of wavelength shorter than 300 nm reaches the surface of the Earth. But some does, and this is life-threatening.

Ozone in the stratosphere also reacts with nitrogen monoxide, the hydroxyl radical and chlorine free radicals. The reactions with nitrogen monoxide and the hydroxyl radical arise from natural causes, such as the dinitrogen monoxide given off from the soil, but the concentration of chlorine has increased rapidly in the second half of the twentieth century. This is mainly due to the chlorofluorocarbons (CFCs) that used to be widely used as solvents, as the coolant liquid in refrigerators, and in aerosol cans, before the damaging consequences were recognized by research.

Once released from its CFC the chlorine radicals reduce ozone concentration by the chain reaction

$$Cl\cdot + O_3 \longrightarrow ClO\cdot + O_2$$
$$ClO\cdot + O\cdot \longrightarrow Cl\cdot + O_2$$

(the dots represent unpaired electrons)

In this chain reaction one chlorine radical can turn many thousands of ozone molecules into oxygen molecules before the chain is terminated.

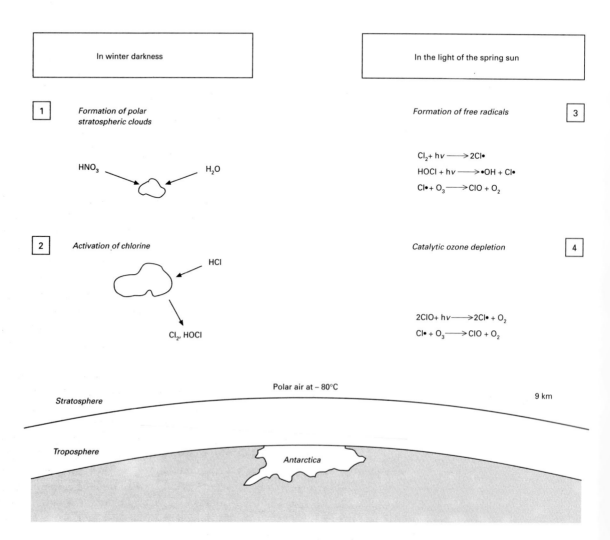

Figure R3.3 Some reactions which cause ozone depletion in the Antarctic atmosphere

There are other processes which can reduce ozone concentration. For instance, the detection of the 'ozone hole' over the Antarctic led to research which identified a set of reactions which occur on the surface of ice crystals in Antarctic stratospheric clouds. In these reactions hydrogen chloride is converted into the photochemically active species ClO. A concentration of 1 part per billion of ClO can destroy nearly all the Antarctic stratospheric ozone in two to three weeks and can be directly attributed to the release of CFCs. This was a surprising and alarming result because 99% of the atmospheric chlorine is in the form of hydrogen chloride and other species which do not normally take part in photochemical reactions.

Skin and photochemistry

In sunlight, our skin undergoes photochemical reactions. These may be beneficial, cosmetic, damaging or even fatal. The Victorians thought sunlight was beneficial; they knew it promoted healing and prevented rickets, the vitamin D deficiency that affected inhabitants of smoky industrial cities. This belief in the blessings of sunbathing lingers on, although most people now do it for cosmetic reasons to get a golden tan, despite the fact that this browning is the body's response to a threat. Ultraviolet light

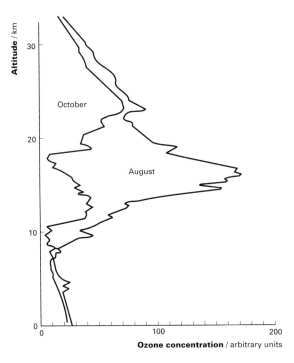

Figure R3.4 Ozone destruction in the Antarctic during September

which causes tanning is very damaging. At best it weakens the skin and leads to excessive wrinkling in old age, at worst it causes skin cancers. Every year, doctors treat more than 25 000 people for this condition in Britain alone. Fortunately, 95% of patients are cured.

Ultraviolet light kills cells, and light of wavelength 260–265 nm is the most deadly. This band is exactly that at which DNA and RNA absorb, and may explain why cells are so vulnerable and strong sunlight is so dangerous. We can identify pyrimidine and purine groups as the components of DNA and RNA which absorb at 260 nm. When these are activated they undergo chemical reactions, either with another group nearby, or with a water molecule, or with oxygen. The danger of heavy damage is that it will swamp the body's repair mechanisms, and allow a mutant cell to survive. If that cell then replicates it starts a cancer.

We have about 100 billion (10^{11}) cells in our body, each of which is damaged 100 times a day. Some of this damage is caused photochemically. Cells have evolved to deal with ultraviolet light. Many cells contain sunscreens to prevent damage occurring, others use repair kits, others employ another photochemical reaction which darkens our skin. Sunbathing starts to redden our skin after about 12 hours, and this process peaks after 24 hours. Cells die or deform, and the longer we are exposed, the greater the damage. Ultraviolet light of around 300 nm penetrates the clear cells on the surface of our skin and reaches the next layer of cells, which die. One defence is tanning in which cells called melanocytes in this lower layer are activated to produce deposits of brown melanin. This compound absorbs ultraviolet light and protects the DNA. A natural suntan cuts down ultraviolet light penetration by 90%. Alternatively, we can apply a sunscreen lotion to our skin. This will contain a substance which absorbs in the 290–320 nm region, such as benzoic acid, menthyl anthranilate, or even zinc oxide.

Most skin cancers are easily noticed and dealt with by a doctor, but one type called melanoma is dangerous and can be fatal if not treated early. In Britain there has been an increase of 50% in cases of malignant melanoma since 1980, and there are about 2700 cases a year (1993 figures). The increase could be due to stratospheric ozone depletion, the popularity of sunbathing, a series of fine summers, or cleaner air which no longer cuts out ultraviolet light over urban areas.

Some skin conditions can be cured by photochemical therapy. Psoriasis, or flaky skin, is caused by overactive skin cells. Sensitize these cells with a compound called psoralen and bathe the skin with ultraviolet light of wavelength 320–400 nm and you can destroy the overactive cells.

Industrial applications

Photochemistry has limited application in the chemicals industry, which still drives most reactions with heat, not light. The mercury arc lamps that drive photochemical processes are not cheap, so they are only economical for making compounds, such as drugs and cosmetics, that are highly valuable even in very small amounts. If other ingredients absorb light, they too may undergo reactions and reduce the yield or contaminate the product.

Cholesterol, a fatty component of blood, can be converted to vitamin D, and one step in the process is best done photochemically. The form of the vitamin this produces, called D_3, is vital to poultry, and it is added to their feed.

Few companies make bulk chemicals photochemically, but one in Japan has devised an economical method of making caprolactam, the material from which nylon is made. The plant can produce up to 135000 tonnes of caprolactam a year from cyclohexane and nitrosyl chloride, NOCl. The company bathes the reactants in light from high-pressure mercury lamps which can operate efficiently for more than a year. The light splits the nitrosyl chloride into chlorine radicals, Cl·, and nitrosyl radicals, NO·. The chlorines snip a hydrogen atom from the cyclohexane, and then the nitrosyl radicals move in to form caprolactam.

Questions

Reread the passage and answer these questions.

1 Write an account in about 150 words about the importance of the ozone layer and how it is being depleted.

2 Could scientists and governments have done more to prevent the development of our problems connected to ozone layer depletion?

TOPIC 9

Intermolecular forces and solubility

Figure 9.1 Harold Lloyd relying on the intermolecular forces in a flagpole

When atoms combine with each other and electrons are exchanged to give ions, or shared between the atoms to give molecules, the resulting forces of attraction are very strong and require quite large amounts of energy to break them. The lattice energies of ionic solids, for example, range in numerical value from about 600 kJ mol^{-1} to over 3000 kJ mol^{-1}. Covalent bonds are not as strong as this usually but the bond energies of most single bonds lie between 200 and 500 kJ mol^{-1}.

When you melt and then boil an ionic compound the forces of attraction between the ions are overcome and the ions actually separate to a considerable extent

$$NaCl(s) \longrightarrow Na^+(g) + Cl^-(g)$$

When a molecular substance is melted or boiled the covalent bonds normally remain intact: the molecules merely separate from each other

$$H_2O(l) \longrightarrow H_2O(g)$$

Since the molecules do not break up when the liquid water is boiled, there must be some forces of attraction between the molecules themselves which are broken when the molecules separate from each other. These are called **intermolecular forces**, and, since every molecular substance can be made into a solid or liquid by suitable adjustments of temperature and pressure, intermolecular forces must always exist, though sometimes they may be very weak.

At the other end of the scale some intermolecular forces must be very large. In this picture the 'hero' is being attracted downwards by the gravitational pull of the whole Earth and this force of attraction is being resisted successfully by the intermolecular forces in the flagpole!

QUESTIONS

You will need the *Book of data*.
1 Look up some lattice energies for a variety of ionic compounds. What patterns are there in the values?
2 Look up some covalent bond energies. What patterns are there in the values?

EXPERIMENT 9.0 Evidence for intermolecular forces

This experiment provides some evidence for the presence of intermolecular forces.

Procedure

1 Floating corks

Put seven new wine-bottle corks in a large beaker of water. They float, of course, but can you make them float standing on their ends in the water? Try floating several of the wet corks as a group. You should be able to manage it with all seven!

2 Floating needles

Steel is obviously more dense than water yet you can float a blunt razor blade or steel sewing needle on the surface of water if you are very careful. Try putting the needle on a small piece of paper on the water's surface and then carefully pushing the paper away from under it.

You might try floating the blunt razor blade on the surface of a non-aqueous liquid such as hexane (CARE: flammable liquid; harmful vapour).

- What conclusions can you draw about the size of the intermolecular forces in water and hexane?

9.1 Van der Waals forces

Helium, which does not form normal covalent or ionic bonds and has symmetrical atoms, condenses to a liquid and ultimately freezes to a solid at very low temperatures. Energy is released in this process, showing that there are cohesive forces operating.

The energy of sublimation of solid helium when it changes from solid to gas is only $0.105 \text{ kJ mol}^{-1}$. This should be compared with the dissociation energy of the hydrogen molecule, 436 kJ mol^{-1}, required to break the one covalent bond.

The weak forces of attraction independent of other bonding forces are known as **van der Waals forces**.

Van der Waals forces are considered to be due to continually changing induced electric charge interactions between atoms, called dipole-dipole interactions.

These interactions are thought to arise because the electron charge-cloud in an atom is in continual motion. In the turmoil it frequently happens that rather more of the charge-cloud is on one side of the atom than on the other. This means that the centres of positive and negative charge do not coincide, and a fluctuating dipole is set up. This dipole induces a dipole in neighbouring atoms. The sign of the induced dipole is opposite to that of the dipole producing it, and consequently a force of attraction results. These **induced dipole-dipole interactions** produce a **cohesive force** between neighbouring atoms and molecules.

The greater the number of electrons in an atom, the greater will be the fluctuation in the asymmetry of the electron charge-cloud and the greater will be the van der Waals attraction set up. The rise in boiling point down Group 7 (fluorine, chlorine, bromine, and iodine) is due to the increasing numbers of electrons in the atoms and the consequent increase in van der Waals attractions, rather than to the increase in the mass of the atoms.

The energy of sublimation is the sum of the enthalpy change of melting and the enthalpy change of vaporization.

Compare these data by plotting a graph.

Entity	Number of electrons	Molar mass /g mol^{-1}	Boiling point /K
F$_2$	18	38	85
Cl$_2$	34	71	238
Br$_2$	70	160	332
I$_2$	106	254	457

Similarly the increase in boiling point up the homologous series of alkanes is due to the increased number of electrons in the molecules and the increased total van der Waals attractions and not to the increase in mass of the molecules. The difference in boiling point between isomers can also be explained in terms of van der Waals attractions.

STUDY TASK

The boiling points of pentane and 2,2-dimethylpropane, two isomers of molecular formula C_5H_{12}, are 36 °C and 9 °C respectively. Their structures are shown in figure 9.2.

Build space-filling models of each of these structures, and compare the shapes of the molecules.

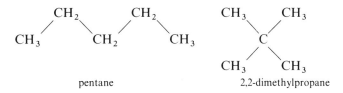

Figure 9.2 Why do these isomers differ in boiling point?

In the case of pentane, the linear molecules can line up beside each other and the van der Waals forces are likely to be comparatively strong, as they can act over the whole of the molecule. In the case of 2,2-dimethylpropane, the spherical molecules can only become close to one another at one point, so the van der Waals forces are likely to be comparatively weak. The isomer with the linear molecules thus has a higher boiling point.

Van der Waals radii

The normal bonding forces in molecules are concentrated within the molecules themselves; they are intramolecular. In molecular crystals individual molecules are held to each other by van der Waals forces. Examples are iodine, solid carbon dioxide, and sulphur. As the forces are weak, the melting points of molecular crystals tend to be low.

In molecular crystals the van der Waals forces draw molecules together until their electron charge clouds repel each other to the extent of balancing the attraction. Thus for argon the atoms are drawn together until the atomic nuclei have a separation of about 0.4 nm (figure 9.3).

The atomic distances **within** simple molecules and **between** simple molecules are not the same.

> **The *covalent radius* is one half of the distance between the nuclei of two atoms in the *same* molecule. The *van der Waals radius* is one half of the distance between the nuclei of two atoms in *adjacent* molecules.**

9 Intermolecular forces and solubility 248

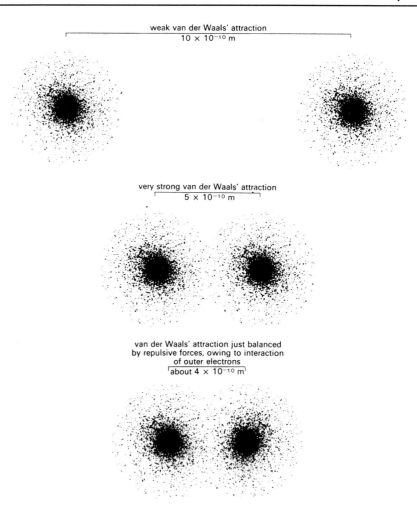

Figure 9.3 Van der Waals attractions between argon atoms

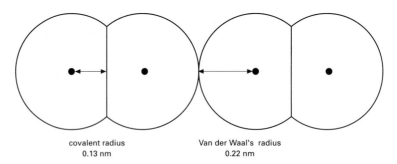

Figure 9.4 The difference between covalent radius and van der Waals radius for I_2

The relative values of these radii for certain elements can be seen in the table.

Atom	Covalent radius/nm	Van der Waals radius/nm
H	0.037	0.12
N	0.075	0.155
O	0.073	0.150
P	0.110	0.185
S	0.102	0.180

249   9   Intermolecular forces and solubility

STUDY TASK

Although the van der Waals forces between individual atoms, as in helium, give only a small bonding energy, the total van der Waals bonding energy can be significant in large molecules with many contacts between atoms.

Use tables 7.7 and 7.2 in the *Book of data* to compare the tensile strength of a substance such as poly(ethene) that has only van der Waals forces between its molecules and the tensile strengths of a range of metals.

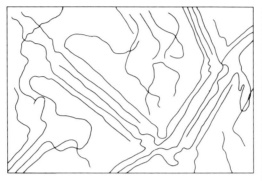 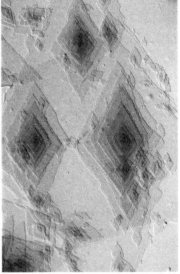

Figure 9.5   **a** Crystalline packing in a plastic polymer   **b** Electron micrograph of poly(ethene) crystals

9.2 Molecules with permanent dipoles

The attraction between the 'flickering dipoles', the van der Waals forces, is universal; all molecules exert some degree of attraction on all other molecules by this means. In some molecules, however, the dipole is not instantaneous and rapidly changing but permanent. A good example of this is the molecule of propanone which has a permanent dipole because of its strongly polarized carbonyl bond.

Figure 9.6   Space-filling model of propanone

Molecules with a permanent dipole are called **polar molecules**. In experiment 9.2 you can investigate how permanent dipoles affect the properties of molecules.

EXPERIMENT 9.2 What is the effect of an electrostatic field on a jet of liquid?

You should plan this experiment in groups and produce an agreed list of liquids to test. Organic liquids provide a good range of compounds which have asymmetric molecules; for reasons of safety and cost it is best to confine your attention to hydrocarbons, alcohols and ketones.

Test **one** of the liquids on your list and watch the other tests.

Make a risk assessment before starting any experiments.

SAFETY ⚠
Read the hazard warnings about the liquids selected for investigation and take appropriate precautions. Make sure you have a fume cupboard if your risk assessment requires it.

Procedure
Fill a burette with a liquid, stand it over a large empty beaker, and turn on the tap so that a stream of the liquid flows into the beaker. Hold a charged plastic rod near the jet of liquid, but do not let it touch the liquid.

A plastic rod (such as a ball-point pen) can be charged by rubbing it vigorously with a piece of cloth, provided both are thoroughly dry.

- What happens to the jet of your liquid?
  What happens with other liquids?
  Why does this happen?

Interpretation of the experiment
The electrostatic field will be interacting with molecules that have an imbalance in charge distribution, that is, the molecules are **polarized**.

You are detecting the polarization of **molecules**, but the polarization of **bonds** can only be inferred from your results. A further complication is that the electrostatic field will create a temporary polarization in the molecules, so the interpretation of these experiments cannot be precise. Nevertheless, you should have obtained some marked effects.

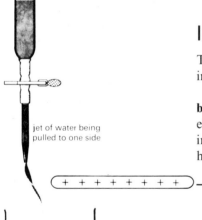

jet of water being pulled to one side

QUESTIONS
1 Which liquids seemed to contain strongly polarized molecules?
2 Which bonds in the molecules might be strongly polarized?
3 Can you suggest an order of polarization for the various bonds involving carbon atoms in the liquids you tested?

Figure 9.7 Apparatus for experiment 9.2

The permanent polarization of molecules can be measured as a **dipole moment**. The direction of the dipole moment in a molecule is represented by the sign ↦, which is placed to point from the positively charged atom to the negatively charged atom in the molecule, as shown in figure 9.8.

Figure 9.8 Representation of the dipole in the hydrogen chloride molecule

	Dipole moment /debye
Ethane C_2H_6	0
Hydrogen chloride HCl	1.1
Ammonia NH_3	1.5
Water H_2O	1.8
Chloroethane C_2H_5Cl	2.0

A difference in electric charge equal to the charge on the electron at a separation of about a bond length (0.1 nm) has a dipole moment of about 5 debye (a measure of charge × distance apart).

We can compare this value of 5 debye with the values for some molecules.

STUDY TASK

Molecules with permanent dipoles attract each other in a way which is not possible with symmetrical molecules. The extent of this can be seen by comparing the boiling points of polar and non–polar substances with approximately the same number of electrons in their molecules. One pair of examples is given; find others from the *Book of data*.

Substance	Number of electrons	Boiling point/°C
Propanone	36	56
2-methylpropane	34	−12

9.3 Hydrogen bonding

In many molecules there is the possibility of a very special kind of intermolecular force which has certain specific requirements. These are:

> **One molecule must have a hydrogen atom which is very highly positively polarized. So highly, in fact, that it is almost ready to be donated as a proton to a base.**

> **The other molecule must have one of the small strongly electronegative atoms of the elements nitrogen, oxygen or fluorine and this atom must have available a lone pair of electrons.**

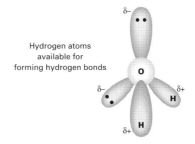

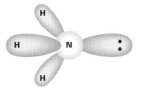 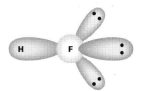

Figure 9.9   Lone pairs available for forming hydrogen bonds

Because of the small size of the hydrogen atom and the comparatively small size of the other atom involved, the two atoms are able to approach one another closely enough for the forces of attraction between them to reach nearly one tenth of the strength of a typical covalent bond. The resulting force of attraction is known as a **hydrogen bond**.

Hydrogen bonds are represented by three dots thus:

$$H_3N \cdots H\text{—}OH, \qquad H\text{—}F \cdots H\text{—}F$$

Bond energies of typical hydrogen bonds are around 30 kJ mol^{-1} or less whereas those of covalent bonds tend to be about 300 kJ mol^{-1}. By contrast van der Waals forces are only of the order of 3 kJ mol^{-1}.

One way of visualizing a hydrogen bond is to think of it as being a proton, H$^+$, between two negative charge clouds, the covalent bond on one side and the lone pair of electrons on the other:

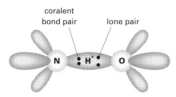

Figure 9.10

The angle between the hydrogen bond and the covalent bond to the hydrogen atom is normally 180° for maximum bond strength.

Evidence for hydrogen bonds

When the boiling points of the four hydrides HF, HCl, HBr, HI are plotted against the halogen period number in the Periodic Table the result is as shown below:

a

	Melting point/K	Boiling point/K
CH$_4$	91	109
SiH$_4$	88	161
GeH$_4$	108	185
SnH$_4$	123	221
NH$_3$	195	240
PH$_3$	140	185
AsH$_3$	157	218
SbH$_3$	185	256
H$_2$O	273	373
H$_2$S	188	212
H$_2$Se	207	232
H$_2$Te	224	271
HF	190	293
HCl	158	188
HBr	185	206
HI	222	238

Figure 9.11 a Data on the hydrides of the p-block elements. **b** Boiling points of the Group 7 hydrides and the noble gases.

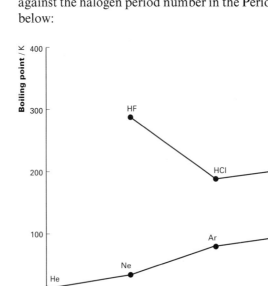

b

The rise in boiling point HCl < HBr < HI is due to an increase in van der Waals forces because of an increase in the number of electrons; but if this were the only type of intermolecular force, HF would boil at about −90 °C whereas in fact it boils at +20 °C due to hydrogen bonding.

STUDY TASK

The hydrogen bonds between water molecules are particularly important.

1 Draw a graph for the hydrides of Group 6 of the Periodic Table similar to the one for the hydrogen halides.
  Without its hydrogen bonding, what would you predict as the boiling point of water?

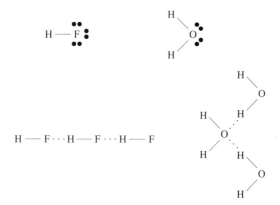

Figure 9.12

2 The molecule of hydrogen fluoride has three lone pairs of electrons but only one hydrogen atom so that each molecule can only use one of its lone pairs on average. The water molecule has two lone pairs and two hydrogen atoms and so can take part in hydrogen bonding to twice the extent.
  What feature of the graph you have just drawn arises from this ability?
3 Plot similar graphs for the hydrides of Group 4 and Group 5 and explain the results.

How strong are the intermolecular forces?

The intermolecular forces in water are largely overcome when the water is vaporized:

$$H_2O(l) \longrightarrow H_2O(g)$$

The next experiment is designed to measure the enthalpy change of vaporization of water.

EXPERIMENT 9.3 Measuring enthalpy changes of vaporization

The aim is to measure the energy required to boil away a known mass of water. The simplest procedure is to use an electric kettle; but good results can be obtained using a Bunsen burner. If the apparatus shown in figure 9.13 is available results can be obtained for a range of liquids.

Procedure 1

Record the power in watts of an electric kettle, fill with about 500 cm³ of water, and weigh with its lid. Remove the lid, plug in and switch on. Start timing when the water starts to boil and steam comes out of the kettle. Boil for a fixed period of time, three minutes is sufficient. Switch off, put the lid on at once and reweigh.

Work out the electrical energy supplied using the conversion kilowatts = kJ s⁻¹

Procedure 2

Set up the apparatus as shown in figure 9.13. Have a second, weighed, 100 cm³ beaker ready. Switch on the electrical supply to the meter and wait until water is distilling steadily. Switch off the electricity, read the joulemeter, and place the weighed beaker to collect the distillate. Switch on the electricity and distil over approximately 10 cm³ of water. Switch off, reweigh the beaker, and read the joulemeter.

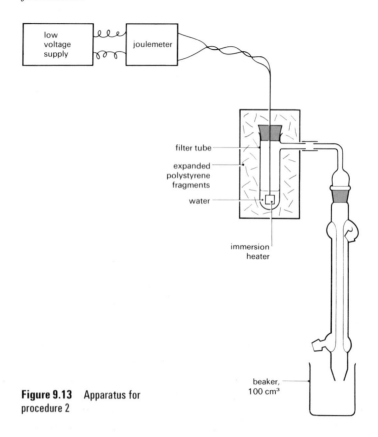

Figure 9.13 Apparatus for procedure 2

Calculations

1 Work out the mass of water evaporated. What is this amount in moles?
2 Calculate the enthalpy change of vaporization (sometimes called the molar latent heat of vaporization) from the equation

$$\Delta H_{vap}/\text{kJ mol}^{-1} = \frac{\text{electrical energy supplied}/\text{kJ}}{\text{amount of water distilled}/\text{mol}}$$

A value from the *Book of data* is 40.7 kJ mol^{-1}. This would represent the total for all the types of intermolecular forces between the water molecules and would include contributions from van der Waals forces and dipole-dipole forces as well as hydrogen bonding.

STUDY TASK

In this study task you will estimate the hydrogen bonding contribution in water.

1 The enthalpy changes of vaporization of the hydrides of the Group 6 elements are as follows:

Compound	ΔH_{vap} /kJ mol^{-1}
H_2O	40.7
H_2S	18.7
H_2Se	19.3
H_2Te	23.2

Plot a graph of ΔH_{vap} on the vertical axis against period number on the horizontal axis. Use it to estimate the value of ΔH_{vap} for water if there were no hydrogen bonding. Subtract this from the actual ΔH_{vap} to obtain a measure of the hydrogen bonding contribution.

Divide your answer by two to get an estimate of the strength of 1 mole of hydrogen bonds in water.

2 Do a similar estimation of the hydrogen bond strength in ammonia. Suitable data are:

Compound	ΔH_{vap} /kJ mol^{-1}
NH_3	23.4
PH_3	14.6
AsH_3	17.5

Hydrogen bonding in ice

Water molecules have **two** hydrogen atoms and **two** non-bonded electron pairs each and so can form an average of **two** hydrogen bonds each. There is, therefore, the possibility of water molecules being bound by hydrogen bonds into a three-dimensional lattice. This is so in ice, which usually has the 'wurtzite' structure (see zinc blende, colour plate 1 in the *Book of data*)

Although this structure accounts for very many of the properties of ice it does not account for all of them, and the structure of ice is not fully understood. The structure of water is even less certain. In water, the strong hydrogen bonding still succeeds in retaining parts of the three-dimensional lattice, and there is a short-range order but no long-range ordered structure.

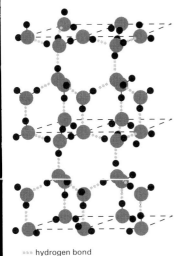

··· hydrogen bond

Figure 9.14 Water molecules in ice

STUDY TASK
What explanation can you offer for the following in terms of hydrogen bonding?
1 The way in which the densities of ice and water vary with temperature between −10 °C and +10 °C.

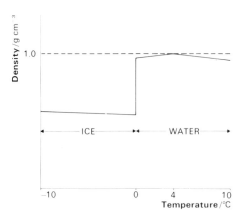

Figure 9.15 The density of water at different temperatures

2 The high surface tension of water which enables insects to 'walk' on the surface, and steel needles to float.

Figure 9.16 Using hydrogen bonding to walk on water

Hydrogen bonding in living organisms

In living processes hydrogen bonds are responsible for a very wide range of structural features and chemical processes, and the full extent is certainly greater than has so far been discovered.

To take one outstanding example, the highly complex structures of proteins are often maintained by means of hydrogen bonds. Enzymes are proteins, and depend for their action on the retention of highly specific molecular shapes. Consequently, the whole range of enzyme-catalysed reactions upon which life depends is determined by hydrogen bonding. The double helical structure of DNA (colour plate 3 in the *Book of data*) also depends on hydrogen bonds.

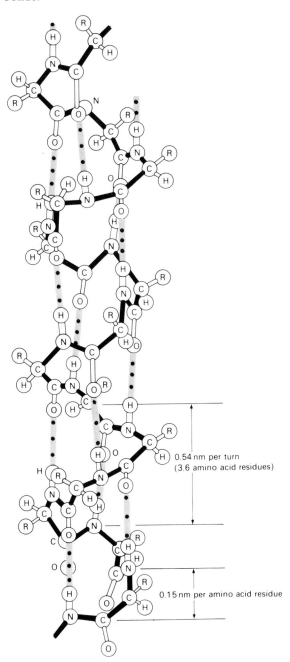

Figure 9.17 Hydrogen bonds hold the α-helix in shape

The solubility in water of organic compounds is due to hydrogen bonding involving electronegative atoms in the organic molecules and the oxygen and hydrogen atoms in water. An example is glucose, the large number of —OH groups per molecule giving rise to the possibility of extensive hydrogen bonding with water molecules, and hence solvation; an animal's ability to transport energy supplies rapidly to those parts of the body where they are needed depends upon the water-solubility of glucose.

Figure 9.18 The protein myoglobin has several lengths of α-helix in its structure. A haem group, with an iron atom at its centre, is held in place between the helices.

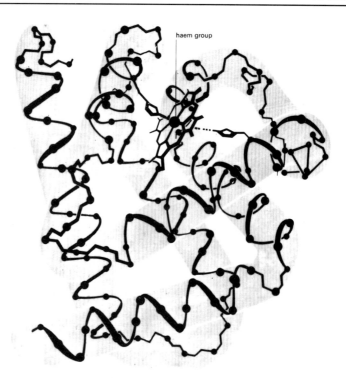

Fats, although relatively richer in energy than sugars, are not suitable for this purpose because they are not soluble in blood.

STUDY TASK

Cotton, hair, and wool absorb moisture while some synthetic materials such as nylon do not. Why should cotton, made of cellulose, be able to absorb moisture? Why should hair and wool which are protein fibres absorb moisture?

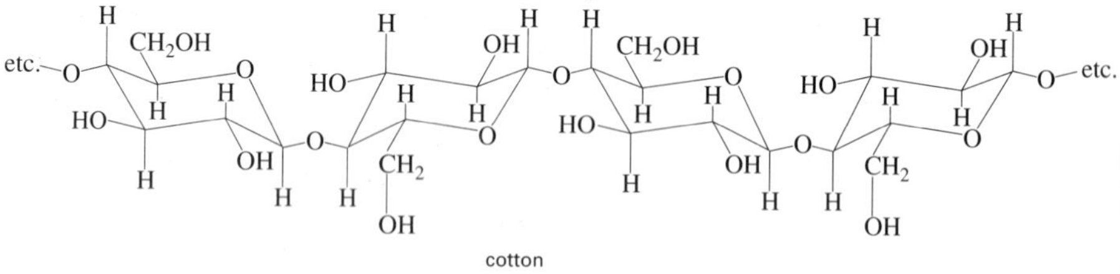

Figure 9.19 Cellulose chains can cross-link by hydrogen bonds, forming a fibrous material

9.4 The solubility of molecular compounds

We are now in a position to look for patterns in the behaviour of solutes and solvents. All types of compounds may be involved and all types of interaction are possible: covalent and ionic compounds; hydrogen bonding, dipole-dipole interactions, and van der Waals forces. We will start by looking at the mixing and dissolving of covalent compounds.

EXPERIMENT 9.4a The solubility of molecular compounds

You will be supplied with the following liquids: methylbenzene, hexane, methanol and water; and the following powdered solids: glucose and iodine.

Procedure

Remember **not** to wash out your apparatus with water during this experiment.
1. Test the solubility of iodine in each of the solvents.
2. Test the solubility of glucose in each of the solvents.
3. Test the solubility of each solvent in water.
4. Test all the other solvent pairs.

- Draw up a table of your results. In general, what type of liquids are good solvents for iodine, and what type are poor solvents? Does this correspond to the solubility of the solvents in each other? What type of solvent dissolves the polar glucose molecule?

SAFETY ⚠
All the organic solvents are flammable; methanol is toxic; iodine is corrosive. Do not dispose of these solvents in the laboratory sinks.

The pattern of solubility

The general pattern emerging from these experiments is that **non-polar liquids are good solvents for non-polar substances**, and **polar liquids are good solvents for polar substances**.

Iodine molecules are held to each other in the solid by weak van der Waals forces, and methylbenzene molecules are held to each other by the same type of forces. These are similar in magnitude to the iodine–methylbenzene van der Waals forces. Thus it is easy for methylbenzene molecules to penetrate into the iodine crystal and to separate the iodine molecules; similarly it is easy for iodine molecules to penetrate between the methylbenzene molecules. In this way the iodine dissolves.

Solvent	Solubility/g per 100 g of solution
Hexane	12
Methylbenzene	3.6
Methanol	8
Water	0.03

Solubility of iodine at 25 °C in various solvents

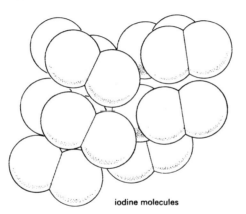
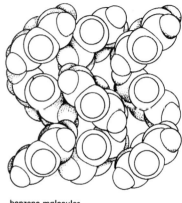

Figure 9.20 The crystal structures of iodine and benzene (benzene is comparable to methylbenzene)

iodine molecules | benzene molecules

Similarly, hexane–hexane, methylbenzene–methylbenzene, and hexane–methylbenzene attractions are weak and comparable to each other, and hexane dissolves in methylbenzene.

In liquids of moderate polarity such as methanol the situation is more complicated and solubility effects cannot be predicted with confidence.

For liquids of high polarity such as water a general pattern of behaviour can be seen. For water, the water–water attraction is much increased by the existence of hydrogen bonding; these water–water attractions are very much stronger than either the iodine–iodine attractions or the iodine–water attractions, and the iodine molecules cannot force their way into the water structure. Iodine is thus almost insoluble in water.

A similar situation exists for water and methylbenzene, and these two do not dissolve in each other.

On the other hand glucose and water can interact by forming hydrogen bonds that are comparable to the water–water and glucose–glucose hydrogen bonds, so glucose dissolves in water.

The dissolving of organic compounds in water

Water is a very good solvent for ionic substances. It is not, in general, a good solvent for dissolving molecular substances. Hydrocarbons, for example, do not mix with water but float on its surface. This is, of course, the reason why the spillage of oil from wrecked tankers into the sea is such a terrible problem.

Organic compounds consisting of small ions are likely to be soluble in water; thus sodium ethanoate, $CH_3CO_2^- Na^+$, and sodium butanoate, $CH_3CH_2CH_2CO_2^- Na^+$ are soluble.

A very wide range of non-ionic organic compounds is also soluble; but examination of their structures shows that they contain at least one polar group, and it is the polar group that confers water solubility.

Thus propylamine, $CH_3CH_2CH_2NH_2$, is very soluble in water; as in ammonia, the nitrogen end of the molecule is negative with respect to the other end, and a dipole exists.

$$CH_3CH_2CH_2\overset{\delta+}{-}\overset{\delta-}{N}\underset{H}{\overset{H}{\diagdown}}:$$

Not only can dipole-dipole interaction with the water result, but hydrogen bonding can also take place.

$$CH_3CH_2CH_2\overset{\delta+}{-}\overset{\delta-}{N}\cdots H - O\underset{H}{\diagup}$$

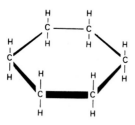

cyclohexane

glucose

Cyclohexane, a six-membered ring compound containing six carbon atoms, is insoluble in water. Yet glucose, a six-membered ring compound containing six carbon atoms, is very soluble in water. (See formulae in margin.) Why is this?

On this planet the solvent upon which life is based is water. Plant fluids, blood, and body fluids are aqueous solutions. Sugars, amino acids, peptides of low molar mass, and amines can be transported rapidly in a living organism because they are water-soluble. Without the polarity conferred upon carbon compounds by a nitrogen or an oxygen atom and the opportunity which this provides for hydrogen bonding with water, life could not have evolved as we know it.

STUDY TASK

To work out why hydrocarbons such as decane $C_{10}H_{22}$ do not mix with water we have to consider three things. What are the forces of attraction:

1 Between water molecules?
2 Between decane molecules?
3 Between water molecules and decane molecules?

If the strength of the attraction in **3** were similar to the attraction in **1** and **2** added together, the decane would be likely to mix with water. It is quite obvious however, that any possible attraction in **3** would be weak compared with **1** and **2** so that mixing does not occur. Apply this line of reasoning to explain the following situations:

a Glucose can be carried around the body in solution in the bloodstream (blood consists mostly of water).
b Iodine does not dissolve very well in water but dissolves freely in decane.
c Ethanol mixes with water in all proportions but cyclohexanol mixes only partially with water.

It must be borne in mind that the pattern of solubility which has emerged from this section, while true for many common substances, is untrue for an important number of special cases. The important criterion is that the new solute-solvent interactions must outweigh the existing solute-solute and solvent-solvent interactions if solution is to occur.

INVESTIGATION 9.4b An investigation of soluble laundry bags

Advertisements claim that laundry bags made of a suitable grade of polyvinyl alcohol, also named poly(ethenol), will dissolve in cold or hot water.

Poly(ethenol) is made from poly(ethenyl ethanoate) by a process known as 'ester exchange':

$$\left[\begin{array}{c}CH_3-CO_2 \quad CH_3-CO_2 \\ | \quad\quad\quad | \\ CH-CH_2-CH-CH_2\end{array}\right]_n + nCH_3OH \longrightarrow \left[\begin{array}{c}CH_3-CO_2 \quad OH \\ | \quad\quad\quad | \\ CH-CH_2-CH-CH_2\end{array}\right]_n + nCH_3CO_2CH_3$$

Solubility	Ester groups replaced /%
Soluble in cold water	<90
Needs warm water	90–96
Needs hot water	97–98
Insoluble in water	99–100

The product's solubility in water depends on the percentage of ester groups replaced.

SOLUBLE LAUNDRY BAG

FOR THE SAFE ISOLATION, TRANSPORTATION, DISINFECTION & CLEANING OF FOUL/INFECTED OR CONTAMINATED LINEN.

These Hot Water Soluble Laundry Bags have been developed and improved using the experience of the last 15 years, to produce the ultimate in efficient and practical laundry bags.

Only this amount of field testing can produce a consistent quality of material that is free from the problems of blocking and gelling and produces utterly reliable operating parameters, freeing the user from continuous monitoring of performance to ensure trouble free operation.

Manufactured from high grade Polyvinyl Alcohol to exacting standards, the bag has a deep embossed finish to ensure the bag does not cling and complicate filling. The attached cold water soluble Tie Tape offers convenient closure of the bag and allows rapid opening during the initial sluice, essential to prevent the setting of stains by unremoved body waste, blood* etc., that could occur during the following wash cycles.

Bacteriological Tests confirm that the bag material is impermeable to bacteria and thus provides an effective barrier to isolate Foul/Infected linen.

★ Hazard warnings can be printed on the bag to customer specifications.

★ Attached cold water soluble tie tape.

★ Embossed for easy handling.

★ PROVEN impermeable to bacteria.

Figure 9.21

Make a risk assessment before starting any experiments.

Procedure

Use samples of a laundry bag to investigate the rate of solution in water and the effect of factors that might have an influence.

Suggest why your bag material behaves the way it does, using the data provided.

9.5 The solubility of ionic compounds

As an aid to discussing this question we will start by investigating the energy changes taking place.

EXPERIMENT 9.5a Enthalpy changes of solution

You will be supplied with the following powdered solutes:
anhydrous potassium chloride, anhydrous calcium chloride, anhydrous iron(III) chloride.

The solids being tested absorb moisture from the atmosphere, so the stoppers of the stock bottles should be replaced securely immediately after use.

SAFETY
Calcium chloride and iron(III) chloride are irritant.

Procedure
Put about 5 cm³ of water in a test-tube and record the temperature. Add one of the solutes heaped on the end of a spatula, replace the thermometer and stir, and note the temperature of the solution. Repeat the experiment with the other two solutes, and record any temperature changes.

■ Draw up a table of your results. Is there any clear pattern in your results?

The dissolving of ionic solids

For an ionic solid to dissolve the crystal lattice must be broken down, and the positive and negative ions separated from each other. This requires a substantial amount of energy. When the ions are separated to an infinite distance apart the energy needed is the lattice energy; thus in the formation of a fairly dilute solution the separation of the ions from the lattice must involve the absorption of a quantity of energy almost equal to the lattice energy.

For the alkali metal halides the lattice energies are in the region of 600–800 kilojoules per mole; the absorption of so much energy from the solvent and its immediate surroundings would freeze most solvents. But the dissolving of many ionic solids in a solvent produces a rise in temperature, and when a drop in temperature is produced it is usually small. Some additional phenomenon which is exothermic must therefore be occurring and the energy released by it must be about the same in value as the lattice energy.

The most likely source of such a large quantity of energy is the occurrence of some form of bonding. You have seen that polar molecules are best for dissolving ionic compounds; the negative end of a polar solvent molecule can be attracted to a positive ion, and the positive end of a solvent molecule can be attracted to a negative ion, and thus ion-solvent molecule bonds are produced with a release of energy. This release of energy enables the ions to become detached from the lattice.

Figure 9.22 represents the process. Copy the diagram into your notes: positive ions may be coloured in red, the negative ions blue, and the polar solvent molecules some other suitable colour.

The attachment of polar solvent molecules to ions is known as **solvation**, and the ions are said to be **solvated**. When the solvent is water the process is known as **hydration**, and the ions are said to be **hydrated**. Figure 9.23 represents two hydrated ions, one positive and the other negative.

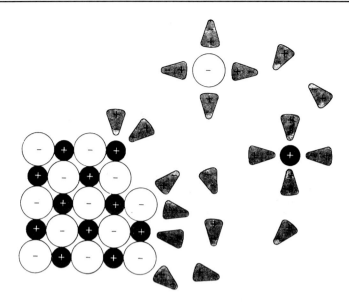

Figure 9.22 Solvent molecules dissolving an ionic solid

Energy changes and solution

When a mole of an ionic substance is formed in solution in water from its **elements** in their standard states the energy change is the **standard molar enthalpy change of *formation* of an aqueous solution**:

$$M(\text{element}) + X(\text{element}) + aq \longrightarrow M^+(aq) + X^-(aq)$$

Table 5.6 in the *Book of data* lists data for individual ions which can be added together to obtain values for ionic substances.

When a mole of an ionic substance **in the form of gaseous ions** dissolves in water the term used is **hydration energy**:

$$M^+(g) + X^-(g) + aq \longrightarrow M^+(aq) + X^-(aq)$$

Like lattice energies, neither of these can be readily measured in the laboratory.

What you can measure easily in the laboratory is the **enthalpy change of solution** when a mole of a substance dissolves:

$$M^+X^-(s) + aq \longrightarrow M^+(aq) + X^-(aq)$$

Enthalpy changes of solution can also be calculated by a Hess cycle (figure 9.23) using table 5.3 and 5.6 in the *Book of data*.

STUDY TASK

1 Using figure 9.23 and data from tables 5.3 and 5.6 in the *Book of data* calculate the enthalpy change of solution of the salts:

KCl   CaCl$_2$

2 Construct a Hess cycle using lattice energies and your results from **1** to calculate the hydration energies of the two salts.

What interpretation can you suggest for your results?

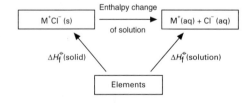

Figure 9.23

The structure of some hydrated ions

Hydrated iron(III) chloride crystals have the formula $FeCl_3.6H_2O$. Structure determinations have shown the compound consists of the ions $Fe(H_2O)_6^{3+}$ and $3Cl^-$, with the six water molecules distributed octahedrally.

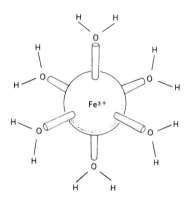

Figure 9.24 A hydrated iron(III) ion

Copy figure 9.24 into your notes and join the oxygen atoms by means of coloured lines to show the octahedral faces and the square plane. Use thick lines for edges close the observer and thinner ones for edges further away from the observer.

When iron(III) chloride is added to water the energy changes can be thought of as taking place in two stages. In the first stage the ions are hydrated

$$FeCl_3(s) + 6H_2O(l) \longrightarrow Fe(H_2O)_6Cl_3(s)$$

and in the second stage the hydrated salt dissolves in more water

$$Fe(H_2O)_6Cl_3(s) + aq \longrightarrow Fe(H_2O)_6Cl_3(aq)$$

6-coordination of water molecules around a positive ion is common in crystals, and 4-coordination is fairly common. Examples are $CrCl_3.6H_2O$ which contains $Cr(H_2O)_6^{3+}$ ions and $CuSO_4.5H_2O$ which contains $Cu(H_2O)_4^{2+}$ ions.

Solvent molecules coordinated around an ion can often be replaced by another type of polar molecule, or by another ion. For instance, when concentrated ammonia solution is added to copper sulphate solution a very deep blue coloured solution results containing $Cu(NH_3)_4^{2+}$ ions. The four ammonia molecules have replaced four water molecules, retaining a planar distribution.

$$Cu(H_2O)_4^{2+}(aq) + 4NH_3(aq) \longrightarrow Cu(NH_3)_4^{2+}(aq) + 4H_2O(l)$$

What shape is an ammonia molecule? Draw a diagram to represent the structure and orientation of the $Cu(NH_3)_4^{2+}$ ion.

INVESTIGATION 9.5b An investigation of an enthalpy change of hydration

Carry out an investigation to measure the enthalpy change of hydration of an anhydrous ionic salt. A suitable example is

$$Na_2S_2O_3(s) + 5H_2O(l) \longrightarrow Na_2S_2O_3.5H_2O(s)$$

Make a risk assessment before starting any experiments.

Summary

In every example in this Topic when a substance has dissolved in a liquid there has been an interaction between the substance and the solvent.

Without a solvent-solute interaction a substance cannot dissolve, but for solution this interaction must outweigh the previously existing forces of attraction **within the solvent** and also those **within the solute**.

Draw up a chart or poster in which you should aim to distinguish between
1 Polar and non-polar substances
2 Polar substances with hydrogen bonds and those without
3 Typical properties you can attribute to molecular interactions.

At the end of this Topic you should be able to

a demonstrate understanding of van der Waals forces and dipole-dipole interactions

b interpret using the concept of van der Waals forces and dipole-dipole interactions
   i properties which imply weak cohesive forces between all molecules
   ii increase in boiling point with increasing size among similar molecules, and with increasing surface area among isomers

c demonstrate understanding of hydrogen bonding, and identify the atoms involved in such bonding in specified cases

d interpret using the concept of hydrogen bonding
   i anomalous physical properties among the hydrides of the period 2 elements
   ii anomalous physical properties among organic compounds (including their infra-red spectra, see Topic 12)

e demonstrate understanding of the importance of hydrogen bonding in determining the structures of some materials, particularly ice and hydrated salts

f predict some of the properties of an unfamiliar substance which contains hydrogen bonds

g demonstrate understanding of solubility patterns in terms of the possible interactions between solute and solvent molecules

h calculate enthalpy changes of hydration and solution from experimental results and data.

Review questions

*9.1 Consider the three organic liquids tetrachloromethane, CCl_4
  octane, C_8H_{18}
  ethanol, C_2H_5OH.

a Which of the three has a permanent dipole?
b i Using the *Book of data*, arrange the three liquids in order of increasing boiling point.
  ii It is often supposed that hydrogen bonding is the strongest of the intermolecular forces; to what extent is this true for these three liquids?

9.2 Urea is soluble in water due to the formation of hydrogen bonds between its molecule and water molecules. Despite being a molecular substance, urea is a solid because of hydrogen bonding between the urea molecules.

a Draw displayed formulae to show
i a hydrogen bond between a water molecule and a urea molecule
ii a hydrogen bond between the same two molecules as in **i** but in a different pattern
iii a hydrogen bond between two molecules of urea.
b Urea and propan-1-ol both have a molar mass of 60 g mol^{-1} yet propan-1-ol is a liquid at room temperature whereas urea is a solid. Explain this in terms of intermolecular forces.

9.3 Arrange the following in the order you should expect for their boiling points, putting the one with the highest boiling point first.

a $CH_3-CH_2-CH_2-CH_3$

b $CH_3-\underset{\underset{CH_3}{|}}{\overset{\overset{CH_3}{|}}{C}}-H$

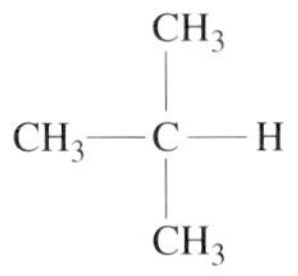

c $CH_3-CH_2-CH_2-CH_2Cl$

d $CH_3-\underset{\underset{CH_3}{|}}{\overset{\overset{CH_3}{|}}{C}}-Cl$

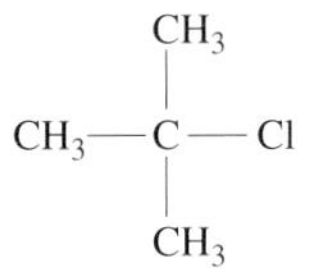

Give reasons for your answer.

9.4 Octane and 2-methylheptane are isomers with molecular formula C_8H_{18}.

a Which would have the higher boiling point? Explain your answer.
b Would either of them mix with water? Explain your answer.
c Would the two mix together? Explain your answer.

9.5 Classify the following mixtures of liquids into

i   Those with van der Waals forces between the two molecules
ii Those with van der Waals forces **and** dipole-dipole attraction between the two molecules
iii Those with van der Waals forces, dipole-dipole attraction **and** hydrogen bonding between the two molecules

a Propan-1-ol and butan-1-ol
b 1,1,2,2-tetrachloroethane and ethanol
c Methylbenzene and 1,2-dimethylbenzene
d Trichloromethane and dichloromethane
e Benzene and cyclohexane
f Bromobutane and ethanol.

9.6 Arrange each of the following groups of substances in the order you would expect for their boiling points, putting the liquid with the highest boiling point first.

a Neon, helium, argon
b Propane, butane, pentane
c Hydrogen fluoride, hydrogen chloride, sodium bromide
d Hydrazine, N_2H_4, disilane, Si_2H_6, diborane, B_2H_4
e Benzoic acid, 4-hydroxybenzoic acid, benzene.

Give reasons for your choice in each case.

9.7 Explain, in terms of intermolecular forces, why:

a Water has a higher boiling point than hydrogen fluoride although oxygen is less electronegative than fluorine.
b propan-1-ol has a higher boiling point than ethanol.
c Ethanol C_2H_5OH has a higher boiling point than thioethanol C_2H_5SH.
d Cyclohexane does not mix with water.
e 1 g of ice at $-1\,°C$ occupies more space than 1 g of water at $+1\,°C$.

Examination questions

9.8 Three liquids have the following properties:

	Water	Propanone	Pentane
Molecular formula	H_2O	C_3H_6O	C_5H_{12}
Molar mass/g mol^{-1}	18	58	72
Enthalpy change of vaporization/kJ mol^{-1}	41.1	31.9	27.7
Boiling point/°C	100	56	36

a What type of intermolecular force predominates in **each** liquid?
b What can you deduce about the relative strength of these forces in the liquids? Justify your conclusions.
c i If the liquids were shaken together in pairs, which pair would be unlikely to mix?
ii Explain this immiscibility in terms of the forces between the molecules.
d Choose ONE of the pairs that mix and say whether the enthalpy change on mixing would be exothermic or endothermic. Justify your answer.

9.9 This question is about hydrogen bonds and their effect on the properties of substances.

a Which of the following hydrides can form hydrogen bonds?
  i CH_4 **ii** NH_3 **iii** H_2O **iv** HF.
b Describe how the physical properties of hydrides alter down a group in the Periodic Table because of the presence or absence of hydrogen bonds.
c State two physical properties of water which can be attributed to hydrogen bonds.
d In what ways would you expect the properties of crystalline gypsum, $CaSO_4.2H_2O$, to differ from the properties of crystalline anhydrite, $CaSO_4$?

9.10 The following table lists the molar masses and boiling points of some organic compounds.

Compound	Molar mass/g mol^{-1}	Boiling point/K
Benzene	78	353
Methylbenzene	92	383
Phenol	94	454
Nitrobenzene	123	483
2-nitrophenol	139	494
4-nitrophenol	139	518

a What reason can you suggest for the **general** rise in boiling point with increasing molar mass of these compounds?
b Why would water not fit into the pattern of the table?
c i Draw the structural formulae of 2-nitrophenol and 4-nitrophenol.
ii Suggest a reason why the boiling point of 4-nitrophenol is higher than that of 2-nitrophenol.
iii Which of the two isomers is likely to be more soluble in water? Give a reason.

9.11 The energy changes which take place when an ionic solid, M^+X^-, dissolves in water can be represented by the energy cycle shown.

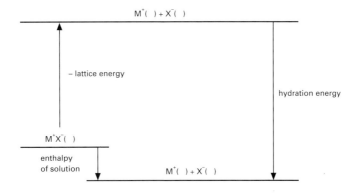

a Copy the diagram and add the appropriate state symbols.
b The lattice energies (in kJ mol^{-1}) of some ionic crystals are given in the table below.

	Anion		
Cation	F$^-$	Cl$^-$	Br$^-$
Li$^+$	-1031	-848	-803
Na$^+$	-918	-780	-742
Mg^{2+}	-2957	-2526	-2440

State the simple relationship which appears to exist between:
i The ionic radii of singly charged cations and anions and the lattice energies of their salts.
ii The size of the charge on cations and the lattice energies of their salts.
c The hydration energies, $\Delta H_{hydration}$, for a series of cations are listed below.

Cation	$\Delta H_{hydration}$/kJ mol^{-1}
Ca^{2+}	-1562
Cu^{2+}	-2069
Fe^{3+}	-4330

Account for the relative magnitudes of the $\Delta H_{hydration}$ values for:
i The Ca^{2+} and Cu^{2+} ions.
ii The Cu^{2+} and Fe^{3+} ions.
d Why are the ionic salts more likely to be soluble in propanone than in cyclohexane?

9.12 This question is about four liquids of which the formulae, molar masses and boiling points are given in the table:

		Molar mass /g mol^{-1}	Boiling point /°C
Butanone	$CH_3COCH_2CH_3$	72	80
1-aminobutane	$CH_3CH_2CH_2CH_2NH_2$	73	78
1,2-dibromoethane	CH_2BrCH_2Br	188	132
Pentane	$CH_3CH_2CH_2CH_2CH_3$	72	36

a i In which liquid would hydrogen bonding occur?
ii Draw the structural formulae of two molecules of this liquid, showing a hydrogen bond between them.
b Suggest a reason for the difference in boiling point between pentane and butanone.

9.13 Clathrates are materials in which molecules are trapped in 'cages' formed within the open crystal lattice structure of such substances as ice and 1,4-dihydroxybenzene, $C_6H_4(OH)_2$. Substances found in clathrates include some noble gases, methane and oxygen.

a i Draw a structural formula for 1,4-dihydroxybenzene.
ii Draw a diagram to show how strong intermolecular forces might arise between two adjacent molecules of 1,4-dihydroxybenzene. Mark and give the values of two different bond angles outside the benzene ring.
iii Explain, in terms of electronic structure, how water molecules combine to form an open crystal lattice structure in ice.
iv Describe how 1,4-dihydroxybenzene molecules might form a 'cage' to trap a molecule such as oxygen.
b Noble gases, except helium, also form clathrates with 1,4-dihydroxybenzene.
i Which type of intermolecular force is likely to be involved between the noble gas and the 1,4-dihydroxybenzene molecules in the clathrate?
ii Suggest why helium does not form such a clathrate.

*9.14 What are the characteristics of a solvent which is capable of dissolving a wide range of ionic compounds? Discuss how these characteristic properties facilitate the solubility of such compounds.

Using tables 5.3 and 5.6 in the *Book of data*, tabulate the standard enthalpy changes of **formation** of the solids and the solutions for fluorides of the Group 1 elements from lithium to caesium.

Use these data to calculate the enthalpy change of solution of each of the fluorides:

$$M^+F^-(s) + aq \longrightarrow M^+F^-(aq)$$

Give a brief explanation for any trends which are visible in the tabulated and calculated values.

Comment, in the light of the above, on the following solubility data:

Formula of substance	LiF	NaF	KF	RbF	CsF
Solubility measured in grams per 100 grams of water at 18 °C	0.27	4.22	92	130	367

9.15 The following substances either mix completely with water or dissolve very readily in it:

    ethanol, glucose, propylamine, sodium ethanedioate

On the other hand, the following substances are either sparingly soluble or insoluble in water:

    cyclohexane, heptan-1-ol, phenylamine, calcium ethanedioate

Discuss, with reference to these or other examples, some of the factors which determine whether an organic compound is likely to be soluble or insoluble in water.

9.16 Give an account of two of the following aspects of hydrogen bonding:

    Evidence for the existence of hydrogen bonds
    Measurement of the strength of a particular hydrogen bond
    Hydrogen bonding in non-metal hydrides.

*9.17 For each of the following substances discuss how the structure and bonding both have a significant influence on the properties of the substance:
    graphite
    ice
    poly(phenylethene)
    potassium bromide
Quote appropriate data from this book and the *Book of data* to illustrate your discussion.

TOPIC 10

Entropy

You may not have thought about it so far, but the chemical reactions you have met have taken place readily. The laboratory gas once lit will burn on and on, as it does at oil fields; as soon as a torch is switched on the chemicals in its batteries will react to produce an electric current.

Figure 10.1 The gas flare on a North Sea oil platform. For safety it burns continuously.

Sometimes addition of a catalyst has speeded up a reaction. Sometimes we have heated the reagents to start the reaction, but once started the reaction continued until the reagents were used up, without any further intervention by us. The reactions all 'go'.

There are many examples of events that happen naturally: cyclists can freewheel downhill; ice-cream melts in a warm room; the smell of cooking spreads through a building as the molecules diffuse naturally by their random chance mixing with the air.

Changes that tend to 'go' naturally are called spontaneous changes; once started they will go to completion. A spontaneous change can be quite slow, spontaneity is not a matter of speed.

10 Entropy

The reason why a chemical reaction 'goes' can be stated in a very short principle:

A reaction occurs when it is overwhelmingly probable by chance alone.

We shall show you that chance in chemistry operates predictably and quantitatively, which is extremely helpful. By knowing the rules by which chance operates in chemistry we can tell whether a reaction we wish to carry out will 'go' or not. We do not have to gamble; if the probability is favourable then provided we get the conditions right we will be able to obtain the product we want.

10.1 How does chance operate in chemistry?

If energy is to be shared by some molecules it **will be** shared, and in all possible ways. We will take a very simple case: an assembly of vibrating atoms making up a crystal. We will assume that each vibrating atom has one quantum of energy. This makes things simple and calculations show that it is a fairly accurate assumption at room temperature.

We will start with a system of 2 atoms and 2 quanta of energy: there are only 3 ways of arranging the quanta. For 3 atoms sharing 3 quanta there are 10 ways.

Ways of arranging quanta

Figure 10.2 Ways of arranging quanta (·) on atoms (□)

What about a system with a larger number of atoms and quanta? 36 atoms sharing 36 quanta will do. An arrangement that only exists in one way is to have one quantum on each atom. Rearranging the energy by chance – on the throw of dice – will eventually produce an arrangement that can exist in many ways.

The initial arrangement and a more probable arrangement, after 100 moves.

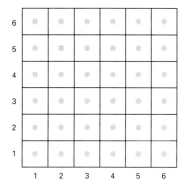 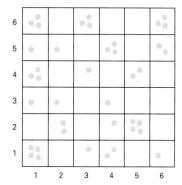

Figure 10.3 The initial arrangement and a more probable arrangement, after 100 moves

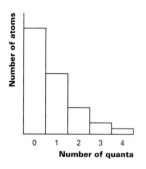

Figure 10.4 The pattern of atoms with differing number of quanta

The energy goes on moving around but the pattern of numbers of atoms with different energy stays much the same. This makes the pattern a probable arrangement whereas the initial arrangement can be described as improbable (it will only occur very rarely as the quanta move around).

The number of ways of arranging the energy will be many, many more if we have more energy levels for each atom than the simple vibrating atoms of our example.

> **MATHEMATICAL COMMENT**
>
> To take our example any further we need a general mathematical formula. One is available that you may meet in a mathematics course. Using W to represent the number of ways of arranging the energy, N to represent the number of atoms, and q to represent the number of quanta of energy, the formula is
>
> $$W = \frac{(N+q-1)!}{(N-1)!\,q!}$$
>
> So that for three atoms sharing three quanta of energy
>
> $$W = \frac{(3+3-1)!}{(3-1)!\,3!} = \frac{5!}{2!\,3!} = \frac{5 \times 4 \times 3 \times 2 \times 1}{2 \times 1 \times 3 \times 2 \times 1} = 10$$
>
> You can check this calculation and try out others with even larger numbers of atoms and quanta by using the factorial button, $x!$, on your calculator.
>
> For 100 atoms sharing 100 quanta
>
> $$W = \frac{199!}{99!\,100!} = 8 \times 10^{59}$$

A chemical system behaves in the same way: it will change from a system with a limited number of ways of arranging its energy to a system with the most ways of arranging that energy. And this will happen inevitably through random changes which occur by chance. The change occurs spontaneously.

Can we predict what type of chemical system is likely to have more ways of arranging its energy than another type? The answer is that it is not too difficult to make shrewd predictions about some systems. For example, gas molecules can move from place to place, vibrate and rotate. All these energies involve a considerable number of quantized energy levels. By contrast in a solid the particles will be restricted to mainly vibrational energy levels. And a liquid or solution is likely to have an intermediate number of ways of arranging its energy.

So chemical reactions which involve the spreading out of energy, as crystalline solids change to liquids or solutions, are probable changes. Reactions which produce gases are even more probable.

The trend is from order to disorder.

'Order' means few arrangements, 'disorder' means many arrangements. Let us see how these ideas might apply to some real reactions.

EXPERIMENT 10.1 Chance in chemical reactions

We are going to examine some chemical reactions. You are asked to estimate whether the possible arrangements of the product particles represents a more ordered or a less ordered system than the original reagents. You should also classify the reactions as exothermic or endothermic.

An increase in temperature is a quite common observation when a chemical change occurs, but decreases in temperature during spontaneous reactions are less frequently observed. Some of the examples are unusual but thinking about unusual examples is often the making or breaking of a chemical theory.

Read the Study Task at the end of these experiments before you begin.

Ethanoic acid is flammable and corrosive; ammonium nitrate is oxidizing; ammonium chloride is harmful; sodium nitrite and soluble barium salts are toxic.

Procedure

1 Dissolving

Measure 5 cm³ of water into a test-tube and record the temperature of the water. Add 5 g of solid ammonium nitrate, stir and note the change in temperature as the ammonium nitrate dissolves.

- Consider the surroundings. Did the surroundings gain or lose energy as the chemicals returned to room temperature after dissolving?
  Consider the ammonium nitrate–water system. Are the particles in ammonium nitrate solid, $NH_4^+ NO_3^-$, likely to be arranged in an orderly or disorderly way?
  Are the particles in a solution normally arranged in a regular or a random way?

2 A neutralization reaction

Measure 5 cm³ of pure ethanoic acid into a test-tube and record the temperature of the acid. Add 3 g of solid ammonium carbonate, stir and note the change in temperature as carbon dioxide is produced.

- Write a balanced equation for the reaction.
  Consider the surroundings. Did the surroundings gain or lose energy as the chemicals returned to room temperature?
  Consider the acid–base system. Do you think the product particles can be classified as more or less ordered than the reagent particles?

3 A combustion reaction

Hold a short length of magnesium ribbon in crucible tongs and light it in a Bunsen burner flame. Does it continue to burn when removed from the flame?

- Write a balanced equation for the reaction. Is the reaction exothermic or endothermic?
  Do you think the product, ionic $Mg^{2+}O^{2-}$, can be classified as more or less ordered than the reagents, metallic Mg and gaseous O_2?

4 A redox reaction

Mix roughly equal small amounts of solid ammonium chloride, NH_4Cl, and solid sodium nitrite, $NaNO_2$, in a small beaker. Add sufficient water to cover the solids and warm to dissolve them. Remove the Bunsen burner when an effervescence starts.

- What are the products? Hint: the gas evolved is inert.
  What changes are occurring in this reaction?
  Why is this reaction classified as a redox reaction?

5 An acid–base reaction

Mix 3 g of solid hydrated barium hydroxide, $Ba(OH)_2.8H_2O$, with 1 g of solid ammonium chloride, NH_4Cl. Note the temperature change as the mixture 'melts'.

- What changes are occurring in this reaction?
  Why is this reaction classified as an acid–base reaction?

6 A thermal decomposition

Heat a small amount of zinc carbonate. Test for the evolution of carbon dioxide and try to estimate the temperature at which decomposition can be detected.

- What changes are occurring in this reaction?

STUDY TASK

Draw up a table in your notes listing all the experiments and classify them as either tending to go or not taking place **at room temperature** (spontaneous or non-spontaneous). Reactions which are slow at room temperature but are helped by a catalyst or 'started off' by heating can be classified as spontaneous. Reactions which only go when **kept** at a high temperature should be classified as non-spontaneous.

Write a description of the spontaneous changes, comparing the degree of order in the reagents to the degree of order in the products.

State whether energy is transferred to the surroundings (an exothermic reaction) or to the reagents – the system (an endothermic reaction).

Can you find a pattern that describes all the spontaneous changes, or are there exceptions to any rule you propose?

Figure 10.5 Ludwig Boltzmann. He first saw the link between the laws of thermodynamics and the underlying behaviour of atoms and molecules. Boltzmann's work was rejected by most other scientists of the time and in 1906 he committed suicide, a disillusioned man.

Counting numbers of ways

The importance of energy distribution as the key to whether changes are spontaneous or not was first recognized by the German physicist Clausius and he introduced the term **entropy** as the unit by which energy distribution is measured.

But it was the Austrian scientist Boltzmann who worked out how to convert the **number of ways** atoms can share energy into **entropy units**. For an ideal gas counting the number of ways can be done, but what matters most to chemists are changes in the number of ways and measuring the amount of change in convenient units.

Boltzmann demonstrated that

$$S = k \ln W$$

where S is called the entropy of a substance
k is Boltzmann's constant
ln W is the natural logarithm of the number of possible arrangements

Our example of 100 atoms sharing 100 quanta is now more manageable, $\ln W$ is only 138. Check that $\ln(8 \times 10^{59})$ equals 138 by using the '**ln**' button on your calculator.

By making entropy, S, proportional to the natural logarithm of the number of ways Boltzmann made the amount of entropy directly proportional to the amount of substance in a system. Thus doubling the number of moles of a substance doubles $\ln W$ and doubles the amount of entropy.

The constant, k, is made equal to 1.381×10^{-23} J K^{-1} so that $\ln W$ is converted to joules and changes in entropy are made inversely proportional to temperature (as was deduced from experimental results). When calculating entropies chemists consider moles of substances.

The units of entropy per mole are therefore J mol^{-1} K^{-1}.

The relationship gives manageable values for entropies. For example, for a mole of atoms sharing 10^{23} quanta the entropy is

$$S = 11.4 \text{ J mol}^{-1} \text{ K}^{-1}$$

For a particular physical or chemical situation, the greater the number of possible arrangements, W, the more likely the situation is to come about. For any physical or chemical change which happens spontaneously, of its own accord, W must increase. This in turn means that **entropy must always increase in spontaneous changes**. Another way of putting this is to say that for any spontaneous change, the entropy change, ΔS, must be positive.

This is one of the cornerstones of our present understanding of how energy behaves and is referred to as the **Second Law of Thermodynamics**:

Spontaneous reactions go in the direction of increasing entropy

Entropy and our experiments

To apply the Second Law correctly to chemical reactions we need to be clear about the idea of a closed system and its surroundings, an idea we have already met in Topic 4. It may be helpful to repeat the main features again.

A **closed system** will consist of our reagents and, after a reaction, our products. A closed system is a theoretical model and does not change in temperature or pressure as a result of a reaction. Energy can transfer in or out, but no matter can transfer – it is 'closed' to matter.

The **surroundings** can transfer energy to or from the system, and are considered to be so large that this takes place without the surroundings altering in temperature or pressure.

We should now be able to see why it was so difficult to see any pattern to describe all the spontaneous changes that were identified in experiment 10.1.

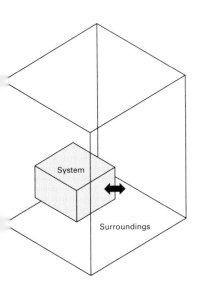

Figure 10.6 A closed system and its surroundings. Only energy transfers take place and only between the system and its surroundings.

Changes in the system

Molar amounts of solids, liquids and gases at the same temperature will in general follow a trend of increasing entropy, with a perfect crystal having relatively few ways of arranging its atoms and energy, a liquid having more ways and a gas having the most ways.

So the chance of a spontaneous reaction is increased when
    solute and solvent ⟶ solution
    solid ⟶ liquid ⟶ gas
    few gas molecules ⟶ many gas molecules

QUESTION
List the results of Experiment 10.1 and decide whether the entropy change in each *system* was positive or negative.

QUESTION
List the results of Experiment 10.1 and decide whether the entropy change in the *surroundings* was positive or negative.

COMMENT
The surroundings are so large that they will always have more ways of arranging energy than any system. Clausius recognised the ultimate size of the surroundings in his remark:

The entropy of the universe tends to increase.

Changes in the surroundings

Most of the entropy changes in the systems of experiment 10.1 were positive but for the combustion of magnesium (a common type of reaction) it looks likely that the entropy change in the system was negative. But nevertheless the combustion occurred. Why?

The clue to the answer lies in the enthalpy change for the combustion: the reaction is exothermic. When the combustion occurred, energy was given out and passed to the **surroundings** – the air, container, or whatever.

Now, this energy will have had an effect on the entropy of the surroundings. Extra quanta of energy were being made available, and of course this increases the number of ways energy quanta can be arranged among the molecules in the surroundings. In other words, an exothermic change increases the entropy of the surroundings, and the change is positive. As we shall see later, entropy changes in the surroundings can be calculated.

To summarize:

1 When a reaction is **exothermic** energy is transferred **to** the surroundings. This increases the number of ways of arranging the energy and corresponds therefore to an **increase in entropy**. This increases the probability of a change occurring spontaneously.

2 When a reaction is **endothermic** energy is transferred **from** the surroundings. This reduces the number of ways of arranging the energy and corresponds to an **decrease in entropy**. This decreases the probability of a change occurring spontaneously.

> **So the chance of a spontaneous reaction is increased when reactions are exothermic**

The total entropy change

You should now have realized that the pattern that describes our spontaneous changes is that they all corresponded to an increase in the number of ways of arranging energy and particles **in the system and surroundings taken together**. In other words, there was always an overall increase in entropy.

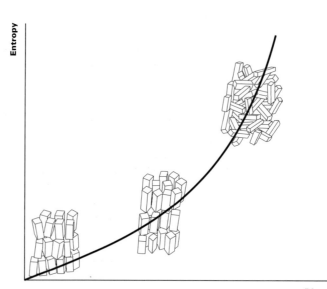

Figure 10.7 Increasing disorder means increasing entropy

When we want to consider the total entropy change of any process, we will always have to take into account what happens to the surroundings, $\Delta S_{surroundings}$, as well as the entropy change of the chemicals themselves, ΔS_{system}. In other words

$$\Delta S_{total} = \Delta S_{system} + \Delta S_{surroundings}$$

This a useful way of looking at changes in entropy when we are interested in chemical reactions. For spontaneous changes the criterion is

ΔS_{total} is positive

For reactions that do not 'go' the opposite is true and ΔS_{total} is negative.

Figure 10.8 How can the total entropy change be positive in the process of transforming a heap of stones from disorder to order?

10.2 Measuring entropy

Before we can make progress in our use of the idea of entropy we need to know how to calculate entropy changes in the surroundings and in systems.

Calculating the entropy change in the surroundings

Chemical reactions usually involve quite large enthalpy changes. Therefore in most reactions $\Delta S_{surroundings}$ is quite substantial and certainly cannot be ignored. But how can we calculate its value? We cannot possibly count all the ways, W, of sharing the energy among the surroundings – for one thing, the surroundings are impossible to define exactly.

Fortunately, there is a simple relation which enables us to calculate $\Delta S_{surroundings}$ exactly. It is

$$\Delta S_{surroundings} = \frac{-\Delta H_{reaction}}{T}$$

where T is the temperature in kelvins of the surroundings. Clausius proposed a form of this relationship on the basis of experimental measurements but its application to reactions is due to the work of Gibbs.

When a reaction is exothermic a quantity of energy, $\Delta H_{\text{reaction}}$, is passed to the surroundings: and the more energy passed to the surroundings, the greater the increase in the number of ways of sharing energy, and so the greater the entropy change.

But why divide by T? Well, when T is high the surroundings are already hot, so giving them more energy will not make much difference to the entropy – there is already plenty of energy to share.

But when T is low the surroundings are cold, so passing energy to them will make a bigger difference to the entropy, and will greatly increase the sharing possibilities. In other words, the entropy **change** will vary **inversely with temperature**. That is why we divide by T.

After all, someone of a low salary appreciates a £10 rise more than someone on a high one. But both would prefer a £20 rise to a £10 one.

Let us apply this relationship to the combustion of magnesium that we looked at in experiment 10.1, part **3**:

$$2\text{Mg(s)} + \text{O}_2(\text{g}) \longrightarrow 2\text{MgO(s)}$$

The standard enthalpy change of formation at 298 K is

$$\Delta H_f^{\ominus}[\text{MgO}] = -601.7 \text{ kJ mol}^{-1}$$

therefore $\Delta H_{\text{reaction}}^{\ominus} = -2 \times 601.7 \text{ kJ mol}^{-1}$

and $\Delta S_{\text{surroundings}}^{\ominus} = \dfrac{-\Delta H_{\text{reaction}}^{\ominus}}{T} = \dfrac{-(-2 \times 601.7 \times 1000)}{298}$ J mol^{-1} K^{-1}

$= +4038$ J mol^{-1} K^{-1}

This large positive value makes it likely that that $\Delta S_{\text{total}}^{\ominus}$ will be positive and the reaction will be spontaneous at 298 K once started, as we actually found.

Figure 10.9 Rudolf Clausius, who invented the term 'entropy' in 1865

Standard entropy of substances

Before we start calculating entropy changes in systems we need to look at the entropy values of some substances.

It has been said that the best way to become acquainted with birds is to look at some birds. And a good way to become acquainted with entropy is to look at some values. The list alongside gives some typical standard entropies at 298 K.

What useful comments can be made about the entropy values in the table?

Notice that hard structures generally have smaller entropies than soft structures:

- $S^{\ominus}[\text{C, diamond}]$ is less than $S^{\ominus}[\text{C, graphite}]$
- $S^{\ominus}[\text{Mg}]$ is less than $S^{\ominus}[\text{Pb}]$

and simple substances generally have smaller entropies than complex substances:

- $S^{\ominus}[\text{MgO}]$ is less than $S^{\ominus}[\text{MgCO}_3]$

and solids have smaller entropies than liquids, which are generally smaller than gases:

- $S^{\ominus}[\text{MgO}]$ is less than $S^{\ominus}[\text{H}_2\text{O}]$, which is less than $S^{\ominus}[\text{O}_2]$

Entity	$S^{\ominus}$/J mol^{-1} K^{-1}
C (diamond)	2.4
C (graphite)	5.7
Mg	32.7
Pb	64.8
MgO	26.9
MgCO$_3$	65.7
H$_2$O(l)	69.9
C$_2$H$_5$OH(l)	160.7
He	126
NH$_3$	192.3
O$_2$	205

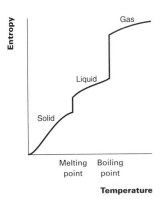

Figure 10.10 The entropy of a substance increases as the temperature rises; and increases in a jump when a substance melts or boils.

In hard solids the crystals have rigid structures with stiff bonds and the thermal motion of the atoms is very restricted: the entropy is correspondingly small. In soft solids there is rather more thermal disorder and entropy is larger.

For liquids and gases there is much more disordered motion and entropy values are much larger.

COMMENT

There are two possible approaches to determining the entropy content of a substance. A statistical approach is to count the actual number of ways, but this approach is only possible for very simple substances. A thermodynamic approach is to measure the effect of transferring energy to the substance. In order to do this we have to have a starting point and this is provided by the **Third Law of Thermodynamics**.

All perfect crystals have the same entropy at absolute zero.

The convention adopted by chemists is that this entropy value is zero. So our starting point is one mole of a perfect crystal at absolute zero. A close approach to this situation would be a flawless diamond weighing 12 g cooled in solid helium:

$S_0[\text{C, diamond}] = 0$

By how much does the entropy increase as the diamond warms up from zero K to 298 K, the standard temperature? This is actually quite easy in principle because we can apply Clausius's relationship to the process:

$\Delta S[\text{C, diamond}] = \dfrac{\dot{q}}{T}$ where q is the energy transferred reversibly from the surroundings

We need to know the energy transferred from the surroundings to the diamond at each stage of the warming process. We have to monitor continuously the temperature rise and the joules being transferred: 1 joule transferred at 10 K has a much greater effect on the number of arrangements than 1 joule transferred at 100 K.

Figure 10.11 A rough diamond crystal (12 g is 1 mol)

Calculating the entropy change in the system

It now becomes possible to calculate entropy changes in the system for chemical reactions, $\Delta S^{\ominus}_{\text{system}}$.

As an example we will again use the reaction of magnesium with oxygen.

Once ignited, the magnesium burnt spontaneously in air. A mole of oxygen molecules, O_2, was used up for every two moles of $Mg^{2+}O^{2-}$ formed. There are fewer possible arrangements of the energy in the solid product than in the gaseous reactant, so we might expect the entropy to decrease, giving a negative value for $\Delta S^\ominus_{system}$:

$$2Mg(s) + O_2(g) \longrightarrow 2MgO(s)$$

To work out the exact value of the entropy change of the system for this reaction, we need to look up the standard entropy values of the reactants and products in tables 5.2 and 5.3 of the *Book of data*.

The standard entropy values at 298 K are:

$S^\ominus[Mg(s)] = +32.7 \text{ J mol}^{-1}\text{ K}^{-1}$
$S^\ominus[\frac{1}{2}O_2(g)] = +102.5 \text{ J mol}^{-1}\text{ K}^{-1}$
$S^\ominus[MgO(s)] = +26.9 \text{ J mol}^{-1}\text{ K}^{-1}$

Notice that the values quoted for elements in table 5.2 are **per mole of atoms**. Notice also how much higher the standard entropy of the gas, oxygen, is than that of the solids, magnesium and magnesium oxide.

We can calculate $\Delta S^\ominus_{system}$ for the reaction in much the same way as we found $\Delta H^\ominus$ values using an energy cycle in Topic 4.

$$2Mg(s) + O_2(g) \longrightarrow 2MgO(s)$$
$$2 \times 32.7 \quad 2 \times 102.5 \quad 2 \times 26.9$$

$$\Delta S^\ominus_{system} = S^\ominus_{products} - S^\ominus_{reactants}$$

Therefore $\Delta S^\ominus_{system} = 2S^\ominus[MgO(s)] - 2S^\ominus[Mg(s)] - 2S^\ominus[\frac{1}{2}O_2(g)]$

$$= +(2 \times 26.9) - (2 \times 32.7) - (2 \times 102.5)$$
$$= -216.6 \text{ J mol}^{-1}\text{ K}^{-1}$$

$\Delta S^\ominus_{system}$ is indeed negative, showing that the entropy of the reaction system has decreased. The reaction nevertheless goes spontaneously. Why? Because there is also the entropy change in the surroundings to be considered.

The total entropy change

We know from experiments that the combustion of magnesium is spontaneous, but do our calculations say the same? We can find out by using the relationship

$$\Delta S^\ominus_{total} = \Delta S^\ominus_{system} + \Delta S^\ominus_{surroundings}$$

with the values we have already calculated for the burning of magnesium

$\Delta S^\ominus_{system} = -216.6 \text{ J K}^{-1}\text{ mol}^{-1}$
$\Delta S^\ominus_{surroundings} = +4038 \text{ J K}^{-1}\text{ mol}^{-1}$

Therefore $\Delta S^\ominus_{total} = -216.6 + 4038 \text{ J mol}^{-1}\text{ K}^{-1}$
$$= +3821.4 \text{ J mol}^{-1}\text{ K}^{-1}$$

and the reaction should go spontaneously at 298 K.

This positive value for $\Delta S^\ominus_{total}$ tells us that the reaction should happen spontaneously. The example warns us of one limitation of 'spontaneous', namely that a spontaneous reaction may be very slow at 298 K.

> **COMMENT**
> The appropriate answer for the total entropy change is 3820 J mol^{-1} K^{-1} (to 3 SF).

The entropy change of an endothermic reaction

You should now be able to calculate the entropy change at 298 K for another reaction you carried out in experiment 10.1 (part **5**).

$$2NH_4Cl(s) + Ba(OH)_2(s) \longrightarrow BaCl_2.2H_2O(s) + 2NH_3(g)$$

You will have to start by looking up the $S^\ominus$ and $\Delta H_f^\ominus$ values at 298 K for all the reactants and products. You then have to work out $\Delta S_{system}^\ominus$ for the reaction using the relationship:

$$\Delta S_{system}^\ominus = S_{products}^\ominus - S_{reactants}^\ominus$$

when you should get the value

$$\Delta S_{system}^\ominus = +298.6 \text{ J mol}^{-1}\text{ K}^{-1}$$

$\Delta H_{reaction}^\ominus$ is calculated using a similar energy cycle:

$$\Delta H_{reaction}^\ominus = \Delta H_{f\ products}^\ominus - \Delta H_{f\ reactants}^\ominus$$

when you should get the value

$$\Delta H_{reaction}^\ominus = +21.2 \text{ kJ mol}^{-1}$$

and from $\Delta H_{reaction}^\ominus$ you calculate $\Delta S_{surroundings}^\ominus$ using the relationship

$$\Delta S_{surroundings}^\ominus = \frac{-\Delta H_{reaction}^\ominus}{T}$$

when you should get the value

$$\Delta S_{surroundings}^\ominus = -71.1 \text{ J mol}^{-1}\text{ K}^{-1}$$

Finally obtain a value (to 3 SF) for $\Delta S_{total}^\ominus$ by using the relationship

$$\Delta S_{total} = \Delta S_{system} + \Delta S_{surroundings}$$

with the values you should have obtained:

$$\Delta S_{total}^\ominus = +298.6 - 71.1 \text{ J mol}^{-1}\text{ K}^{-1}$$
$$= +228 \text{ J mol}^{-1}\text{ K}^{-1}$$

and the reaction will go spontaneously at 298 K.

The experiment you did matches this theoretical prediction. We have demonstrated that entropy calculations match our practical experience, which is a test all scientific theories must pass.

Notice that the production of ammonia gas with its large entropy plays a major role in making this reaction go spontaneously.

The decomposition of carbonates

You studied the decomposition of zinc carbonate in experiment 10.1 (part **6**).

$$ZnCO_3(s) \longrightarrow ZnO(s) + CO_2(g)$$

When you do the necessary calculations you should be able to confirm that

$$\Delta S_{system}^\ominus = +174.8 \text{ J mol}^{-1}\text{ K}^{-1}$$

and $\Delta S_{surroundings}^\ominus = -238 \text{ J mol}^{-1}\text{ K}^{-1}$

so that $\Delta S_{total}^\ominus = -63 \text{ J mol}^{-1}\text{ K}^{-1}$

So the reaction will not go spontaneously at 298 K. But you may remember that most carbonates decompose when heated. How does this happen when we have confirmed that the entropy change is negative? We will leave this question until we next consider entropy.

QUESTIONS

1 Carry out the complete calculation for the endothermic reaction of barium hydroxide with ammonium chloride in order to confirm the values quoted in the text.
2 Carry out the complete calculation for the decomposition of zinc carbonate in order to confirm the values quoted in the text.
3 In Topic 1 you studied the thermal decomposition of iron(II) sulphate and in Topic 4 calculated the standard enthalpy change of reaction for:

$$2FeSO_4 \cdot 7H_2O(s) \longrightarrow Fe_2O_3(s) + SO_2(g) + SO_3(g) + 14H_2O(l)$$
$$\Delta H^\ominus_{reaction} = +194.8 \text{ kJ mol}^{-1}$$

Which entropy values are mainly responsible for this reaction being spontaneous when heated, in spite of it being endothermic? What are the differences between the molecules that might account for this entropy effect?

HISTORY

Chemists were not measuring heating and cooling changes in reactions properly until after the work of Hess, published in 1840. A Dutchman, van't Hoff, was the first chemist whose work on the energy relationships in chemistry, published in 1894, was widely understood and accepted. His theories had already been published by other chemists, especially by the American Willard Gibbs in the period 1876–78, but the subtlety of the ideas resulted in them being not understood and therefore neglected. Gibbs's work became more widely known after being translated into French by Le Châtelier in 1899.

Finally the Austrian scientist Ludwig Boltzmann was able to show how atomic theory and energy theory could be combined to produce an overall interpretation of the direction of spontaneous change.

Many distinguished scientists in that period, such as Kelvin, regarded aspects of atomic theory as useful fictions, not to be taken too seriously. It was Einstein's explanation of Brownian motion, confirmed experimentally by Perrin, that provided support for the atomic theory. Perrin's book on Brownian motion, published in 1913, was perhaps the first scientific best-seller, 30 000 copies being sold.

We will meet the ideas of many of these scientists again in other Topics, as their work is commemorated in the Boltzmann constant, Le Châtelier's principle, and Gibbs free energy.

Figure 10.12 Boltzmann's tombstone

Summary

We can summarize what we have learnt about the values of entropy changes as

$\Delta S_{surroundings}$	ΔS_{system}	ΔS_{total}
+	+	+
+	−	probably +
−	+	probably −
−	−	−

Notice that $\Delta S_{surroundings}$ is the most important factor in determining whether a reaction will 'go' or not. This is why we find that so many reactions that 'go' are exothermic, and endothermic reactions that 'go' are rare.

At the end of this Topic you should be able to
a interpret the natural direction of change (spontaneous change) as the direction of increasing number of ways of sharing energy and therefore of increasing entropy (positive entropy change)
b recall that the entropy change in any reaction is made up of the entropy change in the system added to the entropy change in the surroundings, summarized by the expression

$$\Delta S_{total} = \Delta S_{system} + \Delta S_{surroundings}$$

c recall the factors affecting the standard entropy of a substance, in particular its physical state, and predict the relative entropies of different substances (qualitatively only)
d calculate the standard entropy change in the system for a stated chemical reaction using standard entropy data
e recall the expression $\Delta S_{surroundings} = -\Delta H_{reaction}/T$ and use it to calculate entropy changes in the surroundings and hence calculate ΔS_{total}
f recall that the feasibility of a reaction depends on the balance between ΔS_{system} and $\Delta S_{surroundings}$.

Review questions

* Indicates that the *Book of data* is needed.

*10.1 Consider the reaction $C(s) + O_2(g) \longrightarrow CO_2(g)$

a Use the *Book of data* to find the standard molar entropies of $C(s)$, $O_2(g)$, and $CO_2(g)$. Use these data to calculate the standard molar entropy change, $\Delta S^{\ominus}_{reaction}$, for the reaction.
b Use the *Book of data* to find the standard enthalpy change, $\Delta H^{\ominus}(298)$, for the reaction. Use your answer to calculate the entropy change in the surroundings, $\Delta S^{\ominus}_{surroundings}$, at 298 K for the reaction.
c Calculate the total entropy change, $\Delta S^{\ominus}_{total}$, for the reaction.
d Does your answer to c suggest this reaction will be spontaneous at 298 K?

10.2 Calculate $\Delta S^\ominus_{\text{surroundings}}$, at 298 K, for the following reactions using the values for the enthalpy changes, ΔH, that you calculated in Topic 4, question 4.

a $SO_2(g) + 2H_2S(g) \longrightarrow 3S(s) + 2H_2O(l)$
b $N_2O(g) + Cu(s) \longrightarrow CuO(s) + N_2(g)$
c $NH_4Cl(s) \longrightarrow NH_3(g) + HCl(g)$
d $Mg(s) + \frac{1}{2}O_2(g) \longrightarrow MgO(s)$
e $H_2(g) + S(s) \longrightarrow H_2S(g)$
f $CO_2(g) + 2Mg(s) \longrightarrow 2MgO(s) + C(s)$
g $Na(s) + \frac{1}{2}Cl_2(g) \longrightarrow NaCl(s)$

*10.3 Use the *Book of data* to calculate $\Delta S^\ominus_{\text{system}}$ at 298 K for the reactions in **10.2**.

10.4 a Use your answers to **10.2** and **10.3** to calculate $\Delta S^\ominus_{\text{total}}$ at 298 K.
b Which reactions will be spontaneous at 298 K?

10.5 Suppose you were given the total entropy changes, $\Delta S^\ominus_{\text{total}}$, for the following reactions.

$$2Fe(s) + 1\tfrac{1}{2}O_2(g) \longrightarrow Fe_2O_3(s)$$
$$Ca(s) + \tfrac{1}{2}O_2(g) \longrightarrow CaO(s)$$

What further information, if any, would you require in order to calculate the the total entropy changes of each of the following reactions?

a $3Ca(s) + Fe_2O_3(s) \longrightarrow 3CaO(s) + 2Fe(s)$
b $Ca(s) + CuO(s) \longrightarrow CaO(s) + Cu(s)$
c $2Fe(s) + 3CuO(s) \longrightarrow Fe_2O_3(s) + 3Cu(s)$

Examination questions

10.6 The apparatus shown in the margin was used to measure the enthalpy change for the reaction:

$$NiSO_4(aq) + Zn(s) \longrightarrow Ni(s) + ZnSO_4(aq)$$

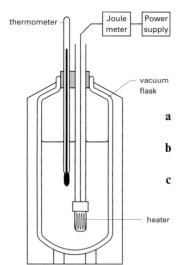

A slight excess of zinc metal was added to 100 cm³ of 0.200 mol dm⁻³ nickel(II) sulphate solution in the flask at room temperature and the maximum temperature rise was noted. The flask was then washed out, filled with 100 cm³ of water at room temperature and electrical energy supplied until the temperature rise was the same as before. 1700 J was recorded on the joulemeter.

a How many moles of nickel(II) sulphate are present in 100 cm³ of the 0.200 mol dm⁻³ solution?
b What mass of zinc must be added if there is to be a 10% excess? (Molar mass: Zn = 65.4 g mol⁻¹)
c Calculate the enthalpy change when one mole of zinc reacts.
(*Continued*)

d i Calculate $\Delta S^\ominus_{\text{surroundings}}$ at 298 K.
ii Calculate $\Delta S^\ominus_{\text{system}}$ at 298 K using the data

$S^\ominus[\text{Zn(s)}] = +41.6 \text{ J mol}^{-1}\text{ K}^{-1}$
$S^\ominus[\text{Zn}^{2+}(\text{aq})] = -112.1 \text{ J mol}^{-1}\text{ K}^{-1}$
$S^\ominus[\text{Ni(s)}] = +29.9 \text{ J mol}^{-1}\text{ K}^{-1}$
$S^\ominus[\text{Ni}^{2+}(\text{aq})] = -128.9 \text{ J mol}^{-1}\text{ K}^{-1}$

iii Suggest why your value for $\Delta S^\ominus_{\text{system}}$ is small.
iv Calculate $\Delta S^\ominus_{\text{total}}$ at 298 K and comment on the significance of your value.

10.7 When 1 mole of rubidium chloride is dissolved in water at 298 K to form a solution of concentration 1 mol dm^{-3}, the enthalpy change is $+19$ kJ mol^{-1}:

$$\text{RbCl(s)} + \text{aq} \longrightarrow \text{Rb}^+(\text{aq}) + \text{Cl}^-(\text{aq}) \qquad \Delta H^\ominus = +19 \text{ kJ mol}^{-1}$$

a Calculate the entropy change in the surroundings when this process takes place.
b Calculate the entropy change in the system for this process from the data:

$S^\ominus[\text{RbCl(s)}] = +95.9 \text{ J mol}^{-1}\text{ K}^{-1}$
$S^\ominus[\text{Rb}^+(\text{aq})] = +121.5 \text{ J mol}^{-1}\text{ K}^{-1}$
$S^\ominus[\text{Cl}^-(\text{aq})] = +56.5 \text{ J mol}^{-1}\text{ K}^{-1}$

c Use the results of your calculations to explain why rubidium chloride dissolves readily in water in spite of this being an endothermic process.

*10.8 This question concerns the standard enthalpy change of atomization of ethanol at 298 K.

$\Delta H^\ominus_{\text{at}}[\text{C}_2\text{H}_5\text{OH(l)}]$

You will need to refer to the *Book of data* for the values of some of the enthalpy changes required. The following diagram may be helpful.

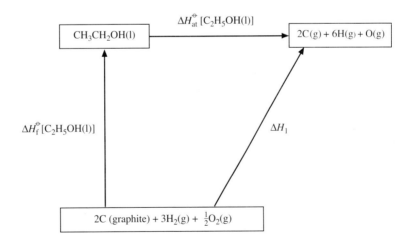

a i What are the conditions for **standard** enthalpy changes of atomization?
ii What does $\Delta H^\ominus_{\text{f}}[\text{C}_2\text{H}_5\text{OH(l)}]$ represent?
iii Calculate ΔH_1 using the standard enthalpy changes of atomization of elements given in the *Book of data*. (Remember these values apply to the production of 1 mol of gaseous atoms of the element.)

iv It is usual to write
$$\Delta H_1 = \Delta H_f^\ominus[C_2H_5OH(l)] + \Delta H_{atl}^\ominus[C_2H_5OH(l)]$$
How would you justify this expression?

v Use the expression to calculate the standard enthalpy change of atomization of ethanol.

b i Calculate $\Delta S^\ominus_{surroundings}$ for the atomization of ethanol at 298 K.

ii Deduce the sign of $\Delta S^\ominus_{system}$ for the atomization of ethanol. Justify your deduction.

iii At 298 K ethanol does not atomize spontaneously as $\Delta S^\ominus_{total}$ for atomization is negative. Suggest conditions in which the atomization of ethanol might occur.

10.9 This question is concerned with the element titanium, Ti, and with some of its compounds.

a i State the electronic configuration of a titanium atom in its ground state.

ii Is titanium likely to display more than one oxidation number in its compounds? Justify your answer.

b Titanium occurs naturally as the mineral rutile, TiO_2. One possible method suggested for the extraction of the metal is to reduce the rutile by heating it with carbon:

$$TiO_2(s) + 2C(s) \longrightarrow Ti(s) + 2CO(g)$$

i Calculate $\Delta H^\ominus_{surroundings}$ for this reaction given that at 298 K
$\Delta H_f^\ominus[TiO_2(s)] = -940 \text{ kJ mol}^{-1}$ and $\Delta H_f^\ominus[CO(g)] = -110 \text{ kJ mol}^{-1}$.

ii Calculate $\Delta S_{surroundings}$ at 2200 K for this reaction using your value for ΔH at 298 K.

iii Using $\Delta S_{system} = +365 \text{ J K}^{-1} \text{ mol}^{-1}$ calculate ΔS_{total} at 2200 K.

iv Is this reaction feasible at 2200 K? Justify your answer.

c Explain the pollution problem which might be caused by this process.

*10.10 Butane is often used as the fuel for gas stoves used by campers. A full gas cylinder contains 2.5 kg of butane. The standard enthalpy change of combustion of butane at 298 K is $\Delta H_c^\ominus(298) = -2877 \text{ kJ mol}^{-1}$.

a i Explain the meaning of the symbol $\Delta H_c^\ominus(298)$.

ii Write a balanced equation for the combustion of butane, including state symbols.

iii Using your *Book of data*, calculate the number of moles of butane in a full gas cylinder.

iv Hence calculate the enthalpy change which should occur if all the butane in a full cylinder were burned.

v Suggest two reasons why this amount of energy is unlikely to be available to heat food on a stove.

vi Suggest another fuel which might be used by campers. Briefly discuss its advantages and disadvantages compared to butane.

c The standard entropy change of combustion of butane at 298 K is
$$\Delta S^\ominus_{system} = +228 \text{ J mol}^{-1} \text{ K}^{-1}$$

i Why does $\Delta S^\ominus_{system}$ have a positive sign?

ii Calculate $\Delta S^\ominus_{surroundings}$.

iii Explain why the reaction is feasible.

TOPIC 11

How far? Reversible reactions

In this Topic we shall explore the common features of reversible reactions and demonstrate how the Equilibrium Law can be used to make quantitative deductions about their behaviour. In particular we shall study the factors that influence the pH of solutions of acids and bases.

Many important reactions used to manufacture useful materials are reversible, for example the production of ammonia for fertilizers and of esters for perfumes and solvents, so a good understanding of reversible reactions is essential in the chemical industry. Also an understanding of the acid–base balance in our bodies helps in the formulation of drugs.

REVIEW TASK
In small groups, write down what you can recall about reversible reactions

Figure 11.1 The manufacture of ammonia is a reversible reaction carried out on a very large scale (Norway).

11.1 Reversible reactions

You are going to start by thinking about some examples of reversible processes in order to be clear about what is happening at the molecular level.

EXPERIMENT 11.1

Reversible processes

SAFETY
Iodine is corrosive; hydrocarbons are flammable; sodium chromate(VI) and dichromate(VI) are irritant; sodium ethanedioate is toxic.

Procedure

1 Iodine, I_2, is only slightly soluble in water but much more soluble in aqueous potassium iodide solution. It is also soluble in hydrocarbons and other organic solvents.

a Use a spatula to put one very small piece of iodine into a test-tube. Add 5 cm³ of a hydrocarbon solvent and shake the test-tube gently (TAKE CARE: do not put your thumb over the mouth of the test-tube). Note the colour of the solution and its intensity.

Now add 5 cm³ of potassium iodide solution and shake again. How is the iodine colour distributed between the two solvents?

b Repeat **a** using a piece of iodine of similar size but add the solvents in the opposite order. Note the colours and their intensity.

- Is the final effect the same in both **a** and **b**? Do you think that iodine molecules are on the move even when no colour changes can be seen? Describe what you think is happening in terms of the movement of iodine molecules between the solvents when they are first added to the test-tubes.

2 Put two samples of aqueous iodine solution (in potassium iodide solution) into test-tubes: to one sample add dilute sulphuric acid and to the other add dilute sodium hydroxide, using a dropping tube.

Look up your notes on Topic 5 if you need help.

- Which sample appears to have reacted?
  Try to reverse the reaction: what are you going to add?
  Write an equation for this reversible reaction.
  Can you achieve a solution that has roughly half the original concentration of iodine?

3 Put two samples of sodium chromate(VI), Na_2CrO_4, solution into test-tubes: to one sample add dilute acid and to the other add dilute alkali.

Repeat the experiment using sodium dichromate(VI), $Na_2Cr_2O_7$, instead of chromate(VI).

- What do you think is happening? Find out if the changes you have seen are reversible.
  Write equations for the two reactions and then combine them in an equation for the reversible process.

4 Prepare three samples of solutions containing the deep red complex ion $Fe(CNS)^{2+}$ by mixing 5 cm³ portions of 0.02 M iron(III) chloride and 0.02 M potassium thiocyanate, KCNS.

To one portion add drops of 0.1 M sodium dihydrogen phosphate(V) until the colour has noticeably changed. To the second portion add 0.1 M sodium ethanedioate and to the third portion add 0.1 M sodium chloride. Note the relative volumes needed to reverse the formation of the complex ion.

COMMENT
You will be using this solution in your study of electrochemical cells in Topic 13.

5 Examine a bottle of saturated potassium nitrate solution.

■ Describe what you imagine is the movement of the potassium ions, K^+, and nitrate ions, NO_3^-, when an excess of solid potassium nitrate is added to a new bottle half-full of water.

Interpretation of the reversible processes

You now know of a number of changes which, when allowed to start, do not proceed to completion. That is, not all the reactants are changed into new forms or new substances.

1 When a small volume of a solvent is added to solid iodine, some of the iodine will dissolve. We could arrive at the same situation by allowing some of the solvent to evaporate from a solution until solid forms.

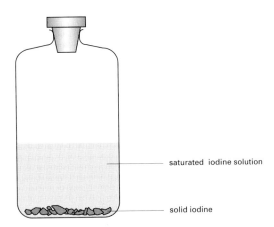

Figure 11.2 Iodine at equilibrium

COMMENT
The symbol ⇌ indicates that we are considering a system in which equilibrium has been reached and also that the equilibrium state can be reached from either direction.

We say that a **state of equilibrium** has been reached between solid iodine and iodine solution. This state of equilibrium is represented by

$$I_2(s) \rightleftharpoons I_2(solvent)$$

With another solvent present the iodine is distributed between the layers until equilibrium is reached

$$I_2(s) \rightleftharpoons I_2(aq) \rightleftharpoons I_2(hydrocarbon)$$

Since the equilibrium state can be approached from either direction it is not unreasonable to suppose that some iodine dissolved in the water may be transferring to the hydrocarbon and vice versa.

If the processes are going on simultaneously, they must also go on at the same rate, otherwise the properties of the system would alter. This possibility can be investigated by using a radioactive isotope which can be detected by a Geiger tube. This is done by adding some iodine in potassium iodide solution containing the radioactive isotope ^{131}I to an equilibrium mixture. After some time has elapsed it is possible to detect radioactivity in both solvents. Thus there must be an interchange of iodine between the aqueous solution and the hydrocarbon solution.

This means that equilibrium in chemical systems must be a **dynamic** state (and not a static state).

11 How far? Reversible reactions

QUESTION
What is the Stock name of the compound with the formula NaIO?

2 As you may remember from Topic 5 iodine reacts with cold sodium hydroxide solution according to the equation

$$I_2(aq) + 2NaOH(aq) \longrightarrow NaI(aq) + NaIO(aq) + H_2O(l)$$

As this reaction proceeds, there is a loss of reactants and a simultaneous formation of products. This process is called the **forward reaction**.

You have seen that the iodine can be recovered by adding acid:

$$NaI(aq) + NaIO(aq) + H_2SO_4(aq) \longrightarrow I_2(aq) + Na_2SO_4(aq) + H_2O(l)$$

This process is called the **reverse reaction**.

The system reaches a state of equilibrium depending on the pH of the mixture. This is more clearly seen in an ionic equation

$$I_2(aq) + 2OH^-(aq) \rightleftharpoons I^-(aq) + IO^-(aq) + H_2O(l)$$

The equation has been turned into an ionic equation, by leaving out the sodium and sulphate ions since these do not undergo chemical change (they are called 'spectator ions').

3 The state of equilibrium between chromate(VI) and dichromate(VI) is another equilibrium that is pH dependent:

$$2CrO_4^{2-}(aq) + 2H^+(aq) \rightleftharpoons Cr_2O_7^{2-}(aq) + H_2O(l)$$

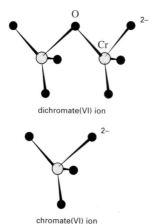

dichromate(VI) ion

chromate(VI) ion

4 The formation of complex ions is a reversible process that depends on competition between two equilibria, thus in the case of the reaction between iron(III) ions and thiocyanate ions the competition (in a simplified equation) is

$$Fe(H_2O)^{3+} + CNS^- \rightleftharpoons Fe(CNS)^{2+} + H_2O$$

When other ions are added they will displace the thiocyanate ion if they form a stable product:

$$Fe(CNS)^{2+} + 3C_2O_4^{2-} \rightleftharpoons Fe(C_2O_4)_3^{3-} + CNS^-$$

5 When a mixture of water and potassium nitrate is shaken, and there is more potassium nitrate than is required to saturate the water at the temperature concerned, some solid will remain undissolved

$$K^+NO_3^-(s) + aq \rightleftharpoons K^+(aq) + NO_3^-(aq)$$

We can approach this equilibrium state from either direction, by dissolving potassium nitrate in water at a given temperature, or by cooling a hot saturated solution of potassium nitrate to the same temperature.

At equilibrium, potassium and nitrate ions will be continuously transferring between solid and solution.

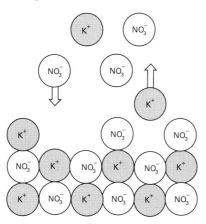

Figure 11.3 Potassium ions and nitrate ions at equilibrium

Characteristics of the equilibrium state

1 You can only have a stable state of equilibrium in a closed system – one that cannot exchange matter with its surroundings. In a system which allows matter to enter or leave, a stable equilibrium is not possible.

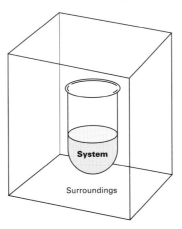

Figure 11.4 A closed system

2 The equilibrium state can be approached from either direction; that is, the products can be used as reactants and the reaction will still take place. Reactions of this kind are called **reversible reactions**.

3 Equilibrium is a dynamic state in that change is continually taking place in opposite directions on the molecular level.

4 The dynamic aspect of equilibrium means that it is stable under fixed conditions but sensitive to alteration in these conditions. The existence of an equilibrium state can be recognized by taking advantage of its sensitivity to changes in conditions. For example, when a change in temperature, pH, or concentration leads to an obvious change in a system it is likely to have been at equilibrium before the change took place.

QUESTIONS

A concentrated solution of bismuth(III) chloride is clear and colourless but when water is added it becomes cloudy. The equation is

$$BiCl_3(aq) + H_2O(l) \rightleftharpoons BiOCl(s) + 2HCl(aq)$$

1 What could you add to the cloudy mixture to make it go clear again?
2 Having made the mixture clear, how could you re-form the precipitate?
3 What do you imagine an intermediate state would look like?
4 Repeating **1** and **2** as a cycle gets progressively more difficult to carry out. Why do you think this happens?

11.2 The Equilibrium Law

From many quantitative investigations of equilibrium reactions, the following general statement emerges. For any system at equilibrium, there is a simple relationship between the concentrations of the substances present:

When a reaction at equilibrium is represented by the equation

$$mA + nB \rightleftharpoons pC + qD$$

the expression

$$\frac{[C]_{eq}^p [D]_{eq}^q}{[A]_{eq}^m [B]_{eq}^n} \text{ is a constant at a given temperature} = K_c$$

This is known as the **Equilibrium Law** and K_c is called the **equilibrium constant**. The subscript c indicates that it is expressed in concentrations. By convention, the concentrations of the substances on the **righthand side** of the equation are always put at the top of the fraction of the equilibrium constant and those of the substances on the **lefthand side** at the bottom.

We can draw some important conclusions from this law:

1 **When K_c is large the equilibrium mixture will contain a high proportion of products; that is, the reaction has gone nearly to completion.**

2 **When K_c is small, the reaction does not proceed very far at the temperature concerned and the concentration of products is low.**

Figure 11.5 Proportions of reactants (in white) and products (in grey) for different values of K_c

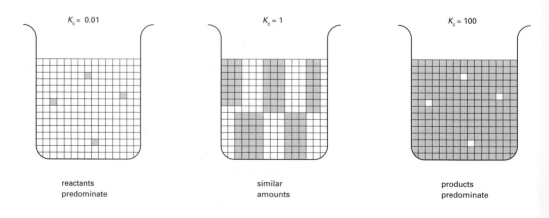

3 The value of K_c is not altered by the addition of more reactants or more products to an equilibrium mixture.

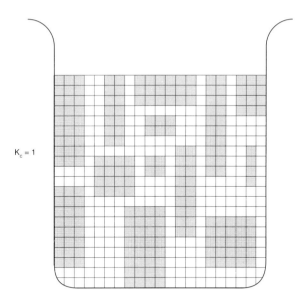

Figure 11.6 Doubling the amounts does not alter the ratios

On the addition of more reactants the system will move in the forward direction until the concentrations again satisfy the value for K_c.

Similarly, on addition of more products, the equilibrium will move in the reverse direction and the concentrations of reactants will increase.

4 When K_c for a reaction is known, the relative proportions of reactants and products at equilibrium can be calculated for any mixture of reactants used initially.

When stating the value of K_c for a particular reaction, it is important to indicate the equation on which the constant is based. For example, in the reaction between ethanoic acid and ethanol to form ethyl ethanoate and water, we can write the equation in the form:

$$\underset{\text{ethanoic acid}}{CH_3CO_2H(l)} + \underset{\text{ethanol}}{C_2H_5OH(l)} \rightleftharpoons \underset{\text{ethyl ethanoate}}{CH_3CO_2C_2H_5(l)} + \underset{\text{water}}{H_2O(l)}$$

The appropriate form of the equilibrium expression is

$$K_c = \frac{[CH_3CO_2C_2H_5(l)]_{eq}[H_2O(l)]_{eq}}{[CH_3CO_2H(l)]_{eq}[C_2H_5OH(l)]_{eq}}$$

and the experimentally determined value of $K_c = 3.7$.

But if we approach the equilibrium from the other direction

$$\underset{\text{ethyl ethanoate}}{CH_3CO_2C_2H_5(l)} + \underset{\text{water}}{H_2O(l)} \rightleftharpoons \underset{\text{ethanoic acid}}{CH_3CO_2H(l)} + \underset{\text{ethanol}}{C_2H_5OH(l)}$$

the appropriate form of the equilibrium expression is

$$K_c = \frac{[CH_3CO_2H(l)]_{eq}[C_2H_5OH(l)]_{eq}}{[CH_3CO_2C_2H_5(l)]_{eq}[H_2O(l)]_{eq}}$$

and $K_c = 0.27$ (which is $1/3.7$).

For reactions in which the number of particles on each side of the equation is the same, as in the example above, the concentration units cancel, and K_c has no units.

> **COMMENT**
>
> In the example K_c is $\dfrac{\text{mol dm}^{-3} \times \text{mol dm}^{-3}}{\text{mol dm}^{-3} \times \text{mol dm}^{-3}}$ which cancels to no units.
>
> For other reactions this may not be the case and units for K_c must be stated. Thus in the equilibrium
>
> $$2NO_2(\text{solvent}) \rightleftharpoons N_2O_4(\text{solvent})$$
>
> $$K_c = \frac{[N_2O_4]_{eq}}{[NO_2]^2_{eq}}$$ and the units of K_c are $\text{mol}^{-1}\,\text{dm}^3$.

Relative concentrations at equilibrium

The reaction of esters with water, for example the reaction of ethyl ethanoate with water, involves an equilibrium state:

$$\underset{\text{ethyl ethanoate}}{CH_3CO_2C_2H_5(l)} + \underset{\text{water}}{H_2O(l)} \rightleftharpoons \underset{\text{ethanoic acid}}{CH_3CO_2H(l)} + \underset{\text{ethanol}}{C_2H_5OH(l)}$$

Equilibrium is reached very slowly indeed at ordinary temperatures in this system. It is reached more rapidly by heating the reagents or by using a catalyst (6% of concentrated hydrochloric acid in the mixture). A mixture of ethyl ethanoate and water alone would take several years to reach equilibrium at room temperature so the catalyst has a remarkable effect.

The equilibrium can be studied by a simple titration technique but has to be set up with special care and needs at least a week for the reaction to reach equilibrium.

Figure 11.7

Concentrations at equilibrium/mol dm^{-3}			
Ethyl ethanoate	Water	Ethanoic acid	Ethanol
15.4	5.27	4.33	4.33
13.6	7.87	5.47	5.47
10.0	17.4	7.00	7.00
6.00	35.4	7.60	7.60

> **QUESTION**
>
> When the ester equilibrium is represented by
>
> $$CH_3CO_2C_2H_5(l) + H_2O(l) \rightleftharpoons CH_3CO_2H(l) + C_2H_5OH(l)$$
>
> $$K_c = \frac{[CH_3CO_2H(l)]_{eq}[C_2H_5OH(l)]_{eq}}{[CH_3CO_2C_2H_5(l)]_{eq}[H_2O(l)]_{eq}}$$
>
> Use your calculator to check that the Equilibrium Law is satisfied and that there is a constant numerical relationship between the equilibrium concentrations of the reactants and products in figure 11.7, within the limits of experimental error.

11 How far? Reversible reactions

As another example let us look at some results from a study of the equilibrium

$$2NO_2(\text{solvent}) \rightleftharpoons N_2O_4(\text{solvent})$$

Equilibrium mixtures can be prepared in chloroform solution at temperatures near 0 °C. The composition of the equilibrium mixture can be calculated from the intensity of colour of the solution because dinitrogen tetroxide, N_2O_4, is colourless but nitrogen dioxide, NO_2, is brown.

In a solution of this kind, at 10 °C, the concentrations in a set of mixtures were

$[NO_2]_{eq}$ /mol dm^{-3}	$[N_2O_4]_{eq}$ /mol dm^{-3}
0.0012	0.13
0.0016	0.28
0.0019	0.32
0.0021	0.42
0.0028	0.78

QUESTION

Calculate the value of K_c for this reaction at 10 °C using an appropriate expression for K_c: with a calculator the work should not take long.

STUDY TASK

1 Write the K_c expressions for the reactions you studied in Experiment 11.1.
2 Write the units for K_c for the reactions you studied in Experiment 11.1.

You now have an opportunity to determine an equilibrium constant in experiment 11.2.

EXPERIMENT 11.2 Measurement of an equilibrium constant K_c

Silver ions and iron(II) ions react in a slow redox reaction which reaches an equilibrium in which both ions are present in measurable concentrations

$$Ag^+(aq) + Fe^{2+}(aq) \rightleftharpoons Ag(s) + Fe^{3+}(aq)$$

The concentration of silver ions can be measured by titration with potassium thiocyanate. The equilibrium constant can then be calculated from the relationship:

$$K_c = \frac{[Fe^{3+}(aq)]_{eq}}{[Ag^+(aq)]_{eq} \times [Fe^{2+}(aq)]_{eq}}$$

COMMENT
[Ag(s)] does not appear in the expression for K_c because the **mass** of solid silver present does not alter the **concentration** of solid silver, which therefore remains constant.

When titrating the reaction mixture with potassium thiocyanate, KCNS, the first reaction is the precipitation of silver thiocyanate.

$$KCNS(aq) + AgNO_3(aq) \longrightarrow AgCNS(s) + KNO_3(aq)$$

When all the silver ions have been removed from solution, thiocyanate ions react with the iron(III) ions in the equilibrium mixture to give the deep red colour with which you should already be familiar. This acts as an indication of the end-point of the titration

$$CNS^-(aq) + Fe^{3+}(aq) \longrightarrow Fe(CNS)^{2+}(aq)$$

This procedure should be successful because the equilibrium changes quite slowly.

Procedure

Using pipettes transfer 25.0 cm³ each of 0.10 M silver nitrate solution and 0.10 M iron(II) sulphate solution into a dry 100 cm³ conical flask, stopper it so that it is air tight and allow to stand undisturbed overnight.

During this time the equilibrium is established. There should be a precipitate of silver, which settles to the bottom of the flask.

Using a pipette, transfer 10.0 cm³ of the solution into another conical flask, disturbing the silver precipitate as little as possible. Titrate the sample with 0.020 M potassium thiocyanate. The end point is marked by the first permanent red colour. Repeat the titration (twice if possible) and calculate the average of your 'good' titres.

> **QUESTION**
> Why do you think it is necessary to make the flask air tight, and leave it overnight?

Calculation

1 When you mix equal volumes of two solutions you are effectively diluting both of them by half so the initial concentrations are:

$$[Fe^{2+}]_{initial} = 0.05 \text{ mol dm}^{-3} \text{ and } [Ag^+]_{initial} = 0.05 \text{ mol dm}^{-3}$$

2 The equation for the titration reaction is:

$$KCNS(aq) + AgNO_3(aq) \longrightarrow AgCNS(s) + KNO_3(aq)$$

Use it to calculate the concentration of $[Ag^+]_{eq}$ from your titration results.

3 Since $[Fe^{2+}]_{initial} = [Ag^+]_{initial}$
it follows that $[Fe^{2+}]_{eq} = [Ag^+]_{eq}$
so $[Fe^{2+}]_{eq}$ = your answer to **2**.
and $[Fe^{3+}]_{eq} = [Fe^{2+}]_{initial} - [Fe^{2+}]_{eq}$

4 Using $K_c = \dfrac{[Fe^{3+}(aq)]_{eq}}{[Ag^+(aq)]_{eq} \times [Fe^{2+}(aq)]_{eq}}$

calculate K_c and give its correct units in your result.

> **QUESTION**
> Can you see why $[Fe^{3+}]_{eq} = [Fe^{2+}]_{initial} - [Fe^{2+}]_{eq}$? If you are unsure, imagine starting with 5 ions of each reagent and letting 2 react. Work it out with labelled pieces of paper.

11.3 Acid–base equilibria

REVIEW TASK
In small groups, write down what you remember about the Brønsted-Lowry theory of acids and bases.

We shall be giving special consideration to two extremely important types of equilibria. They are acid–base equilibria, to be dealt with in this section, and redox equilibria to be tackled in Topic 13.

In Topic 3 we introduced the Brønsted-Lowry theory of acids and bases. In their theory, an acid is a substance which can provide protons (hydrogen ions) in a reaction; a base is a substance which can combine with protons.

Figure 11.8 Corrosion of stone sculpture by acid rain, Cloth Hall, Cracow

We shall now take the theory a little further and look at its quantitative applications.

Experiment 11.3

What is an acid?

We can start by collecting data on the pH of some aqueous solutions. Record the results for all these experiments with particular care. You will need to refer to them while working through the remainder of this Topic.

Procedure

1 Using Full-range Indicator solution, test the pH of dilute solutions of the following sets of compounds:

H_2SO_4, $NaHSO_4$, Na_2SO_4
H_3PO_4, NaH_2PO_4, Na_2HPO_4, Na_3PO_4

Classify the compounds into three groups using their **formulae**:
- those that you would expect to donate protons;
- those that you would expect to accept protons;
- those that you would expect both to **donate** protons and to **accept** protons.

SAFETY ⚠
The chloroethanoic acids are toxic and corrosive; boric acid is toxic; all acidic and alkaline solutions should be handled with care, the majority are corrosive depending on the concentration.

CH$_3$—CO$_2$H
CH$_2$Cl—CO$_2$H
CHCl$_2$—CO$_2$H
CCl$_3$—CO$_2$H
CH$_3$—CO$_2$H
H$_3$BO$_3$
C$_6$H$_5$—CO$_2$H

- Which is more helpful to acid–base classification: the pH value of the solution or the ability to donate or accept protons?

2 Measure the pH of 0.1 M solutions of the following salts: sodium carbonate, ammonium chloride and sodium ethanoate.

3 Measure the pH of 0.1 M solutions of the following acids: ethanoic acid, chloroethanoic acid, dichloroethanoic acid and trichloroethanoic acid (TAKE CARE).

4 Find out whether solutions of the following acids will react with sodium carbonate solution producing carbon dioxide gas: ethanoic acid, boric acid and benzoic acid.

5 Record the colours of the following indicators in both acidic and alkaline solution: methyl orange, phenolphthalein, bromophenol blue, bromothymol blue.

> **COMMENT**
> An acid is a proton donor; a base is a proton acceptor.

Interpretation of acid–base reactions

The chemistry of acid–base systems is concerned with equilibria between ionically-bonded species and covalently-bonded species. Equal sharing of electrons (a true covalent bond) occurs only between like atoms, as in H$_2$, Cl$_2$, etc. Bonding between unlike atoms always results in unequal sharing and polar bonds, for example:

$$\overset{\delta+}{H}—\overset{\delta-}{Cl}$$

The degree of electron sharing in polar molecules is changed when they are dissolved in polar solvents. This often results in the formation of an ionic bond. It happens when hydrogen chloride reacts with water

$$HCl(g) + H_2O(l) \longrightarrow H_3O^+(aq) + Cl^-(aq)$$

a process which can be resolved into two stages:

$$HCl(g) \longrightarrow H^+ + Cl^-$$
$$H^+ + Cl^- + H_2O(l) \longrightarrow H_3O^+(aq) + Cl^-(aq)$$

The hydrogen ion, H$^+$, is a single proton, with no electrons. Thus it is some 50 000 times smaller than the next smallest cation, Li$^+$. The possibility of very close approach between the free proton and the oxygen atom in the water molecule results in a strong bond being formed by the lone pair of electrons on the oxygen atom. Many other substances react with water in this way.

One consequence of the Brønsted-Lowry theory of acids and bases is that when an acid donates a proton in a neutralization reaction it becomes a substance that can, in turn, accept a proton. It becomes, in effect, another base. For example, the equilibria involved in the reaction of hydrogen chloride with water are

$$\underset{\text{acid}_1}{HCl(g)} \rightleftharpoons \underset{\text{base}_1}{Cl^-(aq)} + H^+$$

and

$$\underset{\text{base}_2}{H_2O(l)} + H^+ \rightleftharpoons \underset{\text{acid}_2}{H_3O^+(aq)}$$

These can be combined to give

$$\underset{\text{acid}_1}{HCl(g)} + \underset{\text{base}_2}{H_2O(l)} \rightleftharpoons \underset{\text{base}_1}{Cl^-(aq)} + \underset{\text{acid}_2}{H_3O^+(aq)}$$

In this reaction, water behaves as a base by accepting a proton.

Similarly by accepting a proton a base becomes a substance that can donate a proton: it becomes an acid. In aqueous ammonia, for example, the acid–base equilibrium is

$$\underset{\text{base}_1}{NH_3(aq)} + \underset{\text{acid}_2}{H_2O(l)} \rightleftharpoons \underset{\text{acid}_1}{NH_4^+(aq)} + \underset{\text{base}_2}{OH^-(aq)}$$

In this reaction water acts as an acid by donating a proton.

Acid and base reactions are essentially a competition for protons.

> **COMMENT**
> $H_3O^+(aq)$ can be referred to as the conjugate acid of $H_2O(l)$; and $Cl^-(aq)$ can be referred to as the conjugate base of $HCl(g)$.

> **COMMENT**
> Ammonium chloride, NH_4Cl, is not usually called an acid, but when it dissolves in water the solution is weakly acidic. The Brønsted-Lowry theory copes with this because it views the ammonium ion as a weak acid
>
> $$NH_4^+(aq) + H_2O(l) \rightleftharpoons NH_3(aq) + H_3O^+(aq)$$
>
> The ammonium ion donates a proton to a water molecule. The equilibrium lies to the left but there are sufficient hydrogen ions in solution to make it noticeably acidic.

> **STUDY TASK**
> Use the Brønsted-Lowry theory to interpret the results of Experiment 11.3, part **2**.

pH and the equilibrium constant for the ionization of water

In practical laboratory situations we generally use acids in aqueous solutions, and we must therefore take into account some of the properties of water.

You have seen that water is able to function both as a base, accepting protons

$$H^+ + H_2O \longrightarrow H_3O^+$$

and as an acid, donating protons

$$H_2O \longrightarrow H^+ + OH^-$$

As a consequence, an equilibrium exists in water, with some of the molecules acting as an acid and some as a base

$$2H_2O(l) \rightleftharpoons H_3O^+(aq) + OH^-(aq)$$

or more simply

$$H_2O(l) \rightleftharpoons H^+(aq) + OH^-(aq)$$

and we say that the water is **ionized**. The equilibrium constant for this ionization, K_c, is given by

$$K_c = \frac{[H^+(aq)]_{eq}[OH^-(aq)]_{eq}}{[H_2O(l)]_{eq}}$$

Rearranging the expression, we get

$$K_c \times [H_2O(l)]_{eq} = [H^+(aq)]_{eq}[OH^-(aq)]_{eq}$$

We can treat the concentration of water molecules, $[H_2O(l)]_{eq}$, as constant because the proportion of water molecules that ionizes is very small. The lefthand side of this equation is therefore a constant. It is known as the ionization constant for water, and given the symbol K_w.

At 298 K, $K_w = 10^{-14}$ mol^2 dm^{-6}.

In pure water, and in any absolutely **neutral** solution

$$[H^+(aq)]_{eq} = [OH^-(aq)]_{eq}$$

and so the value of each is $\sqrt{10^{-14}} = 10^{-7}$. This small value shows that our assumption that $[H_2O(l)]_{eq}$ can be treated as constant was correct.

REVIEW
- In 0.1 M HCl

  $[H^+(aq)]_{eq} = 10^{-1}$ mol dm^{-3}

  and since $[H^+(aq)]_{eq}[OH^-(aq)]_{eq} = K_w = 10^{-14}$ mol^2 dm^{-6}

  $[OH^-(aq)]_{eq} = 10^{-14} \div 10^{-1} = 10^{-13}$ mol dm^{-3}

  Therefore in 0.1 M HCl $[OH^-(aq)]_{eq} = 10^{-13}$ mol dm^{-3}

- In 0.01 M NaOH

  $[OH^-(aq)]_{eq} = 10^{-2}$ mol dm^{-3}

  Therefore $[H^+(aq)]_{eq} = 10^{-14} \div 10^{-2} = 10^{-12}$ mol dm^{-3}

COMMENT
The pH scale is a logarithmic scale using logarithms **to base 10**, symbol 'lg'. In other Topics the logarithms that are used are 'natural logarithms' (common logarithms) to base e, where $e = 2.718...$, symbol 'ln'.

Because the range of possible hydrogen ion concentrations in solution is very large, from about 10 to 10^{-15}, chemists find it convenient to use a logarithmic scale. This is the origin of the **pH scale**, and the relationship between the pH value of a solution and the hydrogen ion concentration is

$$pH = -\lg[H^+(aq)]$$

where $[H^+(aq)]$ is measured in mol dm^{-3}. The minus sign is introduced to make pH values positive in almost all cases.

A few examples should help to make the relationship clearer.
When

$[H^+(aq)] = 1 \times 10^{-3}$ mol dm^{-3} pH = 3
$[H^+(aq)] = 1 \times 10^{-8}$ mol dm^{-3} pH = 8
$[H^+(aq)] = 5 \times 10^{-4}$ mol dm^{-3} pH = 3.3

When, with concentrations in mol dm^{-3}:

$[OH^-(aq)]_{eq} = 10^{-2}$ $[H^+(aq)]_{eq} = 10^{-12}$ pH = 12

$[OH^-(aq)]_{eq} = 10^{-4}$ $[H^+(aq)]_{eq} = 10^{-10}$ pH = 10

$[OH^-(aq)]_{eq} = 5 \times 10^{-1}$ $[H^+(aq)]_{eq} = 2 \times 10^{-12}$ pH = 11.7

> **STUDY TASK**
> Use your calculator to check the pH examples; do the calculations from pH to concentration as well.

The range of pH values you will actually encounter in aqueous systems varies between just less than zero (concentrated acid) and a little over 14 (concentrated alkali). The Appendix on mathematics has more examples on how to work out logarithms.

11.4 The strengths of acids and bases

When an acid, represented by HA, is dissolved in water, an equilibrium is established

$$HA(aq) \rightleftharpoons A^-(aq) + H^+(aq)$$

where HA is an acid and A$^-$ is the matching base.

In examples where HA loses protons readily, a high concentration of H$^+$ ions will be produced when the system has reached equilibrium. In this case HA is functioning as a strong acid.

On the other hand, where HA is a weak acid with no pronounced tendency to part with protons to water, the concentration of H$^+$ ions will be smaller.

In effect, the H$^+$ ion can be used as a standard against which to compare the relative strengths of acids. The problem now becomes one of measuring the hydrogen ion concentration, $[H^+]_{eq}$, in aqueous solutions of different acids.

When we measure pH values for progressively diluted hydrochloric acid, we find that there is an increase of about one pH unit per tenfold dilution. For example:

> **COMMENT**
> Instruments known as pH meters can be used to determine the pH of a solution with greater accuracy than can be achieved by using indicators; a glass electrode is immersed in the solution whose pH is to be determined and the value is read off a meter. The pH meter is described in more detail in Topic 18.

pH of 1.0 M   HCl(aq) is approximately 0
pH of 0.1 M   HCl(aq) is approximately 1
pH of 0.01 M   HCl(aq) is approximately 2
pH of 0.001 M HCl(aq) is approximately 3

These observations can be accounted for if we assume that when the gas dissolves in water it is almost completely ionized at all concentrations, so that the equilibrium

$$HCl(g) \rightleftharpoons HCl(aq) \rightleftharpoons H^+(aq) + Cl^-(aq)$$

lies almost completely over to the right. A tenfold dilution will then reduce the value of $[H^+(aq)]$ by $\frac{1}{10}$ and the pH will have increased by 1 unit ($-\lg \frac{1}{10} = 1$).

A few other acid solutions behave in the same way. For most acidic solutions, however, the increase in pH for a dilution factor of ten is less than one unit, and the pH values for comparable concentrations are always greater than for hydrochloric acid solutions. This means that dissociation into ions is incomplete and a considerable proportion of reactants remains when the equilibrium

$$HA(aq) \rightleftharpoons H^+(aq) + A^-(aq)$$

is reached. Thus the hydrogen ion concentration will be smaller than would be expected for complete dissociation and the pH value higher.

Acids which ionize nearly completely at moderate dilutions (0.1 M or 0.01 M) are called *strong acids*.

Those which ionize slightly, or exist mainly as the covalently bonded form under these conditions, are called *weak acids*.

There is no sharp dividing line between strong and weak acids but rather a spectrum of acidic properties.

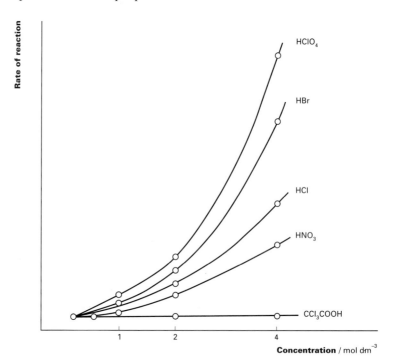

Figure 11.9 The rate of hydrolysis of sugar depends on acid strength

In the Brønsted-Lowry definition, the strength of an acid is measured by the extent to which protons are released and the strength of a base by the extent to which protons are accepted. The strength of an acid, represented by HA, is given by the equilibrium constant, K_c, for the following reaction

$$HA(aq) \rightleftharpoons H^+(aq) + A^-(aq)$$

where HA is an acid and A^- is the matching base. We will use $H^+(aq)$ rather than $H_3O^+(aq)$ to simplify the expressions, so for this equilibrium

$$K_c = \frac{[H^+(aq)]_{eq}[A^-(aq)]_{eq}}{[HA(aq)]_{eq}}$$

When dealing with acid-base equilibria the symbol K_a is often used instead of K_c.

$$\frac{[H^+(aq)]_{eq}[A^-(aq)]_{eq}}{[HA(aq)]_{eq}} = K_a \text{ (the } dissociation\ constant \text{ of the acid)}$$

Table 6.5 in the *Book of data* shows the relative strengths of various acids as represented by their equilibrium constants. The Brønsted-Lowry definition covers almost all the common acid and base reactions and is the one most generally used.

QUESTIONS

Use table 6.5 in the *Book of data* to complete this table:

Name of acid	Equilibrium equation	K_a
Ethanoic		
Chloroethanoic		
Dichloroethanoic		
Trichloroethanoic		
Boric		
Benzoic		
Carbonic		

1 Arrange carbonic, benzoic, boric and ethanoic acids in order of acid strength, putting the strongest acid first.
Which of these acids would you expect to displace carbonic acid (carbonic acid readily decomposes into carbon dioxide and water) from a carbonate and which would not?

2 Write structural formulae for the first four of these acids.
Can you suggest why the acids get stronger? *Hint*: look at the text on dipole-dipole interactions.

3 Use the data to interpret the results of parts **3** and **4** of Experiment 11.3.

Converting K_a values to pH values

As an example we will calculate the value of the pH of a 0.01 mol dm^{-3} solution of methanoic acid, HCO_2H, at 25 °C. The value of K_a from table 6.5 in the *Book of data* is 1.6×10^{-4} mol dm^{-3}.

$$HCO_2H(aq) \rightleftharpoons HCO_2^-(aq) + H^+(aq)$$

$$K_a = \frac{[HCO_2^-(aq)]_{eq}[H^+(aq)]_{eq}}{[HCO_2H(aq)]_{eq}}$$

Neglecting the hydrogen ions which arise from ionization of the water, since the concentration of these will be very small compared with the concentration of those from the acid, we can say that

$$[H^+(aq)]_{eq} = [HCO_2^-(aq)]_{eq}$$

and $[HCO_2H(aq)]_{eq} = (0.01 - [H^+(aq)]_{eq})$ mol dm^{-3}

We will also assume that the concentration of hydrogen ions is small compared to the concentration of methanoic acid. This a less valid assumption but the calculation is much easier and the final pH value will still be reasonably accurate. So we can write

$$K_a = \frac{[H^+(aq)]^2}{0.01} = 1.6 \times 10^{-4}$$

$$[H^+(aq)]^2 = (1.6 \times 10^{-4}) \times 0.01 = 1.6 \times 10^{-6}$$

$$[H^+(aq)] = 1.26 \times 10^{-3}$$

and $pH = -\lg[H^+(aq)]$

so $pH = -\lg(1.26 \times 10^{-3})$

and using your calculator

$$pH = 2.9$$

> **QUESTION**
> The pH of a solution of a weak acid for any molarity can be found if the value of K_a for the acid is known. Calculate the pH of a 0.001 M solution of chloric(I) acid, HClO, given $K_a = 3.7 \times 10^{-8}$ mol dm^{-3}.

EXPERIMENT 11.4 Determination of K_a for a weak acid

You are provided with a pure sample of a weak acid.

To find K_a by experiment you need to make a dilute solution of known concentration and to find its pH using indicators or a pH meter.

> **SAFETY** ⚠
> Make a risk assessment for the acid you are going to use.

Procedure

Weigh a 100 cm^3 standard volumetric flask. Add 1 drop of the weak acid and reweigh. Carefully add pure water to the flask until it is full to the 'ring-mark' on the neck. Mix well by inverting the flask at least 5 times.

From the weighings and the molar mass of the acid, calculate the concentration of the solution.

Measure the pH of the solution with a well washed and calibrated pH electrode.

If a pH meter is not available, pour some of the solution into each of two beakers. Test the contents of the first beaker with Full-range Indicator to get a rough idea of its pH. Then select a suitable narrow-range indicator and use it to test the contents of the second beaker. Record the pH as accurately as you can.

> **COMMENT**
> Amount/mol
> $$= \frac{\text{mass/g}}{\text{molar mass/g mol}^{-1}}$$
> then concentration/mol dm^{-3}
> $$= \frac{\text{amount/mol}}{\text{volume/dm}^3}$$

Calculation

1 You should be able to recall $pH = -\lg[H^+(aq)]$
   from which it follows that $\lg[H^+(aq)] = -pH$
   and $[H^+(aq)] = 10^{-pH}$

Enter the pH of the solution into your calculator and, by using suitable functions, change its sign to '−' and find the value of $[H^+(aq)]$.

2 Calculate a value for K_a from the expression:

$$[H^+(aq)] = (K_a \times [HA(aq)])^{1/2}$$

State the correct units of K_a.

3 Check your answer against the value given in the *Book of data*. Which of your measurements is likely to have been the least accurate?

> **COMMENT**
> [HA(aq)] is the concentration of the acid you calculated from the weighings.

11.5 Acid–base titrations

When a base is added to an acid the reaction

$$H^+(aq) + OH^-(aq) \rightleftharpoons H_2O(l)$$

takes place. As we have already seen, this equilibrium lies far to the right. As a base is added to an acid the hydrogen ion concentration grows progressively less, that is, the pH value of the resulting solution grows progressively greater.

In the following experiment we will investigate, using a pH meter, how the pH changes on addition of a base.

EXPERIMENT 11.5 The change of pH during an acid-base titration

There are four different combinations of acid and base possible, namely
- strong acid and strong base
- strong acid and weak base
- weak acid and strong base
- weak acid and weak base

The best procedure is to use a computer to capture and process the data.

Procedure

Using a pipette and pipette filler, or a burette, put 25.0 cm³ of the 1.0 M acid that you are using, in a 100 cm³ beaker. If a magnetic stirrer is available, stand the beaker on it, and place the stirrer bar in the beaker.

Fill a burette with 1.0 M alkali, and clamp it so that the alkali can be run into the acid in the beaker.

Connect the electrode to the computer via a suitable pH meter or other device, and put it in the acid in the beaker. Clamp it gently in position; if you are using a magnetic stirrer, make sure that the electrode is in a position where it cannot be struck by the stirrer bar when the stirrer is switched on.

Next, switch on the magnetic stirrer, start the computer procedure and run the alkali from the burette as a steady flow into the acid. You should add a total of 35 cm³ of alkali in this way.

Use the computer to plot the results as a graph of pH against time (which will approximate to volume of alkali added).

- Are the shapes of the graphs as you expected?
  Use your results to select an appropriate indicator for each titration. (You looked at the colour change of some indicators in experiment 11.3, part 5.)

COMMENT
pH ranges of indicators are listed in table 6.6 of the *Book of data*

The theory of indicators

You will have used acid–base indicators such as methyl orange, phenolphthalein, and litmus, in previous work. They are used to test for alkalinity and acidity, and for detecting the end-point in acid–base titrations. A particular indicator cannot be used in all circumstances; some are more suitable for use with weak acids and others with weak bases. From a study of the titration curves that you obtained in experiment 11.5 you will see that there is a rapid change of pH in the neutralization reaction near its end-point.

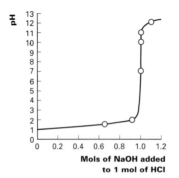

Figure 11.10 Changes of pH in the titration of a solution of hydrochloric acid and sodium hydroxide

However, this rapid change of pH occurs at a comparatively low pH during a titration of a strong acid against a weak base, and at a comparatively high pH during a titration of a weak acid against a strong base.

Now, different indicators change colour at different values of pH. Phenolphthalein, for example, changes colour over the range pH 8–10, and so is suitable for a titration involving a weak acid and a strong base, whose end-point occurs within this range. Methyl orange, however, changes colour over the range pH 4–7, and so can be used to find the end-point in a titration involving a strong acid and a weak base. Phenolphthalein would be unsuitable for this type of titration, as it would only change colour when an excess of alkali was present, and then only gradually, instead of sharply on the addition of one drop of extra alkali.

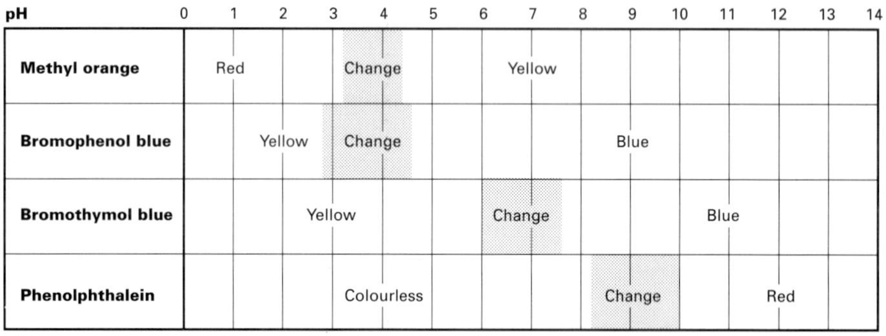

pH	0	1	2	3	4	5	6	7	8	9	10	11	12	13	14
Methyl orange		Red			Change			Yellow							
Bromophenol blue			Yellow		Change					Blue					
Bromothymol blue				Yellow				Change			Blue				
Phenolphthalein					Colourless					Change		Red			

Figure 11.11 Colour change and pH range for some indicators

An indicator may be considered as a weak acid, for which either the acid or the corresponding base, or both, are coloured. We can represent this in a general way, using HIn for the acid form, as

$$\text{HIn(aq)} \rightleftharpoons \text{H}^+\text{(aq)} + \text{In}^-\text{(aq)}$$
colour A   colour B

Addition of acid displaces the equilibrium to the left and increases the intensity of colour A. Addition of base, for example, $NH_3(aq)$, removes hydrogen ions

$$\text{NH}_3\text{(aq)} + \text{H}^+\text{(aq)} \longrightarrow \text{NH}_4^+\text{(aq)}$$

with the result that the equilibrium moves to the right to restore the value of K_a for the indicator and increase the intensity of colour B.

Therefore when an indicator is added to a solution the colour of the system will depend on the relative concentrations, $[\text{HIn(aq)}]_{eq}$ and $[\text{In}^-\text{(aq)}]_{eq}$, which in turn depend on the pH of the solution.

11.6 Buffer solutions

Ethanoic acid is a weak acid, being only slightly ionized in solution ($K_a = 1.7 \times 10^{-5}$ mol dm^{-3}). In an aqueous solution of ethanoic acid the equilibrium

$$\underset{\text{acid}}{CH_3CO_2H(aq)} \rightleftharpoons \underset{\substack{\text{corresponding} \\ \text{base}}}{CH_3CO_2^-(aq)} + H^+(aq) \tag{1}$$

lies well over to the left, and the hydrogen ion concentration is relatively small. There is a second equilibrium involving hydrogen ions in this system

$$H^+(aq) + OH^-(aq) \rightleftharpoons H_2O(l) \tag{2}$$

What happens when extra ethanoate ions are added to the solution? They can be introduced by adding a soluble salt of ethanoic acid, such as sodium ethanoate, which is highly ionized in solution.

EXPERIMENT 11.6a Buffer solutions

In a 100 cm^3 beaker take about 50 cm^3 of 0.1 M ethanoic acid and add a small spatula measure of solid sodium ethanoate. Stir to dissolve and then add Full-range Indicator solution, sufficient to give a recognisable colour.

Determine the pH by comparison with the colour charts provided with the Indicator or by using a pH meter. Stand this beaker on a white tile or piece of white paper.

Label the beaker **buffer solution of pH =**

Take a second beaker and put into it 1 drop of 1 M hydrochloric acid and 50 cm^3 of pure water. Add some Full-range Indicator solution and find the pH of this solution. Now dilute the contents of this beaker with pure water and add more indicator as necessary until you have a solution of hydrochloric acid which has the same appearance (and therefore pH) as the first beaker.

Label the second beaker **unbuffered solution of pH =**

Now add 1 drop of 0.1 M sodium hydroxide to each of the beakers and observe what happen. Follow this with further drops of sodium hydroxide and then drops of 0.1 M hydrochloric acid.

■   Which solution does not change much in pH?
    Which solution changes very easily indeed?

Interpretation of buffer behaviour

The mixture of ethanoic acid and sodium ethanoate contains a relatively high concentration of un-ionized ethanoic acid and a relatively high concentration of ethanoate ion. It therefore contains both an acid and its conjugate base.

When more hydrogen ions are added to this system by adding a small volume of a solution of a strong acid, these hydrogen ions will combine with ethanoate ions to form more un-ionized ethanoic acid. Equilibrium (1) moves to the left, removing nearly all the added hydrogen ions. The concentration of hydrogen ions, and thus the pH of the solution, will alter a little, but not very much.

Adding a strong base, for example sodium hydroxide, to the system disturbs equilibrium (2) so that OH^- ions combine with H^+ ions to form H_2O molecules. This reduces the hydrogen ion concentration, and more CH_3CO_2H molecules ionize to restore it to near its original value.

The two equilibria adjust themselves in this way until nearly all the added hydroxide ions are removed. The pH value of the system will rise a little in consequence, but not very much.

The changes in pH resulting from additions of acid or base are much smaller than they would be if the mixture of weak acid and its salt were not present.

Solutions of this kind, containing a weak acid and its corresponding base, thus provide a 'buffer' against the effects of adding strong acid or strong base. They are therefore known as **buffer solutions**. Essentially they are solutions possessing readily available reserve supplies of both an acid and its conjugate base.

> **COMMENT**
> The compositions of some buffer solutions are listed in table 6.7 in the *Book of data*.

Another example of a buffer solution contains a mixture of ammonium chloride (highly ionized) and ammonia (present mainly as NH_3 molecules). The equilibria present are

$$\underset{\text{acid}}{NH_4^+(aq)} \rightleftharpoons \underset{\text{corresponding base}}{NH_3(aq)} + H^+(aq)$$

and $H^+(aq) + OH^-(aq) \rightleftharpoons H_2O(l)$

The addition of more H^+ ions results in their reacting with the base NH_3 to form NH_4^+ ions. When more OH^- ions are added the following changes occur

$$H^+(aq) + OH^-(aq) \longrightarrow H_2O(l)$$
$$NH_4^+(aq) \longrightarrow NH_3(aq) + H^+(aq)$$

until the two equilibria are again restored. Again, the pH value remains nearly constant.

QUESTION
Use this information about buffers to interpret the titration curves you obtained in Experiment 11.5.

INVESTIGATION 11.6b

Vinegar

Carry out an investigation to measure the concentration of the ethanoic acid in a white-wine vinegar.

Make a risk assessment before starting any experiments.

STUDY TASK
Find out how ethanoic acid is produced from white wine.

11.7 Study task: Acid-base reactions in living materials

QUESTIONS
Read the passage below, and answer these questions based on it.
1 Write an account in not more than 150 words to explain why animals produce acids in their bodies and how they dispose of the waste acid.
2 Describe briefly the problems that plants have to cope with in dealing with acids.

The living cell requires an almost neutral internal environment for its immensely intricate chemical reactions. The pH of the protoplasm in living cells is about 6.9, while the extracellular fluids of vertebrates – blood and lymph – have a slightly alkaline pH of about 7.4. The cells produce acid continuously in the form of carbonic acid (carbon dioxide combined with water); but the production of carbon dioxide does not immediately affect the pH of either the cell or the extracellular fluid because the system is stabilized by a system of buffers.

The carbonic acid–hydrogencarbonate system may be taken as an example to illustrate the action of a buffer:

$$H_2CO_3(aq) \rightleftharpoons H^+(aq) + HCO_3^-(aq)$$

If hydrogen ions are added to the system, tending to cause a fall in pH, the equilibrium is pushed to the left, removing the extra hydrogen ions as un-ionized carbonic acid. The pH is thus kept constant. If the hydrogen ion concentration decreases, bringing about a rise in pH, the equilibrium shifts to the right, and favours the dissociation of the acid to increase the hydrogen ion concentration ions and restore the pH to normal. Buffers in the body are pairs of strong bases with weak acids, so constituted that they respond to changes of pH in the neighbourhood of the neutral point by shifts of equilibrium tending to restore neutrality. The bases consist principally of hydrogencarbonate, hydrogenphosphate, and protein.

However, buffers cannot maintain a constant pH indefinitely if too much carbon dioxide is added to the system. The unicellular animal loses excess carbon dioxide by diffusion through the cell membrane into the surrounding water or air. Multicellular animals have a proportionately smaller amount of surface; as animals increase in size, there has to be an increase of diffusing surface in some way. In insects air is brought to all the tissues by fine tracheal tubes. This diffusion system suffices only for relatively small creatures. In the larger animals, there is an enormous diffusing surface in gills or lungs and a system of extracellular fluids and circulating blood to carry the buffered carbonic acid to them and to carry oxygen to the tissues. As the blood passes through the lungs, the carbonic acid is rapidly converted to carbon dioxide and water by the enzyme carbonic anhydrase. The carbon dioxide diffuses from the blood into the air inside the lungs, and is then removed in the expired air.

Animals also produce strong acids that cannot be excreted in gaseous form. The most important of these is sulphuric acid, formed by oxidation of the

cysteine (one of the amino acids in food)
CH₂(SH)CH(NH₂)CO₂H

urea: O=C(NH₂)(NH₂)

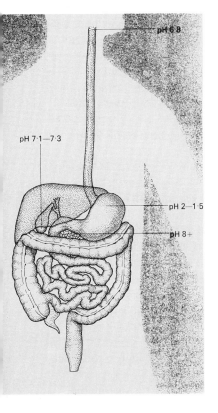

Figure 11.12 pH values in the stomach

Butanoic acid
CH₃CH₂CH₂CO₂H

Lactic acid
CH₃—CH(OH)CO₂H

sulphur in the proteins of the food. In vertebrates, acids of this kind are secreted by the kidney into the urine. Urine may be as acid as pH 4. Additional acids can be excreted as ammonium salts, since the kidney forms ammonium ions from the hydrogen ions and amide groups in proteins. Amino acids are the units that are polymerized to give proteins. The proteins are broken down into amino acids during digestion, and some of the amino acids are used for synthesis of new proteins and other compounds. The amino acids, which are not required as building materials, lose the amine group. This occurs mainly in the liver. The ammonia thus produced exists in equilibrium with ammonium ion

$$NH_3(aq) + H^+(aq) \rightleftharpoons NH_4^+(aq)$$

but at the pH of the cell, equilibrium is well over to the right, and ammonia in more than trace amounts is a tissue poison. It is involved in the liver in a cycle of reactions (the 'ornithine' cycle), from which it emerges in combination with carbon dioxide as the neutral soluble compound urea to be carried by the blood to the kidney for excretion.

The amino acid residues are metabolized, along with carbohydrates and fats, in a way common to practically all cells: two-carbon fragments from the tissue fuels combine with four-carbon carrier molecules, and undergo a series of reactions in which their energy is eventually transferred to organic phosphates, while their hydrogen is oxidized to water and their carbon is set free as carbon dioxide.

The neutrality of the body fluids is further guarded by a centre in the brain. When carbon dioxide accumulates in the blood above the normal level, as in active exercise, the respiratory rate is stepped up to increase the loss of carbon dioxide from the lungs. If the carbon dioxide level falls, as after forced breathing for a minute or two, the brain centre retards the rate of respiration until the carbon dioxide level in the blood has risen to normal. Any condition that alters the pH of the blood beyond what can be readily compensated causes illness.

Although the internal environment is maintained at a nearly neutral pH, organisms can produce acids where required. For example, higher animals' stomachs secrete a fluid containing free hydrochloric acid. Human gastric juice may have a pH as low as 2 or 1.5; the gastric juice of dogs can dissolve bones. This acid solution provides the right conditions for the stomach enzymes to act on food. The acid is neutralized by the food in the process of digestion, and is further neutralized, after leaving the stomach, by the pancreatic juice (pH up to 8.0). Indeed biochemically speaking, the stomach is outside the body, for the digestive canal is open to the external world at both ends, and the digested material has to traverse the cells that line the canal in order to enter the body.

There are other examples of biochemical acids. Some bacteria obtain energy from carbohydrates by a partial breakdown that does not require oxygen. This produces such acids as the lactic and butanoic acids of sour milk. If the bacteria are supplied with oxygen, the acids may be further metabolized. Citric acid is manufactured commercially from molasses or sucrose by taking advantage of the activities of the mould *Aspergillus niger*. The highly aerobic bacterium *Acetobacter* uses atmospheric oxygen to oxidize ethanol to ethanoic acid, turning wine to vinegar. The acids formed by these micro-organisms are by-products of their modes of obtaining energy; the acids accumulate in the media outside the cells, and may prove lethal to them in unbuffered conditions.

Green plants do not have the same need as the animal for protection against acids of internal origin. The plant's oxidative metabolism, and consequently its carbon dioxide production, is much slower, and the carbon dioxide, in daylight, is re-used for the manufacture of carbohydrate in photosynthesis. Similarly, the plant synthesizes its own amino acids for proteins, and does not have the animal's problem of excess ammonia arising from the metabolism of foreign proteins. The plant, therefore, has no need for an excretory system for either carbon dioxide or ammonia. It has, however, been suggested that some of the nitrogenous substances found in plants, such as the chemically complex and pharmaceutically active alkaloids (morphine, strychnine, quinine, etc.), which have no known function in the plant, may serve as a means of storing away excess nitrogen.

The main buffers in the plant cell are the same as in the animal, except that a more important part is played by organic acids, such as citric and malic acids, in equilibrium with their salts. The fluid or sap in the vacuoles that exist in many plant cells is less buffered than the living material, and is often more acid (pH 6.5 to 5.5, or even, in citrus fruits, as low as 2.4).

The chief acid problem of the plant is external, in the soil from which it must draw its supply of mineral salts and nitrogen as ammonium or nitrate ions. The pH of soil ranges from 10 or 11 in alkaline deserts to 3.5 at the most acid. Extremes can be tolerated only by specially adapted plants. The outer ranges of normality in soil, pH 8.0 to 4.5, can be recognized by the different flora characterizing the alkaline chalk and limestone soils on the one hand, and the acid peaty soils on the other. Liming soils to raise the pH is a common agricultural operation.

Besides varying in pH, soils vary in buffering power. Sand contains little buffer, and readily becomes too acid or too alkaline except for specialized plants, while a fertile soil full of decaying organic material is heavily buffered, both by the organic material and by the reactive alumino-silicate particles of clay, which are able to retain ions from the soil water.

The cells of plant root-hairs absorb ions from the soil by an active and selective process. It is active, since it occurs against a concentration gradient, and is accomplished with energy from oxidation; and it is selective, since the different ions absorbed need not be in the same relative proportions as in the soil. The absorption of ions by root-hairs may change the external pH. Thus absorption of nitrate ions from a solution, accompanied and balanced by

Figure 11.13 a Many crops grow better on slightly alkaline soils such as the Wiltshire downs **b** Heather flourishes on acid soil such as the Yorkshire moors

hydrogen ions from water or carbonic acid, leaves the solution more alkaline, because it contains more hydroxide ion or hydrogencarbonate ion. In the case of absorption of potassium ions, accompanied and balanced by hydroxide ion or hydrogencarbonate ion, the solution is left more acid because of increased hydrogen ion concentration.

Plants must cope not only with acids in the soil, but also with those in the atmosphere. When flashes of lightning pass through the atmosphere, they cause the nitrogen and the oxygen of the air to combine. This nitric oxide eventually falls to the earth as nitric acid in rainwater. The carbon dioxide emitted by organisms, and by the combustion of fuels, serves as source of carbon for plants, but also dissolves in rain water to form carbonic acid. As the water percolates through limestone or chalk, the carbonic acid reacts with the calcium carbonate to dissolve calcium as hydrogencarbonate

$$CaCO_3(s) + H_2CO_3(aq) \longrightarrow Ca(HCO_3)_2(aq)$$

so making the water 'hard'.

In the neighbourhood of towns, the air also contains sulphur dioxide, derived from the burning of the sulphur compounds in coal and oil; dissolved in rain, this gas gives sulphurous acid and may be partly oxidized to sulphuric acid. These acids are particularly noticeable in city fogs, and are notorious for irritating the lungs of animals, corroding metal and stonework, and stunting or killing plants.

We can see that any definition of acids has to encompass a vast number of reactions, and, in the case of organic chemistry, some very complex ones. Biochemistry is a rapidly advancing part of present day chemistry.

11.8 Entropy and equilibrium reactions

In Topic 10 on entropy we found out how to calculate the total entropy change for a reaction

$$\Delta S_{total} = \Delta S_{system} + \Delta S_{surroundings}$$

and concluded that for all spontaneous changes

ΔS_{total} is positive

But in this Topic we have been looking at reactions that are spontaneous in both directions:

acid + alcohol $\longrightarrow$ ester + water

is spontaneous but so is

ester + water $\longrightarrow$ acid + alcohol

How can the total entropy change be positive in both directions? The paradox is resolved when we look for the differences between the theoretical world of a system plus its surroundings and the real world of our laboratories.

We have been calculating **standard** entropy changes and to do so we have assumed that in the change from the initial state to the final state:
1 Pure reactants change to pure products with no intermediate mixing.
2 The temperature remained constant at 298 K.
3 The pressure remained constant at 1 atmosphere.

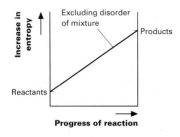

Figure 11.14 Entropy change as a reaction progresses, **excluding** the effect of the mixing of products with reactants

What is the effect on entropy changes when we do real experiments in conditions that are not the standard conditions?

EXPERIMENT 11.8

The $N_2O_4 \rightleftharpoons 2NO_2$ equilibrium

The nitrogen dioxide produced from the thermal decomposition of lead nitrate can be trapped in an ice bath and small amounts sealed in a glass tube if sufficient care is taken. Otherwise samples can be collected in a gas syringe.

Procedure

Have two beakers of water available, one cooled to less than 10 °C with ice and the other warmed to about 50 °C.

Note the colour of the gas mixture at room temperature, and what the changes are on cooling and warming.

If your sample of gas is in a gas syringe note the colour change when the pressure is suddenly altered, both increased and decreased.

- Remembering that N_2O_4 is colourless and NO_2 is brown, how do you interpret the changes in colour? The colour changes with pressure change are not so easy to interpret because the volume change will alter the colour anyhow.

SAFETY
Nitrogen oxides are severely irritant and toxic; this experiment must be carried out in a fume cupboard.

COMMENT
Sudden changes in pressure are approximately 'adiabatic' which means that no energy is exchanged with the surroundings.

The effect of product concentration on entropy change

Consider the conversion of dinitrogen tetroxide to nitrogen dioxide. The molecules of pure dinitrogen tetroxide will be disordered to a certain extent. The disorder will increase when some molecules decompose because there will be two types of molecule instead of one. The disorder will increase further when more molecules decompose, but not by so much because there are already some nitrogen dioxide molecules present.

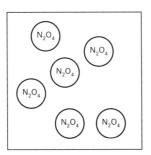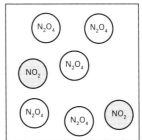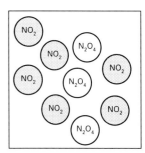

Figure 11.15 Increasing molecular disorder as a reaction progresses

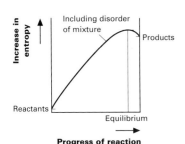

Figure 11.16 Entropy change as a reaction progresses, **including** the effect of mixing of products with reactants

So in the early stages of the reaction we get proportionately greater positive entropy changes in the system than at later stages. You could think of this as an example of the 'law of diminishing returns'. Each increase is not as great as the previous increase.

We can therefore expect that at some stage of the reaction there will be so many product molecules that producing any more will actually reduce the extent of disorder in the system. At this stage the entropy change of the system for the forward reaction will have become negative and as far as the system is concerned the reaction might as well 'stop'.

The amount of entropy change in the surroundings also depends on how much reaction has actually occurred; and the same arguments will apply to the reverse reaction. So we can deduce that at the position of dynamic equilibrium for the **total** entropy change

$$\Delta S_{total(forward)} = \Delta S_{total(reverse)}$$

This means that at equilibrium the overall entropy change is zero.

There is a simple relationship between standard entropy change and extent of reaction.

$$\Delta S^{\ominus}_{total} \propto \ln K_c$$

We can deduce from our discussion and this expression that **all reactions are reversible to some extent**.

When a reaction has a value for its standard total entropy change in the range $+40$ to -40, the reaction will produce a mixture of products and reactants that we regard as an equilibrium.

When a reaction has a total standard entropy change of $+200$ or more the concentration of reactants at 'equilibrium' will be less than the concentration of the traces of impurities that are present in all laboratory reagents, so we regard the reaction as '**complete**'.

When a reaction has a total standard entropy change of -200 or less, the concentration of products at 'equilibrium' will be so small that normal analytical procedures would not detect their presence, so we would normally say that the reaction '**does not go**'.

> **COMMENT**
> The relationship between entropy change and extent of reaction can be stated quantitatively (but you will not be expected to use this relationship):
>
> $\Delta S^{\ominus}_{total} = R \ln K_c$
>
> where $R = 8.31$ J K^{-1} mol^{-1}

The effect of temperature on entropy change.

We have already established in Topic 10 that the entropy change in the surroundings is

$$\Delta S_{surroundings} = \frac{-\Delta H_{reaction}}{T}$$

What happens to the value of the entropy change if we cool our apparatus in an ice bath, or heat it in boiling water, or even heat it in an electric furnace? The entropy change will no longer be a standard entropy but we can still use $\Delta H^{\ominus}(298)$ to estimate a value for the entropy change because $\Delta H^{\ominus}$ does not vary much with temperature.

For our reaction:

$$N_2O_4(g) \longrightarrow 2NO_2(g) \quad \Delta H^{\ominus}(298) = +57.2 \text{ kJ mol}^{-1}$$

so the entropy change in the surroundings (in J mol^{-1} K^{-1}) at different temperatures is

	0 °C	10 °C	100 °C
$\Delta S_{surroundings} = \dfrac{-\Delta H_{reaction}}{T} =$	-210	-202	-153

Figure 11.17 The variation of entropy change in the surroundings with temperature

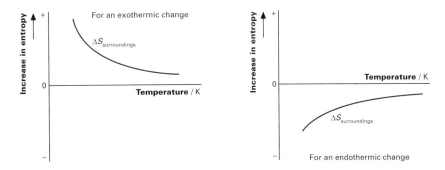

So we see that raising the temperature progressively decreases the extent of the unfavourable entropy change of an endothermic reaction. Since ΔS_{system} does not normally vary significantly with temperature we may even be able to reverse the direction of a reaction by manipulating the temperature.

Figure 11.18 The variation in total entropy change with temperature

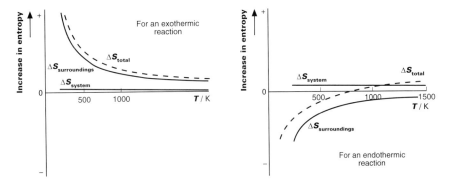

The decomposition of zinc carbonate that we looked at in experiment 10.1 part **6** will provide us with another example

$$ZnCO_3(s) \longrightarrow ZnO(s) + CO_2(g)$$

For this reaction we calculated

$$\Delta S^\ominus_{total} = \Delta S^\ominus_{system} + \Delta S^\ominus_{surroundings}$$
$$-63 \qquad +175 \qquad -238 \qquad \text{J mol}^{-1}\text{K}^{-1}$$

and $\quad \Delta H^\ominus_{reaction}(298) = +71000 \text{ J mol}^{-1}$

We can calculate the effect on the entropy change of raising the temperature, for example to 500 K, by using

$$\Delta S_{surroundings} = \frac{-\Delta H_{reaction}}{T}$$

so $\quad \Delta S_{surroundings}(500) = -71000 \div 500 = -142 \text{ J mol}^{-1}\text{K}^{-1}$

Therefore $\Delta S_{total}(500) \quad = +175 - 142 = +32 \text{ J mol}^{-1}\text{K}^{-1}$

and the reaction is now predicted to be spontaneous. This again matches our experience that zinc carbonate decomposes quite readily when heated.

Figure 11.19 Skater making use of the variation of entropy with pressure

The effect of pressure on entropy change

We should not expect the effect of pressure to be significant for solids and liquids (the melting point of ice changes from 0 °C to only −8 °C even when the pressure is increased to 1000 atmospheres). But the number of ways gas particles can be arranged alters a lot with a change of pressure

When the pressure on a gas is reduced the volume increases and this increases the number of ways of arranging the gas particles, which is an increase in the entropy of the gas.

So a reaction like

$$ZnCO_3(s) \rightleftharpoons ZnO(s) + CO_2(g)$$

will go further in the forward direction when the pressure is reduced.

In general you need to **lower** the pressure to shift a reaction in the direction of **more** molecules, and **raise** the pressure to shift a reaction in the direction of producing **fewer** molecules. For example:

high pressure ⟶
$$N_2 + 3H_2 \rightleftharpoons 2NH_3$$
⟵ low pressure

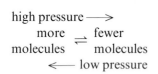

high pressure ⟶
more ⇌ fewer
molecules   molecules
⟵ low pressure

How to make reactions go the way you want

We can shift the position of equilibrium in favour of more product using the general guidelines:

1 Raise the temperature when the reaction is endothermic
2 Lower the temperature when the reaction is exothermic
3 Raise the pressure when there are fewer product molecules than reactant molecules
4 Lower the pressure when there are more reactant molecules than product molecules

The appropriate procedure is to do the opposite to the natural effect of the reaction:

- An endothermic reaction cools the apparatus, so heat it.
- A reduction in molecules causes a pressure drop, so compress it.

These guidelines have been formally stated as **Le Châtelier's Principle**:

When a system in equilibrium is subjected to a change, the processes which take place are such as to tend to counteract the change

STUDY TASK
Apply these ideas to the equilibrium

$$N_2O_4(g) \rightleftharpoons 2NO_2(g)$$

Review of entropy

This Topic has covered a wide variety of ideas and if you are to remember the important points without confusion you need to review the whole Topic now.

Use the summary which follows to plan a chart on which to show the main ideas and their connections. Include examples and illustrations, and even examples of situations to which the ideas do not apply, for example rates of reaction. Try to devise your own mnemonic for Le Châtelier's Principle.

If you can, use an A3 sheet of paper for planning your chart; and keep it visual and simple.

Summary

At the end of this Topic you should be able to:

a demonstrate understanding of the term equilibrium as applied to physical and chemical systems
b apply the Equilibrium Law to a chemical reaction in order to deduce the expression for the equilibrium constant, K_c
c perform simple calculations related to the Equilibrium Law
d understand and use the terms: acid, base, neutral, pH, indicator, buffer
e demonstrate understanding of the term equilibrium as applied to acid–base systems
f apply the Equilibrium Law to an acid–base system in order to deduce the expression for the equilibrium constant, K_a
g demonstrate understanding of practical methods of determining acid and alkali concentrations by titration
h plan an investigation using acid-alkali titrations
i perform calculations from acid-alkali titration data
j demonstrate understanding of the Brønsted-Lowry theory of acid–base behaviour, and use it to interpret the behaviour of strong acids–bases, weak acids–bases, conjugate acid–base pairs, and buffer solutions
k recall the terms pH, K_a, K_w and perform simple calculations using them
l deduce and interpret qualitatively the effect of changes in temperature and pressure on systems at equilibrium in terms of entropy changes.

Review questions

* Indicates that the *Book of data* is needed.

11.1 11 g of ethyl ethanoate were mixed with 18 cm^3 of 1.0 M hydrochloric acid in a flask and allowed to stand at constant temperature until equilibrium had been reached.

$$CH_3CO_2C_2H_5(l) + H_2O(l) \rightleftharpoons CH_3CO_2H(l) + C_2H_5OH(l)$$

The contents of the flask were titrated with 1.0 M sodium hydroxide solution and 106 cm^3 of the alkali were required. Calculate the equilibrium constant K_c. (Assume that 18 cm^3 of 1.0 M hydrochloric acid contain 18 g of water.)

11.2 The equilibrium

$$N_2O_4 \rightleftharpoons 2NO_2$$

can be established in trichloromethane solution at temperatures near 0 °C, and the composition of the equilibrium mixture can be calculated from the density of colour of the solution as N_2O_4 is colourless and NO_2 is brown. In a solution of this kind, at 10 °C, the concentration of NO_2 molecules was found to be 0.0014 mol dm^{-3} and the concentration of N_2O_4 molecules, 0.19 mol dm^{-3}.
Calculate the value of K_c for the reaction at 10 °C.

11.3 Propanone and hydrocyanic acid react in ethanol solution to form a product called 2-hydroxy-2-methylpropanenitrile, according to the equilibrium equation.

$$CH_3COCH_3 + HCN \rightleftharpoons CH_3-\underset{CN}{\overset{OH}{C}}-CH_3$$

At 20 °C, K_c for this equilibrium is 32.8 dm^3 mol^{-1}. When 100 cm^3 of 0.1 M solution of propanone in ethanol are mixed with 100 cm^3 of 0.2 M solution of hydrocyanic acid in ethanol, what mass of the product will be formed at equilibrium? (Molar masses: H = 1, C = 12, N = 14, O = 16 g mol^{-1})

11.4 For the equilibrium

$$C_2H_5OH(l) + \underset{\text{propanoic acid}}{C_2H_5CO_2H(l)} \rightleftharpoons \underset{\text{ethyl propanoate}}{C_2H_5CO_2C_2H_5(l)} + H_2O(l)$$

$K_c = 7.5$ at 50 °C. What mass of ethanol must be mixed with 60 g of propanoic acid at 50 °C in order to obtain 80 g of ethyl propanoate in the equilibrium mixture? (Molar masses: H = 1, C = 12, O = 16 g mol^{-1})

11.5 Calculate the pH of the following solutions at 25 °C. In parts **a** to **d** assume complete ionization.

- **a** 0.2 M HCl
- **b** 0.2 M KOH
- **c** 0.125 M HNO$_3$
- **d** A mixture of 75 cm^3 of 0.1 M HCl and 25 cm^3 of 0.1 M NaOH
- **e** 0.1 M bromoethanoic acid (CH$_2$BrCO$_2$H; $K_a = 1.35 \times 10^{-3}$ mol dm^{-3})

*11.6 What is the concentration of methanoate ion (in mol dm^{-3}) in 0.01 M methanoic acid solution at 25 °C?

11.7 In a 0.1 M solution of an acid HA, $[A^-(aq)]_{eq} = 1.3 \times 10^{-3}$ mol dm^{-3}; calculate K_a for the acid.

11.8 A 0.1 M solution of an acid HA has a pH of 5.1; calculate K_a for the acid.

*11.9 Calculate the pH of a 0.001 M solution of phenylammonium chloride at 25 °C. Assume that phenylammonium chloride, C$_6$H$_5$NH$_3$Cl, is fully ionized.

*11.10 List entropy data for the equilibrium

$$2SO_2(g) + O_2(g) \rightleftharpoons 2SO_3(g) \quad \Delta H^\ominus = -197 \text{ kJ mol}^{-1}$$

Work out the entropy changes for the reaction with the compounds in their standard states, and estimate the position of equilibrium at 298 K.
Calculate $\Delta S^\ominus_{total}$ for the equilibrium at 750 K, at 1250 K and determine the temperature at which $\Delta S^\ominus_{total} = 0$. ($S^\ominus[SO_3(g)] = -256.1$ J mol^{-1} K^{-1})

Examination questions

11.11 Dichloroethanoic acid reacts with pent-1-ene under suitable conditions according to the following equation:

$$CHCl_2CO_2H(l) + C_5H_{10}(l) \rightleftharpoons CHCl_2CO_2C_5H_{11}(l)$$

In an experiment, 0.50 mol of dichloroethanoic acid and 1.15 mol of pent-1-ene were allowed to reach equilibrium. It was found that 0.20 mol of dichloroethanoic acid remained in the equilibrium mixture and that the volume of the mixture remained constant at 0.15 dm³ throughout the experiment.

- **a i** How would you measure out the reactants most conveniently?
- **ii** What other information would you need before measuring out the reactants?
- **iii** Describe how you would measure experimentally the concentration of dichloroethanoic acid in the equilibrium mixture.
- **b i** Draw the displayed formula of pent-1-ene.
- **ii** Into which class of organic compound can $CHCl_2CO_2C_5H_{11}$ be placed?
- **c i** Write down an expression for the equilibrium constant, K_c, for the reaction.
- **ii** Calculate the number of moles of $CHCl_2CO_2C_5H_{11}$ in the equilibrium mixture.
- **iii** Calculate the number of moles of C_5H_{10} left in the equilibrium mixture.
- **iv** Work out the equilibrium concentrations of the three components, and use them to find the value of the equilibrium constant.

11.12 The equilibrium between hydrogen, iodine and hydrogen iodide can be investigated by sealing hydrogen iodide in glass tubes and heating them at known temperatures until equilibrium is reached.

The equation for the reaction is

$$2HI(g) \rightleftharpoons H_2(g) + I_2(g)$$

and the equilibrium constant $K_c = 0.019$ at 698 K.

The tubes are rapidly cooled and then opened under potassium iodide solution when the iodine and hydrogen iodide dissolve.

- **a i** Why are the tubes **rapidly** cooled?
- **ii** Describe how the appearance of the contents of a tube would change as it was cooled.
- **iii** Outline a practical procedure for measuring the amount of iodine dissolved from a tube.
- **b i** Write an expression for the equilibrium constant, K_c.
- **ii** A sample tube is found to contain iodine at a concentration of 4.8×10^{-4} mol dm⁻³. Using the equation, deduce the equilibrium concentration of hydrogen, $[H_2(g)]_{eq}$.
- **iii** Calculate the equilibrium concentration of hydrogen iodide.

11.13 a In this reaction hydrochloric acid is acting as an acid:

$$HCl(aq) + H_2O(l) \rightleftharpoons H_3O^+(aq) + Cl^-(aq)$$

- **i** Identify the Brønsted-Lowry acids and bases.
- **ii** What is the difference between an acid and its conjugate base? Illustrate your answer by means of an example.

- iii Hydrochloric acid is a strong acid. Explain the term **strong acid**.
- iv Calculate the pH of 0.1 mol dm^{-3} hydrochloric acid.
- b Ethanoic acid is a weak acid.

$$CH_3CO_2H(aq) + H_2O(l) \rightleftharpoons CH_3CO_2^-(aq) + H_3O^+(aq)$$
$$K_a = 1.7 \times 10^{-5} \text{ mol dm}^{-3}$$

- i Write down the expression for K_a for ethanoic acid.
- ii When calculating the pH of ethanoic acid it is usual to make two assumptions:
  1. $[CH_3CO_2^-(aq)]_{eq} = [H_3O^+]_{eq}$
  2. The equilibrium concentration of the ethanoic acid is equal to the initial concentration of the ethanoic acid.

  How would you justify these assumptions?
- iii Using these assumptions and your expression for K_a, calculate the pH of 0.1 mol dm^{-3} ethanoic acid.
- c For the reaction between hydrochloric acid and calcium carbonate, the following entropy changes have been calculated at 298 K.

$$\Delta S^\ominus_{system} = +137.5 \text{ J mol}^{-1} \text{ K}^{-1} \qquad \Delta S^\ominus_{surroundings} = +51.0 \text{ J mol}^{-1} \text{ K}^{-1}$$

Calculate $\Delta S^\ominus_{total}$ for the reaction and comment on the value you obtain.

11.14 The ingredients of a typical lemonade are:

> INGREDIENTS: CARBONATED WATER, CITRIC ACID, FLAVOURING
> ACIDITY REGULATOR: SODIUM CITRATE;
> ARTIFICIAL SWEETENER: ASPARTAME
> PRESERVATIVE: SODIUM BENZOATE
> STABILIZER: CMC

- a Carbonated water is an aqueous solution of carbon dioxide. Some of the dissolved carbon dioxide reacts to form carbonic acid, H_2CO_3. The carbonic acid then ionizes

$$H_2CO_3(aq) + H_2O(l) \rightleftharpoons H_3O^+(aq) + HCO_3^-(aq)$$
$$K_a = 2.0 \times 10^{-4} \text{ mol dm}^{-3}$$

- i Explain why carbonic acid is classified as a Brønsted-Lowry acid.
- ii Comment on the strength of carbonic acid as an acid.
- iii Write an expression for the ionization (dissociation) constant for carbonic acid, K_a.
- iv Calculate the pH of 0.1 mol dm^{-3} carbonic acid. State any assumptions you make.
- v Give the name for the mixture of a solution of a weak acid and the salt of the acid, like citric acid and sodium citrate.
- vi Describe and explain the behaviour of a solution of citric acid and sodium citrate when an acid is added.
- b Suggest the signs of the entropy change in the system, $\Delta S^\ominus_{system}$, and the total entropy change, $\Delta S^\ominus_{total}$, for the escape of carbon dioxide from lemonade. Justify your suggestions.

TOPIC 12

Carbon compounds with acidic and basic properties

This third Topic on the organic chemistry of carbon is concerned with some compounds that have acidic properties, the **carboxylic acids**, the compounds that can be made from them, and some compounds that have basic properties, the **amines**. You will also study the properties of phenol.

Ethanoic acid, a carboxylic acid

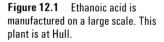

Ethanoic anhydride, an acid anhydride

$CH_3\text{—}CH_2\text{—}NH_2$

Ethylamine, an amine

Phenol

Figure 12.1 Ethanoic acid is manufactured on a large scale. This plant is at Hull.

Carboxylic acids are manufactured on a large scale because they are used in foods and to produce other products as diverse as flavours, perfumes, solvents and polymers.

12.1 Carboxylic acids

alcohol —OH

carbonyl $\diagdown$C=O

carboxylic acid —C$\diagup^{\displaystyle O}_{\displaystyle OH}$

We have already met the hydroxyl group and the carbonyl group in Topic 2 but in the carboxylic acids we meet them on **one** carbon atom, rather than separately.

We might expect some significant alterations in reactivity rather than the sum of the separate reactions of the two groups.

The most significant change is the ionization of the functional group.

$$CH_3-C\underset{OH}{\overset{O}{\diagup\!\!\!\diagdown}} + H_2O \rightleftharpoons CH_3-C\underset{O}{\overset{O^-}{\diagup\!\!\!\diagdown}} + H_3O^+$$

When the functional group ionizes, the resulting anion is stabilized by the π electrons from the double bond and an electron pair from the other oxygen atom, forming a delocalized system. This results in a symmetrical structure with both C—O bonds having the same length. Delocalization was introduced in Topic 6.5 and you should read this section again to refresh your memory on this subject.

The dipole due to the C=O double bond in the carboxylic group means that the carbon atom of the functional group is electrophilic in nature. Reagents such as ammonia, however, react as bases forming a salt

$$CH_3-CO_2H + NH_3 \longrightarrow CH_3-CO_2^- + NH_4^+$$

rather than acting as nucleophiles and attacking the carbon atom.

Very weak nucleophiles, such as ethanol $CH_3CH_2\ddot{O}H$, will react, attacking the carbon atom, but only when catalysed by H^+ ions.

Experiments with ethanoic acid

Pure ethanoic acid is sometimes known as **glacial** ethanoic acid because it freezes at 17 °C and its ice-like crystals were often observed in unheated laboratories. An alternative name for ethanoic acid is acetic acid (from the Latin *acetum* or vinegar, of which it is the active constituent). In these experiments use pure, that is, glacial, ethanoic acid.

EXPERIMENT 12.1a The reactions of ethanoic acid

Procedure

1 Solubility and pH

To 1 cm³ of pure ethanoic acid in a test-tube, add water in drops. Do they mix in all proportions? Add a few drops of Full-range Indicator. Finally add sodium carbonate solution.

- When ethanoic acid and water were mixed, what type of molecular interaction was helping them to dissolve?
  Is it a strong enough acid to displace carbon dioxide from carbonates? Write the equation of the reaction.
  Look up the K_a values of some of the carboxylic acids (in margin overleaf). Is ethanoic acid a strong or a weak acid?

> **SAFETY** ⚠
> Ethanoic acid is corrosive; it has a pungent odour, and can cause painful blisters if left in contact with the skin. Wash it off immediately with plenty of water.

12 Carbon compounds with acidic and basic properties

2 Formation of salts

To 10 cm³ of 0.1 M ethanoic acid in a beaker add a few drops of Full-range Indicator (or use a pH meter); then add, while stirring, 3 cm³ portions of 0.1 M sodium hydroxide until you have added a total of 15 cm³. Note the pH after each addition. Also measure the pH of a solution of sodium ethanoate.

- Explain what happens.

3 Formation of esters

Put 1 cm³ of concentrated ethanoic acid in a test-tube and add 2 cm³ of ethanol and two or three drops of concentrated sulphuric acid (TAKE CARE).

Warm the mixture gently in a hot water bath for five minutes, when an ester will be formed.

What does the product smell like? How does this compare with the smells of the starting materials? Pour the contents of the test-tube into a small beaker of sodium carbonate solution, to neutralize any excess of acid. Stir well and smell again. Is it like the smell of ethyl ethanoate from the bottle?

Cautiously smell a range of esters if they are available.

- Write down what you see and smell. Which ester smells do you recognize? Which esters smell fruity?

You can also attempt to prepare one of the esters listed in figure 12.2 using the same procedure.

Flavour	Alcohol		Carboxylic acid	
Banana	Pentan-2-ol	0.01 mole	Ethanoic acid	0.03 mole
Peach	Benzyl alcohol	0.01 mole	Ethanoic acid	0.03 mole
Pear	Propan-1-ol	0.02 mole	Ethanoic acid	0.04 mole
Pineapple	Ethanol	0.015 mole	Butanoic acid	0.025 mole

Figure 12.2 Flavours of some esters.

Carboxylic acids: acid properties involving the O—H bond

1 Solubility and pH

The carboxylic acids C_1 to C_4 mix with water in all proportions but at C_5 and thereafter solubility rapidly reduces. The molecules will hydrogen bond to each other as well as to water molecules.

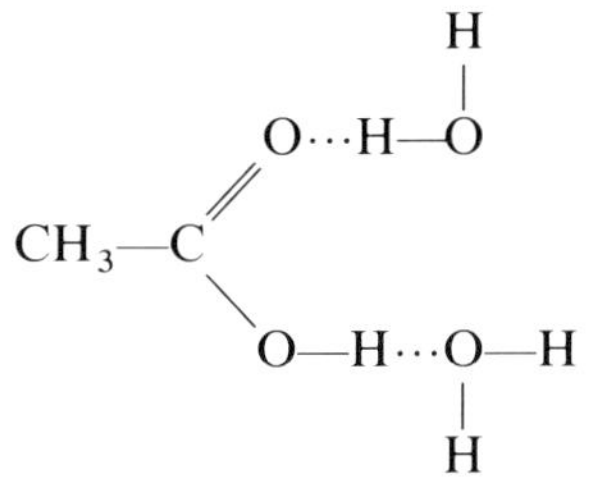

The carboxylic acids are only weak acids in water, as can be seen from their K_a values:

$$RCO_2H(aq) \rightleftharpoons H^+(aq) + RCO_2^-(aq)$$

They are usually less than 1% ionized in water and do not readily produce hydrogen by reaction with metals.

The infra-red spectrum of a carboxylic acid such as ethanoic acid shows a broad absorption due to hydrogen bonding of the O—H group around 3100 cm⁻¹ (figure 12.3). Ethanol has a less pronounced broad absorption also due to hydrogen bonding (figure 12.4).

The infra-red spectrum of a carboxylic acid also shows the characteristic absorption at 1740 cm⁻¹ due to the C═O group, which is again a broader trough. But the hydrogen atoms in propanone are not sufficiently polar to form

Compound	$K_a/10^{-5}$ mol dm⁻³
Methanoic acid	16
Ethanoic acid	1.7
Propanoic acid	1.3
Butanoic acid	1.5

K_a values for some carboxylic acids.

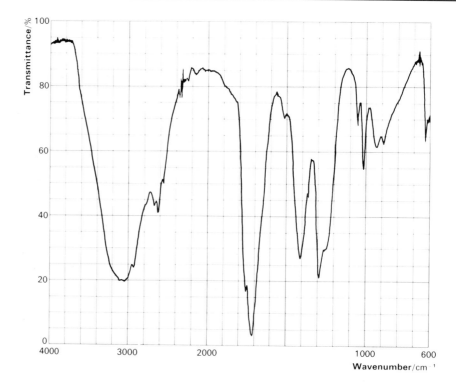

Figure 12.3 The infra-red spectrum of ethanoic acid, showing broad absorptions for both O—H and C=O due to hydrogen bonding.

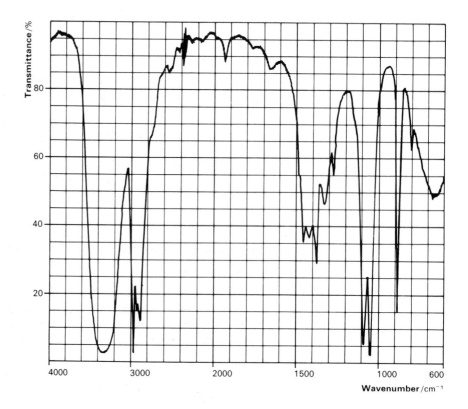

Figure 12.4 The infra-red spectrum of ethanol, showing an absorption at 3300 cm^{-1} that is broadened by hydrogen bonding.

hydrogen bonds so the infra-red spectrum of pure propanone does not have broadened troughs (figure 12.5).

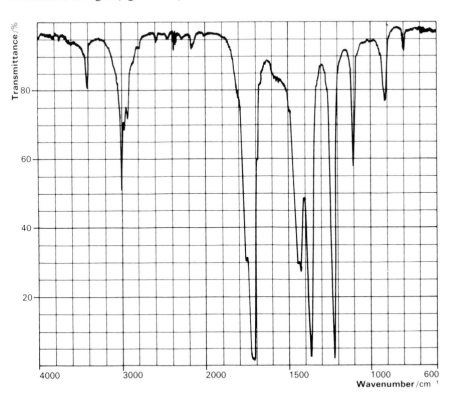

Figure 12.5 The infra-red spectrum of propanone, showing a characteristic sharp absorption at 1740 cm^{-1}

2 Formation of salts

The carboxylic acids are strong enough acids to displace carbon dioxide from sodium carbonate and will neutralize sodium hydroxide, forming a sodium salt.

$$2CH_3CO_2H + Na_2^+CO_3^{2-} \longrightarrow 2CH_3CO_2^-Na^+ + H_2O + CO_2$$
$$CH_3CO_2H + Na^+OH^- \longrightarrow CH_3CO_2^-Na^+ + H_2O$$

Because ethanoic acid is a weak acid, a sodium ethanoate–ethanoic acid mixture can be used as a buffer solution, changing pH only slightly on the addition of a strong acid or a strong base (see Topic 11).

Nucleophilic reactions involving the C=O bond

3 Formation of esters

Carboxylic acids react with alcohols to form **esters**. Alcohols have lone pairs of electrons on their oxygen atom and can therefore act as nucleophiles. The reaction is slow unless the mixture is warmed and an acid is present to act as a catalyst: concentrated sulphuric acid is suitable. The reaction is reversible.

$$\text{C}_6\text{H}_5\text{—CO}_2\text{H} + \text{CH}_3\text{OH} \xrightleftharpoons{\text{H}^+} \text{C}_6\text{H}_5\text{—CO}_2\text{CH}_3 + \text{H}_2\text{O}$$

benzoic acid   methanol   methyl benzoate

From this equation you can see that the water produced may have derived its oxygen atom from either the acid or the alcohol. Check the structural formulae and notice that both molecules have an O—H group. In 1938 two American chemists, Roberts and Urey, found out which compound provided the oxygen for the water, by using methanol containing a high proportion of the isotope ^{18}O.

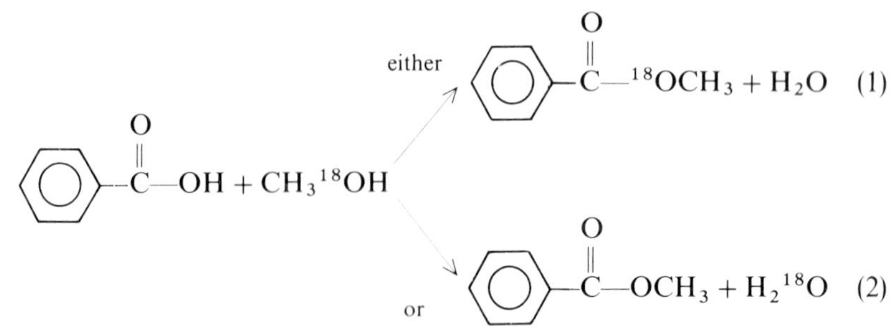

$$\text{C}_6\text{H}_5\text{—CO—OH} + \text{CH}_3{}^{18}\text{OH} \longrightarrow \begin{cases} \text{either} \quad \text{C}_6\text{H}_5\text{—CO—}^{18}\text{OCH}_3 + \text{H}_2\text{O} \quad (1) \\ \text{or} \quad \text{C}_6\text{H}_5\text{—CO—OCH}_3 + \text{H}_2{}^{18}\text{O} \quad (2) \end{cases}$$

Using a mass spectrometer to determine the masses of the products they were able to establish that ^{18}O isotope appears in the ester (equation (1)) and not the water (equation (2)). Thus the new bond was formed between the C atom in the acid and the ^{18}O atom in the alcohol, with the loss of the —OH group from the acid and not from the alcohol. Chemists consider that the first step in an acid-catalysed esterification is the addition of a proton to the C=O group of the acid:

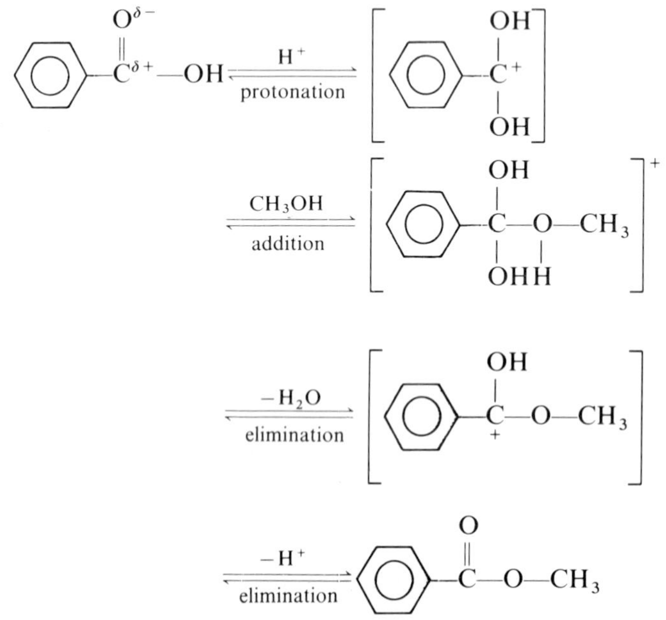

$$\text{C}_6\text{H}_5\text{—}\overset{\text{O}^{\delta-}}{\underset{\|}{\text{C}^{\delta+}}}\text{—OH} \xrightleftharpoons[]{\text{H}^+ \text{ protonation}} \left[\text{C}_6\text{H}_5\text{—}\overset{\text{OH}}{\underset{\text{OH}}{\text{C}^+}} \right]$$

$$\xrightleftharpoons[]{\text{CH}_3\text{OH} \text{ addition}} \left[\text{C}_6\text{H}_5\text{—}\overset{\text{OH}}{\underset{\text{OHH}}{\text{C—O—CH}_3}} \right]^+$$

$$\xrightleftharpoons[]{-\text{H}_2\text{O} \text{ elimination}} \left[\text{C}_6\text{H}_5\text{—}\overset{\text{OH}}{\underset{+}{\text{C—O—CH}_3}} \right]$$

$$\xrightleftharpoons[]{-\text{H}^+ \text{ elimination}} \text{C}_6\text{H}_5\text{—CO—O—CH}_3$$

This is an addition-elimination reaction in which the overall process is substitution of the —OH group by the nucleophilic CH$_3$O— group. You are not expected to learn this set of equations.

4 Reduction

Carboxylic acids can be reduced to alcohols by lithium tetrahydridoaluminate, $LiAlH_4$.

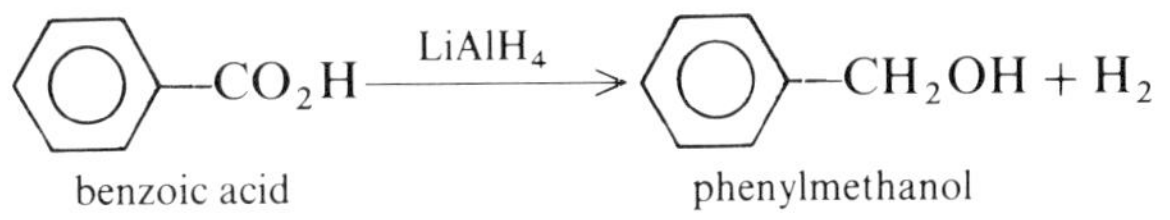

$$\text{C}_6\text{H}_5\text{—CO}_2\text{H} \xrightarrow{LiAlH_4} \text{C}_6\text{H}_5\text{—CH}_2\text{OH} + H_2$$

benzoic acid → phenylmethanol

The reaction proceeds by a nucleophilic attack on the carboxylate ion by the hydride ion, $:H^-$. The same reagent will also reduce aldehydes and ketones to alcohols.

EXPERIMENT 12.1b Oxidation of cyclohexanone

Cyclohexanone can be oxidized to hexanedioic acid by potassium manganate(VII) in a slow ring-opening reaction.

Add 10 cm³ of 2 M sodium hydroxide to 250 cm³ of 0.4 M potassium manganate(VII) (an almost saturated solution) and adjust the temperature to 30 °C. Add 5 g of cyclohexanone (5.3 cm³) and leave for at least 18 hours in a warm room.

Attempt to separate and purify the product, hexanedioic acid, using your knowledge of acid-base reactions and solubility. Helpful data can be found in table 5.5 of the *Book of data*.

Make a risk assessment before starting any separation and purification procedure.

12.2 Carboxylic acid derivatives

The derivatives of the carboxylic acids are considered to be those compounds in which another group appears in the —CO_2H group in the place of the —OH. In this section, we shall be concerned with **acid anhydrides**, **acyl chlorides**, and **amides**.

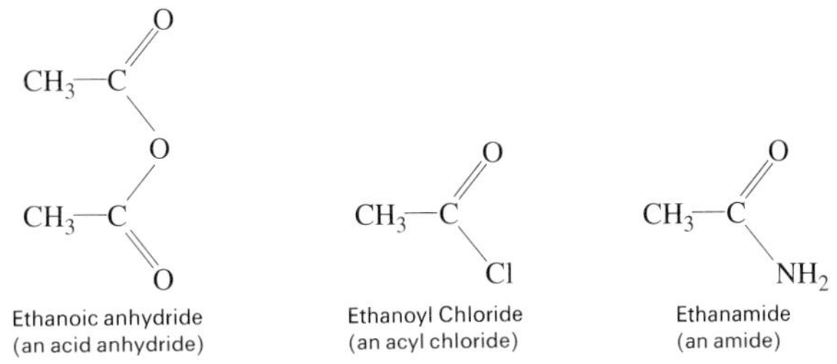

Ethanoic anhydride (an acid anhydride)

Ethanoyl Chloride (an acyl chloride)

Ethanamide (an amide)

EXPERIMENT 12.2 Some reactions of carboxylic acid derivatives

Some of these experiments will be demonstrated to you; you may be able to carry out others for yourself.

12 Carbon compounds with acidic and basic properties

> **SAFETY** ⚠
> Ethanoic anhydride and ethanoyl chloride are corrosive and flammable; they are volatile and form pungent fumes in moist air. Eye protection should be worn throughout this experiment, even when watching a demonstration.

Procedure

1 Demonstration of the reactions of an acid anhydride

a Put 5 cm³ of water in a small (100 cm³) beaker and **very carefully** add a little ethanoic anhydride, one drop at a time. Note the vigour of the reaction.

b Put 1 cm³ of ethanol in a small beaker and **carefully** add 1 cm³ of ethanoic anhydride one drop at a time. When the reaction has subsided, cautiously add sodium carbonate solution to neutralize any acids present and then see if you can detect the presence of an ester by its smell.

c Put 1 cm³ of 8 M ammonia in a small beaker and add 5 **drops** of ethanoic anhydride. Note the fumes produced, then evaporate off the water; a solid product should be obtained.

- For each of the experiments **1a** to **1c**, write an equation showing the structures of the compounds, name the nucleophile involved, and identify the bond which breaks in the ethanoic anhydride.

2 Demonstration of the reactions of an acyl chloride

Repeat **1** using ethanoyl chloride instead of ethanoic anhydride. Be prepared for more vigorous reactions.

- For each of the experiments **2a** to **2c**, write an equation showing the structures of the compounds, name the nucleophile involved, and identify the bond which breaks in the ethanoyl chloride.

3 The reactions of an acid amide

Place 5 cm³ of water in a beaker and add some solid ethanamide (note the smell). Is there any sign of reaction? Now add 5 cm³ of 2 M sodium hydroxide and boil gently. Is any gas evolved? Test by smell and pH.

- Write an equation showing the structures of the compounds, name the nucleophile involved, and identify the bond which breaks in the ethanamide.
  Devise an experiment to see if the same reaction can be catalysed by acid.

Reactions of carboxylic acid derivatives

Nucleophilic reactions

The hydrolysis of the carboxylic acid derivatives amounts to a nucleophilic substitution

$$CH_3-C(=O)W + :OH^- \longrightarrow CH_3-C(=O)OH + :W^-$$

where $—W:$ and $:W^-$ can be $—Cl$ and Cl^-

or $—OCH_3$ and OCH_3^-

or $—NH_2$ and NH_2^-

The ease with which the substitution occurs is related to the power of $—W:$ to attract electrons.

If —W: is strongly electron-attracting then the acid derivative can be hydrolysed by a weak nucleophilic reagent such as water; if —W: is only weakly electron-attracting, then the hydrolysis of the acid derivative will need a strong nucleophilic reagent such as hydroxide ions.

The relative electron-attracting powers of —W: can be summarized as

$$-Cl > -OH > -OCH_3 > -NH_2$$

Acid anhydrides and **acyl chlorides** which mix with water react rapidly, even violently in the case of acyl chlorides, to give the parent acid

$$CH_3-COCl + H_2\ddot{O} \longrightarrow CH_3-CO_2H + HCl$$

Other nucleophiles such as alcohols, ammonia, and amines also react readily and this is an excellent method of preparing esters and amides.

$$CH_3-COCl + CH_3\ddot{O}H \longrightarrow CH_3-CO_2-CH_3 + HCl$$
methyl ethanoate

$$CH_3-COCl + \ddot{N}H_3 \longrightarrow CH_3-CO-\ddot{N}H_2 + HCl$$
ethanamide

$$CH_3-COCl + CH_3NH_2 \longrightarrow CH_3-CONH-CH_3 + HCl$$
N-methyl-ethanamide

The equations for the reactions of ethanoic anhydride are similar except that ethanoic acid is produced instead of hydrogen chloride.

Esters react quite slowly with water and an acid or base catalyst is necessary

$$C_6H_5-CO_2CH_3 + H_2O \xrightleftharpoons{H^+} C_6H_5-CO_2H + CH_3OH$$

If a base is used the product is the anion of the carboxylic acid. The anion is stable to attack by weak nucleophiles such as alcohols and the reaction is therefore not reversible when a base is used.

$$C_6H_5-CO_2CH_3 + OH^- \longrightarrow C_6H_5-CO_2^- + CH_3OH$$

Amides are hydrolysed very slowly. Even with an acid catalyst the reaction is slow but a base is more effective.

$$CH_3CONH_2 + H_2O \xrightarrow{OH^-} CH_3CO_2H + NH_3 \longrightarrow CH_3CO_2^-NH_4^+$$
ammonium ethanoate

Reduction

The reduction of the derivatives of carboxylic acids by lithium tetrahydridoaluminate, $LiAlH_4$, proceeds in the same manner as the reduction of the parent acids. The products from acyl chlorides and esters are alcohols.

$$C_6H_5-CO_2CH_3 \xrightarrow{LiAlH_4} C_6H_5-CH_2OH + CH_3OH$$
phenylmethanol

COMMENT

Some insect sex attractants are unsaturated esters. They are examples of **pheromones**, compounds used by animals to communicate with each other (see Topic 16). An example is *cis*-7-dodecenyl ethanoate, used by the cabbage looper moth.

They are given out by female insects in the most minute amounts in order to attract males. When the synthetic compounds are sprayed the males are confused and fail to find females and mate. This can be an environmentally 'clean' way of reducing insect damage.

12.3 Phenols

In alcohols the hydroxyl group, —OH, is attached to an alkyl group, and in carboxylic acids it is attached to a C=O. You are now going to examine phenols, compounds in which the molecules have a hydroxyl group attached to a benzene ring.

CH₃—CH₂—OH   CH₃—C(=O)OH   C₆H₅—OH
ethanol   ethanoic acid   phenol

EXPERIMENT 12.3a The reactions of the phenolic functional group

In these experiments your principal objective will be to find out how the properties of the —OH group are modified by being attached to a benzene ring. You will thus see some of the characteristic features of the chemistry of phenolic compounds.

Because of the toxic and corrosive nature of many phenols we suggest methyl 4-hydroxybenzoate is used as an example of a phenolic compound.

1 Solubility in water

Put about 5 cm³ of water in a test-tube and add a small quantity of your phenolic compound, methyl 4-hydroxybenzoate. Heat to boiling and allow to cool slowly.

■ Does the compound dissolve in water?

Test the solution with Full–range Indicator.

■ What is the pH of the solution?
Compare the results with the effect of an ethanol–water and ethanoic acid–water mixture on Full–range Indicator.

Methyl 4-hydroxybenzoate

2 Phenolic compounds as acids

a *Action of sodium*   Place a small amount of your phenolic compound in a **dry** test-tube and warm until molten. Add a small cube (2–3 mm side) of sodium (TAKE CARE: wear eye protection) and watch carefully. **Do not heat the tube continuously.** What do you observe? Dispose of the contents with care: do **not** pour down the sink in case some sodium remains but carefully add 1–2 cm³ of ethanol to the **cool** test-tube, wait until all bubbling has ceased, and then pour away.

> **SAFETY** ⚠
> Sodium is corrosive and flammable; ethanoic anhydride and bromine are corrosive; iron(III) chloride is irritant.

- What bond has been broken in the phenol, a C—H, a C—O or an O—H bond?
  Compare this reaction with that of ethanol and sodium.

b *Action of sodium hydroxide* To about 5 cm³ of 2 M sodium hydroxide solution in a test-tube add a small amount of your phenolic compound and warm. Compare the solubility in alkali with the solubility in water. Now add about 2 cm³ of dilute hydrochloric acid.

- What do you observe and what does this tell you? Compare this reaction with that of ethanoic acid.

c *Action of sodium carbonate* To about 5 cm³ of 1 M sodium carbonate solution add a small amount of your phenolic compound.

- Is there an effervescence of carbon dioxide gas? Does this suggest that your phenolic compound is a strong or a weak acid?

3 Formation of an ester

In this experiment you will try to make an ester, using the —OH group of your phenolic compound in the place of the —OH group of an alcohol. Place about 0.5 g of your phenolic compound in a test-tube and add 4 cm³ of 2 M sodium hydroxide to dissolve the phenolic compound. Add 1 cm³ of ethanoic anhydride, cork the test-tube, and shake it for a few minutes.

An emulsion of an ethanoate ester should form. Cool in an ice bath to obtain the crystalline product.

- In what ways did the conditions for this reaction differ from the preparation of esters in experiment 12.1a part **3**?

4 Properties of the benzene ring

a *Combustion* Set fire to a small amount of your phenolic compound on a combustion spoon.

- What sort of flame do you see? Is a similar flame obtained when ethanol and ethanoic acid are burnt?
  Compare these results with your results for the combustions of alkanes and arenes in Topic 7. Does the presence of a hydroxyl group alter the result?

b *Iron(III) chloride* Dissolve two or three small crystals of iron(III) chloride in about 5 cm³ of water, and add about 1 cm³ of a solution of your phenolic compound in water. Note the intense colour that is formed.

- The formation of intense colours is characteristic of several compounds having hydroxyl groups attached to benzene rings. Find out if this also happens with ethanol.

c *Bromine* Shake a small amount of your phenolic compound with 5 cm³ of water and add bromine water.

- How readily does reaction occur?

> **COMMENT**
> The product from **4 d** can be examined by chromatography in 1:1 hexane/ethoxyethane on a silica thin layer; view the result in ultraviolet light. Make a risk assessment if you attempt this experiment.

d *Nitric acid* To a small amount of your phenolic compound add 2 M nitric acid. Heat to boiling, allow to cool then chill rapidly in an ice bath.

- How readily do coloured products appear?
  Compare your results with those obtained using methoxybenzene in experiment 7.7.
  Do you think these are substitution reactions? Has the phenolic compound been attacked by electrophilic reagents?

A comparison of phenols with ethanoic acid and ethanol

We will discuss the properties of phenolic compounds using the simplest compound as our example, phenol itself.

Phenol

1 Phenol as an acid

Phenol loses a proton much less readily than ethanoic acid; they both form sodium salts but phenol will not displace carbon dioxide from hydrogencarbonate ions.

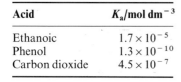

sodium phenoxide

Acid	K_a/mol dm^{-3}
Ethanoic	1.7×10^{-5}
Phenol	1.3×10^{-10}
Carbon dioxide	4.5×10^{-7}

It is suggested that the phenoxide ion is formed because one of the lone pairs of electrons on the oxygen atom in phenol can join the delocalized electrons of the benzene ring. In this way, the negative charge of the ion is stabilized by being spread out over the whole structure (figure 12.6). Nevertheless the delocalization in phenol is not as effective as the stabilization of the charge in a carboxylic acid functional group.

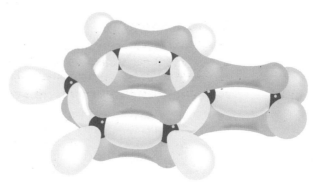

Figure 12.6 A model showing the delocalized orbitals in the phenoxide ring

The ethoxide ion cannot form a delocalized system and so cannot gain the same stability as the phenoxide ion. There is therefore no ionization of ethanol in water.

The delocalization of electrons in the benzene ring was described in Topic 7, and in carboxylic acids in Topic 6.

So phenol is only a weak acid; the pH of its solution in water is only slightly less than 7. (Phenol was formerly known as carbolic acid.)

2 Phenol as an alcohol

a The interaction of a lone pair of electrons on the oxygen atom of phenol with the benzene ring means that phenol normally forms esters much less readily than ethanol.

$$\text{carboxylic acid} + \text{alcohol} \longrightarrow \text{ester} + \text{water}$$

Instead of using a carboxylic acid with an acid catalyst as in the case of ethanol, the more reactive carboxylic acid anhydrides or the acyl chlorides must be used.

However, in the esterification reaction the alcohol is reacting as a nucleophile, $CH_3CH_2\ddot{O}H$. In the case of phenol the reactivity can be enhanced by the addition of sodium hydroxide to form the phenoxide ion, which is a more reactive nucleophile than phenol itself.

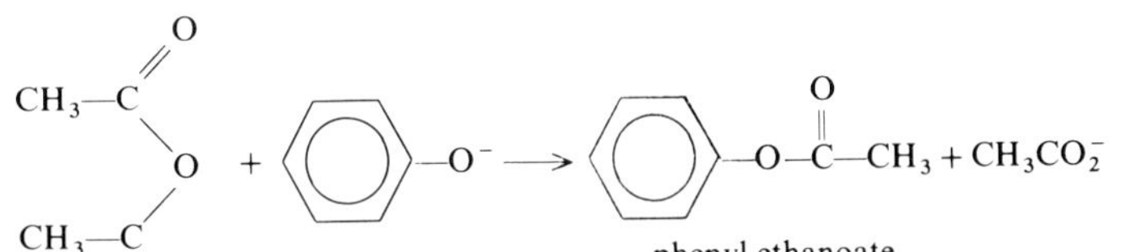

b Phenol will not undergo nucleophilic substitution of the hydroxyl group to form halogenoarenes, whereas ethanol readily reacts with the nucleophile HBr to form bromoethane.

$$CH_3CH_2OH + HBr \longrightarrow CH_3CH_2Br + H_2O$$

In phenol the interaction of the lone pair of electrons on the oxygen atom with the benzene ring makes it more difficult to break the C—O bond.

c Oxidation of the hydroxyl group in phenol to a carbonyl group cannot be carried out because that would involve disruption of the stable benzene ring.

3 Phenol as an arene

The delocalization of electrons from the oxygen in phenol into the benzene ring makes phenol more susceptible than benzene to substitution reactions with electrophilic reagents such as bromine and nitric acid. The reactions occur so readily that multiple substitution often occurs.

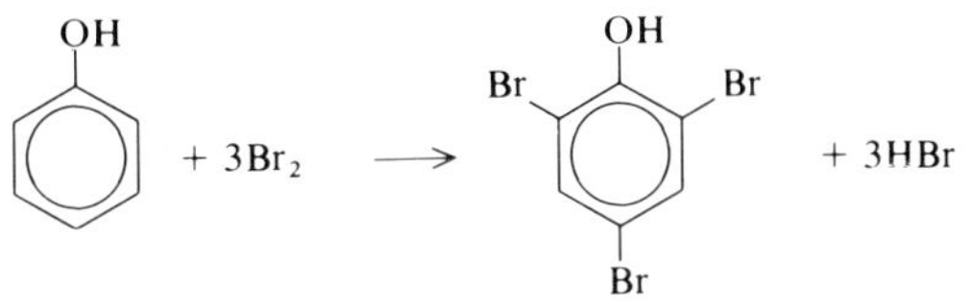

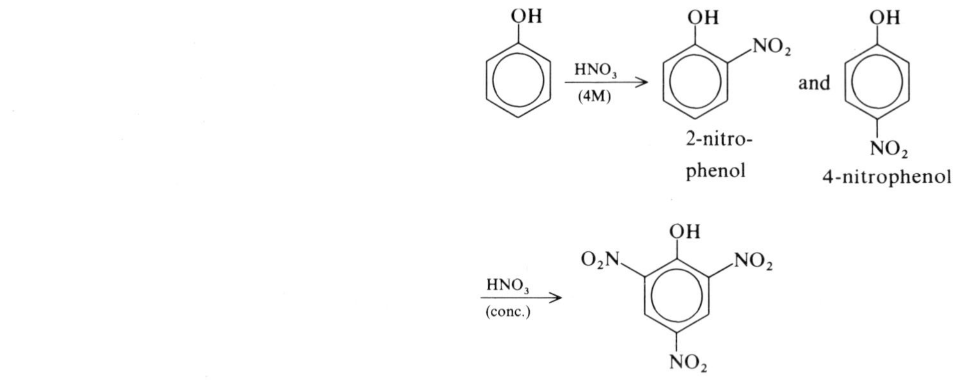

$$\text{phenol} \xrightarrow[\text{(4M)}]{\text{HNO}_3} \text{2-nitrophenol and 4-nitrophenol}$$

$$\xrightarrow[\text{(conc.)}]{\text{HNO}_3} \text{2,4,6-trinitrophenol (picric acid)}$$

COMMENT
The importance of phenol

Phenol was the first **antiseptic** substance to be used in surgery, by Lister in 1857. Although it limits the growth of bacteria in open wounds, it also makes the wounds difficult to heal. The antibacterial properties of phenol can be improved by chlorination and at the same time the healing properties of the antiseptic are also improved. Dettol and TCP are examples of this development from Lister's original discovery.

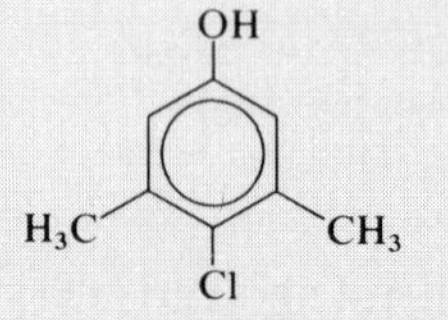

4-chloro-3,5-dimethylphenol (Dettol) 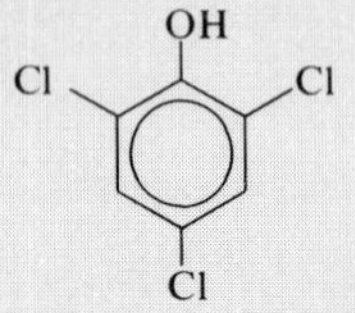   2,4,6-trichlorophenol (TCP)

The effectiveness of germicides relative to phenol in killing the bacterium *Salmonella typhosa* are given below.

Germicide	Relative effectiveness
Phenol	1
TCP	23
Dettol	280

In addition to its use for antiseptics, phenol has been historically important as a component of carbolic soaps in the last century and in the manufacture of plastics such as Bakelite in this century.

The major industrial uses of phenol at the present time include the manufacture of synthetic polymers or fibres, and non-ionic detergents.

Phenolic compounds are also used for making drugs and some of the dyes considered in Topic 17.

Aspirin

Compounds like 2-hydroxybenzoic acid in which the hydroxyl —OH is attached to a benzene ring are phenolic compounds, and you will discover in the next experiment that esters can only be made from them if they are reacted with acid derivatives with enhanced reactivity. If you make aspirin you will be using ethanoic anhydride rather than ethanoic acid.

EXPERIMENT 12.3b Preparations using 2-hydroxybenzoic acid

2-hydroxybenzoic acid (salicylic acid) can be converted by straightforward reactions into two products, both of which find application as medicines. Ethanoylation of the phenolic group produces aspirin

$$\text{2-hydroxybenzoic acid} + (CH_3CO)_2O \xrightarrow{\text{catalyst}} \text{aspirin} + CH_3CO_2H$$

(2-hydroxybenzoic acid — CO₂H, OH; ethanoic anhydride (CH₃CO)₂O; aspirin — CO₂H, O₂C—CH₃; CH₃CO₂H)

and methylation of the carboxylic group produces oil of wintergreen

$$\text{2-hydroxybenzoic acid} + CH_3OH \xrightarrow{\text{catalyst}} \text{oil of wintergreen} + H_2O$$

(methanol CH₃OH; oil of wintergreen — CO₂CH₃, OH)

Aspirin is described as having analgesic (pain killing), anti-inflammatory, and antipyretic (fever reducing) actions. Oil of wintergreen has the same properties, but is applied in liniments and ointments for the relief of pain in lumbago, sciatica, and rheumatism as the oil is readily absorbed through the skin.

The preparation of aspirin uses the same compounds as the industrial method of production.

Procedure

1 Preparation of aspirin

Add to a 50 cm³ pear-shaped flask 2.0 g of 2-hydroxybenzoic acid and 4 cm³ of ethanoic anhydride.

To this mixture add 5 drops of 85% phosphoric(v) acid and swirl to mix. Fit the flask with a reflux condenser and heat the mixture on a boiling water bath for about 5 minutes. Without cooling the mixture, carefully add 2 cm³ of water in one portion down the condenser. The excess ethanoic anhydride will hydrolyse and the contents of the flask will boil.

When the vigorous reaction has ended, pour the mixture into 40 cm³ of cold water in a 100 cm³ beaker, stir and rub the sides of the beaker with a stirring rod if necessary to induce crystallization and, finally, allow the mixture to stand in an ice bath to complete crystallization.

SAFETY ⚠
Ethanoic anhydride is corrosive; 2-hydroxybenzoic acid causes nausea; phosphoric(v) acid and sulphuric acid are corrosive; methanol is flammable and toxic.

Collect the product by suction filtration and wash it with a little water. The product may be recrystallized from water and its melting point recorded. The method of doing this is described in Topic 16.

- What other compounds might have been used to prepare aspirin? Compare their cost with the cost of the compounds actually used. Why do you think ethanoic anhydride and phosphoric(v) acid are used? What further reaction do you think would be necessary to obtain 'soluble aspirin' from aspirin (2-ethanoyloxybenzoic acid)?

2 Preparation of oil of wintergreen

Add to a 50 cm³ pear-shaped flask 9 g of 2-hydroxybenzoic acid, 15 cm³ of methanol, and 2 cm³ of concentrated sulphuric acid. Fit the flask with a reflux condenser and boil the mixture for about an hour.

Cool the mixture to room temperature and pour it into a separating funnel that contains 30 cm³ of cold water. The product now has to be separated by the process known as **solvent extraction**. Rinse the flask with 15 cm³ of ethylethenoate solvent and add this to the separating funnel. Mix the contents of the separating funnel, allow them to settle, and run the lower aqueous layer into a conical flask. Wash the organic solvent layer with 30 cm³ of 0.5 M aqueous sodium carbonate, releasing the pressure in the separating funnel frequently as there is likely to be considerable evolution of carbon dioxide.

Dry the organic solvent extract with anhydrous sodium sulphate and filter.

Remove the organic solvent by distillation (see figure 2.15). Complete the distillation, collecting the distillate boiling above 220 °C as methyl 2-hydroxybenzoate.

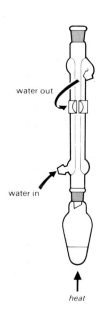

Figure 12.7 Reflux apparatus

- Note the characteristic odour of your product and compare it with a sample of oil of wintergreen. Why does methanol react with the carboxylic acid group and not the phenolic group?
  What is the reason for adding organic solvent to the reaction mixture?
  What is the reason for washing the organic solvent layer with sodium carbonate solution?

Examine the infra-red spectra of 2-hydroxybenzoic acid, aspirin, and oil of wintergreen (figure 12.8) and account for the major differences in the spectra.

Figure 12.8 a The infra-red spectrum of 2-hydroxybenzoic acid

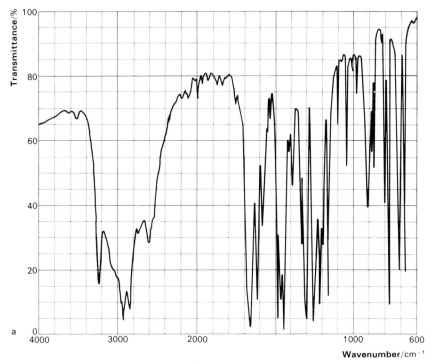

b The infra-red spectrum of aspirin

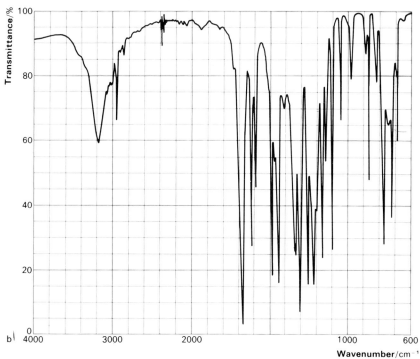

12 Carbon compounds with acidic and basic properties 340

c The infra-red spectrum of oil of wintergreen.

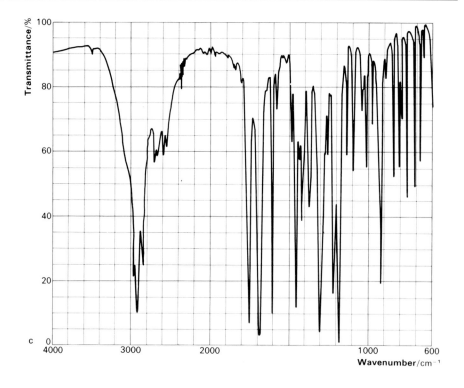

Figure 12.9 The Bill of Mortality for London, 15–22 August 1665

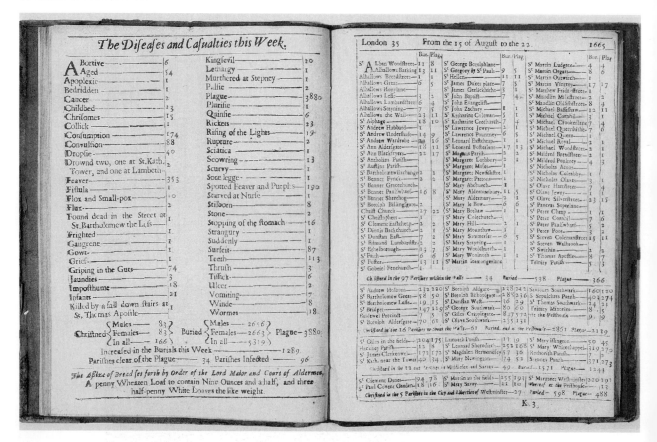

Aspirin, from herbal remedy to modern drug

Look at the Bill of Mortality for London in the week 15–22 August, 1665, the year of the Great Plague, and notice how many of the causes of death could probably be treated successfully in modern London (figure 12.9). It was not that seventeenth-century London lacked doctors and medicines but that the causes of illnesses were very imperfectly understood and many of the herbal and other remedies were ineffective or incorrectly administered.

Herbal remedies are included in the earliest medical writings from ancient Egypt and ancient China but the use of medicines was confused until quite recent times by the inclusion of elements of magic and religion, together with theories that we now know were unsound. Nevertheless, there were many sound remedies, some better known to peasants than to the medical profession, and it has become common practice for the pharmaceutical industry to check all 'old wives' tales' for possible validity. For example, the dried roots of the *Rauwolfia* shrub had been used in Indian folk medicine for over two thousand years but were virtually unknown to Western medical science until 1952; now an alkaloid, reserpine, is extracted from the plant and used in the treatment of hypertension, making it one of the first drugs for the relief of high blood pressure.

The commonest drug of all, aspirin, has a typical history. The use of aspirin as a drug derives from the use of the bark and leaves of willow and poplar trees in a variety of ancient remedies. The active ingredient of the bark and leaves is salicin, the glucoside of 2-hydroxybenzoic acid (salicylic acid).

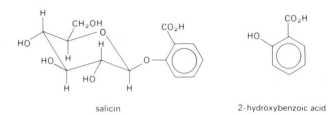

salicin   2-hydroxybenzoic acid

This was not known until 1826 although the bark and leaves had been in use for over 2000 years. From the writings of Hippocrates, the ancient Greek 'Father of Medicine', to those of the Renaissance, willow and poplar extracts had been recommended as remedies in a variety of illnesses – for eye diseases, for the removal of corns, as a diuretic, and in the treatment of sciatica and gout – although such treatments were mainly based on false hopes. In 1763, the Reverend Edmund Stone was the first to describe the value of a willow bark in treating fevers. Even he applied this treatment on the basis of false reasons:

'As this tree delights in a moist or wet soil, where agues (fevers) chiefly abound, I could not help applying the general maxim, that many remedies lie not far off from their causes.'

Natural extract of willow bark was replaced after 1874 by a synthetic process developed by the German chemist, Kolbe:

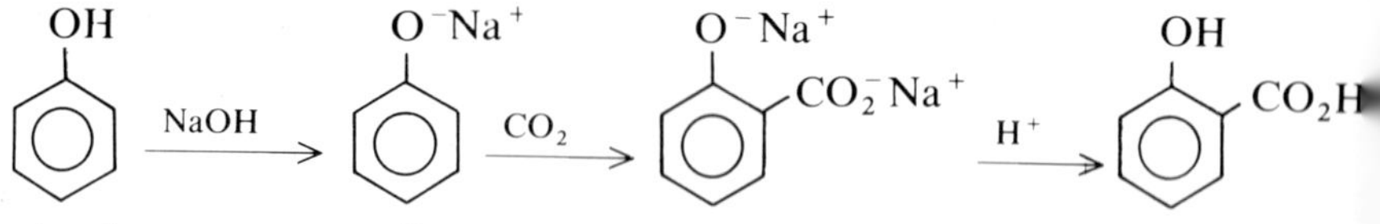

phenol $\xrightarrow{NaOH}$ sodium phenoxide $\xrightarrow{CO_2}$ (intermediate) $\xrightarrow{H^+}$ 2-hydroxybenzoic acid

It was also about this time that the value of willow bark in treating rheumatism was reported by doctors. Although this was new to European medicine, it seems that the Hottentots in Africa had long been familiar with the remedy. However, prolonged treatment of rheumatism by 2-hydroxybenzoic acid has unpleasant side effects and the German chemist, F. Hofmann, supplied his father with a variety of derivatives to try for his rheumatism. This led in a short time to the recognition of the merits of 2-ethanoyloxybenzoic acid (aspirin) and its marketing by the Bayer company in 1899.

The synthetic process developed by Hofmann is still used to manufacture aspirin today, and is based on the reaction between 2-hydroxybenzoic acid and ethanoic anhydride.

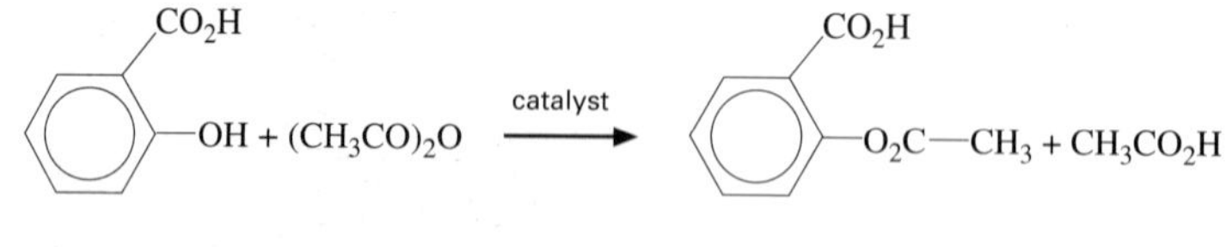

2-hydroxybenzoic acid + ethanoic anhydride $(CH_3CO)_2O$ $\xrightarrow{catalyst}$ aspirin + ethanoic acid CH_3CO_2H

Disadvantages of aspirin

Aspirin is one of the most widely used medicines, with 4000 million tablets produced each year in the United Kingdom. It is mainly taken for feverish colds, headaches, and rheumatism.

However, aspirin is not without its hazards; some 200 people a year die of aspirin poisoning due to deliberate or accidental overdose, with children under five years forming a large proportion of those who die accidentally through eating the tablets. And it has been recognized that aspirin causes internal bleeding. Consumption of aspirin causes a loss of blood of up to 6 cm³ a day in

paracetamol

70 per cent of patients examined and may be the precipitating factor in 50 per cent of patients with gastroduodenal haemorrhage. It has been suggested, since safer drugs are available for headaches and feverish colds, that aspirin should only be available on prescription.

Paracetamol is one possible alternative to aspirin. It is both a useful pain-reducing and fever- reducing drug but lacks some of the side effects shown by aspirin. The physiologically active compound is 4-aminophenol but the potential toxicity of the —NH_2 group must be reduced by conversion to an ethanoyl derivative.

12.4 Amines

Amines contain nitrogen. You can think of the amines as derived from ammonia, with one or more of the hydrogen atoms replaced by an alkyl group.

CH_3NH_2 methylamine,   a primary amine
$(CH_3)_2NH$ dimethylamine,   a secondary amine
$(CH_3)_3N$ trimethylamine,   a tertiary amine

You should compare these structures with the structures of primary, secondary, and tertiary halogenoalkanes and alcohols.

Phenyl groups may also replace the hydrogen atoms:

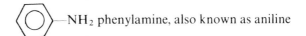—NH_2 phenylamine, also known as aniline

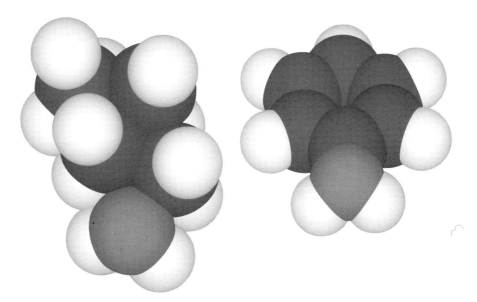

Figure 12.10 Models of butylamine and phenylamine

The properties of ammonia should lead you to expect amines to be basic, to be good nucleophiles, and to form complex ions with metal cations. In the next experiment we shall be testing these expectations.

The N—H bond absorbs in the infra-red in the region 3500–3300 cm^{-1}; the trough is broadened by hydrogen bonding when it occurs (figure 12.11).

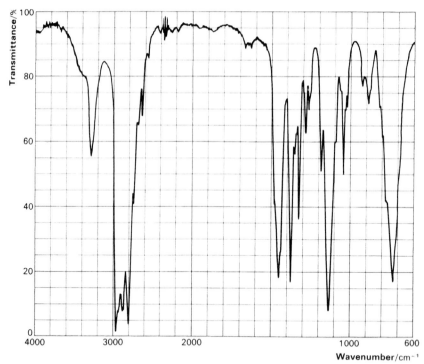

Figure 12.11 The infra-red absorption spectrum of diethylamine

EXPERIMENT 12.4

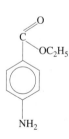

Ethyl 4-aminobenzoate

> **SAFETY** ⚠
> Ammonia is harmful; butylamine is very flammable and an irritant to your skin, eyes and and respiratory system – it should be used in a fume cupboard; ethanoic anhydride is corrosive.

The reactions of amines

In these experiments, you will be looking at the reactions of ammonia, butylamine, and an arylamine. The simplest arylamine, phenylamine, is toxic, so we suggest that you use ethyl 4-aminobenzoate.

Before starting the experiments, make sure that you are familiar with the formulae of the three compounds and the shapes of their molecules, if possible by building models (see figure 12.10).

Procedure

1 Solubility and pH

Prepare or obtain dilute aqueous solutions of the three compounds, warming if necessary. Add drops of Full-range Indicator.

- Are the compounds readily soluble or only sparingly soluble? What type of molecular interaction will be helping them dissolve?
  Are the compounds acidic or basic, strong or weak?

2 Formation of salts

Add small quantities of the three compounds, warming the arylamine, to separate portions of 2 M hydrochloric acid.

- Do the odours disappear?
  Are the amines more soluble in hydrochloric acid than water?
  Could they be reacting with the hydrochloric acid? Write down equations representing any reactions which you consider to be taking place.
  How might you recover the butylamine or arylamine from their mixture with 2 M hydrochloric acid?
  Test your suggestion experimentally.

3 Reaction with transition metal ions

Add small quantities of the three compounds, warming the arylamine, to separate portions of 0.1 M copper(II) sulphate solution until present in excess.

- Are somewhat similar results obtained with the three compounds? What type of interaction do you think is taking place?

4 Preparation of 2-ethanoylaminobenzoic acid

The product of this preparation is a rather unusual ethanoyl compound.

[Reaction scheme: 2-aminobenzoic acid → (intermediate) → 2-ethanoylaminobenzoic acid, with loss of H_2O]

> **SAFETY** ⚠
> Handle ethanoic anhydride with care; it is corrosive.

Place 3.5 g of 2-aminobenzoic acid in a 50 cm³ flask fitted with a reflux condenser and add 10 cm³ of ethanoic anhydride, $(CH_3CO)_2O$, by pouring it carefully down the condenser (TAKE CARE). (See figure 12.7.)

Heat slowly to boiling and reflux for a quarter of an hour.

Allow the solution to cool, add 5 cm³ of water, bring the mixture slowly back to the boil, and then allow it to cool **slowly**.

Collect the crystals of 2-ethanoylaminobenzoic acid by suction filtration, having the apparatus behind a safety screen, and wash them with a small quantity of cold methanol.

The acid (melting point 183–5 °C) may be recrystallized from a mixture of ethanoic acid and water. Allow the crystals to dry thoroughly.

Place several crystals of the product between two watch-glasses. In a darkened room, grind the two glasses together, when flashes of light should be seen. The emission of light has been explained in terms of an electric discharge between the surfaces of the fractured crystal which causes excitation of the molecule with subsequent fluorescence. For the phenomenon to be observed, the crystals must be well formed and free from solvent.

This phenomenon was named **triboluminescence** in 1895 (from the ancient Greek *tribein*, to rub) although the effect had been described by Francis Bacon as long ago as 1605. Bacon noticed that flashes of light are emitted when crystals of sugar are scraped or crushed. Some sugar products also emit light when crushed.

Reactions of amines

1 Basic properties

Amines are readily soluble in water. Solubility is assisted by hydrogen bonding through a lone pair of electrons on the nitrogen atom:

$$CH_3-\underset{H}{\overset{H}{N}}\cdots H-O-H$$

Amines are quite strong bases, as can be seen from the K_a values of their conjugate acids. Phenylamine is an exception; electron pair interaction with π bonds in the molecule of this compound much reduces the acceptance of protons by the NH_2 group. You should compare this interaction with the interaction that occurs in phenol.

Compound	K_a/mol dm^{-3}
Phenylammonium ion	2.0×10^{-5}
Ammonium ion	5.6×10^{-10}
Ethane-1,2-diammonium ion	1.3×10^{-10}
Butylammonium ion	0.15×10^{-10}

K_a values of the conjugate acids of some amines; they refer to K_a for the equilibrium $RNH_3^+(aq) \rightleftharpoons H^+(aq) + RNH_2(aq)$.

In the presence of strong acids the equilibrium is shifted towards the RNH_3^+ ions and the amines form soluble salts:

$$C_6H_5-NH_2(aq) + H^+(aq) \rightleftharpoons C_6H_5-NH_3^+(aq)$$

phenylamine   phenylammonium ion

2 Complex ions

The lone pair of electrons on the nitrogen atom of amines makes the compounds suitable ligands for metal cations, parallel with the familiar deep blue-coloured complex ion formed between ammonia and copper(II) cations

$$4NH_3 + Cu(H_2O)_4^{2+} \rightleftharpoons Cu(NH_3)_4^{2+} + 4H_2O$$

but the stoicheiometry of the reaction may be different:

$$2\,C_6H_5-NH_2 + Cu(H_2O)_4^{2+} \rightleftharpoons Cu(C_6H_5-NH_2)_2(H_2O)_2^{2+} + 2H_2O$$

3 Reaction with electrophiles

Amines, being themselves nucleophiles, will react with the electophilic carbon atom of halogenoalkanes to give secondary and tertiary amines.

$$CH_3CH_2NH_2 + CH_3CH_2Cl \longrightarrow CH_3CH_2-NH-CH_2CH_3 + HCl$$
                                                diethylamine

and then

$$CH_3CH_2-NH-CH_2CH_3 + CH_3CH_2Cl \longrightarrow CH_3CH_2-\underset{\underset{CH_2CH_3}{|}}{\overset{\overset{CH_3}{|}}{N}}-CH_2CH_3 + HCl$$

or $(CH_3CH_2)_3N$ triethylamine

When chloroethane and ammonia react, ethylamine is produced (see Topic 7, section 7.4), but a mixture of products is obtained because the ethylamine will react with more chloroethane, according to the above sequence of reactions.

Amides are produced when amines react with acid chlorides:

$$CH_3COCl + C_6H_5-NH_2 \longrightarrow C_6H_5-NH-CO-CH_3 + HCl$$

ethanoyl chloride   phenylamine   *N*-phenylethanamide

12.5 Study task: The manufacture of citric acid

QUESTIONS

Read the passage below, and answer these questions based on it.

1. Make a summary in 150 words of the submerged fermentation process for the manufacture of citric acid. Try to produce an account that would be useful to visitors to the factory.
2. What uses of citric acid depend on its ability to form salts?
3. What uses of citric acid depend on its ability to form an acidic solution?

Citric acid, a white crystalline material, is a natural product that occurs in most living cells, but is perhaps best known as the acid constituent of citrus fruits. Together with its salts it is used in a wide variety of foods and medicinal products in the UK.

John & E. Sturge Limited, situated at Selby in North Yorkshire, is the only UK producer of citric acid. Founded in 1823, the company began manufacturing citric acid from imported calcium citrate, produced from lime juice, in the 1830s. Production of citric acid is carried out by both a surface fermentation process, developed in the 1920s using pure sugar, and the more important submerged fermentation process. With the increasing cost of pure sugar, beet molasses is now the major carbohydrate feedstock.

The submerged fermentation process for the production of citric acid is described below. In order to achieve a successful fermentation it is necessary to have the right strain of mould, aseptic conditions during all stages of the fermentation and close control of fermentation conditions such as temperature.

The current strain of mould spores used, *Aspergillus niger*, is the result of careful selection and mutation over many years, during which time some 15 000 isolations were obtained and examined. The selected strain is maintained as cultures from which the mould needed to start the fermentation is grown.

A two-stage process is employed by Sturge. The first *inoculum* stage involves the use of a small stainless-steel fermenter which is sterilized by the injection of steam. After sterilization the vessel is filled with a dilute solution of molasses to which have been added various nutrients. This liquid (the 'medium') is sterilized by being passed through a heat exchanger where the temperature of the medium is raised to 130 °C and held at that temperature for up to one minute.

12 Carbon compounds with acidic and basic properties 348

Figure 12.12 Manufacturing plant for citric acid

After filling the fermenter the medium is treated ('sparged') with air and the spores of *Aspergillus niger* introduced. In the agitated, aerated environment the spores germinate over the next few hours and the initial growth phase of the mould takes place. Temperature, dissolved oxygen concentration, pH and evolved CO_2 concentration are carefully monitored, together with the microbiological aspects of this first stage.

Whilst this initial growth phase is proceeding, the larger fermentation vessel, over 200 000 dm^3 in capacity, is prepared in the same manner as the first stage, and filled with a molasses medium. When the inoculum mould growth has reached a satisfactory stage and is judged to be free from infection, approximately 24 hours after the addition of the mould spores, it is transferred to the fermenter where conversion of the sugar to citric acid begins.

As before, temperature, dissolved oxygen and carbon dioxide are monitored and the progress of the fermentation plotted by simple titration of the acid produced. After some seven days the sugar is exhausted, maximum citric acid production achieved and the fermentation is terminated.

The fermentation broth is transferred to a harvest vessel where it is vacuum filtered to remove the unwanted mould. The filtered liquor, which is a crude solution of citric acid, is transferred to the recovery plant for purification. The recovery is started by the addition of slaked lime, $Ca(OH)_2$, which precipitates the acid as calcium citrate. The reaction is exothermic, resulting in a significant rise in the temperature of the liquors which, as calcium citrate exhibits inverse solubility, aids precipitation.

The precipitate is removed by vacuum filtration, during which time it is washed with hot water to remove molasses impurities which would otherwise interfere with the following stages of the process and give quality control problems. Dilute sulphuric acid is then added to the washed calcium citrate. This reaction results in the formation of citric acid solution and insoluble calcium sulphate which is filtered off and currently discarded as no commercial use can be found.

The citric acid solution can be decolorized by the addition of activated carbon before being concentrated by vacuum evaporation. The solution is stirred and cooled, using pure water at 10–12 °C, to promote crystallization of either monohydrate or anhydrous crystals, depending on the temperature at which the crystallization is allowed to take place. The used water is recycled for use elsewhere in the factory.

The crystals are separated from the mother liquor by centrifugation. During this process they are washed with water, and the washings and mother liquor are recycled through the process whilst the crystals are dried in a counter current of hot air.

Finally, the crystals are sieved to produce a range of sizes to meet the varying demands of customers. The finished material is sampled and analysed to ensure that it meets British Pharmacopoeia (BP) standards. Material which fails to meet specification is redissolved to produce salts of citric acid – either trisodium citrate or zinc citrate.

The citric acid crystals are packed into 25 kg bags or into 1 tonne bulk bags. Some is sold in solution form.

All the waste water from the factory is treated in an anaerobic effluent treatment plant before being discharged to the drains. The methane gas produced during the treatment process is used as a fuel to produce steam and electricity for use in the factory.

Citric acid and its salts find a wide variety of uses:

Uses	%
Soft drinks	48
Health salts	21
Other foods	12
Confectionery	9
Chemical cleaning	4
Other uses	6

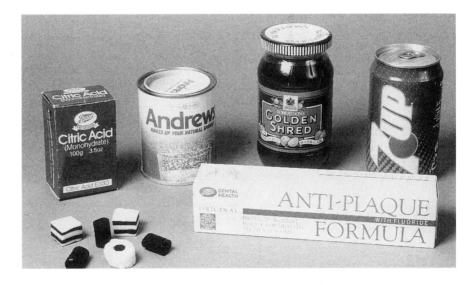

Figure 12.13 The contents lists on these products show that they all contain citric acid

In the soft drinks industry both citric acid and sodium citrate are added to carbonated and non-carbonated drinks. They act as flavour enhancers and help to retain the carbon dioxide. Citric acid can also be added to wine after the completion of fermentation, again as a flavour enhancer.

In the health field, citric acid and sodium carbonate are used in effervescing salts to control the formation of excess hydrochloric acid in the stomach, while citric acid forms an important ingredient in the manufacture of soluble aspirin. Sodium citrate is used as an anticoagulant wherever blood is handled, as, for example, in the blood transfusion service and in slaughterhouses and meat processing. Zinc citrate has been found to be a useful additive to toothpaste, where it is used to inhibit the formation of dental plaque.

In the confectionery field citric acid is added to sweets, cake fillings and jams as a flavouring and it also plays an important part in the setting of jams and jellies by the effect of a lower pH on pectins and gelatine. Citric acid is also used as an antioxidant in the food industry and in the processing of canned foods. In the manufacture of edible oils, fats, dairy products and baby foods citric acid prevents the loss of vitamin C and the formation of a rancid taste by the removal of trace quantities of metals.

Finally, citric acid is used in industrial cleaning as a descaling agent and sodium citrate can be used in the detergent industry as a replacement for phosphates.

Summary

Survey of reactions in Topic 12

The chart below summarizes the main reactions that you have met in this Topic. You should copy the chart into your notes and add appropriate details about the reagents, their chemical nature, and the types of reaction involved. It may help you to learn the material if you make several charts, one for each particular type of information.

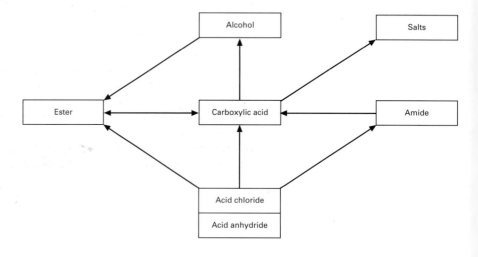

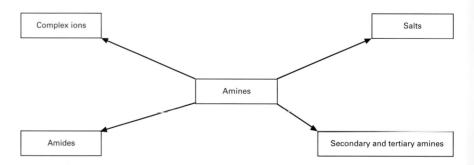

Figure 12.14 Summary of the reactions of **a** carboxylic acids and their derivatives **b** amines

12 Carbon compounds with acidic and basic properties

At the end of this Topic you should be able to:

a demonstrate understanding of the nomenclature and corresponding displayed and structural formulae for carboxylic acids, esters, acid anhydrides, acid chlorides, amides, phenols, and amines (including primary, secondary and tertiary)

b recall the typical behaviour of carboxylic acids limited to:
   solubility in water
   acidity and formation of salts
   reaction with alcohols to form esters
   reduction by lithium tetrahydridoaluminate

c recall the typical behaviour of carboxylic acid derivatives limited to:
   reaction with water
   reaction with alcohols to form esters
   reaction with ammonia to form amides
   reduction by lithium tetrahydridoaluminate

d recall the characteristic behaviour of phenol limited to:
   combustion
   solubility in water and the acidic nature of the mixture
   treatment with **i** sodium
                   ii aqueous sodium hydroxide
                   iii aqueous sodium carbonate
                   iv iron(III) chloride
   ester formation with acid chlorides and acid anhydrides
   comparative reactivity of the ring system with bromine and nitric acid

e recall the typical behaviour of primary alkyl amines and aryl amines, limited to:
   characteristic smell
   solubilty in water and alkaline nature of resulting solutions
   complex ion formation with copper(II) ions
   treatment with **i** halogenoalkanes
                   ii ethanoyl chloride

f interpret the reactions of carboxylic acids, amines and phenols in terms of nucleophilic or electrophilic attack

g evalute information by extraction from text and the *Book of data* about the properties of carboxylic acids, esters, phenols and amines.

12 Carbon compounds with acidic and basic properties

Review questions

Carboxylic acids

12.1

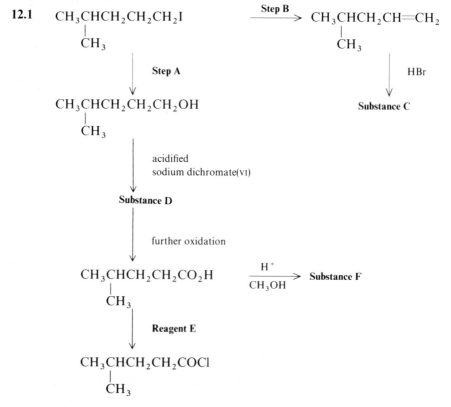

a Name the starting substance.
b Name the reagents and conditions required for step A. Classify this reaction in terms of reaction and reagent type.
c Name the reagents and conditions necessary for step B. What type of reaction is this?
d Give the structure and name of substance C. Classify the reaction for its formation in terms of reaction and reagent type.
e Write the structural formula of substance D.
f Name a suitable reagent E.
g Write the structural formula of substance F.
h Which substance in the scheme would be obtained if the carboxylic acid CH_3—$CH(CH_3)$—CH_2—CH_2CO_2H were treated with $LiAlH_4$?
i Which substance in the scheme would probably fume in moist air?

12 Carbon compounds with acidic and basic properties

12.2 The boiling points of a series of compounds with similar molar masses are given below.

Structure	Molar mass/g mol^{-1}	T_b/K
$CH_3-CH_2-CH_2-CH_3$	58	273
CH_3-CH_2-CHO	58	322
$CH_3-CH_2-CH_2OH$	60	370
CH_3CO_2H	60	391

Name the compounds and suggest reasons for the difference between the boiling points, stating clearly the main types of molecular interaction involved in each of the compounds.

12.3 Arrange the following groups of acids in order of their strength as acids, putting the weakest first.

a (A) $CH_3-\overset{O}{\underset{\|}{C}}-OH$ (B) $ClCH_2-\overset{O}{\underset{\|}{C}}-OH$ (C) $Cl_2CH-\overset{O}{\underset{\|}{C}}-OH$ (D) $Cl_3C-\overset{O}{\underset{\|}{C}}-OH$

b (A) $CH_3-\overset{O}{\underset{\|}{C}}-OH$ (B) $FCH_2-\overset{O}{\underset{\|}{C}}-OH$ (C) $ClCH_2-\overset{O}{\underset{\|}{C}}-OH$ (D) $BrCH_2-\overset{O}{\underset{\|}{C}}-OH$

(E) $ICH_2-\overset{O}{\underset{\|}{C}}-OH$

Amines

12.4 Study the following list of compounds containing nitrogen:

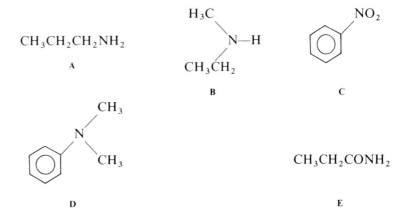

- **a** Which substance is (**i**) a primary amine, (**ii**) a secondary amine, (**iii**) a tertiary amine, (**iv**) an acid amide?
- **b** Which substances react readily with acids to form salts?
- **c** Which amine would be the weakest base?
- **e** Which substance could be hydrolysed to form carboxylic acids?
- **f** Which substances might form complexes with metal cations?
- **g** Which amines could react with ethanoyl chloride?

12.5 Arrange the following compounds of similar molar mass in the order in which you would expect their boiling points to increase, giving reasons for your choice.

$CH_3CH_2CH_2CH_3$, $CH_3CH_2CH_2OH$, $CH_3CH_2CH_2NH_2$, CH_3CONH_2.

12.6 Arrange the following compounds in the order in which you would expect their basic strength to increase, putting the weakest base first.

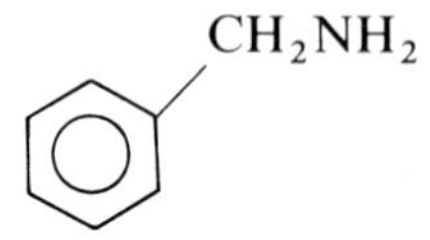

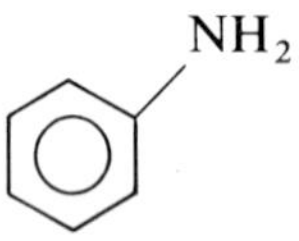

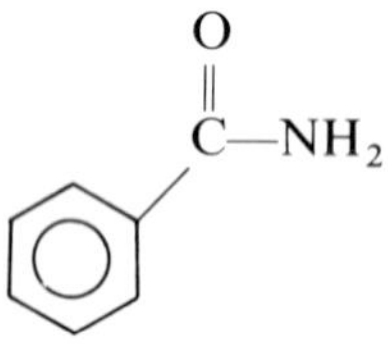

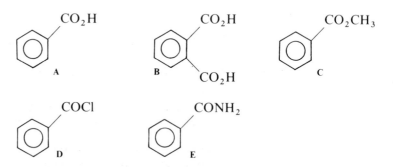

Acid derivatives

12.7 Study the following group of compounds and answer the questions below:

a Which substance should fume in moist air?
b Which substance should have a pleasant smell?
c Which substance is a dibasic acid?
d Write the structural formula of the substance which would be obtained when substance A dissolves in sodium hydroxide solution.
e With what reagent would you treat substance A to form substance C?
f With what reagent would you treat substance D to form substance C?
g With what reagent would you treat substance D to form substance E?
h What products would be obtained if substance C were refluxed with sodium hydroxide solution?
i Which substance might be dehydrated by heating?
j Write the structural formula for the substance you have suggested in **i**.

12.8 Suggest simple **chemical** tests for distinguishing one compound from the other in the following pairs:

a CH_3CO_2H and CH_3CHO

b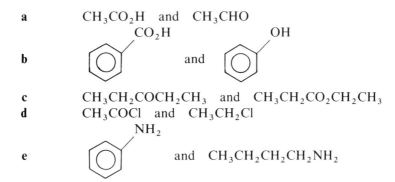

c $CH_3CH_2COCH_2CH_3$ and $CH_3CH_2CO_2CH_2CH_3$
d CH_3COCl and CH_3CH_2Cl

e

12.9 The labels are missing from four bottles known to contain the solids phenol, hexanoic acid, dodecanol, and paraffin wax (a long chain alkane).

Devise a simple series of chemical tests which would enable you to distinguish between them.

Examination questions

12.10 Methyl benzoate can be prepared from methanol and benzoic acid.

$$\underset{\text{benzoic acid}}{C_6H_5CO_2H} + \underset{\text{methanol}}{CH_3OH} \overset{\text{acid}}{\rightleftharpoons} \underset{\text{methyl benzoate}}{C_6H_5CO_2CH_3} + H_2O$$

8.0 g of benzoic acid is added to 15 cm³ of methanol. Then 2 cm³ of concentrated sulphuric acid is added. The mixture is boiled under reflux for 45 minutes, cooled, and put into a separating funnel that contains 30 cm³ of cold water. The reaction flask is rinsed with 15 cm³ of a hydrocarbon solvent and the washings added to the separating funnel.

a i What is the function of the concentrated sulphuric acid?
  ii Why is the concentrated sulphuric acid added last?
  iii Benzoic acid is a **carboxylic acid**. What is the functional group in methanol? What is the functional group in methyl benzoate?
  iv Draw a diagram of the apparatus used for boiling the mixture under reflux.
  v Give reasons why the mixture is not heated in an open beaker.
b Methyl benzoate dissolves in hydrocarbon solvents. After mixing the contents of the separating funnel by vigorous shaking and standing for a few minutes, two layers form. The layers are run into separate conical flasks.

The hydrocarbon layer is returned to the separating funnel and washed first with water and then with 0.5 M sodium carbonate solution. The hydrocarbon extract is dried over a suitable compound and filtered.
  i Why is the mixture washed with water?
  ii Why is the mixture washed with sodium carbonate solution? What precaution should be taken during this step?
  iii Name a suitable compound for drying the hydrocarbon layer.
  iv How would you obtain pure methyl benzoate from the solution in the hydrocarbon? (Boiling points: low boiling hydrocarbon 80–100 °C, methyl benzoate 200 °C)

c   Methanol was present in excess. Calculate the percentage yield from the benzoic acid if 4.0 g of methyl benzoate was obtained.
(Molar masses: benzoic acid = 122.1, methyl benzoate = 136.1 g mol^{-1})

d   One possible mechanism for the reaction is breaking the C—O bond in methanol, and making a new bond to the carbon atom in methanol with electrons from an oxygen in benzoic acid.

  i   Describe an alternative pattern of bond breaking and bond making for the methanol molecule.

  ii How could the correct mechanism be established?

e   Methyl benzoate is a derivative of benzoic acid. Give the name and formula of one other derivative of benzoic acid, with the reagent and conditions needed for its formation.

12.11   The drug **paracetamol**, used to relieve pain and reduce fever, can be made from phenol in three steps:

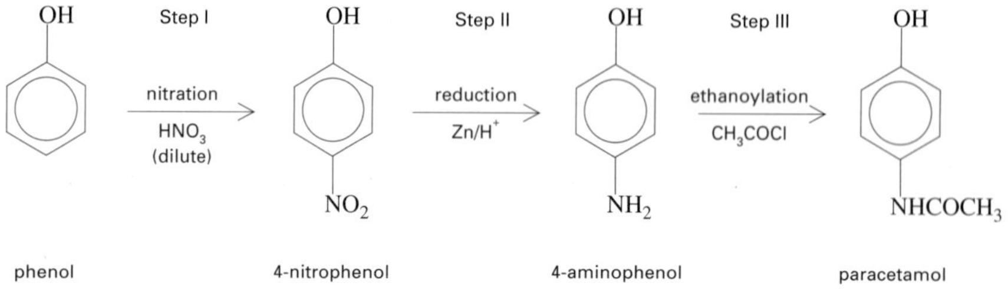

OH — Step I (nitration, HNO$_3$ (dilute)) → OH, NO$_2$ (4-nitrophenol) — Step II (reduction, Zn/H$^+$) → OH, NH$_2$ (4-aminophenol) — Step III (ethanoylation, CH$_3$COCl) → OH, NHCOCH$_3$ (paracetamol)

phenol   4-nitrophenol   4-aminophenol   paracetamol

a i   Explain briefly what is meant by the circle in the diagram of the phenol molecule above.

  ii Draw a displayed formula of the paracetamol molecule. (You can use a circle in your diagram in the same way as in the above diagrams.)

b   What type of reaction is involved in the nitration of phenol in step I? You should indicate the nature of both the reagent and the reaction.

c   Crystals of 4-nitrophenol obtained from step I may be contaminated with 2-nitrophenol. Describe **briefly** how you could show, by a chromatographic method, whether or not any 2-nitrophenol is present, mentioning all the steps in your method.

d   4-aminophenol reacts with both acids and bases. Write equations for the reaction of 4-aminophenol with (i) sodium hydroxide (ii) hydrochloric acid.

e i   Step III is achieved by the use of ethanoyl chloride, CH$_3$COCl, which has a strong dipole. Draw a displayed formula of the ethanoyl chloride molecule, and indicate clearly on it which bonds are likely to be polar, giving the direction of the polarity.

  ii The ethanoyl chloride molecule is said to undergo nucleophilic attack by the 4-aminophenol molecule. Explain this statement as fully as you can.

f   You are asked to carry out the synthesis of paracetamol by this route. State two of the important safety hazards you should consider and state what precautions you would take. (Assume that a laboratory coat and eyeshields are worn as standard procedure.)

12 Carbon compounds with acidic and basic properties

12.12 Methylbenzene can be nitrated using a mixture of nitric acid and sulphuric acid. In an attempt to find out what nitration products result, an experiment was conducted as follows:

'10 drops of concentrated sulphuric acid were carefully added to 10 drops of concentrated nitric acid in a test-tube, while shaking the mixture and cooling the test-tube under a stream of cold water. The mixture of acids was then added to 5 drops of methylbenzene in another test-tube, again shaking this mixture under a stream of cold water. The mixture was poured into a beaker containing 10 cm³ of cold water. The contents of the beaker were transferred to a separating funnel and the organic layer separated off.'

a When methylbenzene is nitrated
  i What **type of reaction** (addition, elimination, etc.) is said to take place?
  ii What **class of reagent** (electrophile, nucleophile, etc) attacks methylbenzene?
  iii What nitrogen-containing ion is believed to be involved in the actual nitration stage of the reaction?
  iv Why is the reaction mixture cooled?
b Suggest briefly how the organic layer resulting from the nitration might be (**i**) treated to remove residual acids (**ii**) dried.
c The methyl group in methylbenzene is said to direct incoming nitro groups to the 2- and 4-positions of the benzene ring. Using this information, draw structural formulae to show three of the possible products of nitration of methylbenzene.
d Give the name of a technique suitable for the separation of the possible products of nitration of methylbenzene which could also serve to identify them.
e How do the conditions for the nitration of phenol differ from the conditions for the nitration of methylbenzene in respect of (**i**) temperature (**ii**) concentration of acids?

12.13 When preparing esters in the laboratory, good yields may be obtained by reaction of an alcohol with an acyl chloride. The reaction is usually rapid and complete.

Plan a reaction scheme for the preparation of butyl ethanoate using this reaction, starting from the appropriate alcohol and carboxylic acid.

Stage I   carboxylic acid + phosphorus pentachloride ⟶ acyl chloride
$$2RCO_2H + PCl_5 \longrightarrow 2RCOCl + POCl_3 + H_2O$$
and $$RCO_2H + PCl_5 \longrightarrow RCOCl + POCl_3 + HCl$$

Stage II acyl chloride + alcohol ⟶ ester

Calculate the quantities of starting materials required to produce 10 g of butyl ethanoate, assuming 60% yield at each of stages I and II. Data relating to the compounds involved may be found in table 5.5 in the *Book of data*.

Using your result, describe how you would carry out this two-stage synthesis in the laboratory, in order to produce a pure sample of about 10 g of butyl ethanoate (boiling point 126 °C; molar mass = 116 g mol⁻¹). Give sufficient details of apparatus, procedures and safety precautions to enable a fellow student to perform the experiment.

TOPIC 13

Redox equilibria and electrochemical cells

STUDY TASK
Make a list of examples of the use of metals in which corrosion is a problem. In each case try to find out how the problem is dealt with.

In this Topic we begin with some simple metal displacement reactions and observe that as well as an enthalpy change we can also obtain electrical energy from the reactions. To make our study quantitative we shall find out how to measure this electrical energy on a systematic basis: as the e.m.f. (electromotive force) of electrochemical cells.

Figure 13.1 The electric eel can produce about 350 volts along its body

COMMENT
An oxidation-reduction reaction can also be referred to as a **redox reaction**.

We shall investigate some oxidation and reduction reactions which involve the transfer of electrons and see how consideration of the electrical energy involved enables us to understand these reactions more fully. We shall also look at some examples of the use of electrical energy data in chemical situations.

An understanding of electrochemistry is essential if we are to control corrosion in metals. Corrosion is an ever-present problem wherever metals are used, for example in ships' hulls, car bodies, steel bridges and pipes for the transport of liquids.

Finally there is a Study Task on the application of these ideas to the design of batteries: from torch batteries to the miniature cells used in watches and hearing aids.

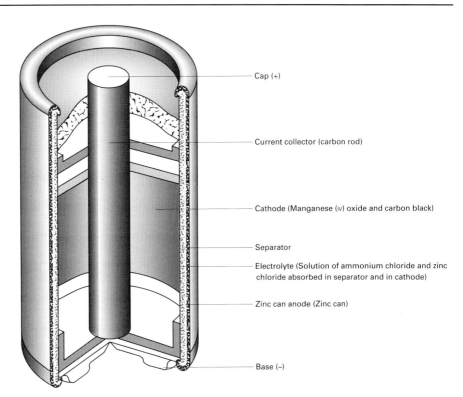

Figure 13.2 The zinc–carbon cell

13.1 Metal/metal ion systems

REVIEW TASK

Working in groups, write down as much as you can recall about oxidation and reduction. Make as long a list as you can of oxidizing agents, with their formulae.

Earlier in the course, you have met a number of oxidation and reduction reactions and have seen, in Topic 5, that these always involve a change of oxidation number of the reacting substances.

We shall start by looking at some reactions between metal **atoms** and metal **ions**. You may have met these before when studying the reactivity series of the metals.

Element	Reduced form	Oxidized form
Most strongly reducing		
Potassium	K	K^+
Sodium	Na	Na^+
Magnesium	Mg	Mg^{2+}
Zinc	Zn	Zn^{2+}
Iron	Fe	Fe^{2+}
Tin	Sn	Sn^{2+}
Lead	Pb	Pb^{2+}
(Hydrogen)	H_2	H^+
Copper	Cu	Cu^{2+}
Mercury	Hg	Hg_2^{2+}
Silver	Ag	Ag^+
Gold	Au	Au^+
Least strongly reducing		

Figure 13.3 The reactivity series of the metals

EXPERIMENT 13.1a

The reactivity series of the metals

All the following reactions can be carried out in test-tubes.

Procedure

1 Add a small amount of zinc powder to 5 cm³ of 0.5 M copper(II) sulphate solution and shake the mixture gently. Measure the temperature change that occurs. Look for other signs of reaction by examining the solid and solution.

2 Repeat the experiment using any metal powders that are available, such as iron, lead and magnesium. Use similar fine powders if you can.

- Was heat evolved or absorbed in the reactions?
  Try to arrange the metals in order of reactivity based on your estimates of the enthalpy changes of the reactions.

3 Repeat part 1, but this time adding magnesium powder to 0.5 M zinc sulphate solution.

- Write down the oxidation numbers of the reactants and the products in each of reactions in **1**, **2** and **3**.
  Which of the reactants has been oxidized and which reduced in each reaction?

> **SAFETY** ⚠
> Copper(II) sulphate is harmful; some metal powders are flammable.

> **COMMENT**
> Some hydrogen may be produced when magnesium powder is used but this is not the effect we are interested in.

> **QUESTION**
> Write ionic equations for the reactions you have observed.

Oxidation and reduction by electron transfer

You should have realized from the equations you have just written that reactions between metals and metal ions involve the transfer of electrons from one reactant to another; and that the reactivity series appears to be connected to the energy change involved in this transfer of electrons.

We described 'half-reactions' in Topic 5 and for the zinc-copper reaction the half-reactions are

$$Zn(s) \longrightarrow Zn^{2+}(aq) + 2e^-$$

$$2e^- + Cu^{2+}(aq) \longrightarrow Cu(s)$$

and you can see that the overall reaction will involve the transfer of electrons.

In the first half-reaction, the oxidation number of zinc increases (0 to +2) so that an **oxidation** is involved; in the second reaction the oxidation number of copper decreases (+2 to 0) thus involving a **reduction**. In the complete reaction zinc is the reducing agent and copper(II) ions the oxidizing agent:

$$\underset{\underset{\text{reducing agent}}{0}}{Zn(s)} + \underset{\underset{\text{oxidizing agent}}{+2}}{Cu^{2+}(aq)} \longrightarrow \underset{+2}{Zn^{2+}(aq)} + \underset{0}{Cu(s)}$$

In experiment 13.1a you will have seen that zinc can act as either an oxidizing agent, $Zn^{2+}(aq)$, or a reducing agent, $Zn(s)$. Each half-reaction can be treated as an equilibrium, so that we have

$$Zn(s) \rightleftharpoons Zn^{2+}(aq) + 2e^-$$

When electrons are added to this system, the equilibrium will move towards the left; removal of electrons will have the opposite effect.

> **QUESTION**
> Write half-reactions for the magnesium/zinc sulphate reaction. Which species is the oxidizing agent and which the reducing agent in this reaction?

13 Redox equilibria and electrochemical cells

QUESTION
Write the equilibrium half-reactions for magnesium and copper.

Thus in the presence of a metal whose tendency to lose electrons and form ions is greater than that of zinc, the reaction moves towards the left and zinc is produced. This is the case with magnesium, which dissolves to form magnesium ions.

In the presence of copper ions the reverse is the case and zinc metal atoms lose electrons to become hydrated ions, while metallic copper is precipitated.

EXPERIMENT 13.1b Electrical energy from a redox reaction

When we keep the reactants of the last experiment apart we can get the electrons to transfer from the metal to the ions via an external wire rather than directly in the test-tube.

Procedure
Set up an electrochemical cell similar to the one in the diagram.

COMMENT
E.m.f. (electromotive force) describes the transformation of the energy of a chemical reaction into electrical energy. It is measured in volts (V).

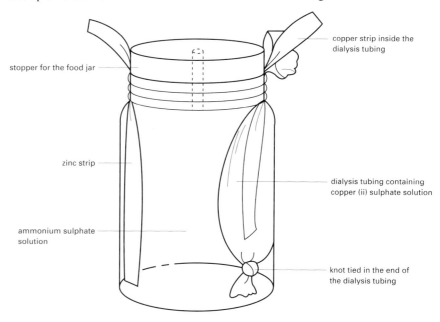

Figure 13.4 A simple cell made from a food jar

Clean the metal strips by rubbing with emery paper if necessary.

Soak the dialysis tubing in water for two minutes until it is soft enough for you to tie the knot. You will find it easier to fill the knotted tubing with copper(II) sulphate solution, using a dropping tube, before you insert the copper metal strip.

Connect the metal strips to a high resistance voltmeter and measure the e.m.f. of your electrochemical cell.

Can you get useful work, such as lighting a bulb or LED, from your cell?

- Which metal was the negative connection in the circuit? Can you explain why?

13 Redox equilibria and electrochemical cells

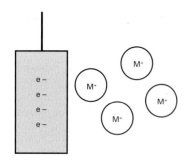

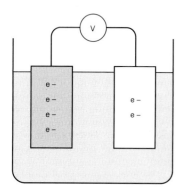

Measuring the tendency of a metal to form ions in solution

When an equilibrium is set up between a piece of metal foil and an aqueous solution of its ions

$$M(s) \rightleftharpoons M^{z+}(aq) + ze^-$$

we can expect the metal to become negatively charged, by electrons building up on it as metal cations are released into the solution. The solution will become positively charged because of the excess of cations.

Thus there should be an electric potential between solution and ions. If the equilibrium position differs for different metals, the potentials set up will differ also. **These potentials (called absolute potentials) cannot be measured** but the **difference** in potential between two metal/metal ion systems can be found by incorporating them into an electrochemical cell and measuring the voltage between the metal electrodes.

Cell diagrams

It is convenient to have an agreed method of representing electrochemical cells and the e.m.f. which they produce. This is called a **cell diagram** and may be illustrated by the Daniell cell, for which the cell diagram is written

$$Zn(s) \mid Zn^{2+}(aq) \parallel Cu^{2+}(aq) \mid Cu(s) \qquad E = +1.1 \text{ V}$$

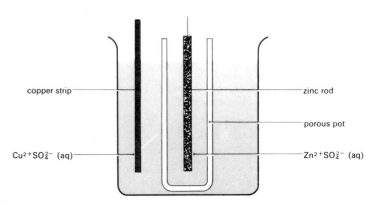

Figure 13.5 A Daniell cell

The solid vertical lines represent boundaries between solids and solutions in each electrode system and the two vertical broken lines represent the porous partition (or other device to ensure a conducting path through the cell). The e.m.f. of the cell is represented by the symbol E, and the value is given in volts, with a sign (+ or −) preceding it which indicates the polarity of the **righthand electrode** in the cell diagram. In the example above the copper plate is the positive terminal of the cell.

Obviously the cell diagram can be written in the reverse order but it is still the righthand electrode whose polarity is indicated. The Daniell cell can thus be written in the alternative form

$$Cu(s) \mid Cu^{2+}(aq) \parallel Zn^{2+}(aq) \mid Zn(s) \qquad E = -1.1 \text{ V}$$

This is the basic pattern for all cell diagrams. Additional conventions are required for more complicated cells. These will be dealt with as they arise.

> **COMMENT**
> We are using the convention agreed by the International Union of Pure and Applied Chemistry (IUPAC).

Contributions made by separate electrode systems to the e.m.f. of a cell

Measurement of the potential of a single electrode system is impossible because two such systems, or **half-cells**, are needed to make a complete cell before an e.m.f. can be measured.

We can, however, assess the **relative** contributions of single electrode systems to cell e.m.f.s by choosing one system as a standard against which all other systems are measured. The standard system is then arbitrarily assigned zero potential and the potentials of all other systems referred to this value. By international agreement the hydrogen electrode is the standard reference electrode.

A hydrogen electrode

From redox reactions such as

$$Mg(s) + 2H^+(aq) \rightleftharpoons Mg^{2+}(aq) + H_2(g)$$

it is clear that an equilibrium can be set up between hydrogen gas and its ions in solution

$$H_2(g) \rightleftharpoons 2H^+(aq) + 2e^-$$

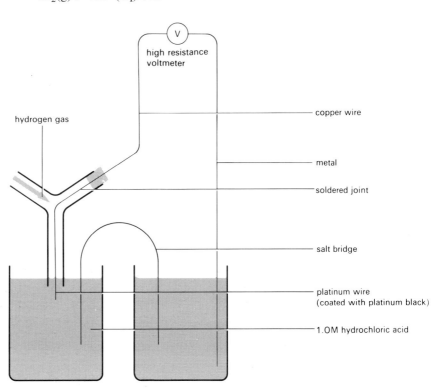

Figure 13.6 A hydrogen electrode

By an arrangement such as the one shown in the figure this reaction can be used in a half-cell. This half-cell is called a **hydrogen electrode**. Essentially it consists of a platinum surface which is coated with finely divided platinum (usually called 'platinum black') which dips into a solution of hydrogen ions. A slow stream of pure hydrogen is bubbled over the platinum black surface. The equilibrium

$$H_2(g) \rightleftharpoons 2H^+(aq) + 2e^-$$

is established rather slowly. The platinum black acts as a catalyst in this process and, being porous, it retains a comparatively large quantity of hydrogen. The platinum metal also serves as a convenient route by which electrons can leave or enter the electrode system. The hydrogen electrode is represented by

Pt[H$_2$(g)] | 2H$^+$(aq)

The potential of the hydrogen electrode under specified conditions, described after the next experiment, is taken as zero. The electrode potential of any other metal is taken as the difference in potential between the metal electrode and the standard hydrogen electrode **with the hydrogen electrode always on the left**:

$$\text{Pt}[H_2(g)] \mid 2H^+(aq) \;\|\; M^{z+}(aq) \mid M(s)$$

If the metal electrode is negative with respect to the hydrogen electrode, the standard electrode potential is given a negative sign. If the metal electrode is positive with respect to the hydrogen electrode, the standard electrode potential is given a positive sign.

Some values for **standard** electrode potentials, $E^\ominus$, are given below. The conditions of temperature and concentration under which these are measured are specified after the next experiment.

$$\text{Pt}[H_2(g)] \mid 2H^+(aq) \;\|\; Mg^{2+}(aq) \mid Mg(s) \qquad E^\ominus = -2.37 \text{ V}$$

$$\text{Pt}[H_2(g)] \mid 2H^+(aq) \;\|\; Zn^{2+}(aq) \mid Zn(s) \qquad E^\ominus = -0.76 \text{ V}$$

COMMENT

When you look at table 6.1 in the *Book of data* under the heading Standard Electrode Potentials you will see that the hydrogen electrode is not usually included in data tables

Pb^{2+}(aq) | Pb(s)	−0.13 V
2H$^+$(aq) | [H$_2$(g)]Pt	0.00 V
Cu^{2+}(aq) | Cu(s)	+0.34 V
Ag$^+$(aq) | Ag(s)	+0.80 V

but each of these values refers to the e.m.f. of a real cell with a standard hydrogen electrode as the lefthand half.

EXPERIMENT 13.1c

To measure the e.m.f. of some electrochemical cells

The circuit for this experiment is shown in figure 13.7**a**.

Several types of cell can be used. The one shown in figure 13.7**b** uses metal strips slotted into pieces of cork as electrodes; they dip into electrode solutions contained in 50 or 100 cm³ beakers. Each beaker contains a half-cell and electrical connection between them is made by a strip of filter paper moistened with saturated potassium nitrate solution (the 'salt bridge').

Connections to the metal electrodes are made with crocodile clips.

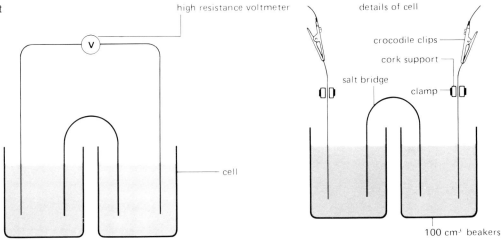

Figure 13.7 **a** The circuit **b** details of the cell

Procedure

1 Set up the cell

$$\text{Cu(s)} \mid \text{Cu}^{2+}(\text{aq}) \mathbin{\|} \text{Zn}^{2+}(\text{aq}) \mid \text{Zn(s)} \tag{1}$$

using 1.0 M ZnSO$_4$(aq) and 1.0 M CuSO$_4$(aq) as the electrode liquids.

Clean metal strips with emery paper before using them. Put a 10 cm × 1 cm strip of filter paper in place and then wet it with saturated potassium nitrate solution using a dropping tube. Connect the cell to the voltmeter and measure the e.m.f.

You should be able to forecast which is the positive pole of this cell, and make the voltmeter connections accordingly.

2 Repeat the measurements for the following cells, using a fresh salt bridge each time a half-cell is changed

$$\text{Mg(s)} \mid \text{Mg}^{2+}(\text{aq}) \mathbin{\|} \text{Cu}^{2+}(\text{aq}) \mid \text{Cu(s)} \tag{2}$$

$$\text{Mg(s)} \mid \text{Mg}^{2+}(\text{aq}) \mathbin{\|} \text{Zn}^{2+}(\text{aq}) \mid \text{Zn(s)} \tag{3}$$

For the magnesium half-cell use magnesium ribbon.

13 Redox equilibria and electrochemical cells 366

QUESTIONS

1 Compare the results of this experiment to the results of the reactivity series reactions in experiment 13.1a.

Is the system with the greatest tendency to lose electrons and form ions the positive or negative pole of each cell?

2 Is the e.m.f. for cell (**3**) what you would expect from the values obtained for cells (**1**) and (**2**)?

You could obtain the equivalent of cell (**3**) by connecting cells (**1**) and (**2**) together as follows

$$\text{Mg(s)} \mid \text{Mg}^+\text{(aq)} \mathop{\|} \text{Cu}^{2+}\text{(aq)} \mid \text{Cu(s)} \quad \text{Cu(s)} \mid \text{Cu}^{2+}\text{(aq)} \mathop{\|} \text{Zn}^{2+}\text{(aq)} \mid \text{Zn(s)}$$

The two copper half-cells would then cancel each other.

COMMENT

You should now realise that the e.m.f. of other cells can be deduced from a list of electrode potentials. Use the *Book of data* to work out the value of the e.m.f., and the sign of the righthand electrode, for the cell

$$\text{Ag(s)} \mid \text{Ag}^+\text{(aq)} \mathop{\|} \text{Cu}^{2+}\text{(aq)} \mid \text{Cu(s)}$$

---

3 Test the effect of temperature on the e.m.f. of the zinc–copper cell by heating a sample of zinc sulphate solution to boiling before setting up a cell.

■ Does the e.m.f. change when the temperature changes?

4 Test the effect of changes in concentration on the e.m.f. of the zinc–copper cell by using a 0.1 M solution instead of one of the 1 M solutions.

■ Does the e.m.f. change when the concentrations are changed?

Standard electrode potentials

In the last experiment you saw that the value of E for an electrode system varies with the concentration of the solution. You also saw that E varies with the temperature. Thus, standardized conditions of temperature and concentration must be used if electrode potential measurements are to be compared. The conditions chosen are:

- **An ion concentration of one mole per cubic decimetre**
- **A temperature of 298 K (25 °C).**

The value of an electrode potential relative to the **standard hydrogen electrode** under these conditions is called the **standard electrode potential**. Standard electrode potentials are denoted by the symbol $E^\ominus$.

By definition, the standard hydrogen electrode consists of hydrogen gas at one atmosphere pressure bubbling over platinized platinum in a solution of hydrogen ion concentration one mole per cubic decimetre, at 298 K.

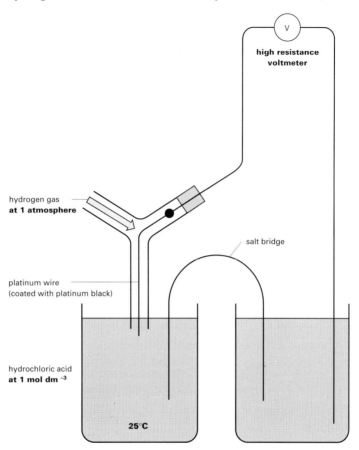

Figure 13.8 The standard hydrogen electrode with hydrogen gas at 1 atmosphere, hydrochloric acid at 1 mol dm^{-3} and the solutions at 25 °C.

A series of $E^\ominus$ values is given in the *Book of data* in table 6.1.

13.2 Redox equilibria extended to other systems

So far in this Topic we have been mainly concerned with redox systems involving a metal in equilibrium with its ions in solution. Earlier in the course, however, you have encountered redox systems of other kinds. Examples of these are

$$2Br^-(aq) \rightleftharpoons Br_2(aq) + 2e^-$$

and $MnO_4^-(aq) + 8H^+(aq) + 5e^- \rightleftharpoons Mn^{2+}(aq) + 4H_2O(l)$

in which the half-cells involve non-metal species or ionic species; and there are other reactions in which the reacting species are ions only:

$$Fe^{3+}(aq) \rightleftharpoons Fe^{2+}(aq) + e^-$$

We shall now see how reactions of this kind can be fitted into the pattern of electrode potentials which we have developed in section 13.1. To begin with, we shall investigate a reaction in which there is no solid metal involved.

EXPERIMENT 13.2a

The reaction between iron(III) ions and iodide ions

To about 2 cm³ of a solution 0.1 M with respect to Fe^{3+}(aq) ions, add an equal volume of 0.1 M potassium iodide solution, noting any changes that you see.

Test separate portions of the original solutions and the final mixture with
1. Starch solution.
2. Potassium hexacyanoferrate(III) solution.

On the basis of your observations write an equation for the reaction.

HINT
Check the behaviour of potassium hexacyanoferrate(III) solution with iron(II) by adding some to a solution containing Fe^{2+}(aq) ions.

Interpretation of the reaction between iron(III) ions and iodide ions

The equation for the reaction between iron(III) ions and iodide ions, studied in experiment 13.2a, is

$$2Fe^{3+}(aq) + 2I^-(aq) \longrightarrow 2Fe^{2+}(aq) + I_2(aq)$$

Each element has undergone a change of oxidation number, which we can write as two separate processes

$$Fe^{3+}(aq) + e^- \longrightarrow Fe^{2+}(aq) \quad \textit{reduction}$$

and $\quad 2I^-(aq) \longrightarrow I_2(aq) + 2e^- \quad \textit{oxidation}$

By analogy with the half-reactions studied earlier we might expect two competing equilibria:

$$Fe^{3+}(aq) + e^- \rightleftharpoons Fe^{2+}(aq)$$

$$2I^-(aq) \rightleftharpoons I_2(aq) + 2e^-$$

with equilibrium positions which differ for each reaction.

When the two systems are brought together the tendency for the iodide/iodine system to lose electrons is greater than that of the iron(II)/iron(III) system. Thus the iodide/iodine equilibrium moves to the right and I_2(aq) is produced.

Gain of electrons by the iron(II)/iron(III) equilibrium causes this to move to the right as well and Fe^{2+}(aq) ions are produced. This process will continue until the two systems are in equilibrium with each other.

> **COMMENT**
> The equilibrium position for the overall reaction
>
> $$2Fe^{3+}(aq) + 2I^-(aq) \rightleftharpoons 2Fe^{2+}(aq) + I_2(aq)$$
>
> lies well over to the righthand side so that we can say it is virtually complete.
>
> $$K_c = \frac{[Fe^{2+}(aq)]^2_{eq}[I_2(aq)]_{eq}}{[Fe^{3+}(aq)]^2_{eq}[I^-(aq)]^2_{eq}} = 10^5 \text{ at } 25\,°C$$

COMMENT

The oxidation numbers of the species in the half-cells are

Fe³⁺(aq), Fe²⁺(aq) | Pt
 +3   +2

and I₂(aq), 2I⁻(aq) | Pt
     0   −1

EXPERIMENT 13.2b

SAFETY
Copper salts are harmful; iodine is corrosive

COMMENT

The platinum electrode need not function as a catalyst, so smooth platinum is suitable. Copper can be used as a substitute for platinum.

It should be possible to use this reaction in an electrochemical cell. The only problem is how the electrons will leave and enter the electrode systems. The solution is to use inert platinum electrodes.

We need to extend our conventions for writing cell diagrams in order to deal with systems such as these. The accepted convention is to put the species with the **lower** oxidation number **nearest** to the electrode, and separate it from the species with the higher oxidation number by a comma. The following electrode systems can therefore be set up and their E values determined by using a reference electrode.

$$Fe^{3+}(aq), Fe^{2+}(aq) \mid Pt$$

and $\quad I_2(aq), 2I^-(aq) \mid Pt$

To measure some electrode potentials

We are going to measure the electrode potentials for the $Fe^{3+}(aq)/Fe^{2+}(aq)$ equilibrium and the $2I^-(aq)/I_2(aq)$ equilibrium.

In order to allow the half-cell reactions to proceed in either direction, we must have both the oxidized and reduced forms present in the half-cells. The righthand half-cell must therefore contain a solution in which both $Fe^{3+}(aq)$ ions and $Fe^{2+}(aq)$ ions are present for the first e.m.f. measurement, and $I^-(aq)$ ions and I_2 molecules for the second e.m.f. measurement.

The $Cu(s) \mid Cu^{2+}(aq)$ system is used as a reference electrode.

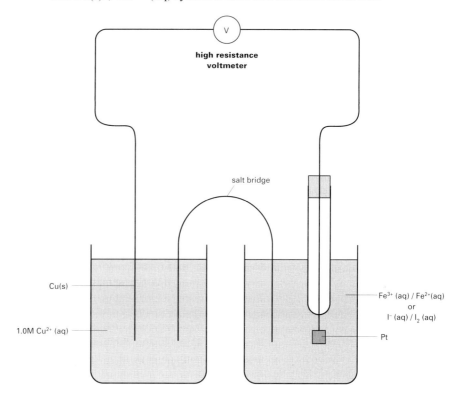

Figure 13.9 The cell for experiment 13.2b.

Procedure

Set up the following cells and measure their e.m.f. The circuit to be used is shown in figure 13.9.

$$\text{Cu(s)} \mid \text{Cu}^{2+}\text{(aq)} \parallel \text{Fe}^{3+}\text{(aq)}, \text{Fe}^{2+}\text{(aq)} \mid \text{Pt}$$

and $\quad \text{Cu(s)} \mid \text{Cu}^{2+}\text{(aq)} \parallel \text{I}_2\text{(aq)}, 2\text{I}^-\text{(aq)} \mid \text{Pt}$

Keep the same cells for the third measurement.
From the results deduce the e.m.f. of the cell

$$\text{Pt} \mid 2\text{I}^-\text{(aq)}, \text{I}_2\text{(aq)} \parallel \text{Fe}^{3+}\text{(aq)}, \text{Fe}^{2+}\text{(aq)} \mid \text{Pt}$$

Check your deduction by measuring the e.m.f. of this cell. A second platinum electrode will be needed for this.

Concentration effects in ion/ion systems

From the results of experiment 13.2b you will have seen that ion/ion systems and non-metal/non-metal ion systems can be used in cell reactions in exactly the same way as metal/metal ion systems.

The equilibrium position is affected by ion concentrations. The greater the relative concentration of $\text{Fe}^{3+}\text{(aq)}$ ions the more the equilibrium

$$\text{Fe}^{3+}\text{(aq)} + \text{e}^- \rightleftharpoons \text{Fe}^{2+}\text{(aq)}$$

moves to the right. This will change the e.m.f. of the cell.

As with metal/metal ion systems, temperature also has an effect on ion/ion equilibria. It is therefore necessary to specify both concentration and temperature conditions for standard electrode potentials involving equilibria between ions. The conditions chosen are:

- **Equal molar concentrations of the reduced and oxidized forms of ion.**
- **A temperature of 298 K (25 °C).**

Some further notes on standard potentials

You should now be in a position to appreciate all the information given in table 6.1 in the *Book of data*, the table of standard electrode potentials. The following notes may, however, be helpful in using this and other similar tables of $E^\ominus$ values.

1 It sometimes happens that the reduced and oxidized parts of an electrode system contain more than one chemical species (ion or molecule) taking part in the cell reaction. For example, the manganate(VII) ion generally exerts its oxidizing power in the presence of hydrogen ions, and water molecules are formed amongst the products of oxidation. These ions and molecules must be included in the oxidized and reduced forms of the equilibrium mixture.

$$\text{MnO}_4^-\text{(aq)} + 8\text{H}^+\text{(aq)} + 5\text{e}^- \rightleftharpoons \text{Mn}^{2+}\text{(aq)} + 4\text{H}_2\text{O(l)}$$

The half-cell diagram for this system is written

$$[\text{MnO}_4^-\text{(aq)} + 8\text{H}^+\text{(aq)}], [\text{Mn}^{2+}\text{(aq)} + 4\text{H}_2\text{O(l)}] \mid \text{Pt}$$

> **COMMENT**
> The square brackets in this and similar diagrams do **not** stand for 'the concentration of' but are merely used to bracket together the oxidized and reduced forms of the equilibrium mixture.

QUESTION
Try converting

$$2IO_3^-(aq) + 12H^+(aq) + 10e^- \rightleftharpoons I_2(aq) + 6H_2O(l) \qquad E^\ominus = +1.19 \text{ V}$$

into a righthand electrode system. Remember the accepted practice is to put the species with the lower oxidation number **nearest** to the electrode, and separate it from the species with the higher oxidation by a comma. Check your answer in the *Book of data*.

2 Metal/metal ion half equations of the general form

$$M^{z+}(aq) + ze^- \rightleftharpoons M(s); \qquad E^\ominus = \pm x \text{ V}$$

such as

$$Al^{3+}(aq) + 3e^- \rightleftharpoons Al(s); \qquad E^\ominus = -1.66 \text{ V}$$
$$Na^+(aq) + e^- \rightleftharpoons Na(s); \qquad E^\ominus = -2.76 \text{ V}$$

QUESTION
Try converting the reaction

$Ag^+(aq) + Fe^{2+}(aq) \rightleftharpoons Ag(s) + Fe^{3+}(aq)$

into a cell diagram.

also convert into electrode systems

$Al^{3+}(aq) \mid Al(s)$

$Na^+(aq) \mid Na(s)$

3 Not all the $E^\ominus$ values given in tables have been obtained by direct measurements. Many of them are calculated from other experimental data.

The chemists' toolkit: some uses of $E^\ominus$ values

We will look at three important uses of tables of $E^\ominus$ values.

1 Predicting whether a reaction is likely to take place

$E^\ominus$ values in order of **increasing negative values** are also in order of increasing tendency for the electrode system to lose electrons. $E^\ominus$ values are listed in figure 13.10 from the most negative value to the least negative value (which is the same as most positive value).

Figure 13.10 Examples of electrode potentials

Electrode system	$E^\ominus$/volt	
$Mg^{2+}(aq) \mid Mg(s)$	-2.37	↑ increasing tendency
$Zn^{2+}(aq) \mid Zn(s)$	-0.76	for electrode to
$S(s), S^{2-}(aq) \mid Pt$	-0.48	release electrons
$Fe^{2+}(aq) \mid Fe(s)$	-0.44	
$Sn^{2+}(aq) \mid Sn(s)$	-0.14	
$2H^+(aq)[H_2(g)] \mid Pt$	0.00	
$Cu^{2+}(aq) \mid Cu(s)$	$+0.34$	
$I_2(aq), 2I^-(aq) \mid Pt$	$+0.54$	
$Fe^{3+}(aq), Fe^{2+}(aq) \mid Pt$	$+0.77$	
$Br_2(aq), 2Br^-(aq) \mid Pt$	$+1.09$	decreasing tendency
$Cl_2(aq), 2Cl^-(aq) \mid Pt$	$+1.36$	for electrode to
$[MnO_4^-(aq)+8H^+(aq)], [Mn^{2+}(aq)+4H_2O(l)] \mid Pt$	$+1.51$	↓ release electrons

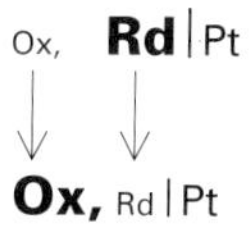

As we go down the series, the oxidizing power of the oxidized forms in the electrode systems **increases** and the reducing power of the reduced forms **decreases**.

This can be illustrated by the $Fe^{3+}(aq)$, $Fe^{2+}(aq)$ and $I_2(aq)$, $2I^-(aq)$ reaction studied earlier. Write down the electrode systems in order of increasing negative value

$I_2(aq), 2I^-(aq) \mid Pt \qquad E^\ominus = +0.54\text{ V}$

$Fe^{3+}(aq), Fe^{2+}(aq) \mid Pt \qquad E^\ominus = +0.77\text{ V}$

Note that $+0.54$ counts as 'more negative' than $+0.77$.

The iodine system will release electrons to the iron system

$2I^-(aq) - e^- \longrightarrow I_2(aq)$

and in the half-cell the reaction will go from right to left

$I_2(aq), 2I^-(aq) \mid Pt$
$\longleftarrow$

For the iron system the opposite is true

$Fe^{3+}(aq) + e^- \longrightarrow Fe^{2+}(aq)$

and

$Fe^{3+}(aq), Fe^{2+}(aq) \mid Pt$
$\longrightarrow$

This is the general situation for all reactions of this kind and can be summed up in a simple procedure:

```
--- I₂(aq), 2I⁻(aq)|Pt ← ---
         above
  --→ Fe³⁺(aq), Fe²⁺(aq)|Pt --
```

1 Write down the reaction in the form of two electrode systems in the order in which they occur in the standard electrode potential series, that is from most negative to most positive.

2 The reaction will go in an anti-clockwise pattern.

Another convenient procedure for predicting the direction a reaction will go in is to write the half-equations for the electrodes in the form of figure 13.11.

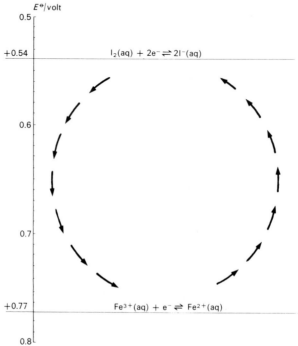

Figure 13.11

2 Balancing the equations of the predicted reactions

The general procedure for balancing redox equations is

1 Predict the direction of reaction using the guidelines given above.
2 Add the number of electrons necessary to balance the charges.
3 Balance the two half-reactions for electron gain and loss.
4 Combine the two half-reactions to give the equation of the reaction.

> **COMMENT**
> Non-standard concentrations or temperatures may alter the equilibrium position and lead to a partial reversal of the predicted reaction. When the difference in $E^\ominus$ values for the electrodes concerned is greater than about 0.6 V, this is unlikely to happen.

Let us apply these rules to the example we used in **1**.

Start from $Fe^{3+}(aq)$ and proceed anti-clockwise

$$Fe^{3+}(aq) + e^- \longrightarrow Fe^{2+}(aq)$$
$$2I^-(aq) - 2e^- \longrightarrow I_2(aq)$$

Balance electron loss and gain

$$2Fe^{3+}(aq) - 2e^- \longrightarrow 2Fe^{2+}(aq)$$
$$2I^-(aq) - 2e^- \longrightarrow I_2(aq)$$

and add

$$2Fe^{3+}(aq) + 2I^-(aq) \longrightarrow 2Fe^{2+}(aq) + I_2(aq)$$

Consider another example. Will potassium manganate(VII) in acid solution be likely to oxidize hydrogen sulphide to sulphur?

From the electrode potential series we find the electrodes in the following order

$$[2H^+(aq) + S(s)], H_2S(aq) \mid Pt \qquad E^\ominus = +0.14 \text{ V}$$
$$[MnO_4^-(aq) + 8H^+(aq)], [Mn^{2+}(aq) + 4H_2O(l)] \mid Pt \qquad E^\ominus = +1.51 \text{ V}$$

and this can be written in a diagram as in figure 13.12.

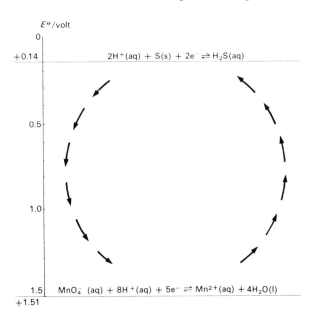

Figure 13.12

The equations are more complicated since more than one species is involved in the oxidized and reduced electrode systems but the half-reactions should proceed

$$MnO_4^-(aq) + 8H^+(aq) \xrightarrow{+5e^-} Mn^{2+}(aq) + 4H_2O(l)$$

$$H_2S(aq) \xrightarrow{-2e^-} 2H^+(aq) + S(s)$$

Balance electron loss and gain

$$2MnO_4^-(aq) + 16H^+(aq) \xrightarrow{+10e^-} 2Mn^{2+}(aq) + 8H_2O(l)$$

$$5H_2S(aq) \xrightarrow{-10e^-} 10H^+(aq) + 5S(s)$$

and add

$$2MnO_4^-(aq) + 6H^+(aq) + 5H_2S(aq) \longrightarrow 2Mn^{2+}(aq) + 8H_2O(l) + 5S(s)$$

The difference between the $E^\ominus$ values is considerable ($1.51 - 0.14 = 1.37$ V), so we should expect this reaction to proceed under all concentration conditions at ordinary temperatures.

The difference in $E^\ominus$ values is, however, no guarantee that it will proceed quickly. In fact, this example is a reasonably fast reaction. Although we can use $E^\ominus$ values to tell us something about the position of equilibrium for a given change, they can never tell us how long it will take for this equilibrium to be attained. Another example is the reaction of aluminium with acids: although the $E^\ominus$ value is large this is quite a slow reaction in many cases.

We have to experiment to find out the rate of a given reaction and, if it is slow, look for a catalyst if we want to make use of the reaction. A table of $E^\ominus$ values enables us to predict whether a search for a catalyst is worth while.

3 Calculating the e.m.f. of electrochemical cells

You have already done this in experiment 13.1c. Now we want to be clear about the general procedure.

$E^\ominus$ values in order of increasing negative values are also in order of increasing tendency for the electrode system to lose electrons as we explained above. Hence when we link two half-cells to form an electrochemical cell, the system which is more negative will become the negative pole and the other system will become the positive pole.

As an example, we will calculate the e.m.f. of the magnesium–lead cell. The cell diagram can be written

$$Mg(s) \mid Mg^{2+}(aq) \parallel Pb^{2+}(aq) \mid Pb(s)$$

1 Look up the $E^\ominus$ values in the *Book of data* and write them down with the more negative value first.

$Mg^{2+}(aq) \mid Mg(s) \qquad E^\ominus = -2.37$ V

$Pb^{2+}(aq) \mid Pb(s) \qquad E^\ominus = -0.13$ V

2 The e.m.f. is the difference between the two values.

e.m.f. $= 2.24$ V

3 The more negative half-cell will be the negative side of the cell.

$Mg^{2+}(aq) \mid Mg(s)$ will be the negative side of the cell.

Therefore the lead electrode will be the positive pole and the e.m.f. of the cell as written above will be $+2.24$ V, under standard conditions.

QUESTION
Use table 6.1 in the *Book of data* to work out the e.m.f. of a lithium–silver cell.

13.3 Entropy changes when metal ions go into solution

REVIEW
Entropy S is related to the number of ways of arranging the particles, and quanta of energy, in a system.

In the last two sections we have looked at redox reactions in cells, particularly cells involving metal/metal ion electrodes. In this section we shall consider changes that take place in these cells, and how they lead to a better understanding of cell reactions.

Consider the reaction

$$Cu(s) \longrightarrow Cu^{2+}(aq) + 2e^-$$

A simplified picture of just one Cu atom becoming a Cu^{2+} ion and entering pure water is shown in figure 13.13. Which has the greater number of possible arrangements – the Cu atom in the solid lattice, or the Cu^{2+} ion in solution? There are more ways of arranging the Cu^{2+} ion, since it could go anywhere among the water molecules.

Figure 13.13

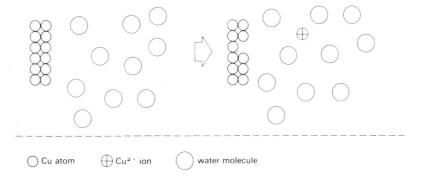

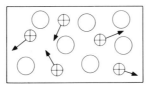 Cu atom 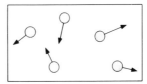 Cu^{2+} ion ◯ water molecule

Ions in solution have a lot of freedom to move about, so the number of arrangements, and therefore the entropy, of ionic solutions is high – much higher than for solid lattices. In fact, ions in solution behave in many ways like molecules in gases.

Figure 13.14

Ionic solution
Ions free to move anywhere among the water molecules

Gas
Molecules free to move anywhere in the container

Of course, this is greatly simplified. For one thing it ignores any interactions between the ions and the water molecules, whereas there are in fact strong attractions. Nevertheless, it is a useful approximate model to think of ions in solutions, especially dilute ones, as behaving like gases.

With this in mind, it is quite clear that whenever a copper atom turns into a copper ion and goes into solution, there must be an increase in the number of arrangements – an entropy increase.

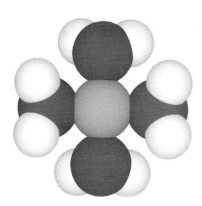

Figure 13.15 Copper ions in water are $Cu(H_2O)_4^{2+}$ entities

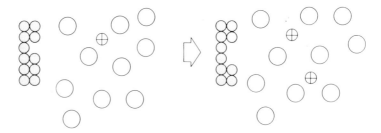

Figure 13.16 What is the entropy change when a second copper atom becomes an aqueous ion?

When a second copper atom becomes an aqueous ion there will be a further entropy increase. But the increase will not be quite as large as last time, because this time there is an ion already present; this slightly reduces the number of arrangements available to the new ion. As more and more ions go into solution, the entropy increases each time – but each time by a little less.

> **COMMENT**
> When there are only a few molecules (or quanta) present, adding one more makes a big difference. When there are lots of molecules or quanta present, adding one more makes a smaller difference. It is the same with money, for that matter: when you are poor an extra pound makes a bigger difference to you than when you are rich.

So if you have a metal electrode surrounded by a solution already containing lots of ions – a concentrated solution – metal atoms turning into aqueous ions will give only a fairly small entropy increase. But if the surrounding solution is dilute, the entropy increase will be bigger.

To summarize, there are two things to bear in mind when considering metal/ion reactions.

1 There is always an entropy increase when metal atoms go into solution as metal ions.
2 The more concentrated the solution of ions surrounding the metal, the smaller the entropy increase when new ions go into solution.

Entropy changes in redox reactions

So far we have looked at the changes that occur when metal atoms turn into ions. This is the kind of change that occurs at a metal electrode in a cell. But to make a cell you need *two* electrodes. What happens when the entropy changes at **both** electrodes are taken into account?

Consider these two simple reactions

$$2Ag^+(aq) + Cu(s) \longrightarrow 2Ag(s) + Cu^{2+}(aq) \quad (1)$$

$$Cu^{2+}(aq) + Zn(s) \longrightarrow Zn^{2+}(aq) + Cu(s) \quad (2)$$

We can use the *Book of data* to look up the entropy changes needed to calculate the entropy changes in the system for these reactions. At standard conditions:

For reaction (**1**) $\Delta S^\ominus_{\text{system}} = -193.0 \text{ J K}^{-1} \text{ mol}^{-1}$

For reaction (**2**) $\Delta S^\ominus_{\text{system}} = -20.9 \text{ J K}^{-1} \text{ mol}^{-1}$

> **COMMENT**
> Collecting standard molar entropy data from tables 5.2 and 5.6 in the *Book of data*, with values in J K^{-1} mol^{-1} at 298 K:
>
> $$2Ag^+(aq) + Cu(s) \longrightarrow 2Ag(s) + Cu^{2+}(aq)$$
> $$2(+72.7) +33.2 2(+42.6) -99.6$$
>
> so
>
> $$\Delta S^\ominus_{system} = -193.0 \text{ J K}^{-1} \text{ mol}^{-1}$$
>
> and
>
> $$Cu^{2+}(aq) + Zn(s) \longrightarrow Zn^{2+}(aq) + Cu(s)$$
> $$-99.6 +41.6 -112.1 +33.2$$
>
> so
>
> $$\Delta S^\ominus_{system} = -20.9 \text{ J K}^{-1} \text{ mol}^{-1}$$

> **COMMENT**
> In the Cu^{2+}/Zn reaction, one ion enters and one leaves. Furthermore, Cu and Zn are very alike as atoms and ions, even down to having similar mass. The entropy of the reactants and the entropy of the products are therefore very similar, so the entropy change for this reaction is small.

We can use the ideas already discussed to try to explain these values. In the Ag^+/Cu reaction, the solution gets only one Cu^{2+} ion but loses **two** Ag^+ ions. We can be sure that the entropy will **decrease** on this account, because aqueous ions have higher entropy than atoms in a solid.

In this way we can rationalize the entropy change for some redox reactions. In many cases however the situation is too complex. But in any case we can **measure** the entropy change for a redox reaction quite easily, by making the reaction occur in a cell, and measuring its e.m.f. We will explore the link between entropy and e.m.f. in the next section.

13.4 Gibbs free energy

Study of the data reveals that there is a relationship between the total entropy of a chemical reaction and the e.m.f. of the corresponding cell.

$$\Delta S_{total} = -\frac{zFE_{cell}}{T}$$

This is why values of E_{cell} can be used with confidence to predict whether reactions are likely to take place or not.

In Topic 10 we introduced a method for calculating the total entropy change for a reaction

$$\Delta S_{total} = \Delta S_{system} + \Delta S_{surroundings}$$

but the method is rather clumsy. For any change there are two entropy changes to consider:

> **COMMENT**
> z is the number of moles of electrons passed round the circuit if the reaction went to completion
> F is the Faraday constant, with the value 96 500 C mol^{-1}
> E_{cell} is the e.m.f. of the cell in volts, J C^{-1}
> You will not be expected to do calculations using this expression.

ΔS_{system}, which we had to calculate from the value of S for each of the reagents and products

and $\Delta S_{surroundings}$, which we had to calculate from the value of ΔH_f for each of the reagents and products, divided by the temperature.

It is rather a nuisance having to bear the surroundings in mind all the time when, as a chemist, what you are really interested in is what is happening in the chemical reaction, or system. Chemists are much more interested in what is happening inside a test-tube than in the test-tube itself, let alone the surrounding laboratory.

To make the situation easier to handle, chemists use a quantity called the **Gibbs free energy change**, or just the **free energy change**, ΔG. It takes its name and the symbol from the American chemist Willard Gibbs. This quantity takes account of both the system and its surroundings in a single expression.

We can write

$$\Delta G = -T\Delta S_{total}$$

For a change in a system to take place of its own accord, that is, for a spontaneous change, we have seen that, quite generally,

ΔS_{total} must be positive

It therefore follows that for a spontaneous change at constant temperature and pressure

ΔG must be negative

The advantage of ΔG is that its values at 298 K for the formation of any compound can be looked up in tables. A calculation then enables us to decide whether a reaction will 'go'. We also have an easy way of measuring ΔG as the e.m.f. of cells.

Since

$$\Delta S_{total} = -\frac{zFE_{cell}}{T}$$

we can deduce

$$\Delta G = -zFE_{cell}$$

but all you need to remember is that

$$\Delta G \propto -E_{cell}$$

Figure 13.17 The American chemist Willard Gibbs who applied the concept of entropy to chemical systems

> **COMMENT**
> Since
>
> $$\Delta S_{total} = \Delta S_{system} + \Delta S_{surroundings}$$
>
> and as $\Delta S_{surroundings} = -\dfrac{\Delta H}{T}$ for a change at constant temperature and pressure, we can substitute in the equation and get
>
> $$\Delta S_{total} = \Delta S_{system} - \frac{\Delta H}{T}$$
>
> When we multiply all through by $-T$, we get
>
> $$-T\Delta S_{total} = -T\Delta S_{system} + \Delta H$$
>
> The quantity $(-T\Delta S_{system} + \Delta H)$ is replaced by the Gibbs free energy change, ΔG, and so we can write
>
> $$-T\Delta S_{total} = \Delta G$$

Gibbs free energy calculations

We can now look at standard free energies of formation of compounds, and see how tabulated values of these quantities can be used to calculate the standard free energy changes for reactions.

The definitions and conventions introduced here are exactly parallel to those concerning the standard enthalpy change of formation, $\Delta H_f^\ominus$. You may find it useful to reread Topic 4.1 as part of your study of this section.

Just as in the case of enthalpies, absolute values of free energies are not known, so it is convenient to choose an arbitrary zero from which to measure the standard free energies of substances. The convention chosen for free energies is the same as that for enthalpies, namely that

at   1 atm pressure
a temperature of 298 K
and the element in the physical state normal under these conditions
the standard free energies of the elements are zero.

It follows necessarily from this convention that
$\Delta G_f^\ominus[\text{element in physical state normal at 1 atm and 298 K}] = 0$
where $\Delta G_f^\ominus[X]$ means 'the standard free energy of X at 298 K'.

It is possible, therefore, using this convention, to tabulate standard free energies of formation of **compounds** rather than standard free energies relating to specific reactions. In fact, of course, the standard free energy of a compound does really refer to a reaction, namely the formation of one mole of the compound from its elements in physical states normal at 1 atm and 298 K. For example,

$$\text{Mg(s)} + \text{O}_2(\text{g}) \longrightarrow \text{MgO(s)} \qquad \Delta G^\ominus = -569.4 \text{ kJ mol}^{-1}$$

and

$$\Delta G_f^\ominus[\text{MgO(s)}] = -569.4 \text{ kJ mol}^{-1}$$

are exactly equivalent statements.

The tabulation of standard free energy data relating to compounds rather than reactions, as in table 5.3 in the *Book of data*, makes for very general and flexible use of the data.

The method of calculating the free energy change of a reaction is by an energy cycle of the same type that you used for enthalpies in Topic 4 and entropies in Topic 10 (see margin).

> **COMMENT**
> We shall write $\Delta G^\ominus$ rather than $\Delta G^\ominus(298)$ since nearly all our calculations will be at 298 K. When we need to draw attention to the temperature we will write $\Delta G^\ominus(298)$ or $\Delta G^\ominus(500)$, etc.

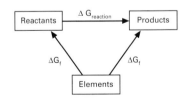

Let us apply this relationship to some examples we have looked at in earlier Topics:

$$\text{ZnCO}_3(\text{s}) \longrightarrow \text{ZnO(s)} + \text{CO}_2(\text{g})$$

$\Delta G_f^\ominus[\text{ZnCO}_3(\text{s})] = -731.6 \text{ kJ mol}^{-1}$

$\Delta G_f^\ominus[\text{ZnO(s)}] \quad = -318.3 \text{ kJ mol}^{-1}$

$\Delta G_f^\ominus[\text{CO}_2(\text{s})] \quad = -394.4 \text{ kJ mol}^{-1}$

$\Delta G_{\text{reaction}}^\ominus \quad\quad\quad = \Delta G_{\text{products}}^\ominus - \Delta G_{\text{reactants}}^\ominus$

$\Delta G_{\text{reaction}}^\ominus(298) \quad = \Delta G_f^\ominus[\text{ZnO(s)}] + \Delta G_f^\ominus[\text{CO}_2(\text{g})] - \Delta G_f^\ominus[\text{ZnCO}_3(\text{s})]$

$\quad\quad\quad\quad\quad\quad\quad\quad = \{-318.3 - 394.4\} - \{-731.6\}$

$\Delta G_{\text{reaction}}^\ominus(298) \quad = +18.9 \text{ kJ mol}^{-1}$

and the reaction is confirmed as one that 'does not go' at 298 K because the Gibbs free energy change is positive.

As another example we will calculate the free energy change for the reaction that is familiar to us from Topic 10:

$$2NH_4Cl(s) + Ba(OH)_2(s) \longrightarrow BaCl_2.2H_2O(s) + 2NH_3(g)$$

using values of Gibbs free energy. Looking up $\Delta G_f^\ominus$ values for all the reactants and products and using the usual energy cycle we find that

$$\Delta G_{reaction}^\ominus(298) = -68.3 \text{ kJ mol}^{-1}$$

So the data predict that the reaction will 'go' at 298 K. We already knew this of course because we did the calculation for $\Delta S_{total}(298)$ in Topic 10 on page 283, and carried out the reaction in the experiment described on page 276.

QUESTION
Carry out your own calculation for the reaction of ammonium chloride and barium hydroxide to confirm that

$$\Delta G_{reaction}^\ominus(298) = -68.3 \text{ kJ mol}^{-1}$$

13.5 Predicting whether reactions will take place: ΔS_{total}, ΔG, E_{cell}, and K_c

Chemists are often involved in predicting whether a chemical reaction will 'go' of its own accord, and if not, what can be done to make it go. We have seen that reactions are always possible if ΔS_{total} is positive – but the surroundings as well as the reacting system itself must be taken into account. A convenient way of doing this is to use ΔG instead of ΔS_{total}. A reaction can occur if it has a negative value for ΔG.

But what does it mean to say a reaction 'goes'? If you put zinc into copper sulphate solution, a reaction certainly goes:

$$Zn(s) + Cu^{2+}(aq) \longrightarrow Zn^{2+}(aq) + Cu(s)$$

The reaction appears to go to completion – equilibrium lies well over to the right. This is the Daniell cell reaction, for which $\Delta G^\ominus = -212.3 \text{ kJ mol}^{-1}$, a large, negative value.

We could measure just how far to the right the equilibrium lay if we knew the equilibrium constant, K_c. This is easy enough to work out if ΔG is known.

You may remember at the end of Topic 11 we arrived at the relationship

$$\Delta S_{total} \propto \ln K_c$$

and as $\Delta G = -T\Delta S_{total}$, by substituting for ΔS_{total} we arrive at the expression

$$\Delta G \propto -\ln K_c$$

A reaction is usually described as 'going to completion' if $K_c = 10^{10}$ or greater. This corresponds to a value of ΔG of about -60 kJ mol^{-1} or a greater negative value. Even with $K_c = 10^{10}$, though, there will always be a very small

COMMENT
For the zinc/copper reaction

$\Delta G^\ominus = -212.3 \text{ kJ mol}^{-1}$

and $K_c = 1.9 \times 10^{37}$.

The reaction will certainly be very nearly complete.

proportion of reactants left unreacted; the reaction will never go **fully** to completion.

Similarly, if K_c has a value less than 10^{-10}, the reaction is considered not to go at all, even though there must in fact be a tiny amount of product formed (otherwise K_c would be zero). When K_c is less than 10^{-10}, ΔG must be greater than $+60\,\text{kJ mol}^{-1}$.

A reaction in equilibrium with **equal amounts** of reactants and products present has $K_c = 1$; this corresponds to $\Delta G = 0$.

Another way of predicting whether reactions can 'go' is to use $E^\ominus$ values. In general, if $E^\ominus$ for a reaction is positive, the reaction 'goes'. Since $E^\ominus$ is related directly to $\Delta G^\ominus$:

$$\Delta G^\ominus \propto -E^\ominus$$

when $\Delta G^\ominus = -60\,\text{kJ mol}^{-1}$, $E^\ominus$ has a value of about 0.6 V.

All this can be summarized as follows:

QUESTION
Copy this table into your notes and add the correct units.

	Reaction 'does not go'	Reactants predominate in an equilibrium	Equal amounts of products and reactants	Products predominate in an equilibrium	Reaction goes to completion
K_c	$<10^{-10}$	$\approx 10^{-2}$	$=1$	$\approx 10^2$	$>10^{10}$
$\Delta G^\ominus$	$>+60$	$\approx +10$	$=0$	≈ -10	<-60
$E^\ominus$	<-0.6	≈ -0.1	$=0$	$\approx +0.1$	>0.6
$\Delta S^\ominus_{\text{total}}$	<-200	≈ -40	$=0$	$\approx +40$	$>+200$

A final word of warning. Many reactions with high K_c values which should apparently go to completion do not do so at room temperature. This is because the rate of reaction is very slow because of a high activation energy barrier. $\Delta G^\ominus$, $E^\ominus$, $\Delta S^\ominus_{\text{total}}$ and K_c only tell you whether a reaction is **feasible**. They say nothing about how fast it will go.

Reaction	$\Delta H^\ominus$ /kJ mol^{-1}	$\Delta G^\ominus$ /kJ mol^{-1}	$E^\ominus$ /V	K_c
$Cu^{2+}(aq)+Zn(s) \longrightarrow Cu(s)+Zn^{2+}(aq)$	-217	-212	$+1.10$	10^{37}
$Zn(s)+2H^+(aq) \longrightarrow Zn^{2+}(aq)+H_2(g)$	-152	-147	$+0.76$	10^{26}
$Pb^{2+}(aq)+Zn(s) \longrightarrow Zn^{2+}(aq)+Pb(s)$	-154	-123	$+0.64$	10^{21}
$Cu^{2+}(aq)+Pb(s) \longrightarrow Cu(s)+Pb^{2+}(aq)$	-62.8	-89.3	$+0.47$	10^{16}
$Tl(s)+H^+(aq) \longrightarrow Tl^+(aq)+\frac{1}{2}H_2(g)$	$+5.9$	-31.8	$+0.34$	10^5

Figure 13.18 Data for some reactions, at 298 K

13.6 Study task: Cells and batteries

QUESTIONS
Read the passage below, and answer the questions based on it.
1. Which cell has the most constant voltage during its hours of service, and which has the most variable voltage?
2. Which cell has the largest capacity per cm^3?
3. Write an equation for one of the reactions in a Leclanché cell.
4. Draw cell diagrams for two button cells.
5. Use the *Book of data* to work out the Gibbs free energy change for the lithium–manganese(IV) oxide cell.

Batteries and cells are available in a bewildering variety of shapes, sizes and names – AA, D, MN1500, G13, dry cells, button cells, car batteries, zinc–carbon, alkaline manganese, mercury, lithium – so that we have probably all had the frustrating experience at some time or other of buying the wrong battery for a calculator or camera. In your reading you will find that the terms battery and cell are used interchangeably, although the term **cell** is preferred for a single anode-cathode system and **battery** for a connected set of anodes and cathodes.

Cells are divided into two groups. **Primary cells** can be discharged once only and then have to be thrown away. Their active reagents are irreversibly converted to products as electrical current is produced. The other group is **secondary cells** which can be recharged by supplying a current to the cell in the reverse direction to the discharge current. In this group the chemical reaction can be reversed; the car battery is the most familiar example of a secondary cell system.

A battery provides electric power from a chemical reaction and the ones we can buy in shops are sealed for convenience and safety. They work on the same principles as redox cells such as the Daniell cell

$$Zn(s) \mid Zn^{2+}(aq) \parallel Cu^{2+}(aq) \mid Cu(s) \qquad E^\ominus = +1.1 \text{ V}$$

with two electrodes, a salt bridge and an electrolyte. The e.m.f. of the cell depends on the Gibbs free energy change of the reaction

$$Zn(s) + Cu^{2+}(aq) \longrightarrow Zn^{2+}(aq) + Cu(s) \qquad \Delta G^\ominus = -212.3 \text{ kJ mol}^{-1}$$

Standard electrode potentials are measured using a voltmeter that takes little or no current from the cell being tested. But a practical battery has to be designed to produce an electric current to light a flashbulb, operate a calculator, a hearing aid or heart pace-maker. So there are some important differences between laboratory standard cells and commercially-produced batteries. In a laboratory standard cell the electrolyte used in the salt bridge is selected so that it takes no part in the cell reaction. But in a commercial battery the electrolyte may play an important part in the reactions because not only does it connect the negative half-cell to the positive half-cell but it may also be expected to absorb the products of the reaction in such a way that the cell can operate efficiently for most of its life.

When we buy the type of battery that is used in torches, toys and radios we will be buying a battery that was first produced over a century ago. It is known as the Leclanché cell and is based on a zinc negative electrode and a manganese dioxide positive electrode, with an electrolyte of ammonium chloride and zinc chloride in water. The manganese dioxide is mixed with carbon to form the positive electrode and the cell is often referred to as the zinc–carbon battery (see figure 13.2). A number of variants on this standard 'dry cell' are produced in which the electrolyte is different. For instance the 'heavy duty' or long-life battery has an electrolyte of zinc chloride and the 'alkaline manganese' battery has a concentrated potassium hydroxide electrolyte.

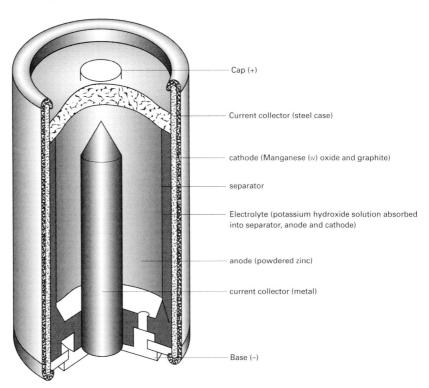

Figure 13.19 The alkaline manganese cell.

What are the chemical reactions on which some of the better known batteries are based?

The Leclanché group of cells all depend on the same basic electrode system

$$Zn(s) \mid Zn^{2+}(s) \parallel MnO(OH)(s) \mid MnO_2(s)$$

The theoretical voltage available ranges from 1.5 to 1.7 volts depending on the reactions that take place with the electrolyte, for example forming $Zn(NH_3)_2Cl_2$ or $Zn(OH)_2$ or ZnO. By connecting the cells in series or parallel batteries of different capacity and voltage are manufactured, ranging from 1.5 to about 500 volts.

The voltage of a Leclanché dry cell tends to drop during its useful life (so your torch gets dimmer rather than suddenly failing) and it performs poorly at lower temperatures (they will work for twice as long on a jungle expedition as on a visit to the South Pole). The heavy duty type can deliver a higher current, which makes them suitable for running small electric motors. The alkaline type has a much greater capacity for the same volume and larger currents can be drawn from the cell, so it is suitable for cassette players.

Although very small Leclanché cells are made, miniature cells are usually used in applications where a very steady voltage is required for the life of the cell. 'Button cells' are manufactured with a wide variety of redox systems but the most commonly available are based on zinc–mercury(II) oxide, zinc–silver(I) oxide, lithium–manganese(IV) oxide and lithium–iodine.

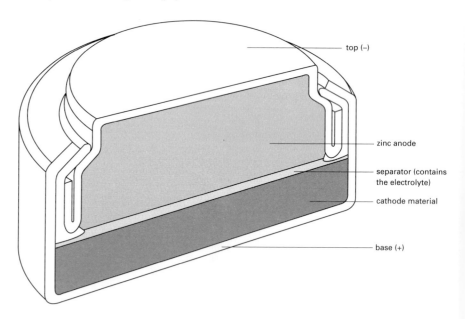

Figure 13.20 A typical button cell.

The reactions in a typical button cell are

$$Zn(s) + Ag_2O(s) \longrightarrow ZnO(s) + 2Ag(s) \qquad E = +1.60 \text{ V}$$

and in a lithium-manganese(IV) oxide cell

$$2Li(s) + 2MnO_2(s) \longrightarrow Li_2O(s) + Mn_2O_3(s) \qquad E = +3.0 \text{ V}$$

Lithium cells use organic solvents because of the reactivity of lithium, and a typical electrolyte is lithium chlorate(VII).

The lithium–iodine cell is a special type because it is a solid–state cell that operates without the need for an electrolyte

$$Li(s) + \tfrac{1}{2}I_2(s) \longrightarrow LiI(s) \qquad E = +2.8 \text{ V} \qquad \Delta G = -270 \text{ kJ mol}^{-1}$$

The cell does not short-circuit internally because a thin layer of lithium iodide forms on manufacture where the two electrodes are in direct contact.

As well as the voltage of the cell, there are other technical features that need to be considered when thinking about batteries. Their **capacity** depends on the amount of reactive chemicals that can be packed into the battery and the number of moles of reaction that will produce a useful flow of electric current; it is measured as **current** × **time** over the life of the battery. Their **power rating** is the power the battery can deliver in normal working conditions and is measured in watts, **current** × **voltage**.

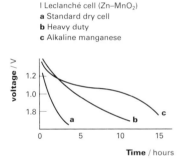
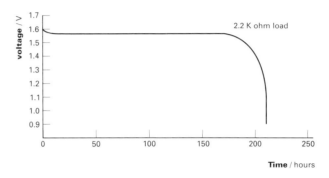

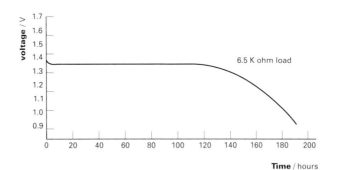
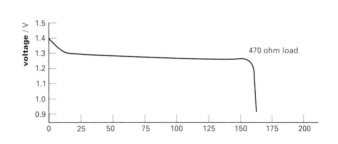

Figure 13.21

	Energy output/ watt-hours kg^{-1}
Primary cells	
Zn–MnO$_2$ (dry cell)	80
Zn–MnO$_2$ (alkaline)	100
Zn–HgO	100
Li–MnO$_2$	260
Li–I$_2$	290
Secondary cells	
Pb–acid	20
Cd–NiO	25
Zn–Ag$_2$O	40

For a torch the mass and volume of the batteries we use does not matter too much and in a car there is room for the large and heavy lead-acid battery. But we now use batteries in a wide range of applications with quite different requirements. One way of comparing batteries is the ratio of their energy output to their mass.

For any battery we can work out what is theoretically possible knowing the amounts of reactants in the battery and the Gibbs free energy change of the reaction, but what can actually be achieved depends on a great deal of attention to the way the battery is constructed. Research into the working of the Leclanché cell followed by technical development has resulted in the capacity being doubled every decade of this century.

As a group batteries represent a remarkable technical achievement. Developing a practical battery at a competitive price takes years, and there is always more work being done to improve the design.

Summary

At the end of this Topic you should be able to:

a demonstrate understanding of the terms: redox, half-reactions, and standard electrode potentials, and use these to interpret reactions involving electron transfer

b construct and interpret cell diagrams using the IUPAC convention, and relate these to the practical arrangements for setting up cells

c plan an investigation to measure a standard electrode potential and justify the procedures involved

d recall that when a metal atom in a solid lattice is converted into an aqueous ion, there is an increase in entropy

e interpret a redox reaction in terms of an appropriate cell by:
  i drawing the cell diagram
  ii selecting appropriate standard electrode potential data
  iii calculating the e.m.f of the cell and the polarity of the electrodes
  iv predicting the direction and extent of reaction

f demonstrate understanding that ΔG is related to the total entropy change and E_{cell} for a reaction

g recall that for all spontaneous chemical changes, ΔG must be negative

h calculate ΔG for a reaction using tables of standard free energy changes and hence predict reaction feasibility

i predict whether a system is at equilibrium, near equilibrium or capable of spontaneous change, using ΔG, E_{cell} and K_c as indicators of thermodynamic feasibility

j demonstrate understanding that some reactions, though predicted as feasible on thermodynamic grounds, do not in practice occur spontaneously

k evaluate information about batteries and cells by extraction from text and the *Book of data*.

Review questions

* Indicates that the *Book of data* is needed.

13.1 Standard electrode potentials are given as the e.m.f. of electrochemical cells with a standard hydrogen electrode forming the left-hand electrode of the system.

Draw a labelled diagram of a standard hydrogen electrode and state the standard conditions under which it is used.

13.2 State which of the reactants is the oxidant and which is the reductant in each of the following reactions.

a $Fe(s) + Cu^{2+}(aq) \longrightarrow Fe^{2+}(aq) + Cu(s)$
b $Al(s) + 3H^{+}(aq) \longrightarrow Al^{3+}(aq) + 1\frac{1}{2}H_2(g)$
c $Zn(s) + Pb^{2+}(aq) \longrightarrow Zn^{2+}(aq) + Pb(s)$
d $2Fe^{3+}(aq) + Sn^{2+}(aq) \longrightarrow 2Fe^{2+}(aq) + Sn^{4+}(aq)$

13.3 Calculate the $E^\ominus$ value and state the polarity of each terminal in the following cells. Assume a temperature of 25 °C and ionic concentration 1.0 M.

a $Pt[H_2(g)] \mid 2H^+(aq) \vdots Fe^{2+}(aq) \mid Fe(s)$
b $Ni(s) \mid Ni^{2+}(aq) \vdots 2H^+(aq) \mid [H_2(g)]Pt$
c $Zn(s) \mid Zn^{2+}(aq) \vdots Ni^{2+}(aq) \mid Ni(s)$
d $Al(s) \mid Al^{3+}(aq) \vdots Cr^{3+}(aq) \mid Cr(s)$

13.4 $E^\ominus_{cell}$ is $+0.62$ volt for the cell:

$$Co(s) \mid Co^{2+}(aq) \vdots Cu^{2+}(aq) \mid Cu(s)$$

Calculate the standard electrode potential for

$$Co^{2+}(aq) \mid Co(s)$$

13.5 $E^\ominus_{cell}$ is $+1.61$ volt for the cell:

$$Zn(s) \mid Zn^{2+}(aq) \vdots Hg^{2+}(aq) \mid Hg(l)$$

Calculate the standard electrode potential for

$$Hg^{2+}(aq) \mid Hg(l)$$

13.6 For each of the following cells construct the two half equations and the whole equations to represent the changes which take place when the cell terminals are connected by a conductor.

a $Al(s) \mid Al^{3+}(aq) \vdots Sn^{2+}(aq) \mid Sn(s)$
b $Ag(s) \mid Ag^+(aq) \vdots Pb^{2+}(aq) \mid Pb(s)$
c $Pt[H_2(g)] \mid 2H^+(aq) \vdots Mg^{2+}(aq) \mid Mg(s)$

13.7 Arrange the following groups of ions in order of their **ability to oxidize**. Put the one with the greatest ability to oxidize first. Assume that they are all of molar concentration.

a $Cu^{2+}(aq), Ag^+(aq), Pb^{2+}(aq), Cr^{3+}(aq)$
b $Mg^{2+}(aq), Zn^{2+}(aq), Fe^{3+}(aq), Sn^{2+}(aq)$

13.8 The equilibrium constant as measured by an analytical method for the reaction

$$Ag^+(aq) + Fe^{2+}(aq) \rightleftharpoons Fe^{3+}(aq) + Ag(s)$$

is 3.2 dm^3 mol^{-1} at 25 °C.

*a Calculate $E^\ominus_{cell}$ for this reaction at 25 °C.
*b Calculate the standard free energy change for this reaction at 25 °C.
c What is the approximate ratio of reactants to products? Do the values K_c, $E^\ominus_{cell}$, and $\Delta G^\ominus$ all support the same conclusion? Justify your answer.

13.9 When an aqueous solution of bromine is added to a solution of potassium iodide, the following reaction takes place:

$$\tfrac{1}{2}Br_2(aq) + I^-(aq) \longrightarrow \tfrac{1}{2}I_2(aq) + Br^-(aq)$$

a Write two half-equations, one for each half-cell reaction.
b Write down a cell diagram for an electrochemical cell in which this reaction could be carried out.

c   Use the values of the appropriate standard electrode potentials given in the *Book of data* to calculate the standard e.m.f. of your cell.
d   Is the position of equilibrium in favour of reactants or products? Justify your answer.

*13.10 a   Copy from table 5.5 in the *Book of data* the standard free energy of formation of methane gas, CH_4.
b   Does your value suggest that the reaction

$$C(s) + 2H_2(g) \longrightarrow CH_4(g)$$

would be expected to take place?
c   In your experience does such a reaction take place at room temperature and atmospheric pressure?
d   How do you explain any difference in your answers to **b** and **c** above?

*13.11 a   Copy from table 5.3 in the *Book of data* the standard free energies of formation of iron(II) oxide and iron(III) oxide.
b   What is the free energy change for the atmospheric oxidation of iron(II) oxide to iron(III) oxide?
c   Which oxide of iron is likely to be the principal naturally occurring ore of the element?

*13.12 a   Calculate both the standard free energy change, $\Delta G^\ominus$, and the standard enthalpy change, $\Delta H^\ominus$, for each of the following reactions:
i $C(s) + O_2(g) \longrightarrow CO_2(g)$
ii $N_2(g) + 3H_2(g) \longrightarrow 2NH_3(g)$
iii $CaCO_3(s) \longrightarrow CaO(s) + CO_2(g)$
iv $Zn(s) + Cu^{2+}(aq) \longrightarrow Zn^{2+}(aq) + Cu(s)$
b   For which reactions do $\Delta G^\ominus$ and $\Delta H^\ominus$ **i** agree closely **ii** differ significantly?
c   For which **type of reactions** are ΔG and ΔH most likely to show close agreement, and for which are they most likely to differ?
d   Explain why ΔH is often a good guide to whether a reaction is likely to go, but not as reliable a guide as ΔG.

Examination questions

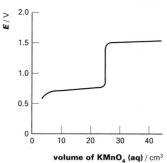

13.13   This question is about the changes in electrode potential that take place when iron(II) ions react with manganate(VII) ions in acidic solution. The half equations are:

$$MnO_4^-(aq) + 8H^+(aq) + 5e^- \longrightarrow Mn^{2+}(aq) + 4H_2O(l) \quad E^\ominus = +1.51\ V$$

$$Fe^{3+}(aq) + e^- \longrightarrow Fe^{2+}(aq) \quad\quad\quad\quad\quad\quad E^\ominus = +0.77\ V$$

25.0 cm^3 of acidified iron(II) sulphate solution, of concentration 0.10 mol dm^{-3}, is placed in a small beaker containing a stirrer and a platinum electrode. The solution is connected to a standard hydrogen electrode with a salt bridge.
Changes in electrode potential, E, are measured with a voltmeter as potassium manganate(VII) solution is added from a burette. The vertical portion of the curve corresponds to the end-point of the reaction.

a i   Write a balanced ionic equation for the reaction between iron(II) ions and manganate(VII) ions.

ii Use this equation and the titration curve to find the concentration of the potassium manganate(VII) solution.

b i Write a cell diagram for the reaction by linking an iron(II)/iron(III) half-cell with a manganate(VII)/manganese(II) half-cell under standard conditions.

ii Which would be the positive electrode in this cell?

iii What would be the e.m.f. of this cell?

13.14 Some standard electrode potentials for iron and vanadium are:

$$Fe^{2+}(aq) \mid Fe(s) \qquad E^\ominus = -0.44 \text{ V}$$

$$Fe^{3+}(aq), Fe^{2+}(aq) \mid Pt \qquad E^\ominus = +0.77 \text{ V}$$

$$V^{3+}(aq), V^{2+}(aq) \mid Pt \qquad E^\ominus = -0.26 \text{ V}$$

a i Draw a diagram to show how you would set up a cell in order to measure the electrode potential for the $Fe^{3+}(aq)/Fe^{2+}(aq)$ system with a standard hydrogen electrode as the reference electrode.

ii At what temperature are standard electrode potentials measured?

iii What substances are used for a 'standard hydrogen electrode'?

iv What reagent pressures and concentrations are used in a standard hydrogen electrode?

v State a suitable combination of materials for the salt bridge.

vi Write out the conventional cell diagram corresponding to your apparatus.

vii Which would be the positive electrode in your cell?

b i In which oxidation state will iron reduce $V^{3+}(aq)$ to $V^{2+}(aq)$?

ii Write an equation for the reaction from **i**.

iii What e.m.f. would you expect if the reaction were set up as a cell, using standard conditions?

13.15 The diagram shows a calomel electrode, which is often used for measuring standard electrode potentials in preference to a hydrogen electrode.

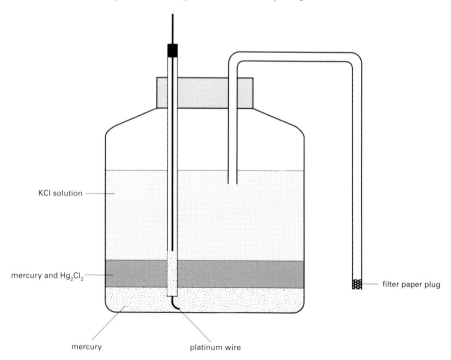

It consists of mercury in contact with a solution of potassium chloride saturated with Hg_2Cl_2. Notice that it contains its own salt bridge in the side tube.

a i What is the oxidation number of mercury in Hg_2Cl_2?
  ii What is the purpose of the filter paper plug?
  iii What do you think are the relative merits of the calomel electrode and the hydrogen electrode for measuring standard electrode potentials? State one advantage and one disadvantage for each electrode.

b i Draw a diagram to show how you would use a calomel electrode to measure the standard e.m.f. of the cell:

$$Pt \mid [2Hg(s) + 2Cl^-(aq)], Hg_2Cl_2(s) \parallel Cu^{2+}(aq) \mid Cu(s)$$

State the concentration of any solution you use and the experimental conditions. You need only draw the calomel electrode in outline.

  ii Predict the e.m.f. of your cell using the following standard electrode potential data:

$$Hg_2Cl_2(s), [2Hg(s) + 2Cl^-(aq)] \mid Pt \qquad E^\ominus = +0.27 \text{ V}$$
$$Cu^{2+}(aq) \mid Cu(s) \qquad E^\ominus = +0.34 \text{ V}$$

13.16 Consider the following pairs of substances:

Potassium iodide and iron(III) sulphate
Manganese(IV) oxide and hydrobromic acid
Tin(II) chloride and mercury(II) chloride

For each pair of substances list the appropriate electrode systems and standard electrode potentials from table 6.1 in the *Book of data*.
Calculate the e.m.f. of each cell that could be formed and deduce the extent to which reaction might occur in each case.
State what the likely products of reaction would be and describe what you would expect to observe on mixing in an appropriate manner each pair of substances.

13.17 Discuss, using suitable examples, what is understood by the terms **oxidation** and **reduction**. Using the *Book of data*, choose an **oxidant** that is likely to bring about the change

$$Fe(CN)_6^{4-}(aq) \longrightarrow Fe(CN)_6^{3-}(aq)$$

and a **reductant** that is likely to bring about the change

$$Cr^{3+}(aq) \longrightarrow Cr^{2+}(aq)$$

Why are some compounds able to act as both oxidants and reductants? Give an example of this type of behaviour.

TOPIC 14

Natural products and polymers

This Topic is about some of the very large molecules whose size depends on the ability of carbon atoms to link together in long chains. When there are more than a thousand atoms in the chain we can call them 'macromolecules' ('macro' means large or long).

We will start with naturally occurring materials – sugars, fats and proteins – and then consider the properties of synthetic materials, such as poly(ethene) and nylon. Except for fats they all occur as polymers and they are of great practical importance to us; as foods, in the structure of our bodies, and as everyday materials so common that they are never out of sight – clothes, furniture, paper.

In this Topic there is more background information than usual to help you appreciate the importance of these substances, but when learning the Topic you should concentrate on the reactions involved.

14.1 The shape of carbon compounds

You are going to smell the contents of three tubes labelled A, B and C and try to decide whether all the tubes have the same smell, or whether the tubes differ in smell from each other.

Remove a stopper and smell the tube, breathing in gently. Try to describe the smell to yourself. Replace the stopper.

Repeat the process with a second tube, and then with the third tube. Does one tube have a different smell? Ignore any differences in the strength of smell.

You can smell a tube again to help you decide, but only open one tube at a time. Record the answer you think is correct:
- A is different from the other two
- B is different from the other two
- C is different from the other two
- All the tubes smell the same
- The tubes all have different smells

Write a short sentence in which you try to describe what you have smelt.

What is odd is that when chemists extract the aroma compounds responsible for the two smells they seem to find only one compound. Both aroma compounds have the molecular formula $C_{10}H_{14}O$, both have a carbonyl group, both decolorize bromine water because they have double bonds, and

they have the same mass spectrum and infra-red spectrum.
They have the same structural formula:

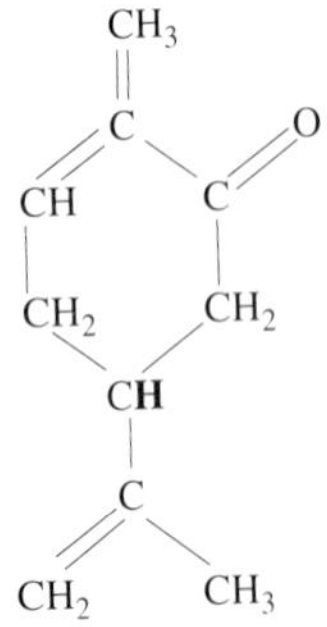

STUDY TASK

1 Use a ball-and-spoke model kit to build a model of the aroma compound. Collect model atoms representing five carbon atoms that are single bonded (tetrahedral) and five carbon atoms that are double bonded (trigonal), plus one double bonded oxygen and one hydrogen atom. Using only one hydrogen atom makes the structure easier to understand.

Build a model representing the molecular structure, using the hydrogen atom where it is shown in bold in the diagram.

Compare your model with models built by other students. Are they all identical? With the rings more or less flat and the solitary hydrogen sticking up, are the oxygen atoms on the left or right when you look at the models?

2 Build a model of 1-bromo-1-chloroethane, CH_3—$CHBrCl$. What do you see when you view your model in a mirror?

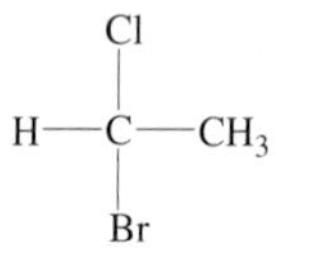

Compare your model with your neighbours' model. You should find that some models are mirror images of other models. Thus 1-bromo-1-chloroethane can exist as two isomers.

If you exchange atoms between models so that two groups are identical, for example making 1,1-dibromoethane and 1,1-dichloroethane, you will find that the new compounds no longer form mirror-image isomers.

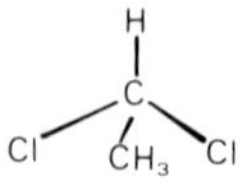

14 Natural products and polymers

The clue to the strange behaviour of the aroma molecules is the arrangement of atoms around the carbon atom drawn in bold. It has four different groups attached:

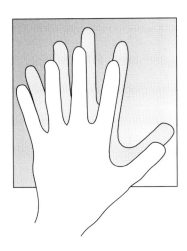

Figure 14.1 Mirror images: two left hands.

This results in two molecular shapes which are **mirror images** of each other and called **optical isomers**. The carbon atom is called a **chiral centre**, pronounced *kiral*.

Our sense of smell works by identifying the exact shape of molecules and therefore responds differently to these two molecules, whereas many chemical tests are 'blind' to this type of molecular difference.

The most common reason for a carbon compound to be a chiral compound is the presence in the molecule of a carbon atom attached to four different atoms or groups of atoms.

The models of 1-bromo-1-chloroethane, CH_3—$CHBrCl$ also exist as two isomers which are mirror images. Thus 1-bromo-1-chloroethane is a chiral compound.

COMMENT
Chiral is from the ancient Greek for 'hand': a left hand is a mirror image of a right hand. Find a mirror and check this claim.

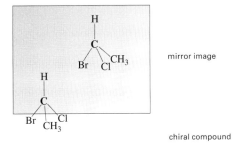

achiral

When you altered your models so that two groups were identical, for example making 1,1-dibromoethane and 1,1-dichloroethane, you should have found that the new compounds were no longer chiral (they are **achiral**, as are most compounds).

Many naturally occurring compounds have very complex structures so it is perhaps not surprising that they are often chiral compounds, sometimes with several chiral centres (see figure 14.2).

Chiral compounds can be detected by an instrument called a **polarimeter**. Plane-polarized light (as produced when sunlight is reflected) is affected by having its plane of polarization rotated by chiral compounds and this effect is detected by a polarimeter (as reflected sunlight is detected and absorbed by some types of sunglasses).

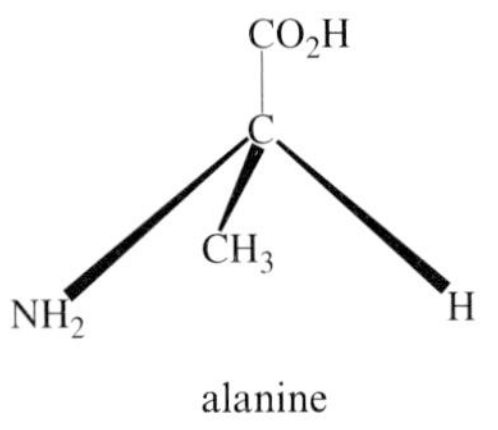

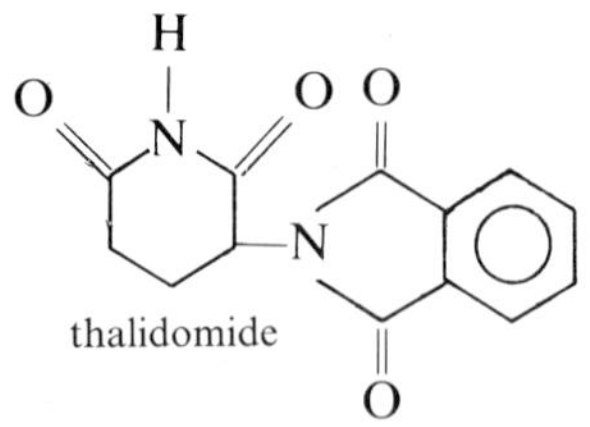

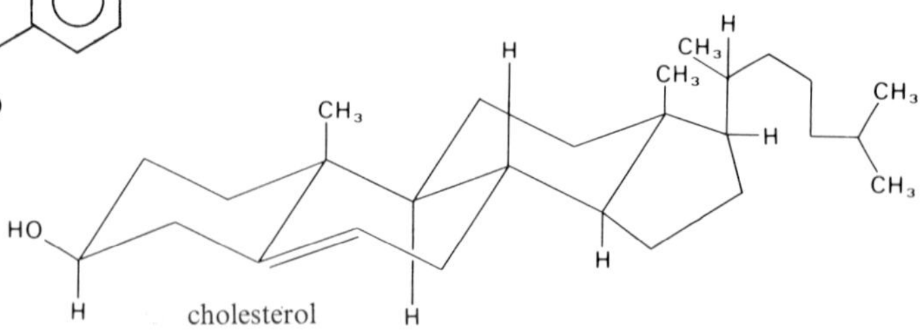

14 Natural products and polymers

Figure 14.2 Examples of chiral compounds.

alanine

thalidomide

cholesterol

EXPERIMENT 14.1 Cholesteryl benzoate, 'liquid crystals'

This experiment is an opportunity to prepare an example of a chiral compound that has the particular properties of 'liquid crystals'.

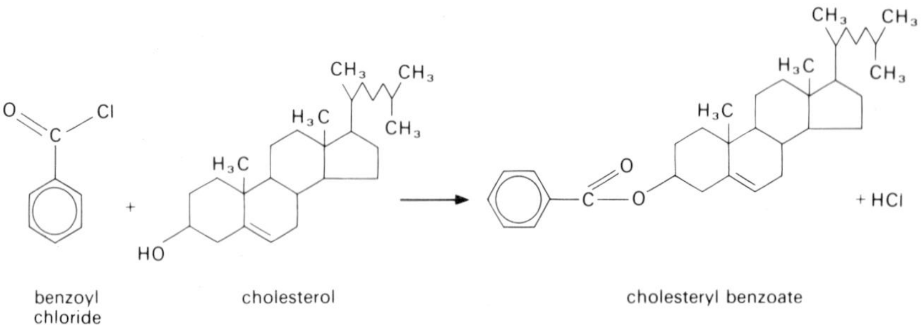

benzoyl chloride + cholesterol → cholesteryl benzoate + HCl

SAFETY ⚠

Two of the reagents are very hazardous: pyridine vapour is harmful, avoid contact with skin and eyes; benzoyl chloride is a lachrymator – that is, it will bring tears to your eyes. Only attempt this experiment in a fume cupboard. Wear gloves and remember to wear your eye protection properly.

Procedure

Place 1 g of cholesterol in a 50 cm³ conical flask. Add 3 cm³ of pyridine (TAKE CARE) and swirl the mixture to dissolve the cholesterol. Then add 0.4 cm³ of benzoyl chloride (TAKE CARE). Heat the resulting mixture on a steam bath for about 10 minutes. At the end of the heating period cool the mixture.

Dilute with 15 cm³ of methanol and collect the solid cholesteryl benzoate by suction filtration, using a little methanol to rinse the flask and to wash the crystals. Recrystallize the cholesteryl benzoate by heating it in a conical flask with ethyl ethanoate (TAKE CARE flammable). Use the minimum volume, about 15–20 cm³, which will completely dissolve the crystals. Cool in an ice bath, and collect the crystals by suction filtration. The yield should be 0.6 to 0.8 g.

The formation of the 'liquid crystal' phase of cholesteryl benzoate can be seen by placing 0.1 g of the compound on the end of a microscope slide and heating the sample by holding the slide with a pair of tongs well above a **small** Bunsen burner flame.

The solid will turn first to a cloudy liquid and then, with further heating, to a clear melt. On cooling, the cloudy liquid crystal phase will appear first, and then it will change to a hard, crystalline solid. With good lighting from the side, for example at a window, a band of colour should be seen at the boundary between the clear melt and the cloudy liquid on both heating and cooling. The more cautious the heating, the better you can see the changes.

You can repeat the heating and cooling many times with the same sample.

Liquid crystal displays

Electronic watches and calculators have made liquid crystal displays commonplace, whereas before the 1980s liquid crystals were only used in specialized applications.

An Austrian botanist, Friedrich Reinitzer, made the first observations of liquid crystal behaviour in 1888. When investigating the formula of cholesterol he prepared cholesteryl benzoate and in determining its melting point noticed that it melted first to a turbid liquid, at 147 °C, and then changed to a clear liquid at 180 °C. The phase between 147 °C and 180 °C is a true liquid phase but the molecules have a degree of orientation and the liquid therefore has some of the properties of a crystal.

Compounds which display liquid crystal behaviour usually have relatively rigid rod-like molecules and in the liquid crystal phase the rod-like molecules take up various parallel arrangements. For example, the molecules may lie parallel but twisted in a helix. The orientation of the molecules can be changed by small electric fields, and this is used as the basis of display units for watches and calculators. A typical display unit consists of a thin film of liquid crystal sandwiched between two glass plates which have a transparent conductive coating on their inner surfaces.

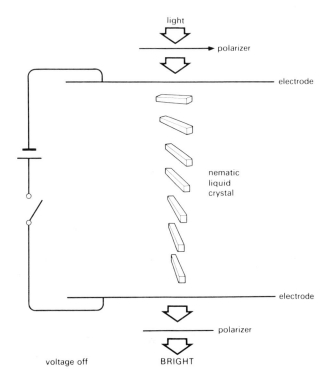

Figure 14.3 The off-state of a liquid crystal display.

Light which enters the display unit first passes through a polarizer. When the plane polarized light passes through the twisted liquid crystal, the plane of polarization is rotated through 90 °.

The light then passes through a second polarizer set at 90 ° to the first polarizer. In this situation the display unit appears bright. If a small voltage is now applied, using the conductive coatings on the glass plates as electrodes, the orientation of the molecules is changed and they no longer rotate the plane of polarization of the light. The plane polarized light therefore does not pass through the second polarizer and the display unit appears dark.

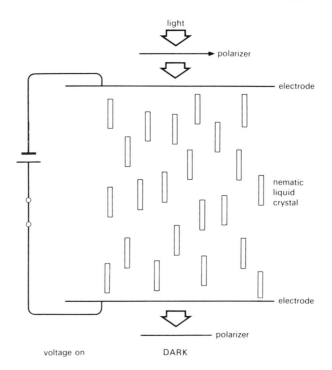

Figure 14.4 The on-state of a liquid crystal display.

On switching off the voltage, the molecules relax back to the original arrangement of figure 14.3, and the cells becomes bright again.

By producing the conductive layer as separate small areas any pattern can be displayed and switched on and off by the application of a small voltage.

Figure 14.5 Liquid crystals used for a computer screen.

As a test of the nature of the light emerging from a liquid crystal display, try the effect of observing the display through a piece of polaroid film or polarizing sunglasses.

14.2 Carbohydrates

Chemists prefer to call sugars carbohydrates because they have the general formula $C_n(H_2O)_m$. As you might expect, this formula does not accurately represent their structures, which usually involve a number of alcoholic hydroxyl groups plus an aldehyde or ketone carbonyl group. We shall be considering the chemistry of four important carbohydrates, glucose, fructose, sucrose and starch.

EXPERIMENT 14.2 Some reactions of carbohydrates

For these experiments we shall use only a small selection of carbohydrates, choosing those that are available economically from natural sources.

SAFETY
Concentrated sulphuric and hydrochloric acids are corrosive; do not inhale any fumes produced.

Procedure

1 Dehydration

To a small portion of a carbohydrate add a few drops of concentrated sulphuric acid in a wide bore test-tube.

2 Benedict's solution

Dissolve a small portion of a carbohydrate in 5 cm³ of warm water and add 5 cm³ of Benedict's solution. Bring to the boil and allow to stand.
　Repeat for the other carbohydrates available.

- Note the colour of any precipitate and note which carbohydrates do not react.

3 Hydrolysis

Dissolve a small portion of sucrose in 5 cm³ of 2 M hydrochloric acid and heat in a water bath for 5 minutes. Neutralize the acid with 2 M ammonia and repeat the test with Benedict's solution.

- Have the properties of the sugar changed?

4 Polarized light

You will need a polarimeter for this experiment. Figure 14.6 shows how the instrument is constructed.
　If solutions of carbohydrates are not available, prepare them by dissolving 15 g in 100 cm³ of warm water.
　Half fill the specimen tube from the polarimeter. **Without** placing the specimen tube in position adjust the polarimeter by rotating the centre of the analyser until, on looking through the analyser and polarizer, you see that the source of light is extinguished. Note the position of the pointer on the scale.
　Put the specimen tube in position and look through the instrument once more. Do you have to alter the setting of the analyser to extinguish the light, and if so, by how much?
　Now fill the specimen tube so as to double the length of liquid through which the light passes. Is a further adjustment of the analyser necessary for extinction of the light?

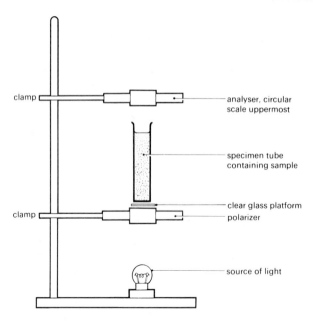

Figure 14.6 A simple polarimeter.

> **COMMENT**
> As an additional experiment if you have time, you can investigate the hydrolysis of sucrose, using the polarimeter.
> Dissolve 100 g of sucrose in 40 cm³ of hot water and leave for 15 minutes to cool and clear. Add 5 cm³ of concentrated hydrochloric acid, mix well, and pour into the polarimeter tube. Take a reading of the setting of the analyser, α_t, and note the time, t. Take further readings every 5 minutes for about 45 minutes until the readings do not change in value (α_0).
> Plot a graph of ($\alpha_t - \alpha_0$) against t and comment on the shape of the graph.

Reactions of carbohydrates

1 Concentrated sulphuric acid is a dehydrating agent and this reaction may remind you of the empirical formula of the carbohydrates

$$C_6H_{12}O_6 \xrightarrow{\text{concentrated } H_2SO_4} 6C + 6H_2O$$

Concentrated sulphuric acid is also an oxidizing agent and the reaction is more complex than the simple equation given.

2 Benedict's solution is a sensitive test for organic reducing agents. When you tested the sugars with Benedict's solution you should have found that glucose and fructose react readily but sucrose did not react. So glucose and fructose are classified as **reducing** sugars.

3 Hydrolysis of sucrose by acid produces equal amounts of glucose and fructose. The reducing property of glucose and fructose is blocked when they form sucrose because they are linked through their functional groups.

14 Natural products and polymers

> **COMMENT**
> Rotation of the plane of polarization in the clockwise sense as viewed by an observer looking towards the source of light is given a (+) sign.

4 Compounds which rotate the plane of polarized light are said to be **optically active**, and this is the standard property by which chiral compounds can be recognized in the laboratory. The two isomers of a chiral compound will produce the same amount of rotation but in opposite directions.

The property can be indicated by adding a prefix to their names: (+) glucose and (−) glucose. When sucrose is hydrolysed it is converted into equimolar amounts of (+) glucose and (−) fructose and the overall rotation changes from + to −. For this reason the mixture obtained from the hydrolysis of sucrose is known as 'invert sugar'.

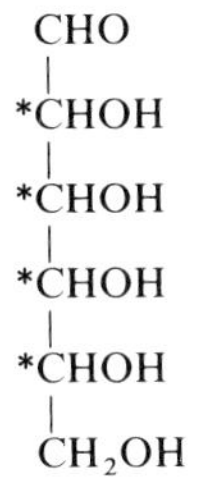

```
   CHO
    |
  *CHOH
    |
  *CHOH
    |
  *CHOH
    |
  *CHOH
    |
   CH₂OH
```

Figure 14.7 Chiral centres in glucose

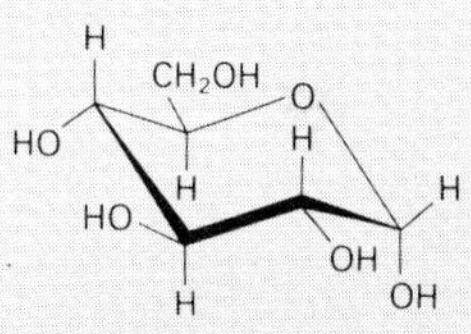

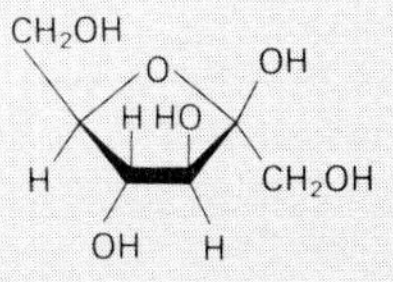

> **COMMENT**
> Glucose, $C_6H_{12}O_6$, is classified as an **aldohexose**: *aldo* denoting an aldehyde functional group, *hex* denoting six carbon atoms, and *ose* denoting that it is a carbohydrate. In its natural form the glucose molecule is wrapped into a ring structure: 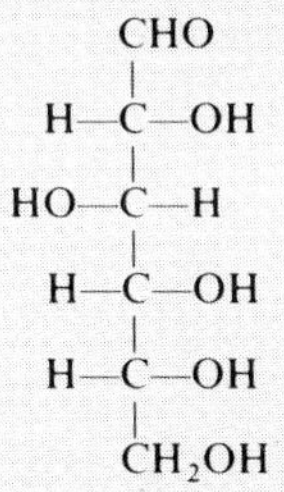
>
> ```
>      CHO
>       |
>   H—C—OH
>       |
>  HO—C—H
>       |
>   H—C—OH
>       |
>   H—C—OH
>       |
>      CH₂OH
>   glucose
>  (chain form)
> ```
>
> α-glucose
>
> Glucose occurs in the blood and other body fluids, and is the monomer for many polysaccharides.
>
> Fructose, $C_6H_{12}O_6$, is a **ketohexose** because it has a ketone functional group. Fructose forms a five-membered ring whereas glucose forms a six-membered ring. 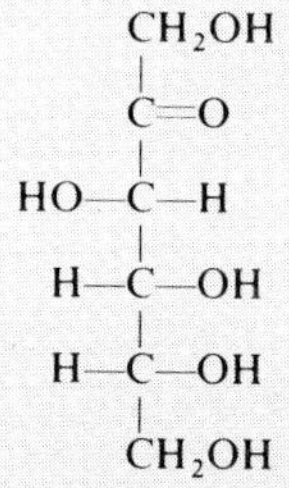
>
> ```
>     CH₂OH
>       |
>      C=O
>       |
>  HO—C—H
>       |
>   H—C—OH
>       |
>   H—C—OH
>       |
>     CH₂OH
>    fructose
>  (chain form)
> ```
>
> fructose
>
> Sucrose, $C_{12}H_{22}O_{11}$, is the common sugar of our diet and is obtained from sugar cane or sugar beet. It can be regarded as the combination of one unit of glucose with one unit of fructose, and with the elimination of one molecule of water. 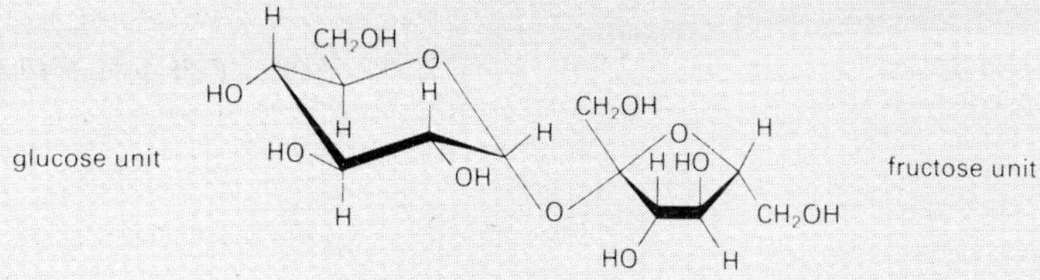
>
> glucose unit ———— fructose unit
>
> sucrose

> Glucose and fructose are classified as **monosaccharides**, while sucrose is a **disaccharide** as it is made of two monosaccharide units. Because the two sugars are linked by their carbonyl functional groups, sucrose has to be hydrolysed before reaction can occur with Benedict's solution.
>
> $$\text{sucrose} + \text{water} \xrightarrow{\text{dilute HCl}} \text{glucose} + \text{fructose}$$
>
> You are not expected to memorize the formulae of these sugars.

Carbohydrates such as cellulose and starch are polysaccharides, and are built up from several hundred monosaccharide units. **Cellulose**, a polymer of as many as five thousand glucose units, forms the framework for cells in plant tissue. It occurs in a fibrous form as cotton and wood. Cotton fibres are long enough to be used as textiles but wood fibres are shorter and are used to manufacture paper. So a piece of paper is mostly a poly(glucose).

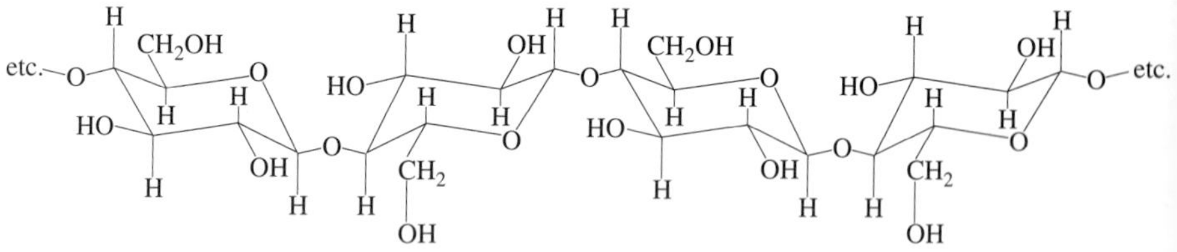

Figure 14.8 Cotton is a type of poly(glucose).

Starch is the main food reserve of plants, forming up to 80% of the mass of seeds, and consists of up to three thousand glucose units. Cellulose and starch are both polymers of glucose, with the monomer molecules linked differently.

Another polysaccharide called **chitin** forms the shell of crabs and the outer covering of beetles.

14.3 Naturally occurring carboxylic acids

Naturally occurring fats and oils are commonly esters of the alcohol propane-1,2,3-triol (glycerol), $CH_2OH-CH(OH)-CH_2OH$, and long chain carboxylic acids such as stearic acid $CH_3-(CH_2)_{16}-CO_2H$. A typical fat molecule therefore has the structure

$$\begin{array}{l} CH_2-O-CO-(CH_2)_{n_1}-CH_3 \\ | \\ CH-O-CO-(CH_2)_{n_2}-CH_3 \\ | \\ CH_2-O-CO-(CH_2)_{n_3}-CH_3 \end{array}$$

where n_1, n_2, and n_3 may be the same, or different, but are almost always **even** numbers. Fats and oils are members of the group of compounds known as **lipids**. Lipids are a class of cell materials that are insoluble in water. Other lipids can be found with more complex structures.

EXPERIMENT 14.3 Hydrolysis of a lipid

Castor oil is a convenient starting point for this experiment but almost any natural plant oil will do.

$$CH_3(CH_2)_5\underset{OH}{CHCH_2CH}=CH(CH_2)_7CHCO_2\overset{CH_2O_2C(CH_2)_7CH=CHCH_2\underset{OH}{CH}(CH_2)_5CH_3}{\underset{CH_2O_2C(CH_2)_7CH=CHCH_2\underset{OH}{CH}(CH_2)_5CH_3}{CH}}$$

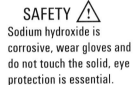

Figure 14.9 The lipid in castor oil, formed from glycerol and ricinoleic acid

SAFETY ⚠
Sodium hydroxide is corrosive, wear gloves and do not touch the solid, eye protection is essential.

Procedure

Dissolve 2 g of sodium hydroxide (TAKE CARE) in 10 cm³ of cold water in an evaporating basin and add 2 cm³ of castor oil. Place the evaporating basin on a water bath. Heat the water bath so that the steam heats the evaporating basin. Continue boiling for about 20 minutes until the oily layer is no longer visible. Add water to the evaporating basin when necessary to maintain the volume at about 10 cm³. Make sure that the water bath does not boil dry.

Dilute the boiled mixture with 10 cm³ of water and saturate with sodium chloride (six spatula measures). Boil again for a minute, then cool and collect the solid product by suction filtration, washing free of alkali with a little water.

Dissolve some of your product in *pure* water and examine its properties in the following test-tube reactions.

1 Add a little of your solution to *pure* water in a conical flask. Shake well. Is a lather formed?

2 Add a little of your solution to a calcium salt solution. Is a precipitate formed?

3 Add a little of your solution to dilute hydrochloric acid. Is a precipitate formed?

■ Is the behaviour of your product that of a soap? Is your product glycerol, ricinoleic acid, or sodium ricinoleate? Write an equation for the reaction (using Ric to represent the ricinoleate group).

Uses of lipids

An important function of lipids is to act as an energy store, as they have the best energy value of all foods; plant seeds are particularly rich in lipids for this reason. Nutmegs, for example, yield up to 40 per cent of fat.

A number of lipids are of economic importance, as they are used to make foodstuffs such as margarine, and soap. Soap is made by boiling fats with akali.

$$\begin{array}{c}CH_2-O-CO-(CH_2)_n-CH_3\\ |\\ CH-O-CO-(CH_2)_n-CH_3 + 3NaOH \longrightarrow 3CH_3-(CH_2)_n-CO_2Na + \begin{array}{c}CH_2OH\\ |\\ CHOH\\ |\\ CH_2OH\end{array}\\ |\\ CH_2-O-CO-(CH_2)_n-CH_3\end{array}$$

$$\text{fat} \quad + \quad \text{alkali} \longrightarrow \quad \text{soap} \quad + \text{glycerol}$$

	% Saturated fatty acids	% Unsaturated fatty acids
Animal		
Beef fat	50	50
Butter	50	50
Cod oil	15	85
Sardine oil	20	80
Vegetable		
Coconut	90	10
Corn	15	85
Olive	15	85
Palm	50	50
Sunflower	10	90
Soybean	15	85

Figure 14.10 Saturated and unsaturated fatty acids in animal and vegetable fats.

Because of the value of lipids for foodstuffs, synthetic detergents based on petrochemicals were introduced, but the original synthetic detergents caused serious pollution in rivers. These problems do not arise with soap because soap has straight-chain alkyl groups and can be degraded by bacteria. The original synthetic detergents had branched-chain alkyl groups and could not be degraded.

Chemists were able to solve this problem by developing biodegradable synthetic detergents which have straight-chain alkyl groups similar to the alkyl chains in natural fats and oils.

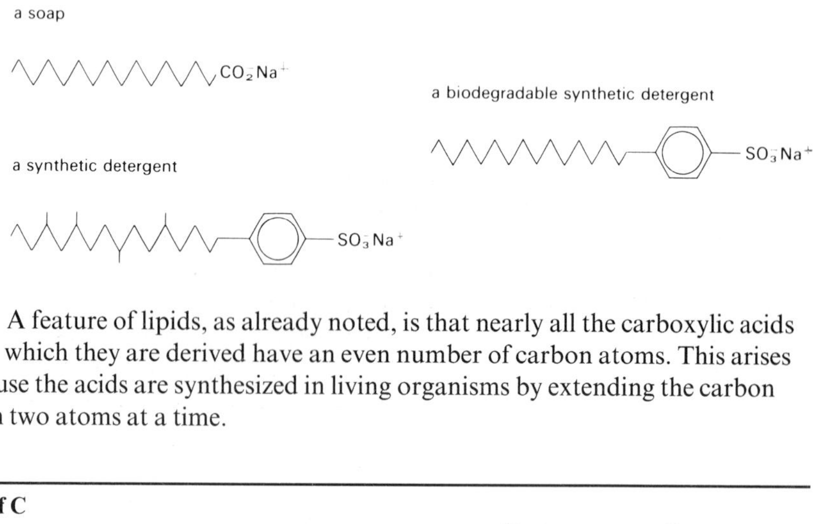

A feature of lipids, as already noted, is that nearly all the carboxylic acids from which they are derived have an even number of carbon atoms. This arises because the acids are synthesized in living organisms by extending the carbon chain two atoms at a time.

No. of C atoms	Naturally occurring compound	Common name	Common source
8	$CH_3(CH_2)_6CO_2H$	caprylic acid	coconut oil
10	$CH_3(CH_2)_8CO_2H$	capric acid	coconut oil
12	$CH_3(CH_2)_{10}CO_2H$	lauric acid	coconut oil
14	$CH_3(CH_2)_{12}CO_2H$	myristic acid	nutmeg seed fat
16	$CH_3(CH_2)_{14}CO_2H$	palmitic acid	palm oil
	$CH_3(CH_2)_{14}CH_2OH$	cetyl alcohol	sperm whale oil
18	$CH_3(CH_2)_{16}CO_2H$	stearic acid	animal fats
	$CH_3(CH_2)_7CH\!=\!CH(CH_2)_7CO_2H$	oleic acid	olive oil
	$CH_3(CH_2)_7CH\!=\!CH(CH_2)_7CH_2OH$	oleyl alcohol	sperm whale oil
	$CH_3(CH_2)_5CH(OH)CH_2CH\!=\!CH(CH_2)_7CO_2H$	ricinoleic acid	castor oil

Figure 14.11 Some naturally occurring carboxylic acids and alcohols

Another use for lipids is to provide carboxylic acids as starting materials for synthesis of organic compounds with eight or more carbon atoms. The synthesis of C_{even} compounds is straightforward but C_{odd} compounds require the introduction of an extra carbon atom (or removal of one atom). So the synthesis of C_{odd} compounds is more costly.

14.4 Amino acids and proteins

Amino acids possess two of the functional groups that you have already studied, namely the amino group, —NH_2, and the carboxylic acid group, —CO_2H. The simplest amino acid is glycine (aminoethanoic acid).

NH_2—CH_2—CO_2H glycine (gly)

As we shall only be concerned with amino acids that occur naturally we shall be using the non-systematic names that are favoured by biochemists. These names are often abbreviated to a 3-lettered 'code', which usually consists of the first three letters of the name. The amino acids that occur naturally are all 2-amino acids with the amino group on the carbon atom adjacent to the acid group.

NH_2—CH—CO_2H
   |
   CH—CH_3
   |
   CH_3
valine (val)

NH_2—CH—CO_2H
   |
   CH_2OH
serine (ser)

The significance of the 2-amino acid structure is that the **naturally occurring amino acids have chiral molecules**. Nature appears to be stereospecific because almost without exception the naturally occurring amino acids in living material have the same configuration. After death a slow conversion from one form to the other occurs until eventually an equilibrium mixture results. But glycine is achiral.

Chemical tests reveal the presence of nitrogen compounds in most animal tissues and, more specifically, the acidic hydrolysis of hair, blood, and muscle tissue shows that they consist almost entirely of amino acids.

NH_2—CH—CO_2H an amino acid
   |
   R

The term **peptide group** is used to describe the amide group —CO—NH— when it links together amino acid residues.

When a number of amino acid residues are connected by peptide groups the molecules are known as **polypeptides** *or* **proteins**. The determination of the amino acid sequence of a protein is considered in the next section.

—NH—CH—CO—NH—CH—CO—NH—CH—CO—   part of a
      | | | protein
      R   R   R

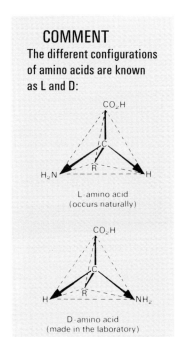

COMMENT
The different configurations of amino acids are known as L and D:

L-amino acid (occurs naturally)

D-amino acid (made in the laboratory)

Figure 14.12 Steel wire of the same mass would be weaker than the filaments of this spider's web, made from proteins.

Experiments with amino acids and proteins

Glycine and glutamic acid can be used in these experiments as examples of simple amino acids and, as an example of a mixture, casein hydrolysate (containing the free amino acids from the protein in milk) might also be used.

In principle every foodstuff and all materials of biological origin are worth testing for protein content. However, as the experiments suggested are carried out with solutions it would be as well to select water-soluble substances.

Fresh milk, egg (white and yolk), or the derived extracts casein and albumin can be tested. Other possibilities are pepsin and trypsin, which are digestive enzymes; or gelatin, from the hydrolysis of the connective tissue of animals.

EXPERIMENT 14.4a

Protein materials

> SAFETY ⚠
> Ninhydrin is harmful and irritant, wear gloves and avoid inhaling the spray; ninhydrin spray is flammable.

Make solutions in water of your samples of protein materials and amino acids, warming if necessary, and use them for the following tests.

1 Acidity and basicity

To 2 cm³ of 0.01 M hydrochloric acid add a few drops of Full-range Indicator and note the effect on the pH of adding 0.01 M sodium hydroxide in 0.5 cm³ portions. Now repeat the experiment, using a solution of 0.01 M glycine or glutamic acid in place of the hydrochloric acid.

- Does the pH change gradually or sharply?
  What type of acid–base behaviour is occurring?

2 Biuret test

Add an equal volume of 2 M sodium hydroxide to one of your protein samples in solution followed by 1 **drop** of 0.1 M copper(II) sulphate. A mauve colour will develop if proteins are present. This is known as the **biuret test** and detects peptide groups.

Carry out a blank test using water as your sample.

- What type of reaction and bonding would you expect between copper(II) ions and peptide groups?

3 Ninhydrin test

On a piece of chromatography paper place small drops of your solutions and allow to dry. Spray lightly with 0.02 M ninhydrin solution in propanone (TAKE CARE **spray in a fume cupboard**) and again allow to dry. Avoid getting the spray on your fingers by wearing gloves. Heat for 10 minutes in an oven at 110 °C or heat cautiously over a Bunsen flame. Red to blue coloured spots will develop if proteins or amino acids are present. Make a note of any unexpected coloured areas.

Ninhydrin is a reagent used as a specific colour test for amino acids. You do not need to be concerned with the formula of the compound (which is complicated) nor with the details of the chemical reaction which produces the colours.

4 Chirality

If there is time, investigate the ability of the amino acid solutions to affect polarized light by examining their solutions in a polarimeter.

EXPERIMENT 14.4b The chromatographic separation of amino acids

The experiment is designed to give you experience and understanding of an important method. To separate and identify naturally occurring amino acids by paper chromatography would require an effort spread over about three days, so this brief experiment can only suggest the potentialities of the method.

To obtain satisfying results you will have to work with care and keep the experimental materials scrupulously clean. Touch the chromatography papers only on their top corners and never lay them down except on a clean sheet of paper.

> **SAFETY** ⚠
> Ninhydrin is harmful and irritant, wear gloves and avoid inhaling the spray; the chromatography solvent mixture is flammable and has an irritant vapour; the ninhydrin and fixing sprays are both flammable; 8 M ammonia is very irritant and pungent

Procedure

Put spots of 0.01 M amino acids in aqueous solution 1.5 cm from the bottom edge of the chromatography paper cut to the dimensions shown in figure 14.13. To do this, dip a **clean** capillary tube in the stock solution and apply a small drop to the chromatography paper, using a quick delicate touch.

Practise on a piece of ordinary filter paper until you can produce spots not more than 0.5 cm in diameter. Apply spots of individual amino acids and also mixtures, making identification marks in **pencil** at the top of the paper. Allow the spots to dry.

Meanwhile prepare a fresh solvent mixture of butan-1-ol (12 cm^3), pure ethanoic acid (3 cm^3), and water (6 cm^3) in a covered 1 dm^3 beaker (TAKE CARE). Cover the beaker to produce a saturated atmosphere.

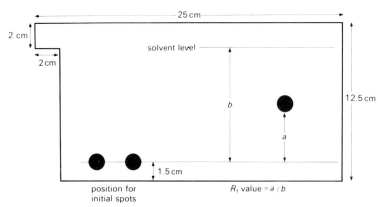

Figure 14.13 Apparatus for simplified chromatography of amino acids.

Now roll the chromatography paper into a cylinder and secure it with a plastic paper clip. Stand the cylinder in the covered solvent beaker and leave it for the solvent to rise to nearly the top of the paper. As 20 minutes are needed to complete the experiment after removing the paper from the beaker, you may not have time to allow the solvent to rise the full distance.

Remove the chromatography paper from the beaker, and mark the solvent level. Dry the paper (without unfastening it), in an oven if possible, but **not** over a Bunsen flame, because the solvent is both pungent and flammable.

Detect the amino acids by spraying the paper sparingly with 0.02 M ninhydrin solution (TAKE CARE) in a fume cupboard and then heating in an oven at 110 °C for 10 minutes. Purple spots should appear at the positions occupied by the amino acids.

Preserve the spots by spraying with a mixture made up of methanol (19 cm^3), M aqueous copper(II) nitrate (1 cm^3), and 2 M nitric acid (a drop), and then expose **in a fume cupboard** to the fumes from 8 M ammonia (TAKE CARE). Determine the R_f value of the amino acid samples (see figure 14.14).

R_f values are obtained using the expression

$$R_f \text{ value} = \frac{\text{distance moved by amino acid}}{\text{distance moved by solvent}}$$

- Have the mixtures separated?

Figure 14.14 The twenty 'standard' amino acids.

Formula	Name	Abbreviation	Nature of side chain	R_f value in butan-1-ol/ ethanoic acid/ water
H₂NCHCO₂H | H	glycine	gly	non-polar	0.26
H₂NCHCO₂H | CH₃	alanine	ala	non-polar	0.38
H₂NCHCO₂H | CHCH₃ | CH₃	valine	val	non-polar	0.60
H₂NCHCO₂H | CH₂ | CH(CH₃)₂	leucine	leu	non-polar	0.73
H₂NCHCO₂H | CHC₂H₅ | CH₃	isoleucine	ile	non-polar	0.72
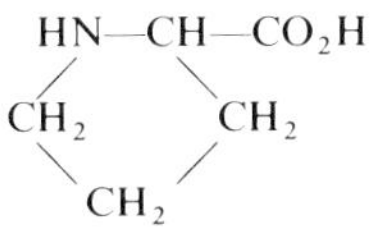 HN—CH—CO₂H / \ CH₂ CH₂ \ / CH₂	proline	pro	non-polar	0.43
H₂N—CH—CO₂H | CH₂ | C (indole ring)	tryptophan	try	non-polar	0.50
H₂NCHCO₂H | CH₂ | CH₂SCH₃	methionine	met	non-polar	0.55
H₂NCHCO₂H | CH₂—C₆H₅	phenyl-alanine	phe	non-polar	0.68
H₂NCHCO₂H | CH₂OH	serine	ser	polar	0.27
H₂NCHCO₂H | CHOH | CH₃	threonine	thr	polar	0.35
H₂NCHCO₂H | CH₂SH	cysteine	cys	polar	0.08
H₂NCHCO₂H | CH₂ | CONH₂	asparagine	asn	polar	0.19

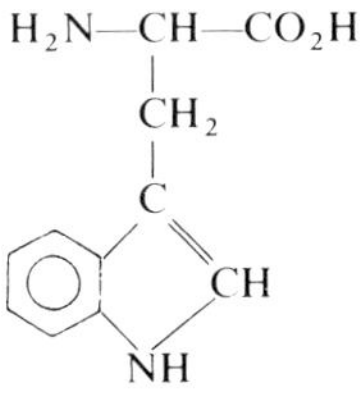

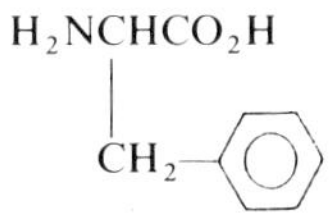

Structure	Name	Abbr.	Type	Value
H_2NCHCO_2H — CH_2 — CH_2CONH_2	glutamine	gln	polar	—
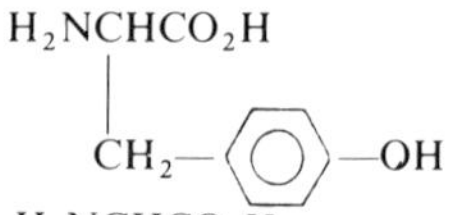 H_2NCHCO_2H — CH_2 — C$_6$H$_4$—OH	tyrosine	tyr	polar	0.50
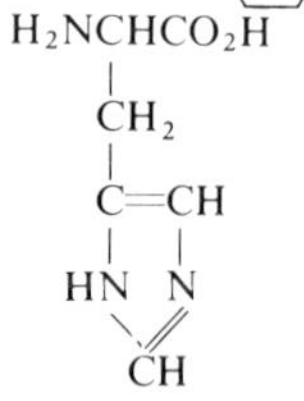 H_2NCHCO_2H — CH_2 — C=CH — HN, N — CH	histidine	his	basic	0.20
H_2NCHCO_2H — $(CH_2)_3$ — NH — HN=C—NH$_2$	arginine	arg	basic	0.16
H_2NCHCO_2H — $(CH_2)_3$ — CH_2NH_2	lysine	lys	basic	0.14
H_2NCHCO_2H — CH_2CO_2H	aspartic acidic	asp	acidic	0.24
H_2NCHCO_2H — CH_2 — CH_2CO_2H	glutamic acid	glu	acidic	0.30

EXPERIMENT 14.4c The enzyme-catalysed hydrolysis of urea

Enzymes are proteins whose function in a living organism is to help to bring about biochemical reactions: enzymes can be said to act as a type of catalyst. Any compound whose reaction occurs through the intervention of an enzyme is known as a **substrate** of the enzyme.

The following experiment is designed to illustrate qualitatively the catalytic properties of an enzyme.

Urease is an enzyme found in plants; jack beans or water melon seeds are convenient sources. It converts urea to ammonia by a hydrolysis reaction:

$$O=C(NH_2)_2 + H_2O \xrightarrow{enzyme} 2NH_3 + CO_2$$

SAFETY ⚠
Ethanamide is harmful.

Procedure

1 To 5 cm³ of a 0.25 M urea solution add five drops of Full-range Indicator, followed by drops of 0.01 M hydrochloric acid until the indicator has just changed to a distinct red colour. Add 1 cm³ of a 1% solution of urease active meal, which has been similarly treated with indicator and acid. Note how quickly the pH of the solution changes.

2 Repeat the experiment with 0.25 M solutions of compounds which have structural similarities to urea and might therefore be hydrolysed by urease to ammonia. Suitable compounds include ethanamide and methylurea.

3 Boil 1 cm^3 of the 1% solution of urease active meal for 30 seconds, then cool to room temperature. Repeat the first experiment, using the boiled urease solution.

■ How specific is the enzyme activity of urease?
What causes the pH of the solution to change?

The properties of enzymes

As a result of your experimental work you should have some ideas about the properties of enzymes. Unlike an inorganic catalyst such as platinum which catalyses a variety of reactions, enzymes are highly specialized and often for a particular enzyme there is only one reaction of one substrate which it can catalyse. That is, enzymes are **specific** catalysts.

As an example of the specific activity of enzymes, fumarase catalyses only the addition of water to **trans**–butenedioic acid; furthermore, of the two possible chiral products only one isomer is produced:

$$\underset{\substack{\text{trans-butenedioic acid} \\ \text{(fumaric acid)}}}{\begin{array}{c} H \quad\quad CO_2H \\ \diagdown \;\;\; \diagup \\ C \\ \| \\ C \\ \diagup \;\;\; \diagdown \\ HO_2C \quad\quad H \end{array}} + H_2O \longrightarrow \underset{\substack{\text{L(−)-2-hydroxybutanedioic} \\ \text{acid (malic acid)}}}{\begin{array}{c} CO_2H \\ | \\ CH_2 \\ | \\ C \\ \diagup | \diagdown \\ HO \;\; H \;\; CO_2H \end{array}}$$

Fumarase is also highly efficient, with a conversion rate at 25 °C of 10^3 molecules of substrate every second per molecule of enzyme! In fact the reaction is so fast, it is controlled by the rate at which the molecules meet by diffusion.

The influence of temperature on catalyst activity is considerable. If the temperature is too high, or too low, enzymes cease to function. So for each enzyme there is an optimum temperature at which, for a given set of conditions, the enzyme will catalyse the greatest amount of chemical change.

Thus a digestive enzyme from a sea squirt was found to have an optimum temperature of 50 °C for a two hour reaction period. Yet being in a sea animal the enzyme will be required to function at a temperature of 15 °C. This does not seem a very efficient digestive situation for the sea squirt until it is realized that it takes up to 60 hours to digest its food. When the optimum temperature over a 60 hour period was investigated it was found to be 20 °C. This illustrates the care needed for the proper investigation of enzyme properties.

Enzymes also function best at a particular pH which is not usually very different from neutrality, being mostly in the range pH 5–7. By the use of buffer solutions the pH dependence of α–amylase from saliva can be studied.

Figure 14.15 The hydrolysis of starch by saliva.

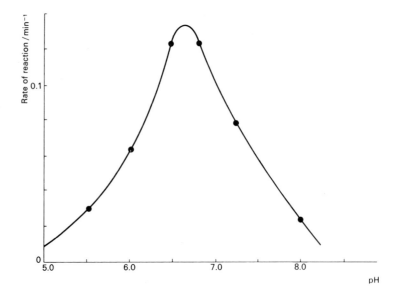

QUESTION
How might an enzyme be influenced by pH? Remember that enzymes are proteins, so amino acids such as lysine and glutamic acid (see figure 14.14) may be present.

An initial understanding of the mechanism of enzyme activity was obtained by kinetic studies. The results suggest that enzymes act catalytically on their substrates and that their reaction, for example for addition of water to the substrate, might be represented as follows:

enzyme + substrate $\longrightarrow$ enzyme – substrate complex

complex + water $\longrightarrow$ enzyme + hydrated substrate

The study of enzymes is taken further in the Special Study *Biochemistry*.

INVESTIGATION 14.4d The pH of amino acid solutions

Carry out an investigation of the change of pH as an amino acid such as glycine is titrated with strong acid and strong alkali.

Make a risk assessment before starting any experiments.

14.5 Study task: The chemical and structural investigation of proteins

QUESTIONS
Read the passage overleaf, and answer these questions based on it.
1 Draw a structural formula for the tripeptide gly-val-ser.
2 Show how two tripeptides can be held together by hydrogen bonds.
3 How would Sanger have separated the seventeen amino acids in the hydrolysate from insulin?

Proteins have remarkably complex molecules, with molar masses of a thousand or more; but at the same time they illustrate how in nature complex ends are often achieved through the infinitely varied use of simple means. All the different naturally occurring proteins are built, not, as might be expected, from many hundreds of different amino acids, but from only about two dozen. The most important of these, together with their structures, are listed in figure 14.14. You will not be expected to remember the names and structures of these compounds.

Protein	Occurrence (examples)	Function	Molar mass	Approximate number of amino acid units
Insulin	animal pancreas	governs sugar metabolism	5700	51
Myoglobin	muscle	oxygen carrier	17000	153
Trypsin	animal pancreas	digests food proteins	23800	180
Haemoglobin	blood	oxygen carrier	66000	574
Urease	soya beans	converts urea to ammonia	480000	4500

Since there are many proteins and few amino acids in nature it is apparent that proteins will be characterized by the sequence in which their amino acids are linked:

  –gly–val–ser–
or –val–gly–ser–
or –val–ser–gly–   etc.

If the correct molecular formula of a substance is to be established it must be available pure, and the necessary experimental techniques must be available. F. Sanger, working at Cambridge, was the first chemist to establish the molecular formula of a protein. When he started work in 1944, he had to develop new experimental techniques because the problems of protein composition were unsolved at that time, and the only simple protein available pure was insulin. Ten year's work was necessary to establish the correct amino acid sequence for insulin.

The amino acid composition of proteins

The molar mass of insulin is about 5700 and its formula is $C_{254}H_{377}N_{65}O_{75}S_6$! As is the case with any protein, the first stage in the investigation of insulin was to discover the nature, and number, of the amino acid residues present.

Hydrolysis of a protein by heating in a sealed tube with 6 M hydrochloric acid for 24 hours produces the free amino acids. Quantitative separation of the hydrolysate will determine which amino acids are present and their relative amounts. Sanger found that the insulin molecule contains 17 different amino acids ranging from six molecules of cysteine and leucine to one molecule of lysine, the molecular formula being accounted for by a total of 51 amino acid units.

Mass spectrometry is becoming increasingly important in the determination of the amino acid sequence of proteins; dihydrofolate reductase in 1979 was the first example. The advantages of the mass spectrometer are speed, the small amount of sample needed (only $5\text{--}30 \times 10^{-9}$ mol are required) and the fact that the polypeptides produced by partial hydrolysis need not be completely separated from each other because analysis is possible on a mixture of two to five peptides.

Sanger used less refined techniques in his work on insulin and the result that he obtained is shown in figure 14.16.

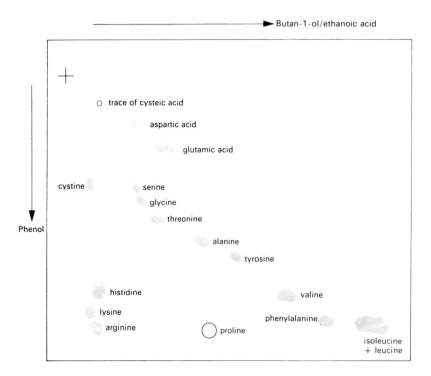

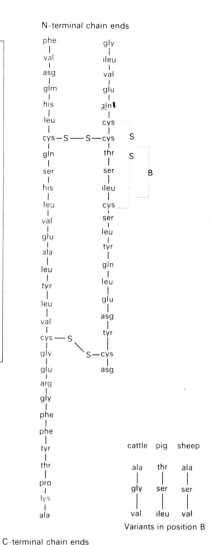

Figure 14.16 **a** The chromatography of the amino acids in insulin. **b** The amino acid sequence in insulin.

The shapes of protein molecules

There are countless possibilities for the shape of a large molecule of a protein such as insulin. These possibilities arise because of the rotation that can take place at the single bonds in the polypeptide chain. A given protein molecule may actually adopt several of these possible shapes, or **conformations**, as they are called. The three-dimensional shape of insulin is shown in figure 14.17.

Hydrogen bonds between the peptide groups play a major role in maintaining the conformation usually adopted by the insulin molecule, as with the other examples mentioned in Topic 9.

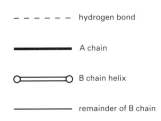

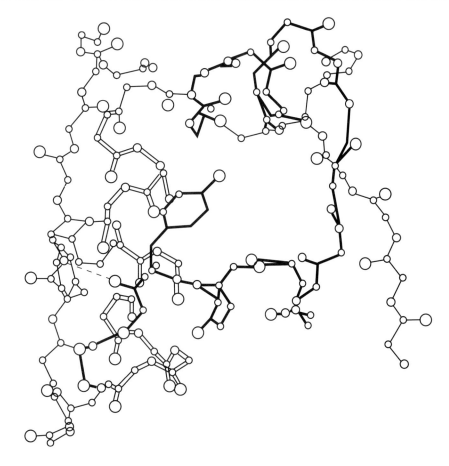

Figure 14.17 The shape of the insulin molecule.

14.6 Polymers

Most technological innovations have been inspired or made possible by the emergence of a new material. Although we tend to take our materials for granted, terms like Stone Age, Bronze Age and Iron Age suggest how significant they have been in our history.

Nowadays synthetic polymers play such an important part in our lives that we could be said to be living in the 'Polymer Age'. It is these materials and their astonishing variety of properties and uses which we shall study in this section.

The nature of polymers

Synthetic polymers can be hard, rigid glasses or soft rubbery solids; they can be made into strong, resilient fibres or tough flexible films. It is not surprising that they have come to be recognized as some of the most useful materials known. You are probably wearing or using something made from a synthetic polymer at this very moment; and if you have recently been to a supermarket, played almost any kind of sport, listened to a tape or CD or watched a video or travelled by bicycle or car you have been using polymers.

Perhaps more important than any one property is the versatility of polymers. The same polymer, PVC, is used to make rigid window frames, credit

> **REVIEW TASK**
> Working in groups, list properties of polymers that make them so useful. Compare each property with the nearest natural material that occurs to you. Also list the formulae of the monomers and corresponding polymers that you have already met in previous Topics.

cards and cling film. Far from being inferior substitutes, they are valuable materials in their own right for which there may often be no substitute.

When Hermann Staudinger, the pioneer of polymer chemistry in the 1920s, first suggested that rubber was a giant molecule with a molar mass of about 100 000, his ideas were ridiculed. He was advised to purify his products properly and obtain an acceptable value. He persisted with his researches however, and established that polymers are very long chain molecules formed by combining a large number of small units or 'monomers'. His views are now accepted and in 1953 he was awarded the Nobel Prize for chemistry.

W.H. Carothers, the discoverer of nylon in 1935, showed that there are two ways in which monomers can be combined.

1 Addition polymerization

In addition polymerization unsaturated monomer molecules add to each other to form a polymer having the same empirical formula as the monomer with no other products, for example

$$n\mathrm{CH_2=CH_2} \longrightarrow \mathrm{-[CH_2-CH_2]}_n$$
ethene → poly(ethene)

Addition polymerization often follows a free radical chain mechanism involving the usual initiation, propagation and termination stages (see Topic 7.2).

Figure 14.18 Rock climbing with a polymer rope.

2 Condensation polymerization

This usually involves two different monomers each having two functional groups at opposite ends of their molecules. Molecules of the two monomers react with each other with the elimination of a small molecule such as water or hydrogen chloride. You have already met this kind of reaction in the formation of esters and peptides. Nylon 66 is a good example:

$$n\mathrm{HO_2C(CH_2)_4CO_2H} + n\mathrm{H_2N(CH_2)_6NH_2} \longrightarrow$$
$$\mathrm{-[OC(CH_2)_4CONH(CH_2)_6NH]}_n + (2n-1)\mathrm{H_2O}$$

STUDY TASK

1 Why are nylons known as polyamides?
2 Try to write an equation for the formation of the polyester formed from ethane-1,2-diol and benzene-1,4-dicarboxylic acid,

HO₂C—⟨benzene⟩—CO₂H

(*Continued*)

14 Natural products and polymers

3 The condensation reaction of phenol with methanal forms a linear polymer at first. These polymers then react with more methanal to form a highly cross-linked product.

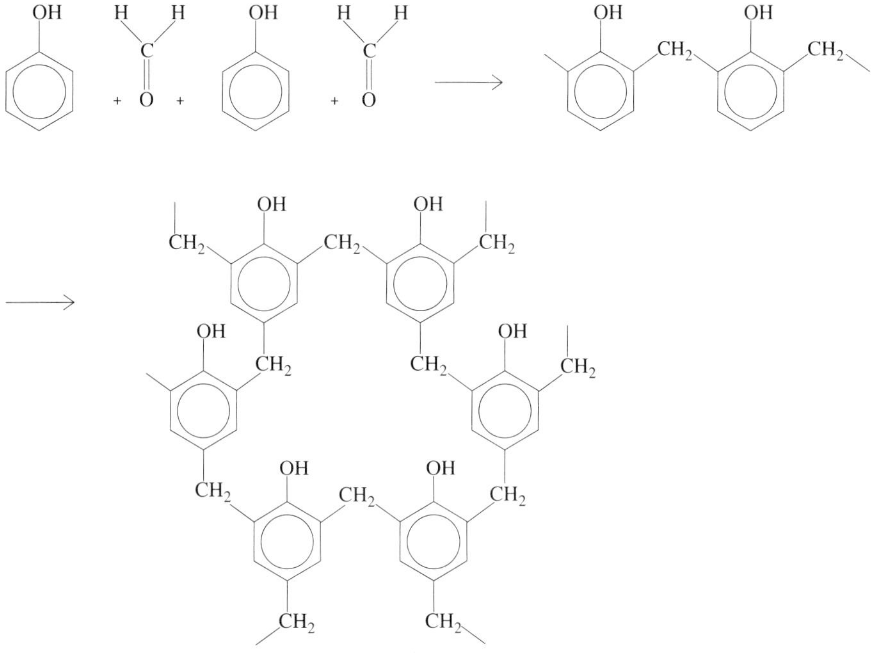

What feature of the methanal molecule enables it to react with the benzene ring?

EXPERIMENT 14.6a Polymerization reactions

You should be able to carry out at least one of the reactions described, but look at them all to identify the common features. Wear gloves to do these experiments and do the work in a fume cupboard if possible.

SAFETY ⚠

Propenamide (acrylamide) is a skin irritant. You should wear protective gloves for this experiment; decanedioyl dichloride is corrosive and hexane-1,6-diamine gives off irritant fumes; benzene-1,2-dicarboxylic anhydride (phthalic anhydride) has a harmful dust; methanal (formaldehyde) is toxic and flammable, sodium hydroxide is corrosive and benzene-1,3-diol (resorcinol) is irritant.

Procedure

1 Poly(propenamide)

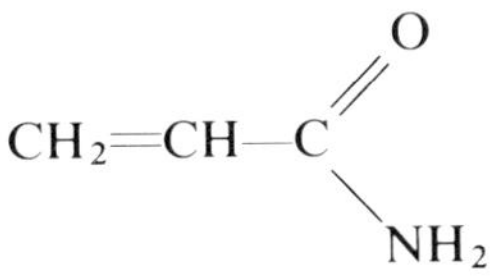

$CH_2{=}CH{-}C(=O)NH_2$ propenamide

Make a solution of 10 g of propenamide (TAKE CARE) in 50 cm^3 of water and warm in a 250 cm^3 beaker to **not more than** 85 °C. Pour the solution into a throw-away container (such as a tin can) on a heat-resistant mat and add about 0.1 g of potassium peroxodisulphate to initiate the polymerization.

- Was the reaction an addition or a condensation polymerization? Try to write an equation for the reaction. Use the table of bond energies in the *Book of data* to estimate the enthalpy change of polymerization.

2 The 'nylon rope trick'

Prepare a solution of 0.5 cm³ of decanedioyl dichloride (TAKE CARE) in 15 cm³ hydrocarbon solvent and, separately, a solution of 0.7 g of hexane-1,6-diamine (TAKE CARE) and 2 g of sodium carbonate in 15 cm³ of water in a 100 cm³ beaker.

Clamp the beaker, and alongside it clamp a pair of glass rods as shown in figure 14.19. If possible allow a drop of about 1 metre from the rod to the receiver.

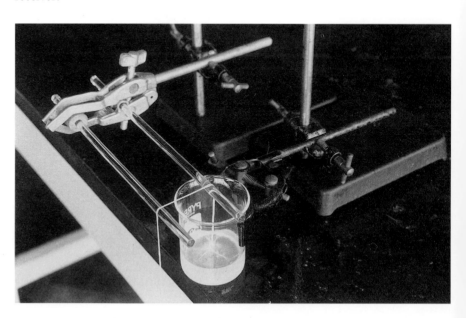

Figure 14.19 The nylon rope trick.

Now pour the hydrocarbon solution carefully onto the aqueous solution and, using crucible tongs, pull the interfacial film out, over the rods, and down towards the receiver. When a long enough rope has formed, the process should go on of its own accord until the reagents are used up, but the rope may need to be pulled out gently, using the crucible tongs. Take care not to get either solvent or reagent on your fingers.

To obtain a dry specimen of nylon polymer, wash it thoroughly in 50% aqueous ethanol and then in water until litmus is not turned blue by the washings. Note that, because of the way in which it is formed, the nylon 'rope' is likely to be a hollow tube, containing solvent and possibly reagent. You should therefore take care when handling it, even after washing in this way.

- Was the reaction an addition or a condensation polymerization? Write an equation for this polymerization reaction.

3 The preparation of a polyester resin

Wearing protective gloves, mix 3 g of benzene-1,2-dicarboxylic anhydride with 2 cm³ of propane-1,2,3-triol in a test-tube. Measure out the propane-1,2,3-triol with a dropping pipette and allow the pipette plenty of time to drain.

Heat to 160 °C and then more slowly to 250 °C in a fume cupboard. When the mixture ceases to bubble allow it to cool. Test the viscosity of your product.

- Was the reaction an addition or a condensation polymerization? Write an equation for a possible polymerization reaction.

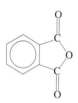

benzene-1,2-dicarboxylic anhydride

4 The preparation of a crosslinked polymer

Measure about 25 cm^3 of a 4% aqueous solution of poly(ethenol) into a disposable container. Rinse out your measuring cylinder thoroughly. Add about 5 cm^3 of 4% sodium borate solution and immediately stir well. Continue stirring as the mixture thickens.

Put on disposable gloves and examine the properties of the 'slime' you have produced. Take care as 'slime' sticks to clothing and removes paint.

- What monomer is needed to make poly(ethenol)? The molar mass of the poly(ethenol) used in this experiment is about 10^5 g mol^{-1}. How many monomer units does this involve?

In 'slime' the poly(ethenol) chains have been crosslinked by hydrogen bonds to borate ions, so the links are not permanent.

hydrogen bonding

Classification of polymers

The properties of all materials depend on their structure and on the bonding which holds that structure together. It may help to appreciate the distinctive properties of polymers if we fit them into the spectrum:
methane – poly(ethene) – diamond.

Methane consists of small separate molecules able to move with complete freedom because only weak, non-directional van der Waals forces act between them.

At the other extreme there is no independent movement at all in diamond, because all the atoms are held in one giant structure by strong covalent bonds.

In poly(ethene) the atoms are covalently bonded into long molecular chains. They therefore have less freedom than methane molecules but much greater freedom than the atoms in diamonds, since only van der Waals forces act between the chains. Furthermore, although the covalent bonds in the linear chains of poly(ethene) are directional, rotation about a bond is still possible. The chains are thus flexible and can slither about in tangled coils like freshly cooked spaghetti. This tangling reduces the freedom of the chains, since any large-scale movement of one chain affects others. Breaking a piece of poly(ethene) therefore involves pulling chains from an entangled mass, which gives the polymer greater strength than it would have simply from van der Waals forces between short sections of neighbouring chains.

While this gives some indication of why polymers differ from other molecular solids, it does not explain the large differences between one polymer and another. They all have long chains, so why the variety? Before tackling this question we need to recognize that there is more than one kind of polymer.

The wide field of polymeric materials can be divided into three broad categories on the basis of their response to heating and to an applied force.

1 Some polymers soften on heating; they can then be shaped and the new shape is retained on cooling. These are **thermoplastic polymers**.

2 Some polymers that soften on heating show rubber-like elasticity rather than plastic behaviour: these are **elastomers**.

3 At the other extreme come the **thermosetting polymers** (or resins), which are hardened by heating into a brittle mass. Subsequent heating does not soften them so they can only be shaped by moulding before the heat treatment (or 'curing').

1 Thermoplastics

This group shows the greatest variety, and includes the familiar commercial plastics such as polythene and nylon. We can divide them into three main groups, amorphous polymers, crystalline polymers and fibres.

Amorphous polymers

These materials are made of linear chains: when molten they are loosely arranged and can move, although their spaghetti-like entanglements make the liquid viscous. When this writhing mass is cooled the disordered arrangement is frozen in to give an amorphous solid. Unlike normal solids, however, some limited movement persists: the chains may be able to twist or even slide past each other segment by segment, like a snake. Consequently, the solid can be bent or stretched: it is rubbery.

On further cooling even this limited movement is lost, and the rubber changes into a rigid, brittle glass-like solid. The glass state differs from the rubber state not in structure but in the freedom of movement of the chains. The temperature at which this change occurs is called the glass temperature, t_g. Although you do not need to remember any details about the glass temperature, it is of great practical importance since a designer needs to know when this marked change in behaviour will occur.

Data on the physical properties of polymers are listed in table 7.7 in the *Book of data*.

Polymer	Repeat unit	t_g / °C
Poly(propene)	—CH$_2$—CH— CH$_3$	−10
poly(chloroethene) rigid	—CH$_2$—CH— Cl	85
flexible		−20 to −30
poly(phenylethene)	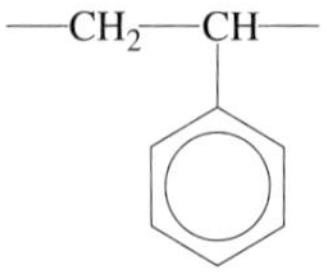 —CH$_2$—CH—C$_6$H$_5$	100

Figure 14.20 Glass temperatures.

STUDY TASK

Anything which restricts the movement of the polymer chains will give a higher t_g – and make for a stronger, stiffer polymer. Three factors are important:
- The flexibility of the polymer chain itself
- The presence of polar groups which give additional dipole-dipole forces between the chains
- Bulky side groups which restrict the rotation of chain segments.

1 Use these principles to explain the relative values of t_g for the polymers shown in figure 14.20. Will the polymers be glassy or rubbery at room temperature and at $-20\,°C$?

2 The polymer Nomex has the repeat unit

[Nomex structure]

Nomex

and has t_g of 270 °C. What factors do you think contribute to this very high value?

3 One way in which a polymer can be made more pliable is by adding a plasticizer, a non-volatile liquid which is absorbed into the solid polymer.

a How do you think these plasticizers work?

b Why are window frames made from unplasticized poly(chloroethene), uPVC?

c Would uPVC be a suitable material for an electric kettle?

Crystalline polymers

When some polymers are cooled the jumbled arrangement of the liquid is not retained and the molecular chains become aligned in rows: the polymer has become crystalline. However the chains are too long and flexible to form the regular crystals of ionic compounds so the crystalline regions are still surrounded by amorphous regions.

	Estimated maximum crystallinity/%
Natural rubber	30
Poly(propene) fibre	60
Nylon fibre	60
Polyester fibre	60
Poly(ethene) – low density	75
– high density	95

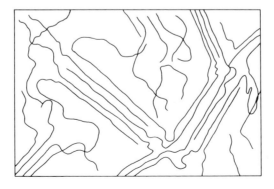

Figure 14.21 Crystalline regions in a polymer.

In the crystalline regions sections of the chains are packed in closer 'contact' so there will be greater intermolecular forces. The crystalline regions will therefore make the polymers harder than rubbers but in the amorphous regions the chains can still move relatively easily (above t_g) so the polymer will be less brittle (tougher) than glass-like polymers. Toughness is a very desirable property and some of the most useful polymers are partially crystalline.

How can this crystallinity be encouraged? The essential feature of crystals is their regular repeating pattern. It is thus not surprising that the more regular the polymer chains themselves the more crystalline is the solid. Low density poly(ethene) has numerous short side chains branching off from the main chain: these obstruct the alignment of the chains, and the material is only about 75% crystalline. The catalytic process used to make high density poly(ethene), however, produces a regular chain with no branching which can give almost 100% crystallinity. High density poly(ethene) is therefore harder, stronger and less flexible than low density poly(ethene).

In polymers with substituent groups replacing some of the hydrogen atoms, as in poly(chloroethene) or poly(propene), a more subtle kind of regularity becomes important. When these groups are regularly arranged the polymer is said to be **isotactic**.

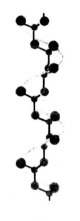

Isotactic chain making one complete turn for every 3 monomer units

Figure 14.22 The regular structure of isotactic poly(propene)

When the methyl groups are randomly oriented, this regularity is lost and the polymer is described as **atactic**; atactic polymers are unable to crystallize. Atactic poly(propene) is an amorphous rubbery polymer of little value, whereas isotactic poly(propene) is highly crystalline and can be drawn into an extremely useful fibre.

The flexibility of the chains is a second factor which affects the ease with which they can be organized into a regular arrangement in the solid.

In polyester the chain is regular and not branched, and the polar C=O groups in the ester linkages increase intermolecular forces, encouraging crystallinity. However the planar benzene rings in the chain produce a larger, less flexible repeating unit and as a result polyester crystallizes rather slowly. It can be cooled below its t_g without crystallizing to form the transparent material familiar as large fizzy drinks bottles. But if the polymer is held at a temperature above t_g the molecules have time to wriggle into the crystalline arrangement which is more suitable for a fibre. By contrast, poly(ethene) with its flexible chain crystallizes so readily that even very rapid cooling fails to form an amorphous material.

A third way of encouraging crystallization is by stretching (drawing) the polymer; this both increases the crystallinity in amorphous regions, and also orientates any existing crystalline regions along the direction of draw. This in turn increases the strength in that direction but usually at the expense of weakness in other directions. Poly(ethene) bags or tubes made by extrusion show this directional strength.

Drawing the polymer film during manufacture can be used to adjust the properties of the polymer: a large amount of draw produces a high strength nylon ideal for tyre cord, while less draw leads to a softer, more elastic fibre which is more suitable for making clothes.

STUDY TASK

Cut several strips, 2 cm × 10 cm, of poly(ethene) film. Strips of film should be cut with a sharp knife; otherwise, the film may tear rather than stretch when under stress. It is worth while cutting two strips, one parallel to the edge of the film and the other at right angles, as the film will have been partially stretched when being processed. Pull the strips gently so as to stretch the film.

What do you think happens to the molecular chains in poly(ethene) when it is stretched?

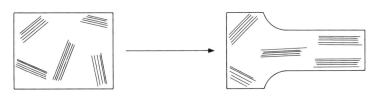

Figure 14.23 Aligning crystalline regions by 'drawing'.

Fibres

Fibres are a special case of crystallinity in which the chains are aligned along the fibre axis. This gives the high strength needed for a successful fibre. A good fibre-forming polymer usually possesses some or all of the following features:
- Regular linear chains favouring high crystallinity.
- Strong intermolecular forces, but no actual cross-linking
- Some, but not too much, chain flexibility.

Thus isotactic poly(propene) forms excellent fibres. 'Acrylic' fibres are based on poly(propenenitrile), which is less regular than poly(propene) but the polar —CN groups compensate for this.

The classic fibre-forming polymers are the polyamides (nylon) and polyesters (Terylene) both of which have regular, linear chains. In nylon 66 crystallinity is further encouraged by hydrogen bonding between adjacent chains.

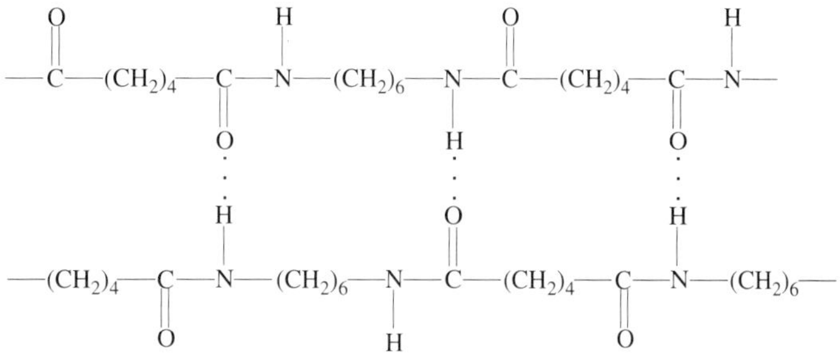

Figure 14.24 Hydrogen bonding between nylon chains

STUDY TASK

Terylene has the structure

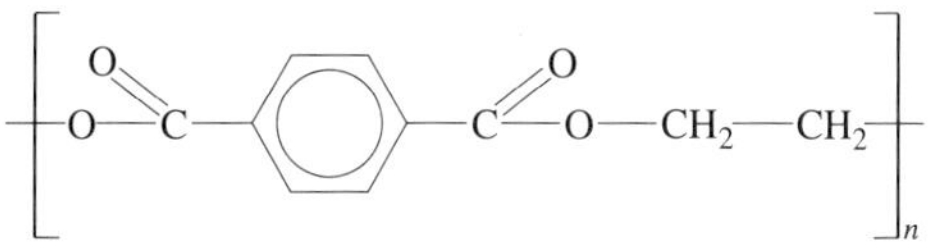

Terylene

1. What forces act between Terylene chains? Are they stronger or weaker than the forces in nylon?
2. What effect will the planar benzene rings have on the Terylene chains?
3. Pleats can be ironed into a Terylene fabric using a hot iron; and they will survive a 50 °C wash, unlike pleats in nylon. Why is this?
4. The elastic recovery of Terylene is noticeably poorer than for nylon. What problem might arise with stockings or trousers made from Terylene?

For clothing in general, a fibre with reasonable moisture retention is more comfortable and attracts less dirt because it is less prone to static. Nylon has poor moisture retention, but this is improved by grafting on poly(epoxyethene): the chemist tailors the polymer before the tailor cuts the cloth.

2 Elastomers

Some rubbery polymers show a quite exceptional elastic behaviour. Natural rubber, for example, can be stretched by 600% and still snap back to its original length when the force is removed. Such a material is an **elastomer**.

To understand this unusual behaviour we need to remember that entropy, rather than energy or force, is the overriding factor for spontaneous changes. A long polymer chain with freely rotating bonds can be coiled into innumerable shapes, all having the same energy. The most unlikely shape is the perfectly ordered straight chain, just as it is most unlikely that a drunken reveller, whose every step is in a randomly chosen direction, will walk straight home.

The natural shape of a chain is that which has the highest entropy: a highly crumpled random tangle in which the two ends may be quite close together. The chain may be readily straightened out by applying a force: and since this involves only bond rotation rather than bond stretching, a relatively small force is needed.

This is why it is easier to stretch a rubber band than a steel wire or a carbon fibre. For this uncoiling to take place, of course, the molecules must have some freedom to move: the polymer must be above its glass temperature. A useful elastomer must therefore have a glass temperature well below room temperature, which implies that the interchain forces must be weak.

These factors do not account fully for the essential feature of 'rubberiness'. Such loosely held molecules should slide past each other, untangling as well as uncoiling: but elastomers seem to have a molecular 'memory' which enables them to return to their original arrangement. The reason for this elastic recovery is that the coiled chains in unstretched rubber are actually looped around each other so that they are virtually knotted together. This effect can be greatly enhanced by replacing these knots with cross-links in the form of covalent bonds which permanently anchor the chains at various points.

The key feature of elastomers is thus limited cross-linking which can successfully combine high local movement of chain segments with low overall movement of chains (see figure 14.25).

Figure 14.25 Movement of the chains in an elastomer.

unstretched   stretched

3 Thermosetting polymers

When these materials are heated, extensive cross-linking takes place between the polymer chains: this causes the material to harden permanently or 'set'. Like diamond, these cross-linked polymers are hard and brittle. Unlike all the other polymers we have discussed, their properties are not sensitive to temperature since there is no possibility of even local movement of the 'chains'.

One of the first thermosetting polymers was Bakelite, formed by the condensation polymerization of phenol and methanal. It is brittle, but it is harder and has better wear and temperature resistance than thermoplastics, keeping its shape even when hot. It is also an excellent insulator and Bakelite is still widely used for plug tops, electrical fittings and heat resistant handles and knobs.

Polymerization of methanal with melamine produces the resins familiar as tableware and laminates for table surfaces. Being almost colourless they can be produced in a much wider and more decorative range of colours. This, together with their greater hardness and chemical resistance, justifies the extra cost.

Polymers are an invaluable and evolving group of materials, continually being adapted to meet new demands. Such developments are based on our growing understanding of the chemistry of polymerization reactions and of the intimate connection between properties and structure.

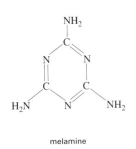

melamine

INVESTIGATION 14.6b

The properties of 'slime'

Carry out an investigation of the reaction between poly(ethenol) and borate ions.

Make a risk assessment before starting any experiments.

Summary

At the end of this Topic you should be able to
a recall the meaning of: chiral and achiral
b identify chiral centres in organic molecules and recall the properties associated with chiral molecules
c recall the characteristic behaviour of the carbohydrates limited to:
    solubility in water
    acid hydrolysis of sucrose
    treatment with Benedict's solution
    effect of aqueous solutions on polarized light
d recall the characteristic behaviour of lipids limited to:
    alkaline hydrolysis of fats

e recall the characteristic behaviour of amino acids limited to:
   acidity and basicity
   separation and identification by chromatography
   effect of aqueous solutions on polarized light
   enzymes as catalysts
   formation of peptide group in proteins
f recall the meaning of the following terms: polyamide, polyester, polysaccharide, protein
g recall the meaning of the following terms: monomer, polymer, addition polymerization, condensation polymerization, initiators; thermosetting polymers, thermoplastic polymers, elastomers
h demonstrate an understanding of the relation between monomer and polymer for both addition polymerization and condensation polymerization, including structural formulae
i recall the formation of poly(ethene), poly(chloroethene) and nylon, and the types of reactions involved
j evaluate information by extraction from text and the *Book of data* about natural products and polymers.

Review questions

14.1 When glucose is dissolved in water an equilibrium is established between the cyclic structure shown below and a small proportion of open chain molecules. Study the structures carefully and then answer the questions:

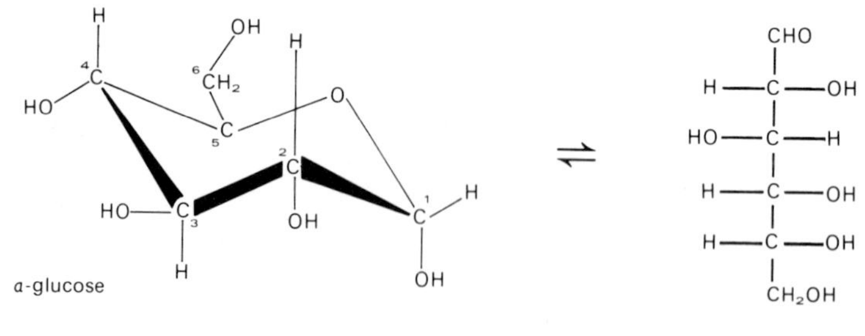

a Which numbered carbon atom is a part of the aldehyde group in the open chain structure?
b Which numbered carbon atoms in the cyclic structure carry (i) primary, (ii) secondary, or (iii) tertiary hydroxyl groups?
c Which numbered carbon atoms are involved when glucose polymerizes to form cotton (see figure 14.8)?
d Explain why you might expect glucose to be a reducing sugar.
e What product would you expect to be obtained if glucose were reacted with hydrogen formed by reacting sodium amalgam with water?

14.2 The formula below represents a molecule of alanine:

$$CH_3—CHNH_2—CO_2H$$

a The infra-red absorption spectrum of alanine has certain features which are very similar to features in the infra-red spectra of aminoethane, $CH_3—CH_2—NH_2$, and chloromethane, CH_3Cl.
 Which is the smallest part of the molecule which could be said to be responsible for these features of the infra-red spectrum?

b Alanine can behave both as an acid and a base.

i Draw the structural formula for the anion formed from an alanine molecule on adding a strong base.

ii Draw the structural formula for the cation formed from an alanine molecule on adding a strong acid.

c The following questions concern the optical isomers of alanine. What differences, if any, would there be in:

i the melting points of the two isomers?
ii the rates of their reaction with ethanoyl chloride?
iii their effect on copper(II) sulphate solution?
iv their effect on plane polarized light?
v their occurrence in natural protein?

14.3 The diagram represents, in two dimensions, a very small part of the structure of a naturally occurring material. The letter R represents not just one group but a variety of relatively small structures.

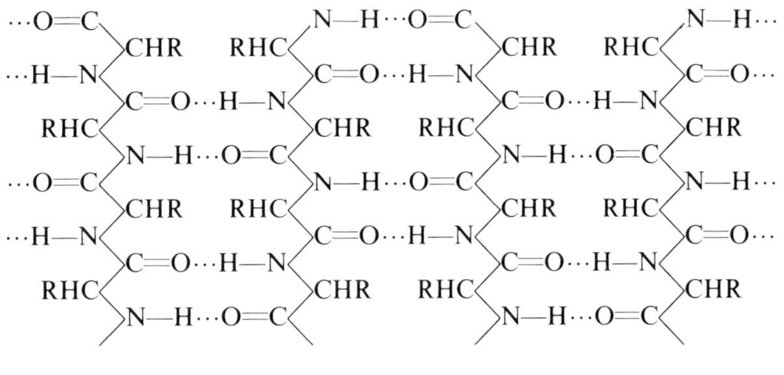

a To what class of naturally occurring material does this substance belong?
b State two physical properties which a substance of this structure would be expected to have.
c Assuming that you swallowed some of this substance and that it could be digested, indicate the chemical reactions by which the process of digestion would probably begin.
d By what practical process would you attempt in the laboratory to break down the substance in order to investigate its structure?
e Having broken down the structure, state the practical procedure you would use next in your investigation.
f After the chemical structure had been worked out, what physical method would be used to help to work out the three-dimensional structure of the molecule?

14.4 The technique of electrophoresis enables a solution of amino acids to be separated and identified by observing their relative movement on chromatography paper under the influence of an applied electric field. At pH 7 little movement of glycine is observed; at pH 2 glycine migrates to the cathode, but at pH 12 glycine migrates to the anode. Suggest reasons for these results.

14.5 A substance, A, of molecular formula $C_5H_{13}N$, exists as two optical isomers. On treatment with hydrochloric acid, A forms B, $C_5H_{14}NCl$. A also reacts with ethanoyl chloride to form C, $C_7H_{15}NO$.

  a Give structural formulae which could apply to A, B, and C.
  b Give the structural formulae of two isomers of A which are not optically active and name them.
  c Suggest a method by which one of your suggestions for part **b** could be synthesized starting from the appropriate alcohol containing five carbon atoms.

14.6 This question is concerned with the synthesis and properties of polymers. A reaction scheme is shown for producing compound Y from readily available materials:

$$\underset{\underset{Br}{|}}{CH_2}\underset{\underset{Br}{|}}{CH_2}CH_2CH_2 \xrightarrow{(i)} CN(CH_2)_4CN \xrightarrow{(ii)} NH_2(CH_2)_6NH_2 \text{ Compound X}$$

$$\underset{OH}{C_6H_5} \xrightarrow{(iii)} \underset{OH}{C_6H_{11}} \xrightarrow[\text{nitric acid}]{\text{conc.}} HO_2C(CH_2)_4CO_2H \xrightarrow{(iv)} ClOC(CH_2)_4COCl \text{ Compound Y}$$

  a Name the principal reagents which could be used to carry out step **(iii)**.
  b What type of reaction takes place at step **(iii)**?
  c Compounds X and Y react together by polymerization. By what type of polymerization reaction do they react?
  d Name the polymer produced.
  e Indicate by means of an equation how a molecule of X and a molecule of Y react together.

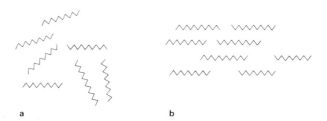

  f The figure represents polymer chains packed together under different conditions. How can the non-crystalline form be converted into the crystalline form?
  g How will the two forms differ
    i in elasticity?
    ii in tensile strength?

Examination questions

14.7 Sucrose, household sugar, is a disaccharide which is a combination of one unit of glucose and one unit of fructose.

a i What is meant by the term **disaccharide**?
ii Describe how you would convert sucrose into glucose and fructose. What type of reaction is involved?
iii Benedict's solution is added to separate solutions of glucose and sucrose and boiled. Describe the results of this test and explain the chemical reactions that have taken place.
b Carbohydrates have chiral centres.
i What is meant by the term **chiral**?
ii Copy the glucose structure and mark the chiral centres.
iii How can chiral compounds be recognized in the laboratory?
c The three main steps in determining the structure of a protein are:
i determination of the amino acids present
ii determination of the amino-acid sequence
iii determination of the arrangement (shape) of the protein chain.
State briefly, the method used in each step.

14.8 This question is about amino acids.

a Most naturally occurring amino acids are chiral compounds. Copy the formulae and mark the chiral centres.

Alanine   Glycine   Aspartic acid

b Glycine is a solid at room temperature.
  i What type of bonding occurs between glycine molecules in the solid state?
  ii Draw the structural formula of glycine in the solid state taking account of the bonding you have suggested.
c Describe an experiment to show that a mixture of amino acids contains glycine and alanine. Your answer should include full practical details and how you would interpret your results. You may wish to include a diagram.

14.9 Study the following reaction sequence and answer the questions below:

$$
\begin{array}{ccc}
\text{CH}_3 & \text{CH}_3 & \text{CH}_3 \\
| & | & | \\
\text{CH}_2 & \text{CH}_2 & \text{CH}_2 \\
| & \xrightarrow{} \quad | \quad \xrightarrow{\text{OH}^-} & | \\
\text{CH} & \text{CHBr} & \text{CHOH} \\
\| & | & | \\
\text{CH}_2 & \text{CH}_3 & \text{CH}_3 \\
\mathbf{A} & \mathbf{B} & \mathbf{C}
\end{array}
$$

a Draw a structural formula to illustrate the addition polymer formed by A, including at least three monomer units in your diagram.
b i What reagent is required to convert A to B?
  ii Why, in converting A to B is a mixture of two isomers obtained, both of which may be represented by formula B above? Use diagrams to assist your explanation.
  iii State one physical property by which the isomers of B may be distinguished.
c For the conversion of B to C the rate expression is found to be:

$$\text{Rate} = k[\text{OH}^-][\text{CH}_3\text{—CH}_2\text{—CHBr—CH}_3]$$

Suggest a mechanism for the reaction that would be consistent with this rate expression.

14.10 Poly(isoprene) is one type of synthetic rubber made from crude oil. The products of thermal cracking of oil include 2-methylbut-1-ene and 2-methylbut-2-ene. These alkenes can be extracted from refinery gases by reaction with sulphuric acid. Catalytic dehydrogenation of both of these alkenes yields isoprene, C_5H_8 which can be polymerized to produce the synthetic rubber.

$$\text{CH}_2\!\!=\!\!C\!\!\begin{array}{l}\diagup\text{CH}_3 \\ \diagdown\text{CH}_2\text{—CH}_3\end{array}$$

2-methylbut-1-ene

a Draw the **displayed formula** of the second hydrocarbon, 2-methylbut-2-ene.
b What type of reaction occurs between sulphuric acid and these hydrocarbons?
c i What is meant by **catalytic dehydrogenation**?
  ii Hence draw the structural formula of isoprene.
d The graph (overleaf) shows the production of natural and synthetic rubbers from the beginning of the century until 1970.

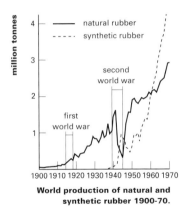

World production of natural and synthetic rubber 1900-70.

i Why do you think that the demand for rubber has increased during this century?
ii Why do you think that there was a substantial replacement of natural rubber by synthetic rubber during the Second World War?
iii Suggest two circumstances which could result in the production of natural rubber overtaking the production of synthetic rubber in the future.

14.11 Epoxy resins are polymers which are used in heavy-duty adhesives. Epoxy resins are made from two monomers, for example:

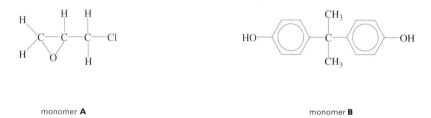

monomer **A**   monomer **B**

a Monomer A is manufactured from propene in three stages:

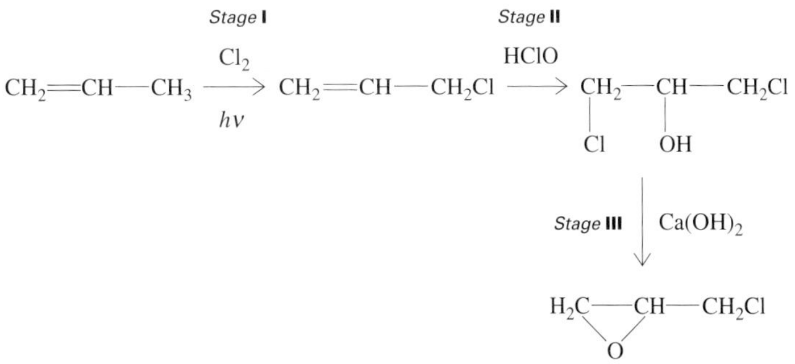

$$CH_2=CH-CH_3 \xrightarrow[hv]{Cl_2} CH_2=CH-CH_2Cl \xrightarrow{HClO} CH_2-CH-CH_2Cl$$
$$\underset{Cl \quad OH}{}$$

Stage III ↓ Ca(OH)$_2$

$$H_2C-CH-CH_2Cl$$
$$\underset{O}{}$$

In stage I the reaction requires ultraviolet radiation.

i Name the type and mechanism of reaction that takes place in stage I.
ii What is the reason for using ultraviolet radiation?
iii Under other conditions a different reaction can occur between propene and chlorine. Write down the name and structural formula of the product of this reaction.

iv In stage II the organic molecule, 3-chloropropene, undergoes **electrophilic** addition with chloric(I) acid. Explain why the 3-chloropropene molecule can undergo electrophilic attack.

v In stage III the reaction taking place is called an **elimination** reaction. Explain why this name can be applied to this reaction.

b Monomer B is made by reacting propanone with phenol. Write a balanced equation, using structural formulae, for this reaction.

c When monomer A reacts with monomer B, a long chain molecule is formed, with monomers A and B forming alternate units in the chain, part of which is shown below:

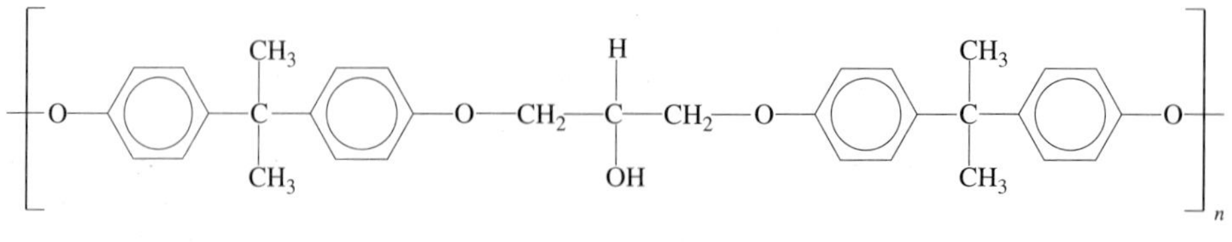

$$\left[-O-\underset{\underset{CH_3}{|}}{\overset{\overset{CH_3}{|}}{C}}-O-CH_2-\underset{\underset{OH}{|}}{\overset{\overset{H}{|}}{C}}-CH_2-O-\underset{\underset{CH_3}{|}}{\overset{\overset{CH_3}{|}}{C}}-O- \right]_n$$

i Copy the displayed formula of monomer A and mark on your copy the bonds which must break in the polymerization reaction.

ii Explain why this reaction is an example of **condensation polymerization**.

d Epoxy resin adhesives are formed by mixing the resin with a **hardener**. The latter contains amines which have more than one amine group per molecule. These react with some of the hydroxy groups in the resin in a process known as **curing** in which the final polymer is hard and strong, as well as a good adhesive for many materials.

i What type of structure would you predict for a hardened epoxy resin adhesive?

ii How does this structure account for the strength and hardness of the adhesive?

14.12 Explain, with examples, the difference between addition polymerization and condensation polymerization. By discussing molecular structure, explain the differences between thermosetting polymers, elastomers and thermoplastic polymers. How would you investigate the physical properties of a new polymer?

READING TASK 4

DIABETES

The sweetness of sugar

For all the flavours that we can distinguish in our foods we are indebted mainly to our sense of smell because it seems that our sense of taste is restricted to four sensations: salt, acid (or sour), sweet, and bitter.

The sensation of sweetness has for a long time been associated with a range of naturally occurring carbohydrates, and nowadays also with a small group of synthetic compounds. Honey as a source of sweetness must have been known in the Stone Age because a cave painting in Spain records a woman gathering honey from a wild bees' nest.

And sucrose from sugar cane has been known in the East for a very long time, being recorded in the fourth century BC. Sugar beet as a source of sucrose came into prominence during the Napoleonic Wars when in 1811 France was cut off from her traditional sources of cane sugar. In the last hundred years several synthetic agents been developed, including saccharin discovered in 1879, cyclamate in 1937, and aspartame in 1965.

saccharin

cyclamate

aspartame (aspartic acid plus the methyl ester of phenylamine)

Figure R4.1 A woman gathering honey in the Neolithic period (a cave painting).

Different compounds have very different degrees of sweetness and this accounts for some of the success of the synthetic sweetening agents. Most sweeteners have been synthesized as part of research programmes quite unrelated to sweetness and owe their discovery to unhygienic laboratory practice. The laboratories became contaminated, and this was noticed because of an unexpected intense sweetness, by casual licking of the fingers or through smoking a contaminated cigarette. The source of sweetness was soon traced to the appropriate compound. The chemists involved were lucky they discovered sweet rather than toxic compounds.

Compound	Source	Relative sweetness
Lactose	milk	0.2
Corn syrup	corn starch	0.3
Glucose	fruit	0.5
Sucrose	cane and beet	1
Invert sugar	honey	1.2
Fructose	fruit	1.7
Cyclamate	synthetic	30–80
Aspartame	synthetic	100–200
Saccharin	synthetic	500–700

Figure R4.2 Relative sweetness of various compounds

Cyclamate quickly became a popular additive to foodstuffs after 1950 because it lacks the bitter aftertaste associated with saccharin and is not destroyed by heat. However, the wisdom of permitting cyclamate to be used as a food additive was questioned when it was found that a small proportion of people converted cyclamate to cyclohexylamine, and large doses of cyclohexylamine, when injected into rats, caused chromosome damage. Cyclamate was banned in the United States and the UK in 1970 after research reporting that rats given large daily doses of a mixture of cyclamate and saccharin developed bladder tumours. Although the doses given to rats were very much higher than might be taken by humans, the use of cyclamates was banned in the United States because a part of the country's food law (the Delaney Amendment) says that if any substance is found to cause cancer in an animal, no matter what the dose level used, that substance shall not be used in food for humans.

You may well wonder why saccharin was not banned as well. The fact is that saccharin has been used for over 80 years by many people, including people with diabetes, who are unable to eat sugar, and there is no evidence of a higher incidence of bladder tumour in these people than in non-users. The cyclamate data have not been used by other countries to impose a ban on its use in foods.

Aspartame has had a similar history, being banned for a time but now generally approved.

Diabetes

The disease *diabetes mellitus* is caused by a failure of the body to produce a protein known as **insulin** in sufficient quantities. The degree to which insulin production fails varies from one individual to another, and for this reason people who have diabetes may suffer from a wide range of ailments.

Insulin is produced in the pancreas, and passed into the bloodstream where it regulates the uptake and release by tissues of glucose, carboxylic acids, and amino acids. We shall briefly consider each of these substances in turn.

The **glucose** in the blood is taken up continuously by tissues such as muscle, heart, brain, and fat depots, and in order to maintain a steady level it must be replaced. Fresh supplies come either from the digestive products of the gut or, if the individual has not eaten recently, from the liver. Glucose is synthesized in the liver from other compounds, and can be stored there in the form of glycogen. After a meal, when much glucose is likely to be available from digestive processes, the rate at which insulin is produced can increase by a factor of ten. The resulting higher level of insulin in the blood reduces the quantity of glucose released from the liver, and increases the amount taken up by the tissues. In this way the blood glucose level is kept nearly constant.

The **carboxylic acids** with which we are concerned are those which are found in a combined form in fats. Fats are esters of these acids and propane-1,2,3-triol (glycerol), and for this reason the acids are taken up continuously by tissues, especially muscle and heart where they are by far the most important fuel. They are replaced continuously by fatty acids released from fat depots. Insulin regulates this process too, in much the same way as it regulates the glucose level. After a meal the relatively high insulin level keeps the release of free fatty acids from fat depots to a minimum, but if the individual has not eaten recently, the lower level of insulin results in a more rapid release of free fatty acids from fat depots. If insulin is very much reduced, and only if this is the case, the excessive amounts of free fatty acids that are released are taken up by the liver, where they are broken down to a variety of compounds including propanone. This compound is eliminated in the urine and, being volatile, may also be exhaled. This gives people with diabetes who fail to control their blood sugar levels a characteristic sweet breath.

Amino acids are also taken up continuously by the tissues. The main source for replacement is the digestive products of the gut, but amino acids are also released by tissues, especially muscle. The quantity that is released is mainly controlled by insulin; if it is present the amino acids are conserved as tissue proteins but if it is reduced, muscle protein is broken down. The bulk of the amino acids so produced is transferred to the liver and converted there to glucose or ketones.

People with diabetes lose this fine regulation. When they eat sugary foods the blood glucose rises to a high level and is eliminated in the urine. There is a constant need to pass large volumes of urine, and the individual becomes very thirsty. If the deficiency of insulin is more severe, not only is glucose lost in the urine, but both fat from the fat depots and amino acids from the muscle protein are mobilized, and the individual loses weight. If insulin production is near zero the free fatty acids are converted by the liver to keto-acids and transferred to the bloodstream as an alternative form of energy. This high level of acid causes nausea and vomiting. The individual rapidly becomes dehydrated and may lapse into a diabetic coma.

Normally, urine does not contain glucose although traces may be present after heavy

Figure R4.3 Insulin was discovered by the Canadian doctors Banting and Best.

meals. Diabetes may therefore be detected during a routine examination by testing the urine for glucose. A positive result, even in an apparently healthy patient during a routine medical, indicates the need for a more detailed clinical examination and further tests.

To test for reducing sugars (glucose) in urine

A laboratory test can be carried out as follows. Add 8 drops of urine to 5 cm^3 of Benedict's solution and boil vigorously for two minutes. If more than 0.2 g of sugar per 100 cm^3 of urine is present, a yellow to brown deposit of copper(I) oxide will be produced and the solution will become colourless; if less sugar than that is present the solution will become green and only a small deposit will form.

To test for ketones (propanone) in urine

A laboratory test can be carried out as follows. To 2 cm³ of urine add a spatula measure of mixed ammonium sulphate (100 parts) and sodium nitroprusside (1 part). Add 2 cm³ of 8 M ammonia solution. A faint purple colour will develop if propanone is present in a proportion of 1 in 20 000 or more.

People with a mild insulin deficiency will have only glucose in the urine, but anyone with severe insulin deficiency will have both substances present in the urine. This kind of patient is seriously ill and needs to seek medical advice urgently.

Once a sufferer has been diagnosed as having diabetes, he or she has to maintain a near normal blood glucose level. Here too, simple colour tests are available which enable patients to check their own levels. Thus, they may begin to care for themselves whilst leading full and active lives.

The measurement of glucose levels

Glucose levels are readily measured by using a test stick containing a dye and two enzymes, glucose oxidase and horseradish peroxidase.

Glucose oxidase catalyses the oxidation of any glucose present in blood or urine to gluconic acid and hydrogen peroxide. The hydrogen peroxide is rapidly decomposed in the presence of the horseradish peroxidase and the oxygen produced oxidizes the dye. During oxidation the dye changes colour, the intensity indicating the concentration of glucose present.

$$\text{glucose} + \text{oxygen} + \text{water} \xrightarrow{\text{glucose oxidase}} \text{gluconic acid} + \text{hydrogen peroxide}$$

$$\text{hydrogen peroxide} + \text{reduced form of dye} \xrightarrow{\text{peroxidase}} \text{water} + \text{oxidized form of dye}$$

This test enables patients to check their own glucose levels and use the information to maintain a near normal blood glucose level by taking appropriate action. The test may be carried out on the urine using a simple dip stick or on a small drop of blood, by pricking a finger. Meters are also available to measure the colour intensity on the strip rather than estimating the result by eye.

Figure R4.4 A specialized colorimeter can be used to measure blood sugar levels

Questions

1 Deduce the equation for the reaction that occurs when sodium cyclamate is converted to cyclohexylamine.

2 The taste-bud receptors for sweetness in the tongue consist of tiny cavities into which the 'sweet' molecule fits and can be held temporarily. This action triggers the sensation of taste.

The surface molecules of the taste-bud receptors have C=O and N—H groups facing out into the cavity.

a What type of bonding is likely to be involved in temporary linking of the 'sweet' molecules to the receptors?

b Are all the 'sweet' molecules you have met capable of forming the type of bonding you have suggested?

c A drink of water will rapidly clear the mouth of any remaining 'sweet' tastes left after eating or drinking. How might this be explained?

3 The discovery of a treatment for diabetes began to save lives within months of being announced in 1922; but the research involved the use of dogs. What are the arguments for and against the use of animals in medical research?

TOPIC 15

The transition elements

In previous work you have probably used the term **transition element** to refer to those elements which come in the d-block of the Periodic Table, between Groups 2 and 3. In this Topic, we shall only be concerned with the elements of the first row. All of these elements are metals.

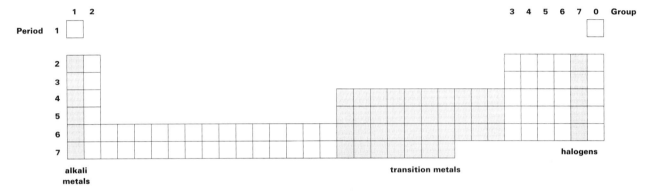

Figure 15.1 Transition elements in the Periodic Table.

As you can see iron is one of the transition elements, so you will be using your knowledge of iron chemistry from Topic 1 to consider in what ways iron is a typical transition element. The transition elements, gold, copper and iron were amongst the earliest materials we produced and their importance has not diminished, because metals have found an ever increasing range of uses as we have learnt more about their properties.

REVIEW TASK
From your general knowledge make a list of what you regard as typical physical and chemical properties of metals. For each property give an example of an application that depends on the property.

Figure 15.2 Uses of some transition elements. **a** Titanium oxide has excellent covering power

15 The transition elements 436

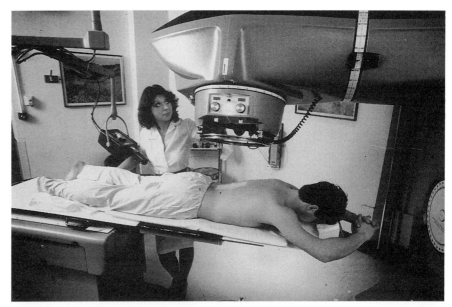
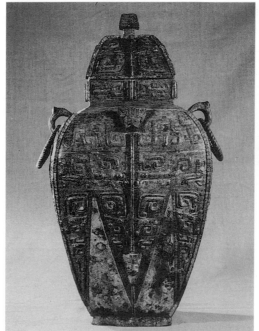

Figure 15.2 Uses of some transition elements. **b** Vanadium is used in this colour display screen **c** Cobalt's radioactive isotope Co-60 is used in medical treatment **d** Copper is alloyed with tin to form bronze, used for this ancient Chinese ritual vessel **e** Nickel occurs in baked beans

Element	Electronic structure
(Sc)	$3d^1 4s^2$
Ti	$3d^2 4s^2$
V	$3d^3 4s^2$
Cr	$3d^5 4s^1$
Mn	$3d^5 4s^2$
Fe	$3d^6 4s^2$
Co	$3d^7 4s^2$
Ni	$3d^8 4s^2$
Cu	$3d^{10} 4s^1$
(Zn)	$3d^{10} 4s^2$

The electronic structures of the first row of transition elements: only the outermost energy levels are shown.

True transition elements and their compounds have a number of characteristic properties, and these properties are usually a consequence of the ions of the elements having d-orbitals which are incompletely filled with electrons. This effectively removes both scandium and zinc from the list, since the only ions that these elements form are Sc^{3+} and Zn^{2+}, neither of which has incomplete d-orbitals. In future, therefore, when we refer to transition elements, we shall mean an element which contains an incomplete d-orbital in at least one compound. In the first row, this means elements Ti to Cu inclusive.

We shall start by looking at how the atoms of metals are packed together in the solid state.

15.1 Metallic bonds

Three significant properties of metals are their high melting points (as contrasted with most non-metals), their high electrical conductivities, and their high thermal conductivities. Any model of the nature of the bonding in metals must be able to account for these properties. A simple model of bonding in a solid metal consists of positive metal ions surrounded by a sea of mobile electrons.

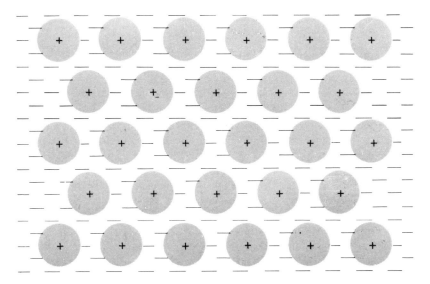

Figure 15.3 A simple representation of bonding in a metal lattice.

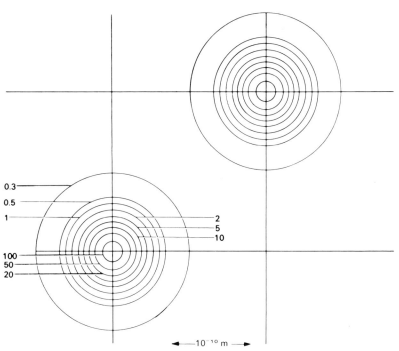

Figure 15.4 An electron density map for aluminium. (Contours are in electrons per 10^{-30} m^3.) Are the ions likely to be spherical?

The sea of electrons bonds the metal ions tightly into the lattice and confers a high boiling point compared to their melting points, which means that metals have a wide temperature range over which they are liquid. The mobility of the electrons provides a means of conducting electricity and heat. The mobile electrons are another example of delocalization.

This model is an oversimplification, and it does not account for all of the properties of metals.

15.2 Crystal structures of metals

There are three patterns in which the atoms of metals usually pack. They are known as **body-centred cubic** (BCC), **face-centred cubic** (FCC) and **hexagonal close-packed** (HCP).

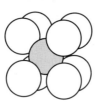
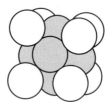
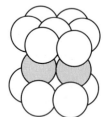

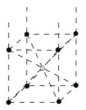

Figure 15.5 Packing patterns of metal atoms.

Body-centred cubic structure (for example, chromium)

Face-centred cubic (ABC) structure (for example, copper)

Close-packed hexagonal (ABAB) structure (for example, zinc)

In this section we are going to explore the nature of these packing patterns.

EXPERIMENT 15.2 ## Models of metallic structures

You will need a collection of polystyrene spheres, either loose or stuck together as small rafts. Keep them in a tray.

15 The transition elements

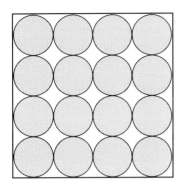

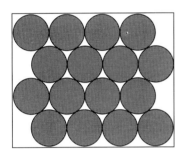

Procedure

1 Construct or examine rafts of sixteen spheres built to match the diagrams in the margin. Measure the sides of the squares that will just enclose the rafts.

■   Which arrangement is more closely packed?
    Make a drawing in your notebook of the arrangement that has a hexagonal pattern.

2 Construct a close-packed raft (A) of sixteen spheres and then use more spheres to build a second layer (B) close-packed on top of the first layer.

A

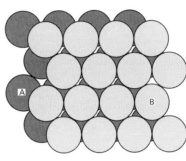

B

Now find out how a third layer can be added. By looking down through the holes in the layers work out how many patterns are possible: a repeat of the first layer (A) or a different arrangement (C).

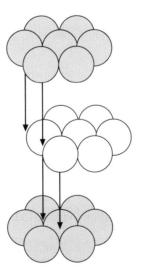

a Close-packed spheres in ABA sequence

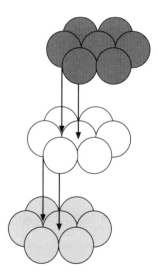

b Close-packed spheres in ABC sequence

Finally add a fourth layer to look at the complete patterns: ABAB and ABCA.

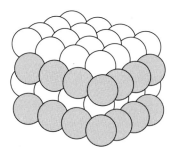

- ABAB... corresponds to the pattern known as the **hexagonal close-packed structure (HCP)**.
- ABCA... corresponds to the pattern known as the **face-centred cubic structure (FCC)**, sometimes referred to as cubic close-packed.

3 Using a FCC model, find out how it can be constructed of close-packed layers. When you suspend the model from a corner sphere you should be able to see that the number of spheres from the layers is 1 from A, 6 from B, 6 from C, 1 from A again.

To convince yourself that the FCC structure consists of close-packed layers, build an array of spheres in which the FCC model is embedded. Start with a standard bottom layer with one sphere removed from the centre: fit a corner sphere of a FCC model into it.

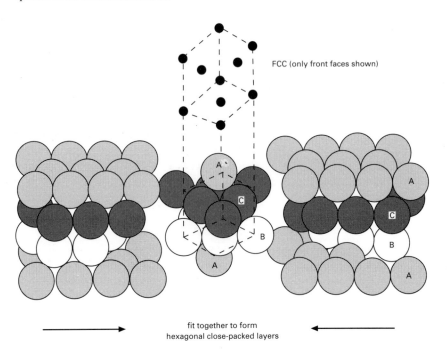

■ Record in your notes the relationship between the hexagonal close-packed layers in the pattern ABCA and the FCC structure.

4 The body-centred cubic structure, BCC, is not as close-packed as the HCP and FCC structures.

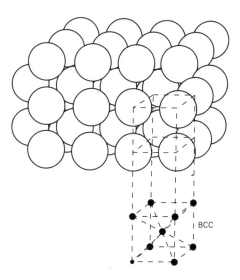

In HCP and FCC structures there is only 26% empty space but in the BCC arrangement there is 32% empty space.

Almost all of the metals crystallize into one or more of these three systems and the information is listed in table 5.2 in the *Book of data*. Iron atoms can, however, be found in different crystal packings: α-iron is BCC and γ-iron, formed above 914 °C is FCC.

QUESTION
What are the relative amounts of empty space in α-iron and γ-iron? Can you calculate the values quoted?

You can work out the proportion of empty space in the structures most readily from their **unit cells**. A unit cell is the smallest unit which, when repeated in three dimensions, gives a whole crystal.

The two close-packed systems account for about fifty metals, and the body-centred system for about twenty metals. There is no obvious relationship between the structural type of a metal and its position in the Periodic Table.

15.3 Variable oxidation number

In this section we shall investigate some chemical reactions of compounds of the transition elements that involve changes of oxidation number. The diagram shows the range of oxidation numbers in transition metal compounds and the common ones you are likely to meet (ringed).

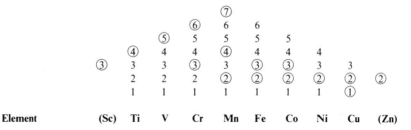

We shall begin our investigation by taking further the study of iron compounds that you started in Topic 1.

The redox chemistry of iron

You should recall that iron has two principal oxidation numbers in its compounds, +2 and +3, as in the ions $Fe^{2+}(aq)$ and $Fe^{3+}(aq)$. The purpose of the experiment in this section is to use standard electrode potentials to predict the likelihood of oxidizing $Fe^{2+}(aq)$, or reducing $Fe^{3+}(aq)$, and then to test your predictions experimentally.

Begin by copying figure 15.6 onto graph paper, and stick it in your notes. The chart is similar to those that you have already met in Topic 13.

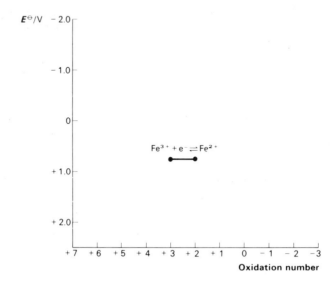

Figure 15.6 Electrode potential chart for experiment 15.3a

Enter on your chart the following equilibria, in the way the Fe^{2+}/Fe^{3+} equilibrium has been entered.

a $\frac{1}{2}Br_2 + e^- \rightleftharpoons Br^-$ $\quad E^\ominus = +1.09$ V
b $MnO_4^- + 8H^+ + 5e^- \rightleftharpoons Mn^{2+} + 4H_2O$ $\quad E^\ominus = +1.51$ V
c $\frac{1}{2}Cl_2 + e^- \rightleftharpoons Cl^-$ $\quad E^\ominus = +1.36$ V
d $Ag^+ + e^- \rightleftharpoons Ag$ $\quad E^\ominus = +0.80$ V
e $SO_4^{2-} + 4H^+ + 2e^- \rightleftharpoons SO_2 + 2H_2O$ $\quad E^\ominus = +0.17$ V
f $Zn^{2+} + 2e^- \rightleftharpoons Zn$ $\quad E^\ominus = -0.76$ V
g $\frac{1}{2}I_2 + e^- \rightleftharpoons I^-$ $\quad E^\ominus = +0.54$ V

Now you are ready to begin the experiment.

EXPERIMENT 15.3a The redox reactions of iron

You will need the following solutions, as far as possible of concentration 0.1 M.

Name	Notes
Iron(II) sulphate solution	Contains Fe^{2+}(aq) ions. These ions have a tendency to react with water (hydrolysis) which eventually makes the solution go brown and become cloudy. This has been minimized by adding some sulphuric acid.
Iron(III) chloride solution	Contains Fe^{3+}(aq) ions which also have a tendency to react with water. In this case the reaction has been suppressed by using hydrochloric acid.
Bromine water	This solution contains Br_2(aq) molecules. Bromine reacts slightly with water but this effect may be ignored.
Potassium manganate(VII) solution	Contains MnO_4^-(aq) ions, and has been acidified with sulphuric acid.
Chlorine solution	This solution contains Cl_2(aq) molecules. Like bromine, chlorine reacts with water to some extent but this effect may be ignored.
Sodium chloride solution	This solution contains Cl^-(aq) ions.
Sulphur dioxide solution	This solution has been made by bubbling sulphur dioxide gas through water. The solution has a strong smell which may be harmful to those who suffer from respiratory complaints.
Silver nitrate solution	This solution contains Ag^+(aq) ions.
Potassium iodide solution	This solution contains I^-(aq) ions.
Powdered zinc	Zinc metal is Zn(0).

SAFETY ⚠
Chlorine and and bromine are very harmful to your lungs; sulphur dioxide is irritant and toxic; zinc powder is flammable.

Use the chart that you have drawn to predict the likelihood of reactions occurring when the following pairs of substances are mixed, most of which are in solution.

> **HINT**
> You may find it helpful to revise the method of predicting redox reactions that was introduced in section 13.2.

Pairs of substances for consideration
a Iron(II) sulphate and bromine water
b Iron(III) chloride and zinc
c Iron(II) sulphate and silver nitrate
d Iron(III) chloride and sodium chloride
e Iron(III) chloride and sulphur dioxide solution
f Iron(II) sulphate and acidified potassium manganate(VII)
g Iron(III) chloride and potassium iodide
h Iron(II) sulphate and chlorine water.

Procedure

For each of the pairs of substances listed, try to confirm your predictions by experiment. Do this by mixing roughly equal volumes of the two solutions in a test-tube, or by adding a spatula measure of the solid to a few cm³ of solution.

In a number of the reactions it should be easy to tell whether a reaction has taken place or not because there are coloured reactants or coloured products. In other cases, however, it may be necessary to test the solution to find out whether the iron has changed in oxidation number. A suitable test is to add sodium hydroxide solution. Iron(II) ions give a green precipitate when sodium hydroxide solution is added, whereas iron(III) ions give a red-brown precipitate.

In your notes draw up a table of results using the following headings:

Substances mixed	Predicted reaction, if any	Observations on mixing
a Iron(II) sulphate and bromine water etc.		

Mention in the 'Observations' column any confirmatory tests you used.

STUDY TASK

Use your table of results to write balanced ionic equations for the redox reactions which took place.

Here is an example of how to do this. In the reaction between Fe^{2+} and Cl_2 the ions $Fe^{2+}(aq)$ are oxidized to $Fe^{3+}(aq)$ and the chlorine molecules $Cl_2(aq)$ are reduced to $Cl^-(aq)$. The equations for the half reactions are

$$Fe^{2+}(aq) \rightleftharpoons Fe^{3+}(aq) + e^-$$

$$Cl_2(aq) + 2e^- \rightleftharpoons 2Cl^-(aq)$$

The first equation involves one electron, whereas the second involves two electrons. The first equation should therefore be doubled throughout, and added to the second, so that the electrons do not appear in the final equation. This is

$$2Fe^{2+}(aq) + Cl_2(aq) \longrightarrow 2Fe^{3+}(aq) + 2Cl^-(aq)$$

EXPERIMENT 15.3b Analysis of 'iron tablets'

Potassium manganate(VII) is a well known oxidizing agent, usually used in solutions acidified with a plentiful supply of dilute sulphuric acid. The following electrode potentials show that manganate(VII) ions should oxidize iron(II) ions:

$$Fe^{3+}(aq) + e^- \rightleftharpoons Fe^{2+}(aq) \qquad E^\ominus = +0.77 \text{ V}$$

$$MnO_4^-(aq) + 8H^+(aq) + 5e^- \rightleftharpoons Mn^{2+}(aq) + 4H_2O(l) \qquad E^\ominus = +1.51 \text{ V}$$

Combining these two equations gives the overall equation for the reaction:

$$MnO_4^-(aq) + 8H^+(aq) + 5Fe^{2+}(aq) \longrightarrow Mn^{2+}(aq) + 5Fe^{3+}(aq) + 4H_2O(l)$$

so that in acid solution:

1 mole of MnO_4^-(aq) reacts with 5 moles of Fe^{2+}(aq)

Solutions containing MnO_4^-(aq) ions have an intense purple colour, whereas those containing Mn^{2+}(aq) ions are almost colourless. Solutions containing Fe^{2+}(aq) ions can be titrated against potassium manganate(VII) solution. The colour of the manganate(VII) is discharged, the end-point of the titration being the point at which the addition of one more drop of potassium manganate(VII) gives a permanent purple colour.

This titration forms the basis of an analytical technique for the estimation of iron, and we will now use the technique for the analysis of a popular 'tonic'.

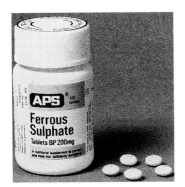

Figure 15.7 Iron is an essential element in our diet. When necessary we can supplement our diet with iron tablets.

Procedure

Weigh accurately two 'ferrous sulphate' tablets. Grind up the tablets with a little M sulphuric acid, using a pestle and mortar. Through a funnel, transfer the resulting paste into a 100 cm³ volumetric flask. Use further small volumes of M sulphuric acid to rinse the ground-up tablets into the flask. During this process, you must take great care to ensure that all the particles of tablet get into the flask.

When this has been done, add sufficient M sulphuric acid to make up the solution to exactly 100 cm³. Stopper the flask and shake it to make sure that all the contents are thoroughly mixed. They will not all be in solution although the Fe^{2+} ions which were present in the tablets will be dissolved.

Titrate 10.0 cm³ portions of the solution with 0.0050 M potassium manganate(VII). The end-point is marked by the first permanent purple colour. Brown or red colours should not be allowed to develop; the remedy is to add more M sulphuric acid.

QUESTIONS

1 What are the reasons for using so much M sulphuric acid during this experiment?
   (There are two reasons, one to do with the behaviour of solutions containing Fe^{2+} ions, and the other to do with the equation for oxidations involving potassium manganate(VII).)

2 Do your results agree with the analysis of the tablets given by the manufacturers on the bottle label?

3 Write a short account of the theory of this analysis, record the practical procedure used, and calculate the percentage of iron in the tablets from your results.

The oxidation numbers of vanadium

Whereas iron has only two readily accessible oxidation numbers in its compounds, vanadium has four: +5, +4, +3, and +2. In this next experiment we shall study this rather more complicated situation.

EXPERIMENT 15.3c

The redox reactions of vanadium

Procedure: Part 1 The reduction of vanadium(V)

The object of this part of the experiment is to start with a solution containing vanadium with oxidation number +5, and to select reducing agents which will reduce it to each of the other oxidation numbers.

The most convenient starting material is solid ammonium vanadate(V), NH_4VO_3. When ammonium vanadate(V) is acidified the vanadium becomes part of a positive ion, $VO_2^+(aq)$, in which the vanadium still has an oxidation number of +5. You are provided with this solution. The other oxidation numbers of vanadium are included with this one in the table:

COMMENT
In solution the vanadate(V) ion is polymerized in a complicated way, so the formula $VO_3^-(aq)$ is a simplification.

Ion	Oxidation number of vanadium	Colour of solution
VO_2^+	+5	yellow
VO^{2+}	+4	blue
V^{3+}	+3	green
V^{2+}	+2	mauve

Copy the table into your notes, and also the chart of electrode potentials and oxidation numbers.

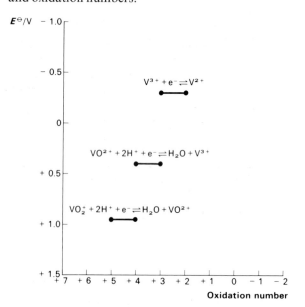

Figure 15.8 Electrode potential chart for vanadium

Enter on the chart the equilibria **a**, **e**, **f**, and **g** from experiment 15.3a, and also the equilibrium

$$Sn^{2+}(aq) + 2e^- \rightleftharpoons Sn(s) \qquad E^\ominus = -0.14 \text{ V}$$

Use the chart to select a reducing agent which should reduce vanadium from +5 in VO_2^+ to +4 in VO^{2+}, but should not reduce the vanadium any further.

Then select a reducing agent which should reduce vanadium from +5 to +4 and also from +4 to +3 but no further.

Finally select a reducing agent which should reduce vanadium all the way from +5 to +2.

Record your selections in a table drawn up in your notes and then try out the reactions, and record your observations.

> **SAFETY** ⚠
> You are using the same reagents as in experiment 15.3a, so you must take the same precautions.

Desired final oxidation number	Selected reducing agent	Observations
+4 (VO^{2+})		
+3 (V^{3+})		
+2 (V^{2+})		

If you decide to use potassium iodide solution as a reducing agent, the iodide will be oxidized to elemental iodine. This will give a colour to the solution which will mask the colour of the vanadium ion. The iodine can be reduced back to colourless iodide by reaction with sodium thiosulphate solution (see Topic 5): add only just enough to discharge the colour due to the iodine. Incidentally, it has been discovered that sodium thiosulphate will itself reduce the VO_2^+ ion. If the reaction involved is

$$2S_2O_3^{2-}(aq) \rightleftharpoons S_4O_6^{2-}(aq) + 2e^- \qquad E^\ominus = +0.09 \text{ V}$$

what oxidation number of vanadium should result?

Keep your samples of vanadium with the four different oxidation numbers for the second part of this experiment.

Procedure: Part 2 Further reactions of vanadium compounds

For the second part of this experiment, try to predict the outcome of mixing the substances listed in the table that follows. Use the standard electrode potentials given in table 6.1 in the *Book of data* where necessary.

Using your samples of vanadium compounds obtained in part 1 of this experiment, try out the various reactions, and interpret the observations that you make. Copy the table into your notebook, and record your observations and comments.

Species to be mixed	Predicted outcome	Observations and comments
VO^{2+} and V^{2+}		
VO_2^+ and V^{3+}		
VO^{2+} and Fe^{3+}		
VO^{2+} and Br^-		
V^{2+} and Cu^{2+}		
V^{3+} and Fe^{3+}		

When making these predictions and interpretations you should bear in mind the appearance of the reactant solutions and that of **all** the products – that is, not only the appearance of the compounds of vanadium but that of any by-products too.

INVESTIGATION 15.3d An investigation of the redox reactions of a transition element

On the laboratory shelves there are likely to be samples of compounds of transition elements of variable oxidation number.

Use the *Book of data* and other reference books to find out in what other oxidation numbers your selected transition element occurs and then carry out an investigation to obtain your selected transition element in its various oxidation numbers.

Make a risk assessment before starting any experiments.

15.4 Complex ion formation

We shall now consider another of the characteristic properties of transition elements, the ability to form complex ions.

A complex is formed when the ion of an element is surrounded by **ligands**. These ligands are either **negatively charged ions**, or **molecules**, and in both cases contain a **lone pair of electrons** which is used to make the bond to the element ion.

Stability constant of copper(II) complexes

The commonest ligand is water, and aqueous solutions of simple compounds of transition elements contain complex ions with formulae such as

$$Cu(H_2O)_4^{2+} \qquad Ni(H_2O)_6^{2+} \qquad Co(H_2O)_6^{2+}$$

> **QUESTION**
> What shapes would you predict for these complex ions?

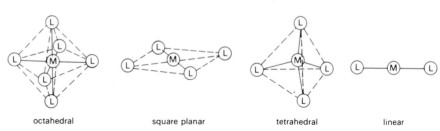

Figure 15.9 Possible shapes of transition metal complexes. M = metal; L = ligand

When a solution containing a different ligand is added to an aqueous solution containing these hydrated ions, an equilibrium is set up in which the water molecules are replaced by the new ligands. For example, in the case of the copper(II) complexes which we shall be investigating in experiment 15.4a, the equilibrium in the presence of chloride ions would be:

$$Cu(H_2O)_4^{2+}(aq) + 4Cl^-(aq) \rightleftharpoons CuCl_4^{2-}(aq) + 4H_2O(l)$$

Application of the equilibrium law gives an equilibrium constant, K, where

$$K = \frac{[CuCl_4^{2-}(aq)]_{eq}}{[Cu(H_2O)_4^{2+}(aq)]_{eq}[Cl^-(aq)]_{eq}^4}$$

$[H_2O(l)]$ is constant and is not included in the expression.

Equilibrium constants such as this are called **stability constants**. They enable us to compare the stabilities of complexes of an element with different ligands; the larger the stability constant, the more stable the complex may be said to be, compared with the water complex.

For convenience, the logarithms of the values of the stability constants are given in the table for various complexes of copper(II):

Ligand		lg K
Cl^-	chloride	5.6
NH_3	ammonia	13.1
(2-hydroxybenzoate structure)	2-hydroxybenzoate	16.9
(1,2-dihydroxybenzene structure)	1,2-dihydroxybenzene	25.0
edta	(ethylenediamine-tetra-acetic acid)	18.8

These stability constants will be referred to in the following experiment. They are 'overall' stability constants for the complete reaction; for example, for the reaction

$$Cu(H_2O)_4^{2+}(aq) + 4Cl^-(aq) \rightleftharpoons CuCl_4^{2-}(aq) + 4H_2O(l)$$

Stability constants can also be found for each stage of the reaction, for example for the reaction

$$Cu(H_2O)_4^{2+}(aq) + Cl^-(aq) \rightleftharpoons Cu(H_2O)_3Cl^+(aq) + H_2O(l)$$

These values are given in table 6.13 in the *Book of data*.

EXPERIMENT 15.4a Some copper(II) complexes

As you write down the answers to the questions in the experiment, you should make it clear what each question was and to which reaction it refers.

Procedure: Part 1

1 Put five or six drops of 0.5 M copper(II) sulphate solution in a test-tube.

■ What complex ion is present? Record its colour.

2 Add ten drops of concentrated hydrochloric acid drop by drop.

■ Record the colour change of the solution. What ligands do you think are now present in the complex ion?

3 Keep half of the solution for **4**. Pour the other half into a test-tube half full of water.

■ What colour is the solution now? What complex ion is now present? In view of what happened in **2** why do you think this reaction occurred?

4 To the solution kept from **3**, add 8 M ammonia solution (TAKE CARE) drop by drop till there is no further colour change. Save this solution for **6**.

SAFETY ⚠
Copper salts are harmful; ammonia is irritant; concentrated hydrochloric acid is corrosive.

- Record the colour of the solution. What ligands to you think are now present in the complex ion?
  What do you think is the order of stability of the complex ions you have seen in these experiments? Compare your order with the data in the table opposite.

5 To 4 or 5 drops of a solution containing $Cu(H_2O)_4^{2+}$(aq) in a test-tube add a solution of the ligand edta until there is no further colour change.

- Record the colour of the edta–Cu(II) complex. Use the stability constants in the table to predict what you think would happen if edta solution were added to the solution obtained in **4**.

6 Test your prediction by adding edta solution drop by drop to the solution obtained in **4** until there is no further colour change.

- Was your prediction in **5** correct?

Procedure: Part 2

You are going to carry out an experiment in which first a solution of ammonia, then one of sodium 2-hydroxybenzoate, then edta, then 1,2-dihydroxybenzene are added in turn to a solution of copper(II) ions. From the stability constants in the table opposite, predict what will happen.

First you need to know the colour of these complexes.

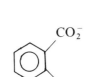

SAFETY ⚠
1,2-dihydroxybenzene is an irritant.

1 Put five or six drops of copper(II) solution in each of four test-tubes. To the first test-tube add edta solution drop by drop until there is no further colour change. To the second similarly add 8 M ammonia solution (TAKE CARE); to the third, sodium 2-hydroxybenzoate solution; and to the fourth, 1,2-dihydroxybenzene solution in 0.5 M sodium hydroxide. Note that the last solution contains sodium hydroxide.

- What are the colours of the four complexes?

2 Now check your predictions by adding 8–10 drops of 8 M ammonia solution (TAKE CARE) to 4–5 drops of copper(II) solution, followed by 10–15 drops of sodium 2-hydroxybenzoate solution, 10–15 drops of edta solution, and 10–15 drops of 1,2-dihydroxybenzene solution in turn.

Add the solutions drop by drop, noting the colours of the mixture.

2-hydroxybenzoate ion

1,2-dihydroxybenzene

edta, ethylenediaminetetra-acetic acid

The 2-hydroxybenzoate ion, 1,2-dihydroxybenzene and edta are **polydentate** ligands. That is, they can form more than one link with the metal ion. Thus in solutions in which there is an excess of ligand present, the predominating species containing the metal are:

$Cu(H_2O)_4^{2+}$ *monodentate* ligands

$Cu(NH_3)_4^{2+}$

$CuCl_4^{2-}$

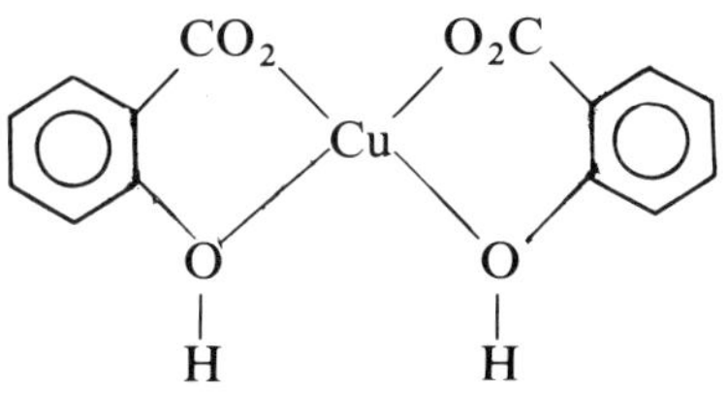

bidentate ligands

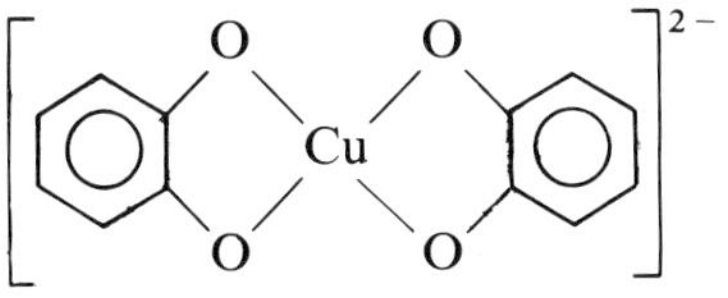

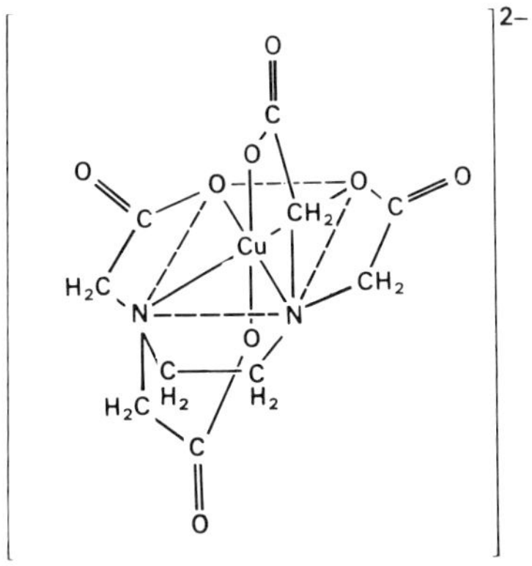

hexadentate ligand

The preparation of compounds containing complex ions

You may have time to carry out the following preparation, or the preparation described in Topic 1 of an iron complex.

EXPERIMENT 15.4b

Stabilizing an unusual oxidation number: chromium(II) ethanoate, $Cr_2(CH_3CO_2)_4(H_2O)_2$

Chromium(II) ethanoate can be regarded as a neutral complex of Cr^{2+} ions with $CH_3CO_2^-$ ions and water molecules as ligands.

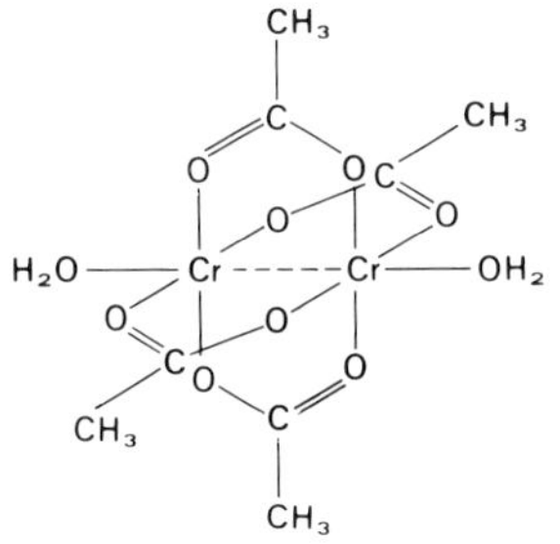

Figure 15.10 The structure of chromium(II) ethanoate.

This compound is interesting because it is an example of the way in which the formation of a complex can sometimes stabilize an oxidation number which would otherwise be unstable. Chromium(II) ions are normally very readily oxidized to chromium(III) by the oxygen of the air.

Write an account of the following method in your notes, and answer the questions at the end, recording your answers in such a way as to make it clear what each question was.

A set of apparatus such as that shown in figure 15.11 can be used, or one which will perform similarly. Because excess hydrogen is needed to force the solution containing the product over into the boiling tube the quantity of zinc and acid you need will depend on the size of your apparatus.

Procedure

Dissolve 1 g of sodium dichromate(VI) in 5 cm³ of water, and put it in the 50 cm³ round-bottomed flask. Add 3 g of granulated zinc. Put about 25 cm³ of approximately 5 M hydrochloric acid in the tap funnel. Put 10 cm³ of saturated sodium ethanoate solution in the boiling-tube. The solubility of sodium ethanoate in water is about 30% by mass.

> **COMMENT**
> Concentrated hydrochloric acid is 10 M; use a mixture of 12.5 cm³ of concentrated hydrochloric acid with 12.5 cm³ of water.

> **SAFETY** ⚠
> Sodium dichromate(VI) is highly irritant to skin, eyes, and respiratory system. Eye protection is essential and you must be careful not to raise any dust when handling the compound. As hydrogen is evolved, naked flames must be kept well clear. Hydrochloric acid is corrosive.

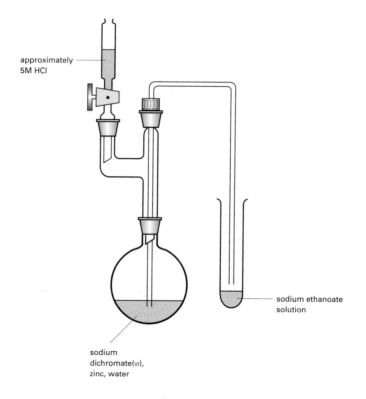

Figure 15.11 Apparatus suitable for the preparation of chromium(II) ethanoate.

> **COMMENT**
> Cr^{3+}(aq) is green and Cr^{2+}(aq) is blue.

Add the hydrochloric acid to the mixture in the flask and leave the tap funnel OPEN to allow the hydrogen which is generated to escape. The reduction of the chromium passes through a green stage and is complete when the solution is blue.

While hydrogen is still being generated close the tap on the funnel so that the pressure of the hydrogen forces the blue solution over into the saturated sodium ethanoate. A red precipitate of chromium(II) ethanoate should be formed, and the solution will contain dissolved red chromium(II) ethanoate.

Dismantle the apparatus and pour the remaining blue solution containing Cr^{2+}(aq) into another boiling-tube. Keep the two boiling-tubes side by side in a rack.

QUESTIONS

1 Show, using electrode potentials, that zinc should reduce chromium from +6 in sodium dichromate(VI) to +2 in chromium(II) ethanoate.
2 What happens to the colour of the Cr^{2+}(aq) solution when it is allowed to stand open to the air?
3 Over the same period of time does the colour of the chromium(II) ethanoate solution also change?
4 Suggest how the complex ion stabilizes chromium(II) to oxidation by the air.

Entropy considerations

When bidentate ligands replace monodentate ligands in a complex, there will be an increase in the entropy of the system because one molecule of ligand is replacing two molecules. See the following example:

$$Ni(NH_3)_6^{2+}(aq) + 3\underset{\text{1,2-diaminoethane}}{NH_2CH_2CH_2NH_2(aq)}$$
$$\rightleftharpoons Ni(NH_2CH_2CH_2NH_2)_3^{2+}(aq) + 6\underset{\text{ammonia}}{NH_3(aq)}$$

In this example there are four particles on the left of the equation but seven particles on the right. Because the entropy of the system depends, amongst other things, on the number of particles present, S_{system} increases when 1,2-diaminoethane molecules replace ammonia molecules.

> **COMMENT**
> ΔS_{total} must always be positive if a change is to occur spontaneously; there are two entropy changes to consider, ΔS_{system} and $\Delta S_{surroundings}$ connected by the expression
> $$\Delta S_{total} = \Delta S_{system} + \Delta S_{surroundings}$$

A similar, but larger, increase of S_{system} takes place when edta replaces monodentate or bidentate ligands in a complex, as, for example

$$Ni(NH_3)_6^{2+}(aq) + edta(aq) \rightleftharpoons Ni(edta)^{2+}(aq) + 6NH_3(aq)$$

Comparison of the numbers of particles involved shows a large increase, corresponding to an increase of S_{system}.

Because of this entropy-increasing effect, complexes with a hexadentate ligand such as edta are usually much more stable than those with a monodentate ligand. A good example of this effect is shown by the stability constants of nickel complexes.

In these complexes the ligands are bonded to the nickel ion by means of lone pairs of electrons on oxygen atoms in water, nitrogen atoms in ammonia, and both oxygen and nitrogen atoms in edta. Ignoring the entropy effect, therefore, the edta complex might be expected to be comparable in stability with the ammonia complex. It is the increase in S_{system} accompanying its formation which accounts for the high stability of the edta complex compared with the other two.

Complex	lg K
$Ni(H_2O)_6^{2+}$	0
$Ni(NH_3)_6^{2+}$	8.01
$Ni(edta)^{2+}$	18.6

INVESTIGATION 15.4c

The formula of a complex

Carry out an investigation to determine the formula of a complex ion. The complex formed between aqueous nickel(II) ions and edta is a suitable complex to investigate. Make a risk assessment before starting any experiments.

15.5 Transition elements as catalysts

The essential feature of a catalyst is that it increases the rate of a chemical reaction without itself becoming permanently involved in the reaction. It does, however, become temporarily involved, by providing a route from reactants to products which has a lower activation energy. (See section 8.5.)

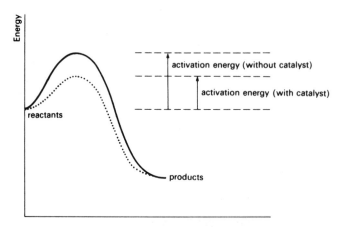

Figure 15.12 The effect on activation energy of a catalyst

The following experiment is about catalysis, though it treats it in an unusual way. Carry out the instructions and try to answer the questions which follow.

EXPERIMENT 15.5

A kinetic study of the reaction between manganate(VII) ions and ethanedioic acid

Potassium manganate(VII) will oxidize ethanedioic acid (oxalic acid) to carbon dioxide and water, in the presence of an excess of acid:

$$2MnO_4^-(aq) + 6H^+(aq) + 5(CO_2H)_2(aq) \longrightarrow 2Mn^{2+}(aq) + 10CO_2(g) + 8H_2O(l)$$

Procedure

1 Carry out a trial experiment by mixing one tenth quantities of Mixtures 1 and 2, but without adding any water. Record the colour changes you see on adding 5 cm³ of 0.02 M potassium manganate(VII).

> **SAFETY** ⚠
> Ethanedioic acid (oxalic acid) is poisonous.

Solution	Mixture 1	Mixture 2
0.2 M ethanedioic acid	100 cm³	100 cm³
0.2 M manganese(II) sulphate	—	15 cm³
1 M sulphuric acid	10 cm³	10 cm³
Water	90 cm³	75 cm³

2 Prepare a reaction mixture according to the table, using measuring cylinders. Some members of your group should use Mixture 1 and some Mixture 2. The results should then be shared.

3 Add 50 cm³ of 0.02 M potassium manganate(VII) and start timing. Shake the mixture for about half a minute to mix it well.

4 After about a minute use a pipette and safety pipette filler to withdraw a 10.0 cm³ portion of the reaction mixture and run it into a conical flask.

5 Note the time and add about 10 cm³ of 0.1 M potassium iodide solution. This stops the reaction and releases iodine equivalent to the residual manganate(VII) ions. Titrate the liberated iodine with 0.01 M sodium thiosulphate, adding a little starch solution near the end point. Record the titre of sodium thiosulphate.

6 Remove further portions every 3 or 4 minutes and titrate them in the same way. Continue until the titre is less than 3 cm³.

QUESTIONS

1 What set of figures gives a measure of the reactant concentration at the various time intervals?

2 Plot an appropriate graph for each experiment. Refer to Topic 8 if you need help.

3 Try to answer these questions about the graphs:

a How does the rate of reaction vary with time?

b What is unusual about the graph for Mixture 1? What explanation can you offer for this abnormal behaviour?

c Can you suggest an experiment that would test your explanation?

Homogeneous catalysis

Although heterogeneous catalysis is of more widespread significance commercially, homogeneous catalysis is important too and often involves transition element ions, because of their ability to change oxidation number readily. It is sometimes possible to identify a possible catalyst by using electrode potentials. Consider the electrode potential chart in figure 15.13.

COMMENT
You will remember from Topic 8 that catalysts can be of two types, *heterogeneous* (described in section 8.5) and *homogeneous*.

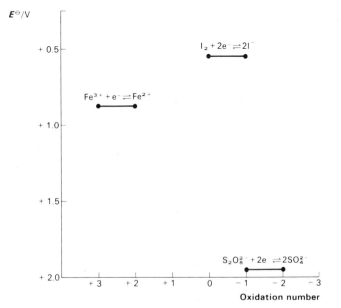

Figure 15.13

Peroxodisulphate ions, $S_2O_8^{2-}$(aq), are capable of oxidizing iodide ions, I^-(aq), to iodine, but the reaction is a very slow one.

$$S_2O_8^{2-}(aq) + 2I^-(aq) \longrightarrow 2SO_4^{2-}(aq) + I_2(aq)$$

If iron(II) ions are added to the mixture, these could be oxidized by the peroxodisulphate ions rather more quickly (perhaps because the negative peroxodisulphate ions would then be reacting with positive ions instead of with negative ones). The resulting iron(III) ions could oxidize iodide ions to iodine and the iron(II) ions would be re-formed.

Clearly, in order for this catalysis to work, the electrode potential for the reaction involving the catalyst must lie between the electrode potentials involving the two reactants.

This predictive method only shows that catalysis should be possible. It does not guarantee that an improvement in rate of reaction will actually take place.

STUDY TASK

Throughout this course there are a number of references to catalysis and you should now attempt to make brief notes on all the different situations in which catalysis is important. Apart from the account given in this section, there are references to catalysis in the following sections:

> section 7.6 (catalysis in the polymerization of alkenes)
> section 8.5 (heterogeneous catalysis)
> section 11.2 (ester hydrolysis)
> section 14.4 (catalysis involving enzymes)

15.6 A study of the chlorides of iron

You are going to react chlorine gas, Cl_2(g), or hydrogen chloride gas, HCl(g), with iron in order to examine the properties of the iron chlorides produced in the reactions.

Iron can form two chlorides, iron(II) chloride and iron(III) chloride. You should be able to find out whether the same chloride is formed in both reactions, or different chlorides are formed. You should also be able to form an opinion as to whether your products are ionic or covalent.

This is an opportunity for you to gain experience of apparatus suitable for reacting gases with solids.

EXPERIMENT 15.6a Preparation of the chlorides of iron

To do this reaction properly you need to exclude air from the apparatus, otherwise the iron may react with oxygen. You also need to exclude moisture as it may react with your product.

Procedure

A suitable apparatus is shown in figure 15.14: the glass joints are gas tight when properly assembled but use rubber bungs and short lengths of plastic tubing in the rest of the apparatus because of the reactivity of the gases you are using.

15 The transition elements

Iron is available as a powder, as coarse filings, as steel wool (impure) or wire. For this experiment iron wire, with its large surface area, is best so that the gases can easily reach all the iron: clean two and a half metres of fine iron wire (0.25 mm diameter) by pulling it through a folded piece of fine emery paper. Using bright iron wire is especially important with hydrogen chloride gas as any surface film of oxide inhibits reaction. Check that your length of wire weighs about 1 g.

Now assemble your combustion tube. The iron wire does not get very hot except where it touches the glass.

Fill your drying tube with roughly 3 g of anhydrous calcium chloride granules. Calcium chloride has a large capacity for absorbing moisture.

Now follow the instructions for **either** chlorine **or** hydrogen chloride; **take care not to muddle the reagents**!

1 Reacting iron with chlorine gas

> **QUESTION**
> Can you think of another way of heating the iron wire?

$$CaCl_2(s) + 6H_2O(g) \longrightarrow CaCl_2 \cdot 6H_2O(s)$$

> **SAFETY** ⚠
> Bleaching powder, hydrochloric acid and sodium hydroxide are all corrosive; chlorine is toxic: it destroys lung tissue and must not be breathed; calcium chloride is an irritant.

> **COMMENT**
> Concentrated hydrochloric acid, HCl(aq), is 10 M. You therefore need 5 cm³ of concentrated acid plus 5 cm³ of water to make 10 cm³ of 5 M hydrochloric acid.

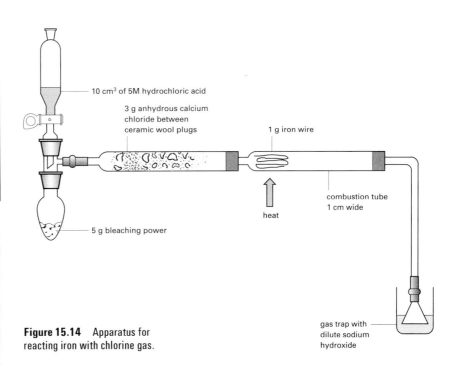

Figure 15.14 Apparatus for reacting iron with chlorine gas.

Put 5 g of bleaching powder, mainly calcium chlorate(I), $Ca(OCl)_2$, with 65% available chlorine, in the pear-shaped flask and moisten with 2 cm³ of water. Fully assemble your apparatus using the diagram as a guide. When everything is firm and stable add sodium hydroxide solution to the gas trap until the whole rim of the funnel is **well under** the surface of the alkali. Make sure the tap is closed, then put 10 cm³ of 5 M hydrochloric acid in the tap funnel.

Now read these instructions right through to the end before attempting to start the reaction.

Allow the hydrochloric acid to drip slowly into the pear-shaped flask. If necessary rock the apparatus **very** gently to mix the reagents. Chlorine gas should be evolved readily and air will bubble out of the gas trap. The sodium hydroxide solution will react with the excess of chlorine and minimize the escape of this toxic gas.

COMMENT

Swimming pools contain only about 1.5 parts **per million** of free chlorine so be careful that no chlorine escapes.

The reaction in the gas trap is

$2NaOH(aq) + Cl_2(g) \longrightarrow NaOCl(aq) + NaCl(aq) + H_2O(l)$

When the bubbling of air stops or slows down at the gas trap heat the iron wire. Make sure that the hydrochloric acid is dripping slowly and steadily into the pear-shaped flask.

Stop heating when the iron wire bursts into flame but heat again if the reaction slows down. When all the hydrochloric acid has been added leave the tap OPEN as this will prevent alkali flowing back into the combustion tube.

- What could you see happening, if anything, in the pear-shaped flask, in the drying tube and in the gas trap?
  Have there been any exothermic or endothermic changes?
  Also, write an account of what happened in the combustion tube. Do this while the apparatus is cooling down.

When the apparatus is cool disconnect the combustion tube (TAKE CARE, the apparatus will still be full of chlorine gas) and dismantle the rest of the apparatus in a fume cupboard. Collect your product in a petri dish. Scrape out any product with a metal spatula and hook out any unreacted iron wire.

The amount of product should be enough for the next set of experiments, even though some material may have escaped into the delivery tube and gas trap.

2 Reacting iron with hydrogen chloride gas

SAFETY ⚠
Concentrated sulphuric acid, concentrated hydrochloric acid and hydrogen chloride gas are all corrosive.

COMMENT
Concentrated hydrochloric acid, HCl(aq), is 10 M. You therefore need 3 cm³ of concentrated acid plus 3 cm³ of water to make 6 cm³ of 5 M hydrochloric acid.

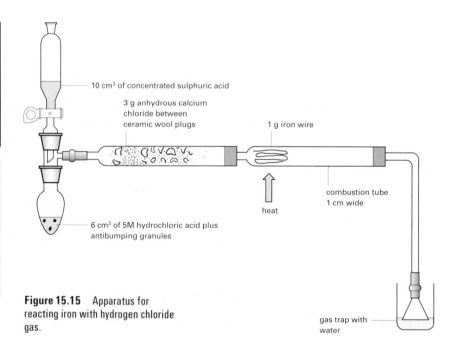

Figure 15.15 Apparatus for reacting iron with hydrogen chloride gas.

COMMENT
1 cm³ of water will dissolve 500 cm³ hydrogen chloride gas at 0 °C.

Put a few pieces of pumice stone and 6 cm³ of 5 M hydrochloric acid in the pear-shaped flask.

Fully assemble your apparatus using the diagram as a guide. When everything is firm and stable add water to the gas trap until the whole rim of the funnel is **just touching** the surface of the water. This will absorb any excess hydrogen chloride gas you produce but you must avoid any risk of water rushing back into the hot combustion tube.

Make sure the tap is closed, then put 10 cm³ of concentrated sulphuric acid in the tap funnel.

Now read these instructions right through to the end before attempting to start the reaction.

Allow the concentrated sulphuric acid to drip slowly into the pear-shaped flask. With some pumice stone in the flask hydrogen chloride gas should be evolved readily and air will bubble out of the gas trap. When the bubbles of air stop or slow down heat the iron wire. Check that the sulphuric acid is dripping steadily into the pear-shaped flask and keeping up the flow of hydrogen chloride gas.

Stop heating when all the sulphuric acid has run into the flask: leave the tap OPEN as this will reduce the risk of water rushing back from the gas trap into the combustion tube. To be safe remove the beaker that formed the gas trap.

- What could you see happening, if anything, in the pear-shaped flask, in the drying tube and in the gas trap?
  Have there been any exothermic or endothermic changes?
  In your notes, write an account of what happened in the combustion tube. Do this while the apparatus is cooling down.

When the apparatus is cool disconnect the combustion tube and move the rest of the apparatus to a fume cupboard. Collect your product in a petri dish. Scrape out any product with a metal spatula and hook out any unreacted iron wire.

The yield is often poor but you should have enough for the next set of experiments.

EXPERIMENT 15.6b

The properties of the chlorides of iron

Copy the headings below into your notebook and carry out the tests on your product from experiment 15.6a. In order to complete the table exchange some product, or results, with students who reacted iron with the other gas.

Notes:
(i) Is the chloride crystalline?
(ii) Leave in a covered petri dish for three days. The words hygroscopic, deliquescent or efflorescent will be needed to describe the results. Look up their meanings in a dictionary if necessary.
(iii) These experiments should enable you to decide which chloride is an iron(II) and which an iron(III) compound.

Procedure

Test	Result with the chloride of iron from $HCl(g)$	Result with the chloride of iron from $Cl_2(g)$
1a Colour and appearance (i)		
b Behaviour in moist air (ii)		
c Solubility in hydrocarbon solvent		

2 Prepare a solution in 10 cm³ of pure water, and divide into three portions (iii).
a Add silver nitrate solution, $AgNO_3(aq)$ to one portion.
b Add sodium hydroxide solution, $NaOH(aq)$ to another.
c Add potassium thiocyanate solution, $KCNS(aq)$ to the third.

STUDY TASK

Complete your study of the chlorides of iron by using the *Book of data* to make a table comparing and contrasting the properties of iron(II) and iron(III) chloride. You should record T_m, T_b, ΔH_f, $S^\ominus$ and ΔG_f.

Write an interpretation of your results, and justify your answers to the following questions:

1 Do the chlorides have the usual properties of salts?
2 Do they seem to be ionic or covalent in structure?
3 Do the energy changes suggest the appropriate products have been formed in each reaction? Why do both reactions not produce iron(III) chloride?

To answer these questions first construct a balanced energy cycle for the reactions involved:

$$Fe(s) + 2HCl(g) \longrightarrow FeCl_2(s) + H_2(g) \qquad \text{i}$$

$$FeCl_2(s) + HCl(g) \longrightarrow FeCl_3(s) + \tfrac{1}{2}H_2(g) \qquad \text{ii}$$

$$Fe(s) + 1\tfrac{1}{2}Cl_2(g) \longrightarrow FeCl_3(s) \qquad \text{iii}$$

Then look up the standard Gibbs free energy changes of formation at 298 K:

$$Fe(s) + Cl_2(g) \longrightarrow FeCl_2(s)$$

$$Fe(s) + 1\tfrac{1}{2}Cl_2(g) \longrightarrow FeCl_3(s)$$

$$\tfrac{1}{2}H_2(g) + \tfrac{1}{2}Cl_2 \longrightarrow HCl(g)$$

Now use the energy cycle to calculate $\Delta G^\ominus_{reaction}[\text{i}]$ and $\Delta G^\ominus_{reaction}[\text{ii}]$.

What conclusions can you draw from your calculations?

15.7 The special properties of the transition elements

Write a general account of transition element chemistry in your notes, using the outline below as a basis. You should begin by explaining what a transition element is, and then list the characteristic properties with well selected examples in each case. Use the results from your experiments, together with the *Book of data*, reference books and textbooks as sources of examples.

The special properties regarded as typical of transition elements are as follows.

1   Similarity of physical properties

The physical properties of the transition elements, which are all metals, show very little variation across the row. Such properties include melting point, boiling point, density, and first ionization energy.

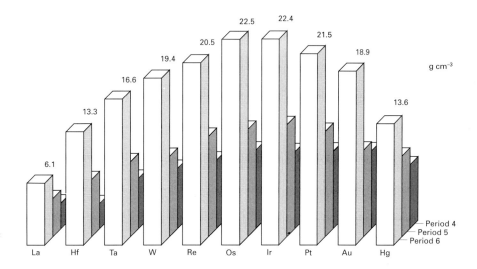

Figure 15.16 Variation in density of the transition elements.

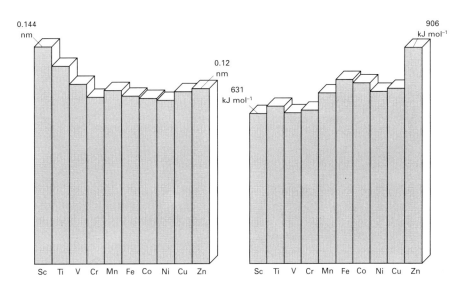

Figure 15.17 Variation in atomic radius of the first row transition elements.

Figure 15.18 Variation in first ionization energy of the first row transition elements.

Include in your survey data from table 7.2 in the *Book of data* on crystal structure and electrical resistivity. Notice that there is no obvious pattern in these data.

2 Variable oxidation number

Most of the transition elements show a range of oxidation numbers in their compounds. To review how these numbers vary in the first-row transition elements, look again at the chart in section 15.3 of the oxidation numbers known to exist, with the more common ones marked.

Find examples in the *Book of data* of compounds with the more common oxidation numbers, and add any uncommon examples that you met as you worked through this Topic.

A novel application of redox equilibria is in **photochromic glass**. The photochromic glass contains tiny crystals of silver and copper halides: in sunlight the high energy ultraviolet light causes clusters of silver atoms to form and the glass darkens. In daylight with reduced ultraviolet light the reaction is reversed and the glass lightens

$$\underset{\text{clear glass}}{Cu^+(s) + Ag^+(s)} \rightleftharpoons \underset{\text{dark glass}}{Cu^{2+}(s) + Ag(s)}$$

One type of this glass is used in sunglasses.

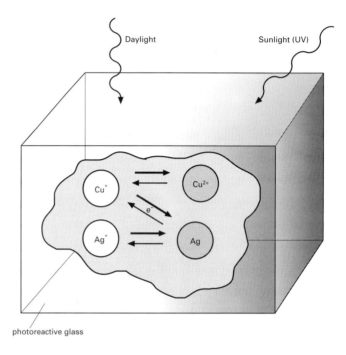

Figure 15.19 Photochromic sunglasses.

3 Ability to form complex ions

Ions of the transition elements, and sometimes the atoms themselves, can be surrounded by, and bonded to, a number of molecules or ions called **ligands**. The result is a molecule or ion called a **complex**. You looked at complex ions in section 15.4.

These complexes generally have structures in which the ligand molecules or ions are arranged around the ion or atom of the transition element in one of the ways shown in figure 15.9.

4 Colour

Many of the compounds of the transition elements are coloured, both in the solid state and in solution. The phenomenon is not, of course, confined to the compounds of the transition elements but it is relatively rare for **metal** ions outside the transition series to form coloured solutions.

5 Catalytic activity

Many slow reactions are accelerated by the presence of transition elements or their ions. This subject was treated in section 15.5.

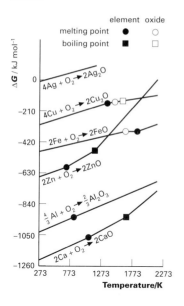

Figure 15.20 An Ellingham diagram.

6 Energy changes

Energy changes in the redox reactions of the transition metals can be recorded as enthalpy changes, e.m.f.s of cells or Gibbs free energy changes depending on which is most suitable. For reactions between solids at high temperatures diagrams based on Gibbs free energy changes are often particularly useful. Such diagrams are known as Ellingham diagrams and one applicable to Thermit type reactions is shown in figure 15.20.

QUESTION

Remembering that reactions only 'go' when ΔG is negative, answer the following questions:

1 What is the approximate value of the Gibbs free energy change for the Thermit reaction between aluminium and iron(II) oxide at the melting temperature of aluminium?
2 List two metal-metal oxide pairs that will react at 1000 K, and two other pairs that will not react at 1000 K.
3 At what temperature does copper reduce zinc oxide? What changes do you think might be responsible for this reaction being reversed at lower temperatures?

15.8 Study task: Micronutrients

QUESTION

Read the passage below, and then summarize it in the form of a table listing six metals, why they are needed in plants and what deficiency diseases occur in animals if they are missing from their diet. Iron has been filled in for you as an example.

Element	Why is it needed in plants?	What is the deficiency disease in animals?
Iron, Fe	to make chlorophyll	anaemia

Plants take up many elements from the soil in the form of soluble compounds. The best known of these are nitrogen, phosphorus, and potassium, all usually abundant in soil and the common stuff of ordinary fertilizers. They are known as **macronutrients** because of the large quantities needed for normal growth. Other elements are required in smaller amounts, and are therefore known as **micronutrients**. Many, though not all, micronutrients are transition elements. Micronutrients are sometimes referred to as trace elements, but this term is misleading because substantially more than a trace is needed of some of them. The table shows the average uptake per hectare of some elements needed by barley to yield 5 tonnes of grain and also to produce 5 tonnes of straw, different amounts of nutrients being needed for the seed and for the leafy parts of the plant.

Produce /5 t ha^{-1}	Uptake in kg ha^{-1}						Na	Fe	Uptake in g ha^{-1}				
	N	P	K	Ca	Mg	S			Mn	Zn	Cu	B	Mo
Grain	85	15	25	4	6	25–7	1	2	100	200	50	30	1
Straw	25	2.5	4	1.5	2.5	10–6	2.5	2	350	200	35	30	1

The first six elements are classed as macronutrients and the last five as micronutrients, sodium and iron falling between the groups.

Other plants need different amounts and some crops need further elements such as cobalt and chlorine. In addition, plants take up elements which, although they do not seem to be needed for the growth of the plant, are vitally necessary for animals feeding on that plant. For example, mammals need all the elements so far mentioned except (as far as is known) boron, plus iodine and for some animals selenium and chromium.

Complete absence of micronutrients prevents any plant growth at all. Shortages produce stunted plants and symptoms such as chlorosis, yellowing of leaves through the absence of chlorophyll; and necrosis, decay of tissues. Animals, including humans, nourished by such plants may themselves suffer from deficiency conditions, often grave.

The position can be dramatically reversed by making good any deficiency. In Australia land which was once desert now grows crops of legumes, thanks to the application of as little as 70 grams of molybdenum per hectare, as sodium molybdate(VI). Even more striking has been the wiping out of iodine deficiency diseases among humans living in regions lacking that element.

Soil dressing with micronutrients needs to be done only once or at most very occasionally. Excessive intake of some of these elements can cause poisoning in humans and animals.

The study of micronutrients depends on analytical techniques sensitive enough to detect, say, 1 gram of molybdenum in 5 tonnes of grain. It is hard to establish whether an element is essential to a plant in conditions resembling its natural habitat, since most of the stable elements are present, however minutely, in the rocks which weather to form soil. Nor is it easy to create an artificial growing medium completely lacking micronutrients.

Let us now consider the role of some transition metal micronutrients in more detail.

Iron

Iron was the first non-major nutrient discovered to be essential for plant growth. It is necessary for the production of chlorophyll although not present in it; deficiency results in chlorosis. In turn, uptake of iron is blocked by excessive alkalinity in the soil, a condition which affects about a third of the World's land surface. In Britain iron deficiency is a common problem in fruit trees on chalk soils.

Many animals including humans need iron for the blood protein haemoglobin: 60 to 70% of the iron in the human body is found here, most of the remainder in other proteins.

Haemoglobin contains only about 0.34% iron by mass, but this iron is the key to the blood's transport of oxygen. When red blood corpuscles come to the end of their usefulness and are broken up, their haemoglobin is destroyed but about 90% of the iron is liberated and reused. Therefore the daily iron requirement of a healthy animal is small. Deficiency causes anaemia, which is most often seen in rapidly growing sucklings since milk contains little iron. Piglets raised in pens without access to soil or pasture often suffer from anaemia. They need about 7 mg of iron a day of which only 1 mg comes from sow's milk. The rest must be provided by dosing or injection.

Manganese

Manganese is needed in several enzyme systems, for example by plants in the reduction of nitrates to amino acids. Uptake of manganese from the soil is linked to that of iron, and is affected by soil pH. In alkaline soils little manganese enters the plant. In highly acid soils too much is taken up, blocking the absorption of iron. Manganese deficiency in plants tends to occur on naturally acid soils which have been limed to over pH 7. The usual symptom is chlorosis. In cabbages severe deficiency causes the leaves to become almost bleached except for the veins which remain green.

Manganese deficiency in animals causes sterility and crippling deformities of the skeleton. However, the element is widely found in food plants. Most pastures contain 40 to 200 mg of manganese per kilogram of dry matter. There is a wide margin of safety. Hens have been given feed with levels as high as 1 g kg^{-1} dry matter without apparent ill effects.

Zinc

Zinc is used by plants in the production of both carbohydrates and fats. Plants lacking zinc accumulate high levels of phenolic compounds such as 1,2-dihydroxybenzene (catechol) in their leaf cells at the expense of substances normally present. Again, the most usual visible symptom is yellowing of the leaves between the leaf veins.

In animals zinc is found in every tissue. Like many micronutrients it is stored mainly in the liver.

Plants vary greatly in their ability to extract zinc from the soil. In Florida, USA, where much of the land is low in zinc, native weed species are particularly efficient extractors compared with most crop plants. Crops can be successfully grown by first planting lucerne (alfalfa), which is also a good extractor, and ploughing it in as a green manure.

Most food plants provide adequate zinc. Yeast, too, is a rich source. Most animals can tolerate high levels of zinc, though a few cases of zinc poisoning have been reported.

Figure 15.21 Round leaved pennycress extracts zinc from the soil. 15% of its ash can be zinc.

Copper

Copper is essential not only to plants, but also to animals in many bodily processes. Low levels of copper restrict the growth of plants; very low levels prevent any growth. There is little copper in peaty heathlands such as those of East Anglia, or in recently ploughed old grassland on chalk soils as in Wiltshire. Deficiency symptoms are different in each place, but in both small dressings of copper compounds give striking improvements in crop yields.

It has been known that copper is needed in the diet of animals since 1924, when experiments with rats showed that it has a role in haemoglobin formation. In some molluscs such as snails oxygen is carried by a blue substance, haemocyanin, in which copper plays a role like that of iron in haemoglobin.

Copper is a constituent of several enzymes including ascorbic acid oxidase, and lactase. Another copper-containing enzyme catalyses the change of the amino acid lysine to a protein that provides cross-links in the main protein in the wall of the aorta and other blood vessels, to which they owe their elasticity and toughness. Copper has been found to influence the rate of growth of pigs and other animals.

Deficiency of copper leads to anaemia and bone disorders. Low copper levels of 2–4 mg kg^{-1} dry matter in some Australian pastures have caused lambs to suffer from a nervous disorder involving lack of coordination or even complete inability to stand up. A British term for a similar disorder of lambs is swayback. In Somerset, soils locally called 'teart', found on the alkaline Lias Clay, have unusually high levels of molybdenum which give pasture a level of 20–100 mg kg^{-1} dry matter compared with 3–5 mg on normal soils. The excess of this element reduces the availability of copper, so that cattle suffer from scouring, a severe diarrhoea. Horses and pigs are not usually affected.

Copper accumulates in the liver. Too much can be poisonous. Sheep are particularly susceptible, so much so that dosing with copper compounds to balance a copper deficiency in pasture can easily kill them. Chronic copper poisoning has occurred naturally among sheep in parts of Australia where the copper content of pasture is high.

Molybdenum

Molybdenum's function in plants appears to be in the reduction of nitrates to amino acids, for lack of it causes nitrate ions to accumulate in the plants. Where nitrogen is supplied as ammonium rather than nitrate ions, much less molybdenum is needed. Legumes require molybdenum to allow their root nodules to fix atmospheric nitrogen. General signs of deficiency in plants are pale or decayed areas in the leaves.

Molybdenum is also needed by animals. It is a constituent of the enzyme xanthine oxidase which plays a part in the metabolism of two of the four bases of the DNA code. In practice the weight of young lambs has been increased by adding molybdenum to a diet known to be low in the element. It is thought to have worked by stimulating the breakdown of cellulose by micro-organisms in the rumen, part of the animal's complex series of 'stomachs'.

Cobalt

Cobalt is not essential to the higher plants, but is certainly needed by at least some animals including sheep. A wasting disease known as pining which affects sheep grazing on pasture on the granitic soils of the West of England has been traced to lack of cobalt in the soil and plants. Pining has also been observed in New Zealand, Australia, and the USA. The condition is due to lack of cobalt preventing micro-organisms in the rumen from synthesizing vitamin B_{12} (a compound containing cobalt). It is corrected by implanting a slow-release cobalt bullet (90% cobalt(III) oxide) in the reticulum (another stomach). Vitamin B_{12} is needed in our diet to prevent pernicious anaemia.

Cobalt can be toxic but safety margins are high, for an excess is soon excreted. In cattle no harm is done unless intake exceeds 9–11 mg cobalt per 10 kg body mass daily.

Summary

At the end of this Topic you should be able to:
a demonstrate understanding of the terms: transition element; complex, ligand, stability constant; homogeneous catalysis
b recall the characteristic properties of transition elements limited to:
   typical physical properties
   variable oxidation number
   complex formation
   coloured compounds
   catalytic activity
c interpret simple structures presented in diagrammatic form, including recalling the three common types of crystal structures in metals
d construct electrode potential/oxidation number charts and use these to predict the feasibility of redox reactions
e plan an investigation using potassium manganate(VII) as titrant, including calculation of the results and justification of the procedures involved
f identify for complex compounds
   i   their relative stability by reference to stability constants (qualitative only)
   ii   their ligands as monodentate, bidentate or hexadentate
   iii from diagrams, their shape as linear, tetrahedral, square planar or octahedral
   iv   their relative stability in terms of a predicted entropy change in ligand replacement reactions at the same central atom (qualitative only)
g demonstrate understanding of methods of preparation of complex compounds
h interpret the catalytic behaviour of transition elements and their compounds in terms of alternative reaction pathways in homogeneous catalysis
i evaluate information by extraction from text and the *Book of data* about the transition elements as micronutrients.

Review questions

*Indicates that the *Book of data* is needed.

*15.1 This question concerns the following electrode potentials for manganese:

I $MnO_4^- + e^- \rightleftharpoons MnO_4^{2-}$ $E^\ominus = +0.56$ V

II $MnO_4^{2-} + 2H_2O + 2e^- \rightleftharpoons MnO_2 + 4OH^-$ $E^\ominus = +0.60$ V

III $MnO_2 + 4H^+ + e^- \rightleftharpoons Mn^{3+} + 2H_2O$ $E^\ominus = +0.95$ V

IV $Mn^{3+} + e^- \rightleftharpoons Mn^{2+}$ $E^\ominus = +1.15$ V

V $MnO_4^- + 8H^+ + 5e^- \rightleftharpoons Mn^{2+} + 4H_2O$ $E^\ominus = +1.51$ V

a Referring to these equilibria by their numbers (I–V), which of these electrode potentials are independent of pH?
b Explain, using the electrode potentials, why a neutral solution of the green manganate(VI) ion changes, on standing, to a purple solution and a black precipitate.

What name is given to a reaction like this in which an element changes both upwards and downwards in oxidation number?

c What would you expect to happen to the red ion Mn^{3+} in neutral solution?

d Solutions of potassium manganate(VII) can be used to estimate the concentrations of other ions by titration.

i Which two of the following ions cannot, for reasons related to electrode potentials, be estimated using potassium manganate(VII)?

$$Sn^{2+}, Zn^{2+}, Co^{2+}, Tl^{+}$$

Justify your answer.

ii Why might it be very difficult to estimate the concentration of $Cr^{2+}(aq)$ accurately by titration with potassium manganate(VII)?

iii What colour change would mark the end point if potassium manganate(VII) were used to titrate a solution containing the ion $V^{3+}(aq)$?

*15.2 2.41 g of a salt containing iron(III) ions were dissolved in dilute sulphuric acid and zinc was added to reduce the $Fe^{3+}(aq)$ to $Fe^{2+}(aq)$. The resulting solution was made up to 100 cm³ with dilute sulphuric acid and 10.0 cm³ portions were titrated with potassium manganate(VII). Exactly 10.0 cm³ of 0.0200 M $KMnO_4$ were required.

a Show by using electrode potentials that zinc should reduce $Fe^{3+}(aq)$ to $Fe^{2+}(aq)$.

b Why was the solution made up to 100 cm³ by using dilute sulphuric acid rather than water?

c What is the percentage of iron by mass in the original salt?

*15.3 When iron(III) ions, $Fe^{3+}(aq)$, are added to iodide ions, $I^-(aq)$, iodine is produced, but if sodium fluoride is first added, the reaction does not take place. The complexes FeF_6^{3-} and FeF_6^{4-} both exist.

a The atomic number of iron is 26. State the electronic configuration of the ion Fe^{3+}.

b Sketch the shape you would expect the complex FeF_6^{3-} to have.

c What is the oxidation number of iron in FeF_6^{4-}?

d Show by means of electrode potentials that iron(III) ions should oxidize iodide ions.

15.4 The stability constants for some octahedral cobalt(II) complexes are listed in the table.

Ligand	lg K(overall)	Colour
Cl^-	(no complex)	
H_2O	—	pink
NH_3	4.39	green
edta	16.3	pink

a What would you expect to see happen when ammonia solution is added to a solution of cobalt(II) chloride?

b What would you expect to see happen when the solution resulting from **a** is added to hydrochloric acid? Give a reason for your answer.

c What would you expect to see happen when edta solution is added to a solution of cobalt(II) chloride? Give a reason for your answer.

d Complexing by edta involves nitrogen atoms and oxygen atoms such as those in ammonia and water, so why is the stability constant of Co(edta) so much larger than that of $Co(NH_3)_6^{2+}$?

15.5 The equation and data for the formation of the diamminosilver ion are

$$Ag^+(aq) + 2NH_3(aq) \longrightarrow [Ag(NH_3)_2]^+(aq)$$

$$\Delta G_f^\ominus(298)[Ag^+(aq)] = +77.1 \text{ kJ mol}^{-1}$$

$$\Delta G_f^\ominus(298)[NH_3(aq)] = +26.6 \text{ kJ mol}^{-1}$$

$$\Delta G_f^\ominus(298)[Ag(NH_3)_2]^+(aq)] = -17.2 \text{ kJ mol}^{-1}$$

a Calculate $\Delta G_f^\ominus(298)$ for this reaction.

b Does your value suggest the equilibrium is in favour of the formation of the complex?

c Can you recall any experimental evidence for **b**?

Examination questions

*15.6 a Construct a list of oxidation numbers found in manganese compounds, based on a representative range of ions and compounds that contain manganese, and including manganese itself.

Using table 6.1 of standard electrode potentials in the *Book of data*, display the possible half-reactions of these oxidation states of manganese in acidic solution at 298 K on a diagram of $E^\ominus$ against oxidation number. You are recommended to use graph paper for this diagram.

b Chromium has two common oxidation states in its compounds, $+3$ and $+6$. The former may be obtained in aqueous solution as $Cr^{3+}(aq)$ ions, and the latter in acid solution as $Cr_2O_7^{2-}(aq)$ ions. Add the data for possible half-reactions for chromium and these chromium ions to your diagram in **a**.

Predict the possible sequence of reactions that could occur in **acidic** conditions starting with manganese metal and excess 1 M sodium dichromate(VI). Support your prediction with equations and $E^\ominus$ values for the stages in the reaction sequence, and describe some of the colour changes you would expect to observe.

c Indicate briefly the relative merits of potassium manganate(VII) and potassium dichromate(VI) as oxidizing agents for practical use in the oxidation of $Ce^{4+}(aq)$ to $Ce^{3+}(aq)$ ions, given:

$$Ce^{4+}(aq), Ce^{3+}(aq) \mid Pt \quad E^\ominus = +1.45 \text{ V}$$

15.7 Iron is the world's most used metal, but it rusts easily. Rusting begins as a redox process involving iron, oxygen and water. The relevant standard electrode potentials are:

$$Fe^{2+}(aq) \mid Fe(s) \quad E^\ominus = -0.44 \text{ V}$$

$$[O_2(g) + 2H_2O(l)], 4OH^-(aq) \mid Pt \quad E^\ominus = +0.40 \text{ V}$$

a i Write the cell diagram for a cell using the two redox systems above.

ii What is the $E^\ominus$ of the cell you have written?

  iii Write a balanced equation, including state symbols, for the reaction between iron, oxygen and water in the cell.
  iv Suggest two ways in which the actual conditions under which iron rusts in winter in the UK are likely to differ from the standard conditions of the cell.
 b The water involved in the rusting process is often rainwater, which has a pH of about 5, even in areas not suffering from atmospheric pollution.
  i Why does rainwater have a slightly acidic pH compared to pure water? Write an equation for any reaction you suggest.
  ii Calculate the concentration of hydroxide ions in rainwater given that

$$K_w = [H^+][OH^-] = 10^{-14} \text{ mol}^2 \text{ dm}^{-6}$$

 c Iron(II) ions are readily oxidized to iron(III) ions, which in aqueous solution form a complex with water molecules: $Fe(H_2O)_6^{3+}$(aq) ions. These ions show acidic properties.
  i Draw the structure of the $Fe(H_2O)_6^{3+}$ ion.
  ii Write an equation for the equilibrium established when an $Fe(H_2O)_6^{3+}$ ion donates a proton to a molecule of water.

*15.8 Copper has two common oxidation numbers ($+1$ and $+2$) whereas vanadium has four.

   $+2$ as V^{2+} (violet)
   $+3$ as V^{3+} (green)
   $+4$ as VO^{2+} (blue)
   $+5$ as VO_2^+ (yellow)

 Use table 6.1 of standard electrode potentials in the *Book of data* to display all the possible half-reactions of these two elements and their compounds on a diagram of $E^\ominus$ against oxidation number.
 What is the highest oxidation number of vanadium which could be obtained starting with vanadium metal and aqueous copper ions, Cu^{2+}(aq)? Give equations and $E^\ominus$ values for each of the reactions which could occur on the way to making this vanadium compound. Describe what you would expect to see as the reactions proceeded.
 Can you make any predictions about the rate of any of these reactions or the extent to which they will occur? Explain your answers.

15.9 Use $E^\ominus$ values to predict what you would see if 5 cm³ of 1 M vanadium(V) ions, VO_2^+, dissolved in a solution containing 1 M H^+, was shaken with excess powdered tin. Give equations for all the reactions you predict.
 Describe with the aid of a diagram how you would set up a cell experimentally with an $E^\ominus$ value of about 0.5 V using electrodes of platinum, tin or vanadium together with tin and vanadium compounds.

15.10 The complex compound cis-dichloro-diammineplatinum(II), also known as 'cis-diplatin', shows activity against cancer cells.

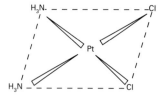

a i What type of bonding exists between the ammonia molecules and the platinum atom?
ii What feature of the ammonia molecule enables this type of bond to form?
iii What name is given to molecules such as ammonia when they are involved in forming this type of compound?
b i What is the value of the N–Pt–Cl bond angle in cis-diplatin?
ii Draw the structural formula of an isomer of cis-diplatin which will have similar bond angles.
c Cis-diplatin reacts to form new complexes with compounds in the nuclei of cells. One such compound is adenine.

i Which atoms in the adenine molecule are most likely to bond to the platinum atom in these complexes?
ii What ions or molecules are likely to be released in this reaction?
iii Suggest a reason why cis-diplatin shows anti-cancer cell activity, and **not** the isomer you have drawn in **b**.

15.11 The following table gives the colour and stability constants, K, for some copper(II) complexes.

Formula	Name	Colour of aqueous solution	lg K
$[Cu(H_2O)_4]^{2+}$	tetraaquacopper(II) ion	pale blue	1
$[CuCl_4]^{2-}$	tetrachlorocuprate(II) ion	yellow	5.6
$[Cu(NH_3)_4]^{2+}$	?	royal blue	13.1

a State the two properties of a typical transition metal which are illustrated in the table.
b Name the ligand present in tetraaquacopper(II) and classify it according to the number of bonds it makes to the metal ion.
c Draw two possible shapes for the tetrachlorocuprate(II) ion.
d Give the Stock (systematic) name for $[Cu(NH_3)_4]^{2+}$.
e Describe and explain the changes you would expect to observe when concentrated hydrochloric acid is added to copper(II) sulphate solution, and then concentrated ammonia solution is added. Use the lg K values in your explanation.

15.12 This question is about the compound 1,2-diaminoethane, $H_2N-CH_2-CH_2-NH_2$.

a **i** Draw a dot and cross diagram for half of the molecule $-CH_2-NH_2$ showing only the outermost shell of electrons for each atom.

ii Estimate the bond angles: H—C—H and H—N—H.

b The molar masses and boiling temperatures of 1,2-diaminoethane and 1,2-dichloroethane are

	Molar mass/g mol^{-1}	T_b/K
1,2-diaminoethane	60	389
1,2-dichloroethane	99	356

Explain why 1,2-diaminoethane has the higher boiling temperature even though it has the lighter molecule.

c 1,2-diaminoethane can act as a ligand.

i How many bonds would one 1,2-diaminoethane molecule be able to form when acting as a ligand?

ii When aqueous 1,2-diaminoethane solution is added to aqueous copper(II) sulphate solution a dark blue/violet complex is formed in solution. Suggest an experiment you could do to find the ratio of 1,2-diaminoethane molecules to copper(II) ions that react to form the complex.

In your description, state clearly the solutions and apparatus you would plan to use as well as the method.

15.13 When cobalt(II) carbonate is treated at 90 °C with pentane-2,4-dione, $CH_3COCH_2COCH_3$, and hydrogen peroxide, a vigorous reaction occurs. Carbon dioxide is evolved because the pentane-2,4-dione acts as an acid.

When the reaction mixture is cooled, green crystals are obtained, melting point 213 °C. The crystals are insoluble in water but soluble in benzene.

Some crystals prepared as above were analysed:

i The percentage of cobalt was found to be 16.25% by mass.

ii On ignition, 1.53 g of the green crystals produced 2.81 g of carbon dioxide and 0.81 g of water.

iii 0.88 g of the green crystals were dissolved in dilute sulphuric acid and treated with excess potassium iodide solution, which reduced the cobalt back to cobalt(II). Iodine was liberated and was found by titration to be equivalent to 24.4 cm³ of 0.05 M sodium thiosulphate(VI) solution.

From the data given, calculate
the formula of the green compound,
the oxidation number of cobalt in the green compound.
Draw a diagram to show a possible structure for the green compound.

*15.14 Use the *Book of data* to establish the range of oxidation numbers found in the compounds of the d-block elements Sc to Fe. Explain how the electronic configurations of these elements can be related to the pattern of the oxidation numbers you have found.

Tabulate the values of $\Delta G_f^\ominus$ for the oxides, M_2O_3, of the d-block elements Sc to Fe. What reasons can you suggest for the variation in these values?

TOPIC 16

Organic synthesis

The purpose of organic synthesis is the preparation of useful substances from simpler materials which are readily available from natural products or the petrochemical industry. The design of the often complex molecular structures required for use as pharmaceuticals, pesticides, perfumes or polymers depends upon a knowledge of the bond-forming reactions of carbon compounds and an understanding of the mechanisms by which they occur.

The search for more effective drugs often requires the synthesis of hundreds of compounds which are variations on a basic structure. For example cocaine is an effective local anaesthetic but has dangerous side effects which prohibit its general use; however other compounds with a similar structure have been developed which are safe enough to be available without prescription.

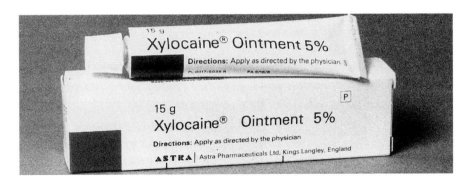

Figure 16.1 A commercial product containing benzocaine.

Carbon chemistry is described in Topics 2, 7, 12 and 14

In this Topic you are going to develop strategies for making new carbon compounds using the reaction schemes you have assembled at the end of each of the previous Topics. You will therefore be revising the reactions of carbon compounds by using them to work out synthetic routes for the conversion of compounds having simple molecules into more complex and useful substances.

There are three main activities in this topic:

1 You will first assemble **flow sheets of synthetic routes** based upon the summary reaction schemes you have constructed at the end of each organic topic. You can then use your flow sheets to devise multi-step syntheses of molecules of increasing complexity.

2 In experiment 16.2 you can perform the **two-step synthesis** of methyl 3-nitrobenzoate which will remind you of the techniques needed to obtain a reasonable yield of pure product. In experiment 16.3 you are asked to use the results of modern instrumental techniques to help you identify three unknown substances.

3 Finally, you can revise the carbon chemistry in the course, using the revision and examination questions that are available at the end of this Topic.

16.1 Devising a synthesis

The enormous range of carbon compounds that are known to exist are the result of the ability of atoms of carbon to join together to form extensive chain or ring structures and to establish covalent bonds to atoms of hydrogen, oxygen and nitrogen which are of similar strength to the bonds that they form with each other.

There are over seven million reported organic compounds, but fortunately their behaviour can be understood in terms of the **functional groups** which they contain. These functional groups usually follow the same type of reactions even though they may be attached to different carbon skeletons. For example, the functional groups of an alcohol, an alkene, or a ketone, behave in a characteristic manner whether they are found in simple molecules such as those of ethanol, ethene and propanone or in a complex molecule such as that of the steroid hormone cortisone.

COMMENT

Bond	Bond energy /kJ mol^{-1}
C—H	413
C—O	336
C—N	286
C—C	347

Use table 4.6 in the *Book of data* to compare these values with other single bond values.

CH_3—CH_2OH CH_2=CH_2 CH_3—CO—CH_3
ethanol   ethene   propanone

cortisone

Figure 16.2 *Sir Robert Robinson* was one of the finest organic chemists of the 20th century. He worked out the structure of many naturally occurring compounds, including the alkaloids morphine and strychnine, for which he was awarded a Nobel Prize in 1947.

Figure 16.3 *Robert Burns Woodward* was born in Boston, USA, in 1917. He hated physical exercise and refused to take any of the compulsory courses in physical education that formed part of the undergraduate curiculum. He was awarded his degree only because of the intervention of one of the professors. He led a group of researchers at Harvard University who achieved many notable firsts in organic chemistry, including the total synthesis of cholesterol in 1951, of strychnine in 1954 and of chlorophyll in 1960.

The structure of cortisone was confirmed in 1951 by Woodward who completed its synthesis in more than forty steps! This achievement clearly involved a remarkable manipulation of the bond-making and bond-breaking processes of carbon compounds. In order to assemble structures of this degree of complexity, where the basic carbon framework must be built from simpler structures, the synthetic chemist needs to deploy suitable **carbon–carbon bond-forming reactions** as well as **functional group interconversions** which alter the groups attached to the basic carbon skeleton.

Carbon–carbon bond-forming reactions

You have already been introduced to the Friedel-Crafts reaction which joins alkyl chains to arene rings (Topic 7).

$$CH_3CH_2Cl + AlCl_3 \longrightarrow CH_3CH_2^+ AlCl_4^-$$

$$CH_3CH_2^+ AlCl_4^- + C_6H_6 \longrightarrow C_6H_5CH_2CH_3 \text{ (ethylbenzene)} + AlCl_3 + HCl$$

Another way of forming new carbon-carbon bonds is to introduce the nitrile functional group, —CN, into organic molecules. This increases the number of carbon atoms in the molecular chain and is therefore a valuable reaction for the synthesis of new compounds.

Methods of producing a nitrile

1 Halogenoalkanes react with cyanide ions in a nucleophilic substitution reaction

$$CH_3(CH_2)_3Br + Na^+CN^- \longrightarrow CH_3(CH_2)_3CN + Na^+Br^-$$
$$\text{pentanenitrile}$$

Good yields of nitriles are obtained by refluxing halogenoalkanes with sodium cyanide in ethanol. The nucleophile is the cyanide ion, $:C\equiv N^-$, with the lone pair of electrons on the carbon atom forming the new bond.

2 Carboxylic acids will form ammonium salts which can be dehydrated in two stages to nitriles via acid amides:

$$CH_3CO_2H \xrightarrow{NH_3} CH_3CO_2^- NH_4^+ \xrightarrow{heat} CH_3CONH_2 \xrightarrow{P_2O_5} CH_3CN$$
$$\text{ethanoic acid} \qquad\qquad\qquad\qquad \text{ethanamide} \qquad \text{ethanenitrile}$$

The ammonium salt of a carboxylic acid is first heated, when the amide will be formed with loss of water; then the amide can be dehydrated by heating with phosphorus(V) oxide.

Reactions of nitriles

1 *Hydrolysis*
Nitriles will react with water if a strong acid or strong base catalyst is used. The reaction can be stopped at the stage when an amide has been formed or carried right through to the parent acid.

$$C_6H_5-CN \xrightarrow[H^+]{H_2O} C_6H_5-CONH_2 \xrightarrow[H^+]{H_2O} C_6H_5-CO_2H$$
$$\text{benzonitrile} \qquad\qquad \text{benzamide} \qquad\qquad \text{benzoic acid}$$

2 *Reduction*

The reduction of the nitriles is best carried out using lithium tetrahydridoaluminate, $LiAlH_4$. The reduction produces primary amines.

$$C_6H_5-CN \xrightarrow{LiAlH_4} C_6H_5-CH_2NH_2$$
$$\text{phenylmethylamine}$$

The general approach

The natural starting point for planning a synthesis is an examination of the desired end-product which is conveniently referred to as the **target molecule**. Consideration of its functional groups will enable a plan to be made for its synthesis from simpler and more readily available **starting materials**. Several steps may well be necessary and a number of **intermediates** may have to be formed first before a synthesis of the target molecule can be achieved. However, the strategy always involves **working backwards** in a logical sequence from the target molecule until suitable starting materials can be identified.

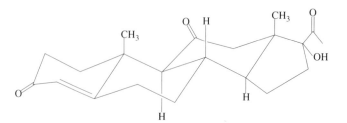

Figure 16.4 Cortisone (Woodward's target molecule).

Starting materials must be cheap and easily available. In practice this means that the majority will be derived from the simpler hydrocarbons obtained by petroleum refining, although natural products from plant and animal sources may also be useful.

Occasionally it may be possible to make the desired product in a single step, but usually at least one intermediate must be formed before the desired target molecule can be obtained. For example, nitriles are useful intermediates in synthesis, but they cannot be formed by the direct reaction between cyanide ions and alcohols, which are often convenient starting materials. Halogenoalkane intermediates have to be prepared

$$\underset{\underset{\textbf{starting material}}{\text{ethanol}}}{CH_3CH_2OH} \xrightarrow{HBr} \underset{\underset{\textbf{intermediate}}{\text{bromoethane}}}{CH_3CH_2Br} \xrightarrow{NaCN} \underset{\underset{\textbf{target molecule}}{\text{propanenitrile}}}{CH_3CH_2CN}$$

The first step in this synthesis is a **functional group interconversion**, with no change in the carbon chain. The second step is a **carbon-carbon bond-forming reaction** in which the carbon chain is extended. This explains the importance of nitriles in synthesis for, although they are hazardous materials which find little use in themselves, they are readily hydrolysed by dilute acids to the carboxylic acids, containing an extra carbon atom compared to the starting material

$$\underset{\text{ethanol}}{CH_3CH_2OH} \xrightarrow{\text{2 steps}} \underset{\text{propanenitrile}}{CH_3CH_2CN} \xrightarrow{\text{hydrolysis}} \underset{\text{propanoic acid}}{CH_3CH_2CO_2H}$$

STUDY TASK
Some synthetic routes

Use your reaction summaries from the end of each organic chemistry topic to help you with these questions. When you present your synthetic route in the final part of each question, use structural formulae and give the reaction conditions over the arrows between compounds, but do not give any details of mechanisms.

1 Your problem is to find a way of converting 1-bromobutane into butanoic acid.
  a What are the structural formulae of these compounds?
  b Is there the same number of carbon atoms in the target molecule as in the starting material?
  c What functional group does the target molecule contain?
  d The target molecule can be prepared by an oxidation reaction.
    What would be a suitable compound to treat in this manner?
  e Can this compound **d** be made from 1-bromobutane, and if so, how?
Write down the synthetic route for the changes you have suggested.

2 Your problem is to make butanoic acid from propan-1-ol.
  a What are the structural formulae of these compounds?
  b Is there the same number of carbon atoms in the target molecule as in the starting material?
  c Therefore, what type of reaction must the desired conversion involve?
  d Carboxylic acid functional groups can be made by oxidation of —CH_2OH groups or by hydrolysis of —CN groups.
    What intermediate does this suggest for the preparation of butanoic acid? Your answer to **b** is a clue.
    What is its structural formula?
  e Think of a way by which the compound **d** can be made from something easily obtained from propan-1-ol.
Write down your suggested route, using structural formulae.

3 Some 1,2-dibromobutane has to be made from butan-1-ol.
  a What functional group does butan-1-ol contain?
  b Write down the structural formula of the target molecule.
  c What sort of reaction can place two bromine atoms on neighbouring carbon atoms?
  d What intermediate compound must you have in order to make 1,2-dibromobutane by reaction **c**?
    Write down its name and its structural formula.
Write down your suggested route, using structural formulae.

Problems in devising a synthesis

In more complex syntheses there will often be more than one possible route to the target molecule and choices will have to be made.

The most desirable route is likely to be the one requiring the fewest steps. Most reactions of carbon compounds are accompanied by side reactions which result in the formation of minor products as well as the major product. A reaction yield of 90% would be considered excellent, 80% is extremely good and even 50% is often considered adequate.

The overall yield in a reaction of several steps is the product of the yields of the individual steps. So in a three-step synthesis where each step had a 90% yield the final overall yield would be only 73%. Clearly in complex syntheses of many stages the effect of low-yield steps can be disastrous.

> **COMMENT**
> **Step 1** 90% of 100 g ⟶ **Step 2** 90% of (90% of 100 g)
>
> ⟶ **Step 3** 90% of 90% of(90% of 100 g)] $= \dfrac{90 \times 90 \times 90}{100 \times 100 \times 100} \times 100 \text{ g} = 73 \text{ g}$
>
> Now calculate the overall yield when each step has a yield of only 50%.

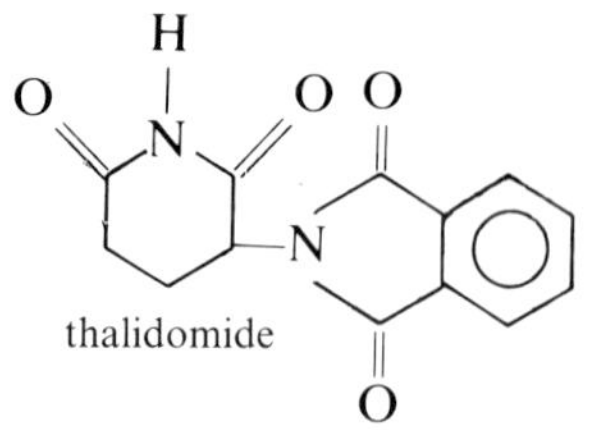

In most synthetic problems it will be essential to check that the reagents required for any one step in the synthesis do not react with other groups present in the molecule. When they do it may be necessary to 'protect' these groups while the desired reaction is carried out. Also, it will often be important to perform the synthesis in a definitely ordered sequence, because changing the functional group may result in changes of reactivity that might block the next step.

Factors associated with health and safety as well as economics will always be important in devising a commercial synthetic route. There is often no single correct solution to a synthetic problem, several routes may be equally possible, and an 'elegant' synthesis will achieve the target molecule by an optimum route which involves the minimum number of steps each with maximum yields, at an acceptable cost.

Figure 16.5 The drug thalidomide exists as two stereoisomers, one is an effective drug but the other is a mutagen. Which is the chiral centre?

> **STUDY TASK**
> **Mapping synthetic routes**
>
> You should now prepare three large **flow sheets** which summarize all the reactions you have met in carbon chemistry. These will provide you with a map with which you should be able to work out synthetic routes.
>
> The three flow sheets should be:
>
> 1 **Alkanes, alkenes and halogenoalkanes**
> 2 **Alcohols, carboxylic acids and their derivatives**
> 3 **Arenes**
>
> You may choose to link flow sheets **1** and **2** together if you have a large enough sheet of paper.
>
> Use your summary reaction schemes from the end of each Topic to draw up the flow sheets in the following way:
>
> **a** Include the conditions needed for each **functional group interconversion**, which you should write above the arrow; below the arrow you may like to remind yourself of the reaction type e.g. substitution or oxidation.
> **b** Illustrate the functional groups in the flow sheets by using the structural formula of ethene for **Flow Sheet 1**, ethanol for **Flow Sheet 2** and benzene for **Flow Sheet 3**.
> **c** Highlight any **carbon-carbon bond forming** reactions on your flow sheets.

You may prefer to create these maps in your own way, but three possible frameworks are given if you would rather follow a standard scheme.

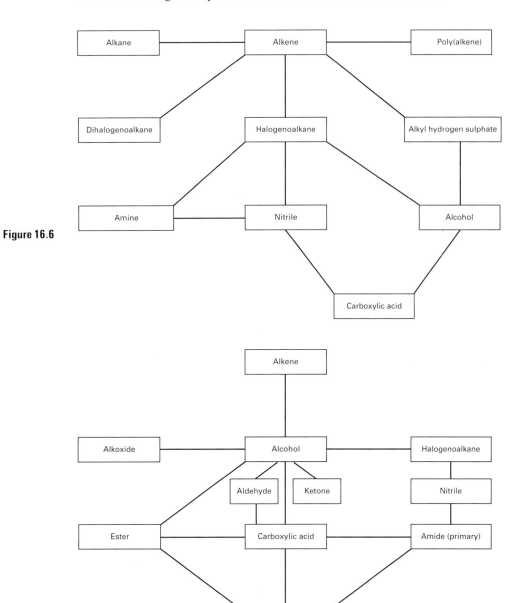

Figure 16.6

Figure 16.7

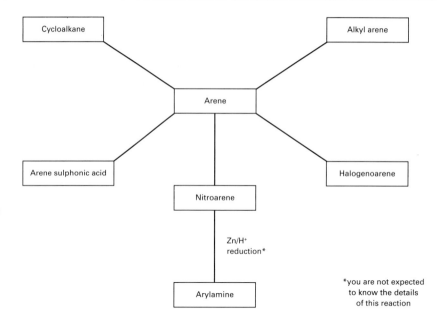

Figure 16.8

*you are not expected to know the details of this reaction

COMMENT
Definite rules govern the position of substitution in benzene and these can be found in more advanced texts.

STUDY TASK
Planning synthetic routes

Here are a selection of synthetic problems for you to try to solve. Use your **flow sheets** to gain familiarity with the synthetic routes available to you, first for simple conversions and then for the more complex syntheses which lead to useful products. In each case write structural formulae for the target molecule, starting material, and any intermediates with the reaction conditions over the arrow.

1 How would you synthesize the following compounds using ethanol alone as the starting material:
  a Ethane
  b Ethyl ethanoate
  c Ethanamide
  d Ethylamine?

2 How would you synthesize the following compounds using benzene alone as the starting material:
  a 4-bromo-ethylbenzene
  b 4-amino-propylbenzene?
  Assume that the reactions you choose place the new functional group in the position you require in the benzene ring.

3 How would you obtain the following target compounds?
  a 2-oxopropanoic acid (pyruvic acid) CH_3—CO—CO_2H, from 2-bromopropanoic acid, CH_3—CHBr—CO_2H
  b Alanine CH_3—$CHNH_2$—CO_2H from lactic acid CH_3—CHOH—CO_2H
  c N-ethanoylpropylamine CH_3—CH_2—CH_2—NH—CO—CH_3 from 1-bromopropane, CH_3—CH_2—CH_2Br
  d Propan-2-ol from propan-1-ol.

16 Organic synthesis

4 These questions are concerned with the synthesis of a number of **pheromones**. Many animals communicate by means of chemical signals. Substances used to convey information in this way are called pheromones and they include sex attractants, alarm signals, trail marking and aggregation pheromones. Aggregation pheromones attract either male or female but not both, and are therefore potentially useful in controlling insect populations.

a 4-methylheptan-3-ol is the aggregation pheromone of the European elm beetle. How might this be synthesized from 4-methylhept-3-ene which is available from cracked petroleum distillates?

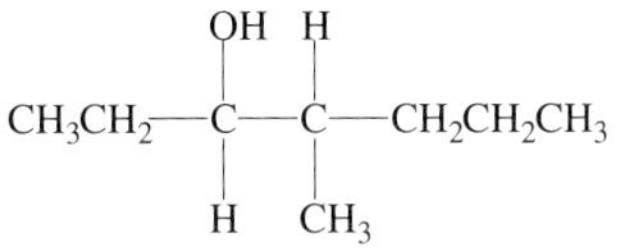

$CH_3CH_2-\underset{H}{\overset{OH}{C}}-\underset{CH_3}{\overset{H}{C}}-CH_2CH_2CH_3$

4-methylheptan-3-ol

$CH_3CH_2-CH=\underset{CH_3}{C}-CH_2CH_2CH_3$

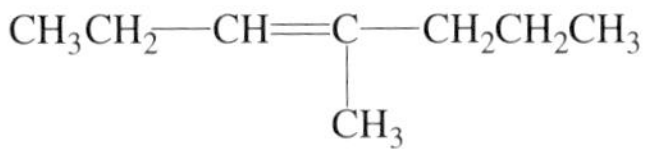

4-methylhept-3-ene

b Pentanoic acid has an unpleasant penetrating smell which attracts the male sugarbeet wire worm. How could this pheromone be obtained from pentan-1-ol?

c Phenylethanoic acid is a primary component of the smell of the stinkpot turtle. How could you synthesize this pheromone from (bromomethyl)benzene?

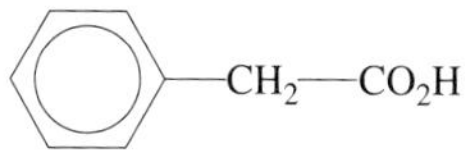

Ph—CH_2—CO_2H

phenylethanoic acid

Ph—CH_2Br

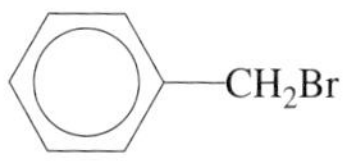

bromomethylbenzene

d The last part of this question concerns pheromones that are all esters.

i 3-methylbutyl ethanoate is a component of the alarm pheromone of the honey bee. Write the structural formulae and name the starting materials needed to make this substance in one step.

ii Heptyl butanoate is used as a bait in wasp traps. Suggest a synthesis for this substance in one step.

iii The ester phenylmethyl benzoate is used as an insect repellent. Suggest a synthesis of this substance using benzoic acid as the only starting material.

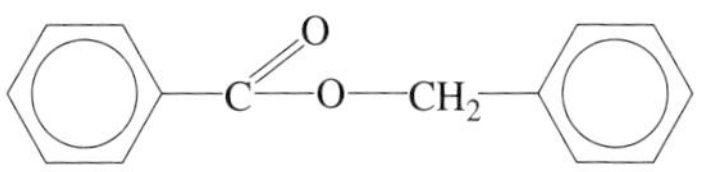

Ph—C(=O)—O—CH_2—Ph

phenylmethyl benzoate

5 These questions are concerned with substances with pleasant smells used in perfume preparations.

a Diphenylmethane is used to impart a geranium odour to soap. What two starting materials would be needed to make this substance in a one-step synthesis and what catalyst would you use?

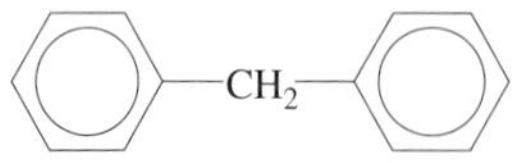

diphenylmethane

b The valuable musk range of perfumes are only available naturally from a small gland in the male musk deer. Musk ambrette (**i**) is a synthetic musk used to enhance and retain the musk odour in perfumes and can be made from the starting material 3-methylmethoxybenzene (**ii**) in a two-step synthesis.

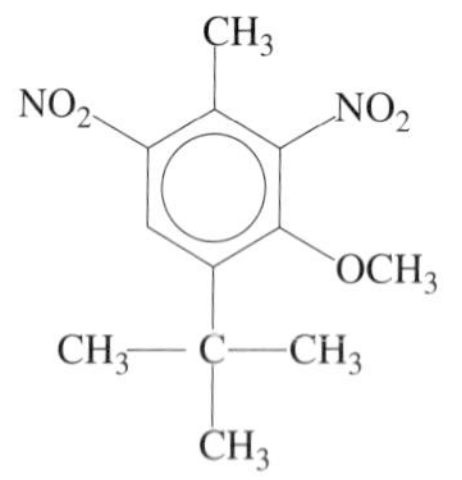

(i) musk ambrette

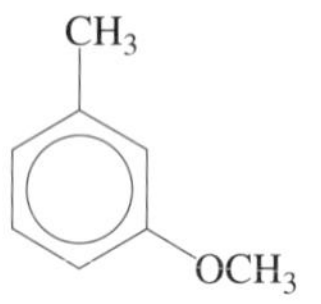

(ii) 3-methylmethoxybenzene

What other carbon compound will be needed for the synthesis? Write down your two-step synthesis assuming that the functional groups are added to the benzene ring in the correct positions.

c 2-phenylethanol is one of the most important substances in perfumery; it has a pleasant mild rose odour. How could it be synthesized from (bromomethyl)benzene?

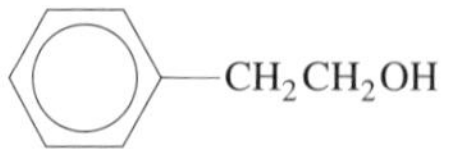

2-phenylethanol

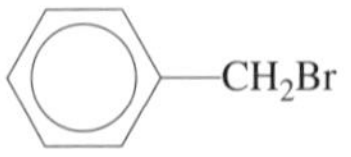

bromomethylbenzene

d 1-phenylethanol has a powerful floral odour and is also used in perfumes. How could it be synthesized from phenylethene in 2 steps?

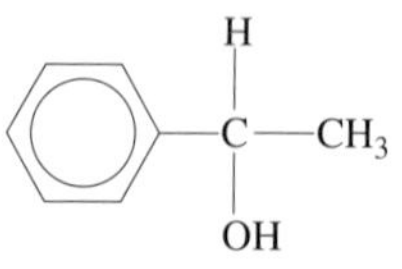

1-phenylethanol

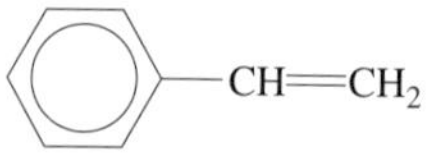

phenylethene

e What steps would be necessary to convert the synthetic perfume (**i**) into the appetite suppressant obesity drug, phentamine (**ii**)?

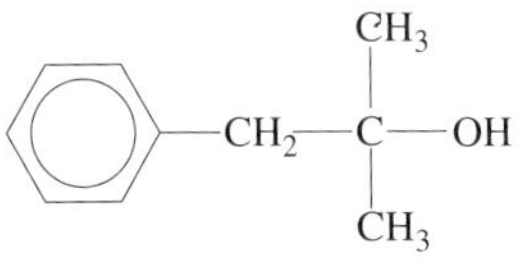

(i) a carbinol perfume

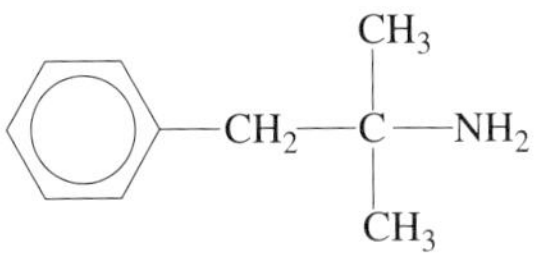

(ii) phentamine

f Suggest how the following two perfumes might be obtained from methylbenzene in multi-step syntheses:

   i Phenylmethyl ethanoate, a constituent of jasmine flower oil widely used for soap aromas.

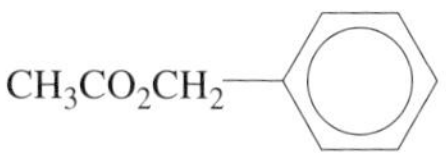

   phenylmethyl ethanoate

   ii Methyl 2-aminobenzoate which occurs in a large number of flowers and is used in making floral blends for soaps and deodorants.

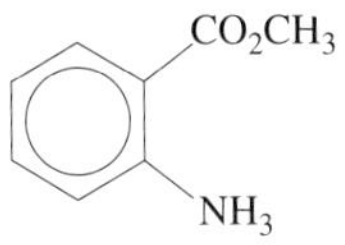

   methyl 2-aminobenzoate

   (Hint: methylbenzene can be oxidized to benzoic acid by powerful oxidizing agents).

16.2 A two-step synthesis

You are now going to carry out a full-scale laboratory synthesis in two steps, followed by some simple tests to check the purity of your product. You should calculate your yield at each step, and also your overall yield.

EXPERIMENT 16.2 The synthesis of methyl 3-nitrobenzoate in two steps

The first step is the preparation of the intermediate ester, methyl benzoate. You will be using a standard procedure for the preparation of esters.

$$CH_3OH + HO-\underset{\text{benzoic acid}}{C(=O)-C_6H_5} \xrightleftharpoons{\text{acid}} CH_3-O-\underset{\text{methyl benzoate}}{C(=O)-C_6H_5} + H_2O$$

methanol   benzoic acid   methyl benzoate

Procedure: Step 1 Formation of the ester, methyl benzoate

To a 50 cm³ pear-shaped flask add 8 g of benzoic acid, 15 cm³ of methanol, and 2 cm³ of concentrated sulphuric acid. Fit the flask with a reflux condenser and boil the mixture for about 45 minutes (see figure 2.10).

Cool the mixture to room temperature and pour it into a separating funnel that contains 30 cm³ of cold water. Rinse the flask with 15 cm³ of hydrocarbon solvent and pour this into the separating funnel.

Mix the contents of the separating funnel by vigorous shaking, releasing the pressure carefully from time to time, allow them to settle, and run the lower aqueous layer into a conical flask.

Wash the hydrocarbon solvent layer in the separating funnel with 15 cm³ of water and then 15 cm³ of 0.5 M aqueous sodium carbonate solution.

Dry the hydrocarbon solvent extract over anhydrous sodium sulphate, and filter.

Remove the hydrocarbon solvent by careful distillation (see figure 2.15). Complete the distillation, collecting the distillate boiling above 190 °C as methyl benzoate.

Weigh your product.

> **SAFETY** ⚠
> Concentrated sulphuric acid is corrosive, other materials are flammable and methanol is toxic.

- What is the percentage yield, based on the mass of benzoic acid you used in your preparation? The yield should be about 70%.
  Which bonds could have broken in the reaction? You should find that two patterns of bond-breaking are possible.

The methyl benzoate is to be used for the next step.

methyl benzoate + HNO_3 $\xrightarrow{H_2SO_4}$ methyl 3-nitrobenzoate + H_2O

The nitration of your methyl benzoate is again a standard procedure, but the wrong conditions could reverse the first step and the temperature and concentration of the nitric acid will influence the number and position of the nitro groups substituted into the benzene ring.

Procedure: Step 2 The nitration of methyl benzoate

1 Measure 9 cm³ of concentrated sulphuric acid (TAKE CARE) into a 100 cm³ conical flask and cool it to below 10 °C in an ice bath. Add 4 cm³ of methyl benzoate while swirling the flask. Prepare a mixture of 3 cm³ of concentrated nitric acid with 3 cm³ of concentrated sulphuric acid in another small flask and cool the mixture in the ice bath.

2 Use a dropping pipette to add the nitric acid–sulphuric acid mixture a drop at a time to the methyl benzoate solution. Swirl the conical flask and control the rate of addition so that the temperature stays in the range 5 to 15 °C. The addition should take about 15 minutes.

3 When the addition is complete, remove the flask from the ice bath and allow it to stand at room temperature for 10 minutes. Pour the reaction mixture over 40 g of crushed ice and stir until the product solidifies. Collect the product by suction filtration after waiting until all the ice melts. Wash with three portions of water, sucking dry and disconnecting the suction pump before each addition of washing water.

4 Change the Buchner flask for a smaller clean dry Buchner flask and wash the product with two portions of 5 cm³ of **ice cold** ethanol. **Keep this wash liquid for examination by chromatography**.

5 To recrystallize the product, transfer it to a 100 cm³ conical flask and add about 15 cm³ of ethanol, the minimum volume that will dissolve the solid when hot. Heat a water bath to boiling and **turn out the Bunsen burner** before putting the conical flask containing the ethanol in the water bath. When the solid has dissolved, it can be recovered by cooling the solution in an ice bath and collecting by suction filtration the crystals which form. Methyl 3-nitrobenzoate is a pale yellow solid of melting point 78 °C.

When your product is dry, weigh it.

For **chromatography** evaporate the wash liquid to 1 cm³ in an evaporating basin, either by standing it overnight or by heating it on a hot water bath. Use a fine capillary tube to put a spot of the solution 2 cm from the bottom of a thin layer of silica on an inert support. Some of the product can be dissolved to make a second separate spot on the plate. Allow the solvent to evaporate and develop with an ethoxyethane–hexane mixture containing 1 volume of ethoxyethane to 9 volumes of hexane.

Methyl 2-nitrobenzoate, a minor product, should be visible on the silica sheet as a yellow spot, while methyl 3-nitrobenzoate can be seen under ultra-violet light or by exposing the sheet to iodine vapour.

The **melting point** of the product can now be found.

Put a sample into a small thin walled capillary tube sealed at one end, and by gentle tapping, or rubbing with the milled edge of a coin, transfer it to the closed end. Fix the tube in the position shown in figure 16.9 by means of a rubber band. Slowly heat the tube by means of a very low Bunsen burner flame so as to maintain an even rise of temperature. Watch the crystals in the melting-point tube carefully, and the moment they melt, note the temperature.

SAFETY
Concentrated sulphuric and nitric acids are corrosive; ethanol and the chromatography solvents are flammable; iodine is corrosive. Remember to wear safety glasses during this experiment.

SAFETY
This mixture is highly flammable.

SAFETY
Do not look directly at the ultra-violet light.

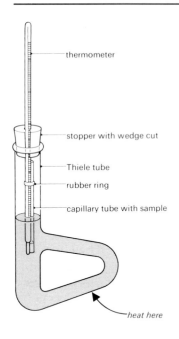

Repeat the process with a fresh melting-point tube containing another portion of the compound, in order to obtain a more accurate value for the melting point. The temperature may now be raised rapidly to within 10 °C of the melting point previously obtained, but must then be raised very slowly (about 2 °C rise per minute) until the crystals melt. Note the temperature at which the crystals first melt and also the temperature at which melting is complete. For pure substances these temperatures are close together and the melting point is called 'sharp'.

If the compound under examination is then recrystallized and dried, and the melting point again determined, it may be a little higher than before. This is because the melting point of a pure compound is always lowered by the presence of impurities. The compound can be made completely pure by repeated recrystallization until the melting point is constant.

- Calculate the yield for this step of the preparation. Work out the overall yield from the 2 steps.

Figure 16.9 The Thiele melting-point apparatus

16.3 The identification of organic compounds

Knowledge of the composition and structure of organic compounds is now largely based on instrumental methods making use of mass spectrometers, infra-red spectrometers, and nuclear magnetic resonance spectrometers (see Topic 18). Examination by these techniques will give precise information on the composition of a compound and the presence of various functional groups in the molecule. Nevertheless, older-established techniques are still useful and necessary.

Combustion analysis of organic compounds

One method for the determination of the elemental composition of organic compounds involves their complete combustion in pure, dry oxygen. An exact mass (0.1–0.3 g) of the compound is burned in a stream of oxygen diluted with helium gas and the combustion products are passed through a complex sequence of chemicals to ensure that the only gaseous products are carbon dioxide, water vapour, and nitrogen, mixed with helium. The amount of carbon dioxide is used to calculate the carbon content, the amount of water vapour is used to calculate the hydrogen content, and the amount of nitrogen gas any nitrogen content of the compound.

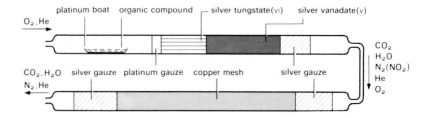

Figure 16.10 Apparatus for the quantitative combustion of organic compounds

Volatile compounds of phosphorus, sulphur, and the halogens are all removed from the gas stream by reaction with chemicals that convert them to involatile substances, while the excess oxygen is removed by reaction with copper. The copper also serves to reduce any oxides of nitrogen to nitrogen gas. The mass of water vapour in the gas stream is determined by comparing the thermal conductivity of the gas stream before and after passing through magnesium chlorate(VII). Magnesium chlorate(VII) absorbs water vapour.

A similar procedure is used to determine the mass of carbon dioxide in the gas stream, using soda lime to absorb carbon dioxide, while the mass of nitrogen is determined by comparing the thermal conductivity of pure helium with the nitrogen-helium gas stream (under the same conditions).

To obtain a pure sample for analysis, or to separate a mixture before analysis, gas chromatography is commonly used when the compounds are sufficiently volatile and stable.

The results of combustion analysis are used to determine the empirical formulae of organic compounds. The mass of carbon dioxide is converted into the mass of carbon in the sample of the compound, and then the mass of carbon is converted to the number of moles of carbon in the sample.

$$\text{mass of carbon dioxide/g} \longrightarrow \text{mass of carbon/g}$$
$$\longrightarrow \text{amount of carbon/mol}$$

The same procedure enables the number of moles of hydrogen and nitrogen in the sample to be determined.

To find out how much oxygen, if any, was present in the sample of the compound the original mass of the sample must be compared with the combined masses of carbon, hydrogen, and nitrogen (and other elements if present), as determined by the combustion analysis. Any original mass of sample not accounted for is attributed to an oxygen content and used to calculate the number of moles of oxygen in the sample of the compound.

$$(\text{mass of sample/g}) - (\text{mass of C, H, N/g}) \longrightarrow \text{mass of oxygen/g}$$
$$\longrightarrow \text{amount of oxygen/mol}$$

The empirical formula of the compound is determined as the whole number ratio of the moles of the elements present.

The empirical formula of a compound can only be converted to a molecular formula if the molar mass of the compound is known.

$n \times$ empirical formula = molecular formula
$n \times$ empirical mass = molar mass

The molar mass can be obtained from a low resolution mass spectrum, provided the spectrum contains a peak corresponding to positively charged ions which are complete molecules of the compound. Without a **parent ion** peak a mass spectrum is not really sufficient for the determination of the molar mass of an unknown compound.

Sample problem 1

0.205 g of the liquid, A, on complete combustion produced 0.660 g of carbon dioxide and 0.225 g water. The mass spectrometer parent ion peak is 81.0. What is the molecular formula of A?

Molar mass of
$C = 12 \text{ g mol}^{-1}$,
$H = 1 \text{ g mol}^{-1}$

Calculation

Part 1: Determination of the empirical formula of A

0.660 g of carbon dioxide contains $0.660 \times \dfrac{12}{44}$ g = 0.180 g of carbon

0.225 g of water contains $0.225 \times \dfrac{2}{18}$ g = 0.025 g of hydrogen

0.180 g + 0.025 g = 0.205 g which is the mass of the original sample

So A consists of carbon and hydrogen only.

0.180 g of carbon is $\dfrac{0.180}{12}$ mol = 0.015 mol of carbon atoms

0.025 g of hydrogen is $\dfrac{0.025}{1}$ mol = 0.025 mol of hydrogen atoms

0.015 mol carbon atoms / 0.025 mol hydrogen atoms = 3C/5H

So the empirical formula of A is C_3H_5.

Part 2: Determination of the molar mass of A

The parent ion peak from the mass spectrometer indicates

Molar mass = 81.0 g mol^{-1}

Part 3: Determination of the molecular formula of A

$n \times$ empirical mass = molar mass

$n \times (C_3H_5) = 81.0$ g mol^{-1}

$n \times 41 = 81.0$

so $n = 2$, to the nearest whole number

Answer

Therefore the molecular formula of A is $(C_3H_5)_2$ or C_6H_{10}.

Sample problem 2

Molar masses of C = 12, H = 1, N = 14, O = 16 g mol^{-1}

0.220 g of the liquid B on complete combustion produced 0.472 g of carbon dioxide, 0.080 g of water, and 0.025 g of nitrogen. The mass spectrometer parent ion peak is 123.0. What is the molecular formula of B?

Answer

You should find that the empirical formula of B is the same as the molecular formula and is $C_6H_5NO_2$, nitrobenzene.

EXPERIMENT 16.3 The identification of organic compounds.

You are asked to try to identify three unknown organic compounds.

You will be provided with the three compounds labelled A, B, and C. A and B contain carbon, hydrogen, and oxygen only.

The problem is to confirm their identity with as much certainty as possible. You should check, therefore, that all the data available are consistent with any identity you propose and not 'jump to conclusions' using only part of the data.

SAFETY
The organic substances are flammable.

Procedure: compound A

1 The quantitative analysis of A gives C, 68.9%; H, 4.9%; O, 26.2%. Use these data to calculate the empirical formula of A.

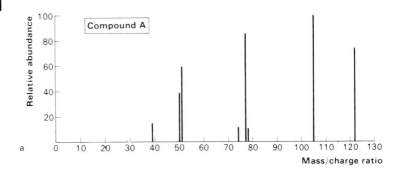

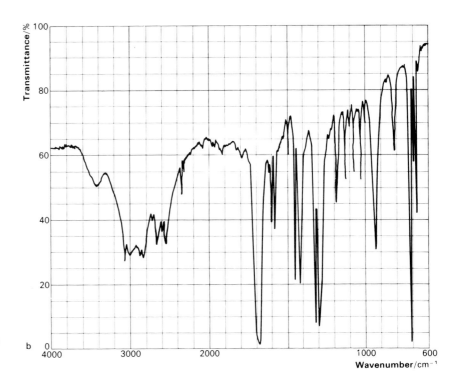

Figure 16.11 a The mass spectrum of A b The infra-red spectrum of A.

2 The mass spectrum of A is given in figure 16.11a. Deduce the molar mass of A from the mass spectrum, assuming that a parent ion peak is present, and determine the masses of the fragment ions from A. Use the molar mass and empirical formula of A to calculate the molecular formula of A. What are the likely formulae of the seven fragment ions?

3 The infra-red spectrum of A is given in figure 16.11b. Use the correlation chart in the *Book of data* to identify the functional group(s) and nature of the hydrocarbon group in A.

4 Carry out the following experiments with A.

a Burn a small amount on a combustion spoon. What type of flame is obtained? What can you deduce about A?

b Test the solubility of A in water by shaking a small amount with 5 cm³ of water. If it does not dissolve in cold water, see if it will dissolve in hot water. What can you deduce about A?

c Test a warm solution of A with 5 cm³ of 1 M sodium carbonate solution. What can you deduce about A?

d Determine the melting point of A.

5 What is your conclusion about the identity of A?

Procedure: compound B

1–3 Follow the same procedure as for A. The analysis of B gives C, 60.0%; H, 13.3%; O, 26.7%; the mass spectrum and infra-red spectrum are given in figure 16.12.

4 Carry out the following experiments with B.

a Burn a small volume on a combustion spoon. What type of flame is obtained? What can you deduce about B?

b Test the solubility of B by shaking 1–2 cm³ of B with 5 cm³ of water, hot and cold. What can you deduce about B?

c Warm a few drops of B with a mixture of aqueous sodium dichromate(VI) and 1 M sulphuric acid. What can you deduce about B?

5 What is your conclusion about the identity of B?

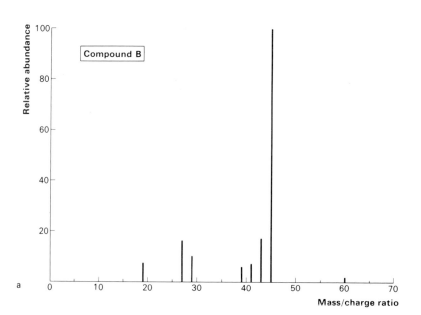

Figure 16.12 a The mass spectrum of B

16 Organic synthesis

Figure 16.12 b The infra-red spectrum of B.

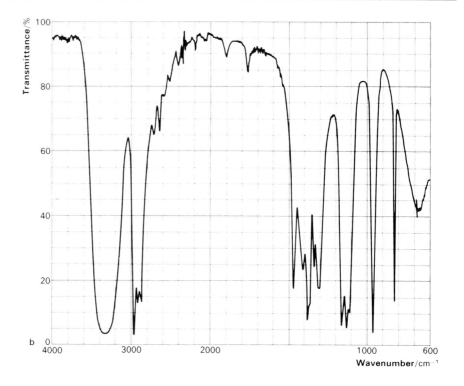

Procedure: compound C

1–3 Follow the same procedure as for A and B. The analysis of C gives C, 71.1%; H, 6.7%; O, 11.9%; N, 10.4%, and the mass spectrum and infra-red spectrum are given in figure 16.13.

4 Carry out the following experiments with C.

a Burn a small quantity on a combustion spoon. What type of flame is obtained? What can you deduce about C?

b Test the solubility of C in water by shaking a small quantity with 5 cm³ of water, cold and hot. Allow to cool slowly. Test the pH of the solution. What can you deduce about C?

c To 5 cm³ of bromine water add a small quantity of C. What can you deduce about C?

d To 5 cm³ of dilute acid or alkali add a small quantity of C and warm. Allow to cool slowly. What can you deduce about C?

5 What is your conclusion about the identity of C?

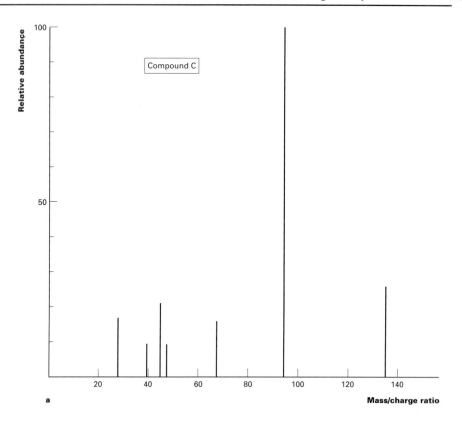

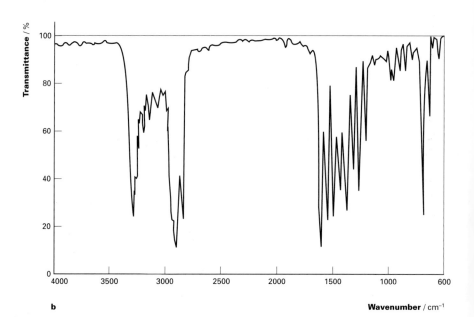

Figure 16.13 a The mass spectrum of C b The infra-red spectrum of C

Summary

At the end of this Topic you should be able to:
a   recall the behaviour of nitriles limited to
         formation from halogenoalkanes by substitution
         formation from amides by dehydration
         hydrolysis by water
         reduction to amines
b   predict the properties of unfamiliar compounds containing the functional groups included in this book and to explain such predictions
c   plan reaction schemes of up to three steps recalling familiar reactions, and using unfamiliar reactions given sufficient information
d   select and describe suitable practical procedures for carrying out reactions involving compounds with functional groups included in this book; justify the selection of such procedures, including consideration of laboratory safety
e   deduce empirical formulae, molecular formulae and structural formulae from data drawn from elemental percentage composition, infra-red spectra, mass spectra, and physical and chemical properties.

Review questions

16.1 Remind yourself of the pattern of carbon chemistry by copying and completing the scheme below.

Covalent bonds may break in 2 different ways.

i ...   ii ...

These bond-breaking processes give rise to 3 different types of reagent.

i ...   ii ...   iii ...

The reactions which result are of 3 main types.

i ...   ii ...   iii ...

Now find suitable examples to illustrate each of the statements you have made.

16.2 Classify the reactions given below, indicating clearly the reaction type and the reagent type in each case:

a $C_6H_{14} + Br_2 \xrightarrow[\text{light}]{\text{ultraviolet}} C_6H_{13}Br + HBr$

b $CH_3CH=CH_2 + Br_2 \longrightarrow CH_3CHBrCH_2Br$

c $C_6H_5\text{—}CH_3 + Br_2 \xrightarrow{Fe} Br\text{—}C_6H_4\text{—}CH_3 + HBr$

d $3CH_3CH(OH)CH_3 + Cr_2O_7^{2-} + 8H^+ \longrightarrow 3CH_3COCH_3 + 2Cr^{3+} + 7H_2O$

e $CH_3CH_2CHO + HCN \longrightarrow CH_3CH_2CH(CN)OH$

f $\langle\rangle$—OH $\xrightarrow{H_3PO_4}$ $\langle\rangle$ + H_2O

g $CH_3COCl + CH_3OH \longrightarrow CH_3CO_2CH_3 + HCl$

16.3 A plan for a sequence of organic reactions is given below.

$$CH_3CH_2CH_2OH \longrightarrow CH_3CH_2CH_2Br \longrightarrow CH_3CH=CH_2$$
$$ABC$$

$$CH_3CH_2CH_2CN$$
$$D$$

$$CH_3CH_2CH_2CO_2CH_3 \longleftarrow CH_3CH_2CH_2CO_2H$$
$$FE$$

$$CH_3CH_2CH_2CH_2OH$$
$$G$$

What reagents would you use and what type of a reaction is involved in each of the following conversions?

a A ⟶ B   b B ⟶ C   c B ⟶ D
d E ⟶ F   e E ⟶ G

16.4 This question concerns the carbon compound M which has the structure:

$$\begin{array}{c}CH_3\\|\\CH_3-CH_2-C-CH_2-OH\\|\\H\end{array}$$

a   Name the substance M.
b i   Draw a diagram to show the structure of substance N obtained by reacting M with hydrogen iodide.
   ii   What alternative reagent(s) could be added to M in order to obtain N?
   iii   Draw a diagram to show the structure of the product first formed when substance N reacts with ethanolic ammonia.
   iv   Suggest another substance that might also be formed if an excess of N were added to ethanolic ammonia.
c i   State how you would convert M into the substance P, of formula:

$$\begin{array}{c}CH_3\\|\\CH_3-CH_2-C-CO_2H\\|\\H\end{array}$$

   ii   What reagent(s) would you use to re-convert P into M?
d i   M exists as a pair of optically active isomers. Indicate by means of suitable diagrams the structures of the two forms.
   ii   What feature of the molecule of M gives rise to these isomers?

16.5 Some of the products which can be obtained from 2-hydroxypropanoic acid (lactic acid) by a single step synthesis are shown below. The reagents for some of the reactions are also shown.

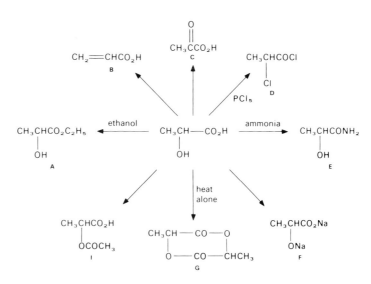

a Suggest suitable reagents for the reactions which form: B; C; F; I.
b Classify the type of reaction leading to: A; B; C; I.
c Which one of the reactions gives an ionic compound?
d With what reagent would you treat 2-hydroxypropanoic acid if you wished to form CH_3—$CH(OH)$—CO_2Na rather than F?
e What substance would first be formed when 2-hydroxypropanoic acid reacted with ammonia? What further operation would be necessary in order to obtain E?
f How can you account for the formation of substance G?
g Give one test which would enable you to confirm that substance C was a ketone rather than an aldehyde.
h Substance B is found to react with hydrogen bromide.
i Write structural formulae for the two possible products.
ii Write a mechanism for the formation of the product you think is most likely to be formed.

16.6 A substance A, had the molecular formula $C_4H_{10}O$. On oxidation it gave B (C_4H_8O) which gave a precipitate of copper(I) oxide with Benedict's solution.
On passing the vapour of A over heated silica it formed C (C_4H_8). C reacted with with hydrogen iodide to form D (C_4H_9I). D after hydrolysis and then oxidation formed E which was neutral and had no reactions with Benedict's solution. Name and give the structural formulae of A, B, C, D and E.

16.7 Compound A is a colourless acidic liquid with the molecular formula $C_2H_4O_2$. When heated with ethanol and sulphuric acid it gives a colourless neutral liquid B, $C_4H_8O_2$. When A is treated with ammonia solution, and the resulting solution is evaporated, a solid is formed, which on heating gives colourless crystals C, C_2H_5NO. Heating with phosphorus pentoxide converts C into D, C_2H_3N, and treatment of D with a powerful reducing agent converts it into E, C_2H_7N. This last compound, when added to copper sulphate solution, causes a deepening in its colour. Write structural formula for A to E.

Examination questions

16.8 The production of aspirin and Disprin from phenol by two alternative methods is shown below. During the synthesis, substances B and C must be purified before being used for the next stage.

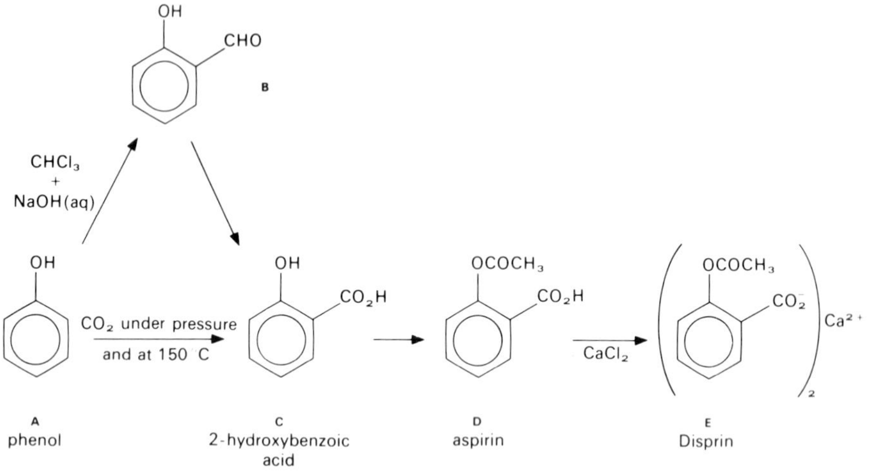

[Scheme: Phenol (A) reacts via CHCl$_3$ + NaOH(aq) to give 2-hydroxybenzaldehyde (B, with OH and CHO groups). Phenol (A) also reacts with CO$_2$ under pressure and at 150 °C to give 2-hydroxybenzoic acid (C, with OH and CO$_2$H). C reacts to give aspirin (D, with OCOCH$_3$ and CO$_2$H). D reacts with CaCl$_2$ to give Disprin (E, calcium salt with OCOCH$_3$ and CO$_2^-$, Ca^{2+}, subscript 2).]

- **a** Name one reagent for carrying out the following stages, and state the type of reaction involved.
  - **i** B ⟶ C
  - **ii** C ⟶ D
- **b** Describe tests (one for each group) to identify the two groups attached to the benzene ring in compound B, and give the results of the tests.
- **c** How could the solid C be purified?
- **d** Of the two methods shown above, which do you think would be the more economic for manufacturing 2-hydroxybenzoic acid from phenol? Give your reasons.
- **e** Disprin is said to be more effective than aspirin for relieving pain because it is more soluble in water. Why is it more soluble?
- **f** Name one other useful class of substances (other than drugs) which is obtained from phenol.

16.9 Limonene, which occurs in lemons and oranges, in peppermint oils, and in oil of turpentine, has the structure shown below and belongs to the class of natural products known as terpenes.

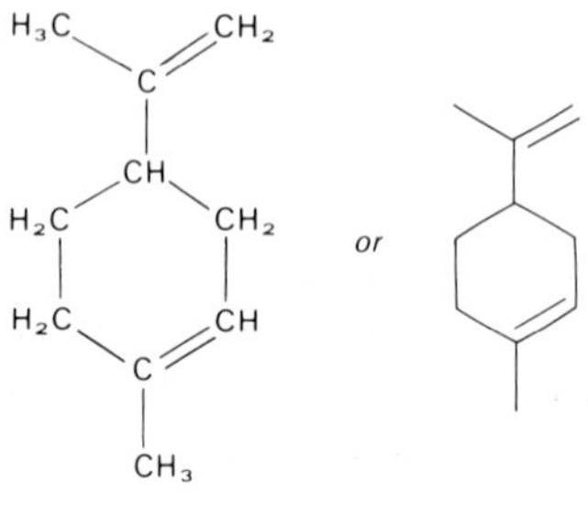

[Structure of limonene shown in two representations.]

16 Organic synthesis

a How many moles of bromine molecules would react with one mole of limonene?
b What would be obtained if limonene were reacted with hydrogen in the presence of a nickel catalyst?
c State how limonene could be converted into

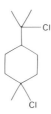

d State how the halide of **c** could be converted into the terpineol

e What product, if any, would be likely to be obtained if the terpineol of **d** were reacted
i with a mild oxidizing agent?
ii with a dehydrating agent?

16.10 The structures of two female sex hormones, oestrone and oestradiol, are given below. They belong to the class of natural products known as steroids.

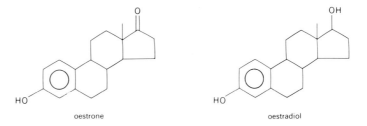

oestrone oestradiol

a Suggest one physical and one chemical method of distinguishing between the two structures.
b What reagent could you use to convert oestrone into oestradiol?
c How would you attempt to remove the double bonds in the ring structure of oestradiol?
d List as many reactions as you can in which the two —OH groups in oestradiol behave:
i in a similar way;
ii differently.
e It is thought that the biosynthesis of steroid structures derives from ethanoic acid. What kind of technique could be used to determine whether the individual carbon atoms in the steroid structures originate in the methyl group or the carboxyl group of ethanoic acid?

16.11 Two organic compounds, A and B, are isomers with the composition by mass of carbon, 70.5%; hydrogen, 5.9%; oxygen, 23.6%. A is moderately soluble in water and B is a pleasant-smelling liquid. Their mass spectra are shown below.

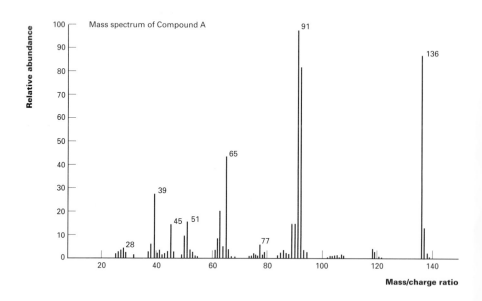

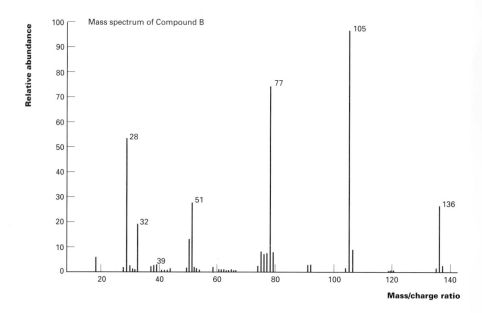

- **a i** What is the empirical formula of A and B?
- **ii** What is the molecular formula of A and B? Justify your answer.
- **b** Give the formulae of the molecular fragments corresponding to the following peaks: Mass/charge ratio: 136, 105, 91, 77.
- **c** What structural formulae would you predict for A and B?
- **d** Describe two tests or chemical reactions in which the behaviour of A and B would differ.

16.12 Give a brief account of how a compound gives rise to a characteristic infra-red absorption spectrum.

An organic compound, A, of molecular formula $C_4H_5O_2N$ and molar mass of 99 g mol^{-1} has the following properties:

i   A is fairly soluble in water, yielding a weakly acidic solution.
ii When refluxed with dilute acid, A forms a compound B of composition $C = 40.7\%$, $H = 5.1\%$, $O = 54.2\%$ and molar mass = 118 g mol^{-1}. B gives a more strongly acidic solution than A.
iii A is optically active.

Suggest structures for A and B, explaining how you reach your conclusions.

Use the *Book of data* to predict the main features of the infra-red spectrum of A.

16.13 A liquid hydrocarbon X is found to contain 85.7% carbon by mass, and to have a molar mass of 70 g mol^{-1}. The infra-red spectrum for this compound includes peaks at 3085 cm^{-1} and 1650 cm^{-1}.

Calculate the molecular formula of X from the percentage composition. Draw the structural formulae of five possible isomers which might produce these two peaks in the infra-red spectrum.

Suggest the structural formula of another isomer, Y, which would not have these two peaks in the infra-red spectrum.

Predict some of the chemical properties of one of your isomers of X. For each prediction write an equation or reaction scheme, and indicate the necessary reaction conditions.

Comment on whether all the isomers you have drawn of X would be expected to have identical physical and chemical properties.

16.14 The drug Salvarsan has been used in the treatment of parasitic infections for over eighty years. It has the formula

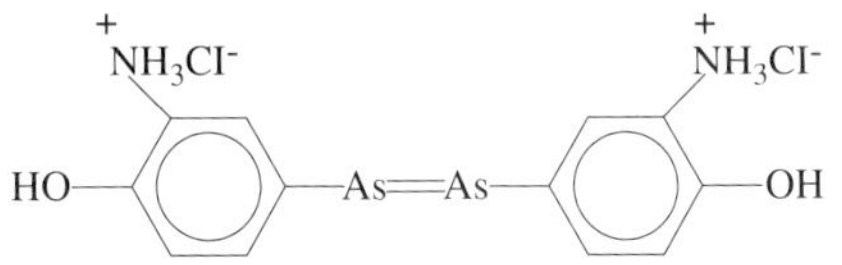

Suggest the physical and chemical properties the drug is likely to have, giving equations and formulae where appropriate.

Properties you might like to consider, include:
solubility in water
reactions with acids and alkalis
features of the infra-red spectrum
reactions with halogens
combustion
reaction with copper(II) ions, Cu^{2+}

16.15 A compound containing an amide group was hydrolysed and gave two organic compounds as products.

One compound was sparingly soluble in water, burnt with a smoky flame and was a monocarboxylic acid; 0.21 g of this monobasic acid when titrated with 0.10 mol dm^{-3} sodium hydroxide solution gave an end-point of 17.2 cm^3. Calculate the molar mass of the organic acid and suggest a structural formula.

The second compound was readily soluble in water and gave a purple coloration when warmed with ninhydrin; but the solution was not optically active when tested in a polarimeter. Analysis by mass of a sample produced the result C, 32.0%; H, 6.7%; O, 42.7%; N, 18.7%. Calculate the empirical formula of the second compound and suggest a possible structural formula.

Draw a displayed formula of the original compound.

Give **two** examples of chemical reactions by which simple amides can be formed.

READING TASK 5

DRUGS

The most remarkable applications of organic chemistry during the last 100 years have occurred in the synthesis of materials used in the successful treatment of disease. In spite of this development over three-quarters of the world's population still relies mainly on plants and plant extracts for their health care, and many of our drugs are based on plant derivatives.

About a third of the world's plant species have been used for medicinal purposes at one time or another. The Chinese have been the most systematic users of herbal remedies with a tradition that goes back 5000 years and is still in regular and effective use today. An example of a herbal remedy now widely adopted as an effective drug is ephedrine. For thousands of years the Chinese have been treating asthma and hay fever with an extract from their shrub Ma Huang, but it was not introduced into western medicine until 1924. The structure of ephedrine, the active ingredient, has been worked out so the drug can now be manufactured and is available pure and in the quantities needed.

Another example is the development of the South American arrow poison, called curare, into effective muscle relaxants. Curare is used for hunting by the indigenous people of the Amazon river. Animals hit by an arrow treated with curare are rapidly paralysed and easily captured. The first medical use of curare for muscle relaxation was in 1942. Its derivatives are now the most frequently used drugs in operating theatres to produce muscle relaxation without the need for high doses of anaesthetics.

The discovery and development of new drugs is a difficult and expensive process. Until recently drug discovery was largely based upon the random screening of numerous chemical compounds obtained from natural sources and by laboratory synthesis; but only one potential drug in 10000 compounds is likely to survive rigorous testing procedures and become commercially available for medical use.

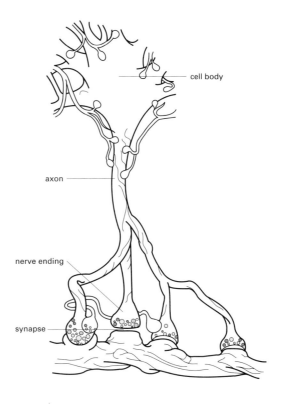

Figure R5.1 The nerve structure of the brain, first identified by the Spanish scientist Santiago Ramón y Cajal

There have been huge developments in our understanding in recent years of how our bodies work which has led to a more rational approach to drug design. For example, the study of the structure of the brain and how it works has led to the successful design of drugs to aid the treatment of people with a wide range of mental disorders such as schizophrenia or clinical depression. These drugs may be prescribed to control some of the symptoms of these illnesses.

The brain consists mainly of two types of cells, neurons and glia. The neurons transmit information; the functions of the glia are still unknown although they comprise 85% of the brain. One neuron communicates with another from specialized nerve endings. At a nerve ending an incoming electrical impulse will trigger the release of chemicals called neurotransmitters.

The neurotransmitters diffuse to receptors on the next neuron and stimulate its electrical activity. At any stage of the process there may be a problem resulting in the release of excess neurotransmitters and over-stimulation of the neurons, or alternatively there may be a shortage of neurotransmitter molecules.

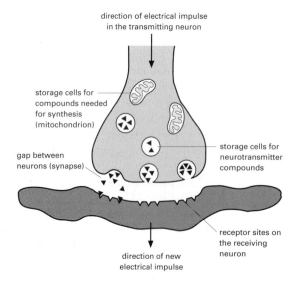

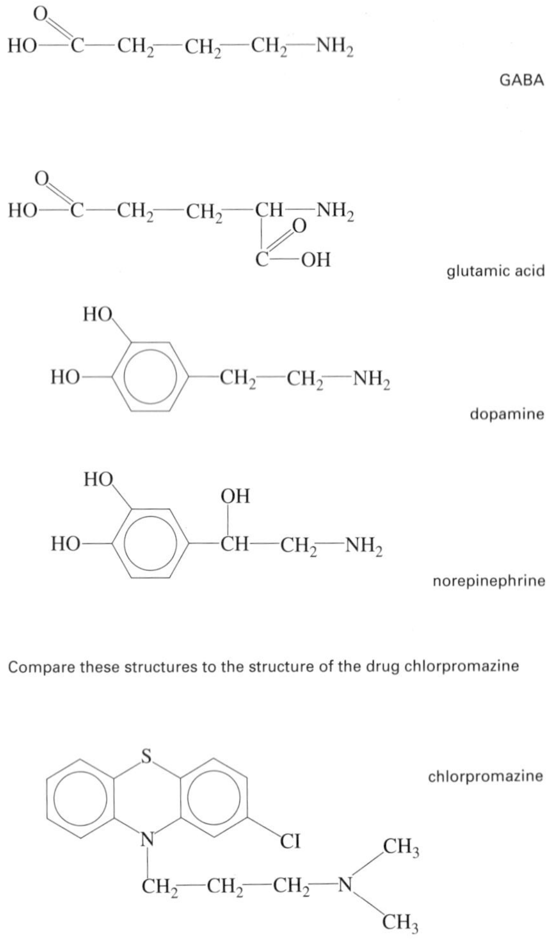

Figure R5.3 Compounds that can act as neurotransmitters

One aspect of depression is that there are reduced supplies of a particular neurotransmitter because the molecules are being broken down by an enzyme; a drug is available that blocks the enzyme activity so that normal amounts of the neurotransmitter are available for communication with the next nerve. Other drugs block the return of neurotransmitter molecules to the original neuron and again the supply of the neurons can be at a normal level.

Figure R5.2 Method of communication between neurons in the brain

Figure R5.4 Drug intervention in neurotransmission I

Figure R5.5 Drug intervention in neurotransmission II

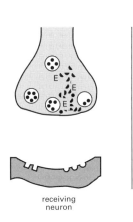
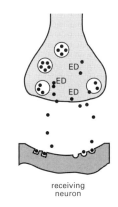

norepinephrine neurotransmitter molecules being degraded by monoamine oxidase enzyme (E) so that normal transmission does not take place

monoamine oxidase activity blocked by antidepressant drugs (D), such as phenelzine

excessive receptor site activity with the neurotransmitter dopamine is partly responsible for schizophrenia

receptor sites blocked by a drug such as chlorpromazine reduces receptor site activity and controls symptoms of schizophrenia

Other drugs that are available can stimulate the release of a neurotransmitter from storage; this again results in more neurotransmitter being available for neuron communication.

Research has shown that sometimes a mental disorder is due to excess receptor site activity. A method which will reduce excess activity is to block some of the receptors on the receiving neuron.

All of these approaches benefit from detailed knowledge of the brain chemicals involved so the drugs of appropriate molecular shape and properties can be synthesized and their effectiveness tested. Nevertheless, most of the original discoveries so far in this area have been due to 'serendipity': research studies that produced totally unplanned results, but the value of the results was nevertheless realized.

Analgesics

Some chemical molecules are able to relieve pain by modifying the pain signals as they approach the brain. A wide spectrum of substances is available which range from the relatively mild, widely used, aspirin and paracetamol to the powerful drugs of the morphine group: codeine is the mildest of these.

All the powerful analgesics in clinical use are related to morphine, which belongs to a group of naturally occurring substances called alkaloids. They have been available for many years from the opium obtained from poppy plants. The chief medical effects are all due to morphine: it has a remarkable ability to relieve pain, but repeated usage leads to addiction.

The molecular structure looks complex since the nitrogen atom is a part of a six-membered ring which is in a different plane to the rest of the molecule: however, the structural relationship between codeine, morphine and heroin should be clear.

The morphine alkaloids act upon the neuron transmission system in the brain. The action of morphine is related to its ability to fit into and block a specific receptor site: communication between the neurons is blocked and the sensation of pain is removed.

The codeine molecule only differs from morphine by a methyl group and, although it is

[morphine structure] morphine

[codeine structure] codeine

[heroin structure] heroin

converted back to morphine in the body, it is significantly less potent and can be used for long periods with little danger of addiction.

In heroin the two alcohol groups of morphine have been converted to esters. Esterification of the two alcohol groups reduces solubility in water because the potential for hydrogen bonding is removed, and the molecule is now more soluble in the hydrocarbon chains of fats. The blood-brain barrier is composed of fatty tissue which prevents the passage of water soluble and large molecules between the blood and the brain. Heroin is able to diffuse across this barrier a hundred times faster than morphine and this is responsible for its enhanced analgesic and euphoric effects: it is so strongly habit-forming that it is the most dangerous of the so called hard drugs.

Hallucinogenic drugs

Hallucinogenic or psychedelic drugs are little used in medicine, but can have a profound effect on the user's mood, memory, or perception. Marijuana (cannabis) has been known for nearly 5000 years, but its active compound has only now been isolated and its structure determined. Cannabis, however, is being used increasingly in medicine because of its ability to stop people from vomiting. It is used particularly for patients who have been given drugs to treat cancer which tend to make people vomit badly.

Cocaine and the amphetamines are perhaps the principal stimulants used in the western world today and they are examples of drugs which can bring both medical advantage and a potential for abuse. Cocaine is extracted from the leaves of the coca plant which grows on the high slopes of the Andes in Bolivia and Peru. Its potential for altering mood has long been known to the South American Indians. Chewing the leaves of the coca plant reduces fatigue and increases endurance; the practice was known to the ancient Inca civilization and is still followed today by people who live in the Andes. The medical use of cocaine in Europe was pioneered by a Viennese physician, Dr Koller, who used cocaine as a local anaesthetic in 1884 for operations on the eye.

Cocaine was the first effective local anaesthetic for use in minor surgery, and some of the important modern local anaesthetics are synthetic substitutes based on the same general structure. The skill of chemists is first to establish the structure of the naturally occurring compound and then to synthesize as many compounds as possible with a similar structural pattern in the search for safer and more effective alternatives. Benzocaine is one of the group of similar compounds which are now used as local anaesthetics in preference to cocaine.

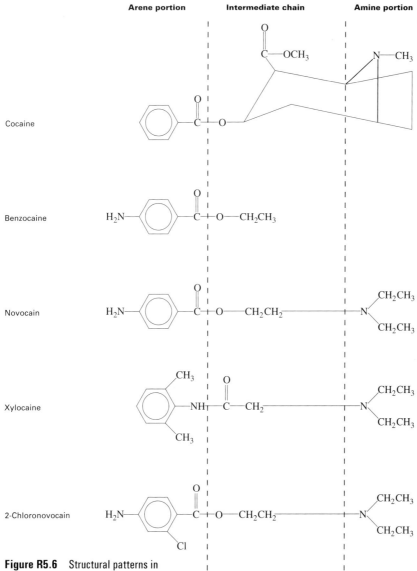

Figure R5.6 Structural patterns in some local anaesthetics

These compounds were developed in order to find substitutes for cocaine, which has dangerously addictive properties. Before the danger of addiction to cocaine was recognized coca extracts were used quite casually as a stimulant and in a range of products. Coca Cola was originally marketed with extracts from coca leaves and kola nuts. After much argument between the Coca Cola company and the United States government the company was obliged to leave coca extract out of its drink, which is now an acceptable product.

Conan Doyle described Sherlock Holmes using cocaine when he was relaxing from his work of detection. Here is an extract from Conan Doyle's story 'The Sign of Four'.

'Sherlock Holmes took his bottle from the corner of the mantelpiece, and his hypodermic syringe from its neat morocco case. With his long, white, nervous fingers he adjusted the delicate needle and rolled back his left shirt-cuff. For some little time his eyes rested thoughtfully upon the sinewy forearm and wrist, all dotted and scarred with innumerable puncture-marks. Finally, he thrust the sharp point home, pressed down the tiny piston, and sank back into the velvet-lined arm-chair with a long sigh of satisfaction.

Three times a day for many months I had witnessed this performance, but custom had not reconciled my mind to it. On the contrary, from day to day I had become more irritable at the sight, and my conscience swelled nightly within me at the thought that I had lacked the courage to protest...

"Which is it to-day," I asked, "morphine or cocaine?" He raised his eyes languidly from the old black-letter volume which he had opened.

"It is cocaine," he said, "a seven-per-cent solution. Would you care to try it?"

"No, indeed," I answered, brusquely. "My constitution has not got over the Afghan campaign yet. I cannot afford to throw any extra strain upon it."'

Drug dangers and dependence

For as long as we have used chemical substances to relieve pain and cure illness we have used other chemical substances to alter our moods and produce feelings of well-being. The social drugs such as alcohol and caffeine are usually only harmful when used to excess, but it should be remembered that alcoholism is a serious health problem, and that even small amounts of alcohol can cause damage to an unborn child. And smoking, which involves the social drug nicotine, is regarded by the medical profession as harmful in any amount, even to non-smokers by exposure to tobacco smoke. The 'soft' drugs, like cannabis and the amphetamines, may not lead to physical dependence, but an increase in availability has lead to an increase in abuse and has involved some users in serious crime. Addiction is inevitable with the 'hard' drugs of the morphine group: these narcotic drugs lead to complete dependence and eventually physical and mental damage. In order to obtain the desired effect, the addict finds it necessary to continually increase the amount of the drug until it reaches a level which is many times higher than might be administered for medical therapy.

The possibilities of drug abuse should not obscure the enormous advantages to be gained from the careful medical application of chemical substances in the treatment of disease. All drugs carry the risk of possible side effects, for the greater the effect a drug has on one part of the body the more likely it is to affect another part: these risks must be weighed against the advantages, which will vary with the patient and the nature of the illness. No drug or medicine can now be marketed without the approval of the expert Committee on Safety of Medicines, which assesses the evidence in support of the safety and effectiveness of new drugs. Nevertheless some hazards may only be identified when a drug is in widespread use, and then the drug company can be forced to stop selling the drug.

Questions

Reread the passage and answer these questions.
1 Describe the structural similarities and differences of morphine, codeine and heroin. Refer to their functional groups and chiral centres.
2 Describe the similarities and differences of benzocaine to the other local anaesthetics illustrated in the passage.
3 The sale of items containing alcohol and nicotine is legal to adults over 18, the sale or use of items containing cannabis, cocaine or heroin is illegal for the general public. Can these distinctions be justified scientifically and socially?

TOPIC 17

Nitrogen compounds

In this Topic we shall study some of the reactions and properties of compounds of nitrogen. The intention of this is to convey some impression of the behaviour of elements in the p-block of the Periodic Table and to provide opportunities for the revision of ideas you have studied in earlier Topics.

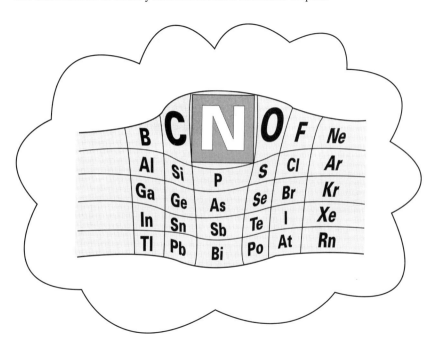

NITROGEN
'O' is for Oxygen
so gregarious
whereas I am
colourless
odourless
and tasteless
unattractive you might say
unreactive in every way
nitrogen: the night
to oxygen's day

I am 75%
of the air you breathe
so keep me clean.
For when I latch on
to fumes that cars exhaust
I am poison
Nitro-glycerin
that's me as well   Dynamite
I can blow you all to hell

But I'm not without
a sense of humour
N_2O is the proof, nitrous oxide
Inhale some laughing gas
and see my funny side
N is my symbol
N for nebulous
necessary
and nondescript.

Roger McGough

In spite of the lack of reactivity of nitrogen itself, its compounds are extremely useful to us: not only as proteins and amino acids in foods but in manufactured products that range from drugs and dyes to fertilizers and explosives.

Figure 17.1 Nitric acid can be manufactured from natural gas, air and water by the correct application of chemical knowledge. This is a catalyst of rhodium metal gauze being put in place.

17.1 Structure and energetics in nitrogen compounds

The simplest nitrogen compounds contain nitrogen atoms each forming three covalent bonds by using three of the five electrons in the highest energy level of the atom. In the ammonia molecule for example, three electrons are used to make covalent bonds and the remaining two constitute a **lone pair**.

$$H \overset{x}{\underset{\underset{H}{\overset{x\bullet}{}}}{\overset{xx}{N}}} \overset{x}{} H$$

In the nitrate ion, the three electrons are again used for covalent bonding and the lone pair is used to form a **dative bond**.

The resulting ion is a structure with delocalized electrons so that all the N—O bond lengths are equal.

In both the ammonia molecule and the nitrate ion, there are eight electrons in the highest energy level of the nitrogen atom. This group of eight electrons, or **octet of electrons**, corresponds to the filling of the 2s and the 2p sub-levels of the nitrogen atom.

Nitrogen does not form five ordinary, single covalent bonds because the energy required for the formation of five unpaired electrons, by promoting an electron to the 3s level, is about 2900 kJ mol^{-1}; and the energy obtained by forming two more covalent bonds, is only about 1300 kJ.

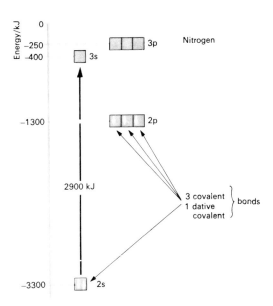

Figure 17.2 Energy levels of electrons in a nitrogen atom.

There are several cases where nitrogen is capable of forming more than one compound with a second element. For example, there are three compounds of nitrogen with hydrogen: ammonia, hydrazine and hydrazoic acid (*azote* is French for nitrogen).

Ammonia is the most familiar of these compounds but is it also the most stable? One way of making a comparison is to look at the value of the Gibbs free energy of formation, ΔG_f, for each compound.

Name	Entity	ΔH_f/kJ mol^{-1}	ΔG_f/kJ mol^{-1}
Ammonia	NH$_3$(g)	−46.1	−16.5
Hydrazine	N$_2$H$_4$(l)	+50.6	+149.2
Hydrazoic acid	HN$_3$(l)	—	—

From the sign of the Gibbs free energies of formation we can deduce that the decomposition of hydrazine to its elements should be a spontaneous process, but ammonia should be thermodynamically stable. Ammonia is said to be **stable with respect to its elements**.

It is possible for a compound to be unstable not because it decomposes into its elements but because it changes into other compounds. We can investigate this possibility by calculating the energy change for any such reaction. For example, might hydrazine decompose into ammonia and nitrogen as an alternative to decomposing into its elements?

We have to remember that even when the decomposition of a compound is expected to go the reaction may be so slow in normal conditions that we can regard the compound as 'stable'.

QUESTIONS

1. Draw dot-and-cross diagrams of the electronic structures of ammonia, hydrazine and hydrazoic acid.
2. Draw displayed formulae, showing your estimate of their bond angles.
3. Calculate the Gibbs free energy change for the decomposition of hydrazine to ammonia and nitrogen. Is the reaction likely to go at 298 K?
4a. Use bond energies from table 4.6 in the *Book of data* to calculate the enthalpy change of formation of hydrazoic acid at 298 K.
 b. Is ΔS_{system} likely to be positive or negative for the formation of hydrazoic acid?
 c. From your answers to **a** and **b** predict whether the formation of hydrazoic acid from its elements at 298 K is likely to be feasible or not.
5a. Use the *Book of data* to calculate the Gibbs free energy changes for the reaction of NO with O_2 to give NO_2 and for the dimerization of NO_2 to N_2O_4.
 b. Comment on the feasibility of these reactions of the oxides.

17.2 The properties of nitric acid

Nitric acid has three sets of properties.
- As a dilute solution it behaves as a **strong acid**.
- Concentrated nitric acid is an **oxidizing agent**. In studying its reactions you will need to refer to the appropriate electrode potentials.
- In its substitution reactions with arenes it acts as an electrophile in the presence of concentrated sulphuric acid.

STUDY TASK
Draw up a chart showing the oxidation numbers of nitrogen in all the compounds listed under **Nitrogen** in table 5.3 in the *Book of data*. Include ammonia and the ammonium ion. The notes overleaf will help you and will be useful for the experiments which follow.

NO_3^- is the formula for the nitrate ion. It is present in solutions of nitric acid and in nitrates. Nitric acid is a strong acid ($K_a = 40$) and so it may be regarded as being completely ionized in aqueous solution.

N_2O_4 is a dimer of NO_2. When nitrogen has this oxidation number in an oxide, an equilibrium mixture of NO_2 and N_2O_4 is present. It is the nitrogen dioxide, NO_2, which has the brown colour. This oxide is acidic and reacts with alkalis to give mixtures of nitrate, NO_3^-, and nitrite, NO_2^-, ions.

HNO_2 is the formula for nitrous acid. At pH = 0, nitrous acid is mostly present as HNO_2 molecules and not H^+ and NO_2^- because it is a weak acid ($K_a = 4.7 \times 10^{-4}$).

NO is the formula for nitrogen monoxide. It is a colourless, neutral gas which reacts immediately on contact with the oxygen of the air to give nitrogen dioxide, NO_2.

N_2O is the formula for dinitrogen oxide, a colourless neutral gas.

NH_4^+ is the formula for the ammonium ion.

EXPERIMENT 17.2a Study of nitric acid as an acid

Use dilute nitric acid for this experiment so that redox reactions are less likely to interfere with your interpretation of your results.

Procedure

Devise your own experiments to study the behaviour of dilute nitric acid with indicators, oxides, carbonates and metals.

EXPERIMENT 17.2b Study of nitric acid as an oxidizing agent

For this experiment you will need to use concentrated nitric acid, which is usually about 16 M.

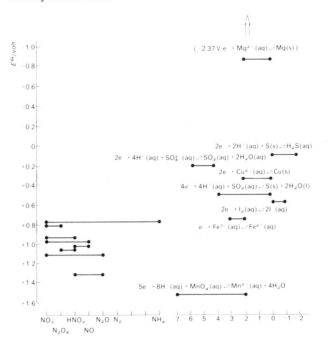

Figure 17.3 Electrode potentials at pH = 0. As these potentials are given at pH = 0, ammonia itself is not shown, only the ammonium ion.

> **SAFETY** ⚠
> Concentrated nitric acid is corrosive and the oxides of nitrogen produced in some of its reactions are severely irritant to the lungs and eyes; sulphur dioxide is irritant and toxic.

Procedure

Use figure 17.3 to predict the outcome of reactions between concentrated nitric acid and:

1. Copper metal
2. Sulphur dioxide solution
3. Potassium iodide solution
4. Iron(II) sulphate solution.

In each case confirm your prediction experimentally, working on a small scale.

- For each experiment record in your notes your prediction, your experimental method, and your result, together with an ionic equation.

INVESTIGATION 17.2c

Study of nitrous acid

Devise your own experiments to study the acid–base and redox reactions of nitrous acid. You will have to start with a solution of **sodium nitrite** because nitrous acid decomposes quite fast at room temperature. Carry out a risk assessment before starting any experiments.

> **SAFETY** ⚠
> Sodium nitrite is toxic and the oxides of nitrogen produced in some of its reactions are severely irritant to the lungs and eyes.

17.3 The properties of ammonia

Ammonia, NH_3, is a gas which is very soluble in water. The concentration of a saturated solution is so high that its density is only 0.880 g cm^{-3} at ordinary temperatures. For this reason saturated ammonia solution is sometimes referred to as 'eight-eighty ammonia'.

Ammonia is a weak base which reacts with the water in which it is dissolved to give an alkaline solution

$$NH_3(aq) + H_2O(l) \rightleftharpoons NH_4^+(aq) + OH^-(aq)$$

Since the solution contains ammonium ions and hydroxide ions it is sometimes loosely referred to as 'ammonium hydroxide' but it should always be remembered that the solution contains many more molecules of ammonia than ions.

In contrast sodium hydroxide is completely ionized in solution and a 0.1 M solution will have a hydroxide ion concentration of 0.1 mol dm^{-3}.

When ammonium chloride, a completely ionized salt, is added to a solution of ammonia, the concentration of ammonium ions is increased. This has the effect of suppressing the ionization of the ammonia still further and producing what you should recognize as a buffer solution.

$$NH_4^+(aq) + OH^-(aq) \rightleftharpoons NH_3(aq) + H_2O(l)$$
addition of
ammonium ions

In such a solution the hydroxide ion concentration would be even lower than it was in ammonia solution.

COMMENT

K_b is the dissociation constant for a base. For any weak base MOH,

$$MOH \rightleftharpoons M^+ + OH^-$$

and

$$K_b = \frac{[M^+]_{eq}[OH^-]_{eq}}{[MOH]_{eq}}$$

QUESTIONS

1 Use the expression and value for the equilibrium constant of the reaction of ammonia with water, K_b, in table 6.5 in the *Book of data* to calculate the equilibrium concentration of hydroxide ion in a solution of ammonia of concentration 0.1 mol dm^{-3}.

What sensible approximation do you need to make to simplify your calculation?

2 What will be the concentration of hydroxide ion in a solution in which ammonia and ammonium chloride both have a concentration of 0.1 mol dm^{-3}?

3 Are organic amines stronger or weaker bases than ammonia? What explanations can you propose for the data you can find?

EXPERIMENT 17.3a — Reactions of ammonia solution

Solution	pH
0.1 M NaOH	
0.1 M NH$_3$	
0.1 M NH$_3$/0.1 M NH$_4$Cl	
0.1 M CH$_3$(CH$_2$)$_3$NH$_2$	

SAFETY ⚠
Ammonia gas is very irritating and toxic; butylamine is an irritant and flammable.

1 Using a pH meter or Full-range Indicator measure the pH of the solutions listed in the table alongside.

- Is the pattern of pH variation consistent with the values of $[OH^-]_{eq}$ you calculated above?

2 Study the buffer action of a solution containing both ammonia and ammonium ions. Look up your notes on Topic 11 for help.

- Does the ammonia/ammonium ion system show buffer action? Explain how this works.

3 Add magnesium to warm ammonium chloride solution.

- Explain your result.

4 Study the action of ammonia solution and the other alkaline solutions on 0.1 M copper(II) sulphate and 0.1 M zinc sulphate.

- Explain your results.

5 In a fume cupboard examine the catalytic oxidation of ammonia by arranging a coil of copper wire over 8 M ammonia solution in a beaker. Heat the coil of copper wire to dull red heat and suspend it in the beaker.

- What happens? Try to explain your results and write equations for the reactions.

EXPERIMENT 17.3b — To estimate the percentage of nitrogen in a fertilizer

Most nitrogenous fertilizers contain ammonium nitrate, NH$_4$NO$_3$. The principle of this method is to convert all the nitrogen in a weighed sample of fertilizer into ammonia, absorb this ammonia in a known amount of hydrochloric acid and then find out by titration how much hydrochloric acid remains unreacted. The procedure is known as 'back titration'.

Nitrate ions in the fertilizer are reduced to ammonia by using an alloy containing 45% aluminium, 50% copper and 5% zinc, known as Devarda's alloy; in alkaline solution all the ammonia is given off as a gas.

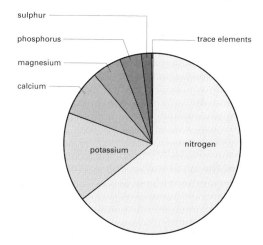

Figure 17.4 The elements that plants need from the soil.

	Element	Ion	Function of ion in plant
Major elements	Nitrogen	NO_3^-	Provides the nitrogen that is an essential raw material in the manufacture of amino acids
	Potassium	K^+	Makes many enzymes active; used in the operation of stomata
	Phosphorus	PO_4^{3-}	Essential in every energy transfer within the cell
	Calcium	Ca^{2+}	Raw material for cell walls
	Magnesium	Mg^{2+}	Raw material needed for making chlorophyll
	Sulphur	SO_4^{2-}	Raw material for making certain amino acids

Figure 17.5 The function of the major elements in plants

Procedure

Set up the apparatus as in figure 17.6. Using a pipette and safety filler put exactly 25 cm³ of 0.100 M hydrochloric acid in the conical flask.

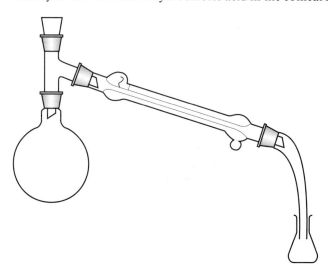

Figure 17.6 Apparatus for distillation

Weigh out accurately about 0.2 g of fertilizer and put it into the 100 cm^3 distillation flask. Add to the flask about 0.5 g of Devarda's alloy powder and about 20 cm^3 of 1 M sodium hydroxide solution.

Gently heat the contents of the flask. There will be a good deal of frothing-up of the mixture so this heating will have to be very gentle indeed.

The sodium hydroxide/Devarda's alloy mixture converts the nitrogen from the ammonium ion and from the nitrate ion into ammonia which is boiled over into the acid in the conical flask.

During the boiling beware that the acid from the conical flask does not 'suck back' into the delivery tube. If it shows signs of doing so remove the stopper at the top of the apparatus briefly, replacing it again as quickly as possible.

Boil for five minutes, then stop heating, remove the stopper from the top of the apparatus to stop sucking back and test the gas inside the top of the apparatus with red litmus paper to check that no more ammonia is being given off. If necessary replace the stopper and continue heating until this test is negative.

When the reaction is over, take the contents of the conical flask, cool and transfer to a 100 cm^3 volumetric flask. Make up the total volume to exactly 100 cm^3 with pure water, rinsing the conical flask and adding the rinsings to the volumetric flask. Mix well and titrate 10.0 cm^3 samples with 0.100 M sodium hydroxide solution using phenolphthalein as indicator. Find the average of your accurate titres in the usual way.

QUESTIONS

1 Calculate the percentage of nitrogen in the fertilizer. Compare your answer with the one given on the packaging of the fertilizer.
2 The aluminium in the Devarda's alloy reduces the nitrate ion to ammonia in the presence of hydroxide ions. The aluminium is oxidized to the aluminate ion, a simple formula for which is AlO_2^-, with water as a by-product. Construct an equation for the reaction. The balancing of this equation involves some surprisingly high numbers so it is best attempted using oxidation numbers.
3 List any errors which might have affected the accuracy of your result.
4 Justify the choice of phenolphthalein as an indicator in this titration. Would any other indicator have done instead?

Altering conditions – the Haber process

The conversion of nitrogen and hydrogen into ammonia in the Haber process is a vitally important part of the chemical industry. If the cost of the operation is to be kept as low as possible, it must be carried out quickly and efficiently. In seeing how chemists do this we shall be looking at how Le Châtelier's principle is applied in industry.

The equation for the Haber process is

$$N_2(g) + 3H_2(g) \rightleftharpoons 2NH_3(g) \qquad \Delta H = -92.2 \text{ kJ mol}^{-1}$$

As can be seen the formation of ammonia is an exothermic reaction. As it proceeds, energy is given to the surroundings, so that the entropy of these surroundings steadily increases.

At the same time, the entropy of the system is decreasing, as four moles of gas are being converted into two. If we follow the reaction at 400 K as one mole of nitrogen, N_2, is converted into ammonia, we can calculate the standard entropy changes involved, and plot them on a graph. If you look carefully at the graph you will see that ΔS_{total} reaches a maximum at about 75% conversion, at which stage the system will have reached equilibrium.

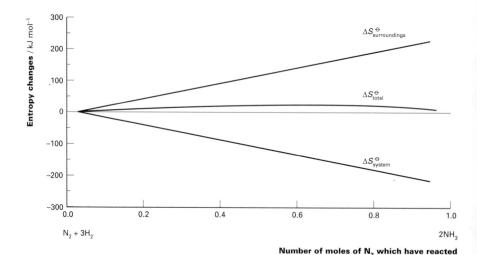

Figure 17.7 How the entropy changes alter as the reaction progresses (at 400 K and constant pressure).

At room temperature, K_c for this reaction is quite large. However, the reaction is very slow, and we should have to wait thousands of years before much ammonia was formed!

One way of speeding up the reaction is to heat it. But this carries a penalty – the more we heat the mixture the less ammonia we get.

How can we understand this decrease? It helps to look at the entropy changes involved. You should remember that

$$\Delta S_{total} = \Delta S_{system} + \Delta S_{surroundings}$$

and that

$$\Delta S_{surroundings} = \frac{-\Delta H}{T}$$

Therefore

$$\Delta S_{total} = \Delta S_{system} - \frac{\Delta H}{T}$$

Suppose that in order to speed up the reaction the temperature is raised. What will happen to these entropy changes? Although ΔS_{system} does alter a little, the big effect is on $-\Delta H/T$. Doubling the temperature (in kelvin) will halve the entropy change in the surroundings (assuming ΔH is approximately constant). As a result the total entropy change will become less favourable, so the equilibrium yield of ammonia will be lower.

The effect is perhaps easiest to see graphically. In figure 17.8, ΔS_{system}, $\Delta S_{surroundings}$, and ΔS_{total} are plotted at different temperatures.

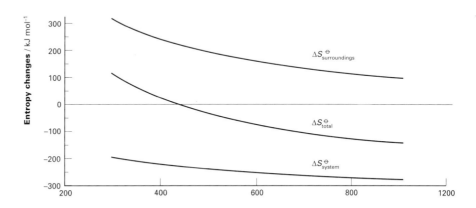

Figure 17.8 How the entropy changes are affected by altering the temperature (at constant pressure and assuming complete conversion to ammonia).

We can now see why Le Châtelier's principle works, at least as far as temperature changes are concerned. As the temperature is increased, the entropy change of the surroundings, $-\Delta H/T$, gets smaller. When ΔH is negative as it is in the Haber process, this means that $\Delta S_{surroundings}$ becomes less favourable, so ΔS_{total} also becomes less favourable, and less of the products are formed.

Indeed at temperatures above 450 K (about 200 °C) ammonia spontaneously decomposes into nitrogen and hydrogen.

We can now see the problem with the Haber process. The reaction goes too slowly to be of much use, and if we heat it to make it go faster it moves in the direction we do not want. Fortunately we can solve the problem by raising the pressure, which is the discovery that made Haber famous and probably doubled the length of the first World War.

Why does changing the pressure alter the equilibrium concentrations? If the pressure is raised, the molecules of gas have less room in which to move, so their entropies decrease. In the Haber process reaction there are more gaseous reactant molecules (four) than gaseous product molecules (two). So when the pressure increases, the entropies of the reactants decrease more than the entropies of the products. The decrease in entropy in going from reactants to products is therefore smaller at higher pressures. But the value of the equilibrium constant is not altered. In short, as the graph in figure 17.9 shows, the higher the pressure, the more ammonia is present at equilibrium. Which is exactly what Le Châtelier's principle predicts.

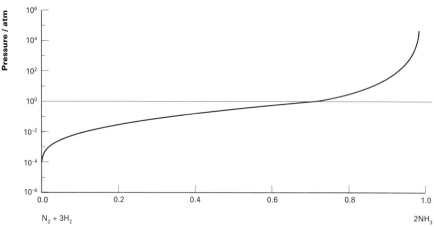

Figure 17.9 How the yield of ammonia is affected by altering the pressure (at 400 K).

Therefore, by increasing the pressure, the equilibrium concentration of ammonia can be increased enough to make the reaction worthwhile, even at the relatively high temperatures required to make it go at a reasonable speed. In the industrial process, pressures of 100–1000 atmospheres are used with a temperature of about 800 K (550 °C). The choice of catalyst for the Haber process was discussed in Topic 8.5.

17.4 Azo-dyes

You should be able to recall from Topic 12 that amines are bases, and react with acids to form salts.

$$CH_3(CH_2)_3NH_2(aq) + HCl(aq) \longrightarrow CH_3(CH_2)_3NH_3^+Cl^-(aq)$$

However nitrous acid, HNO_2, behaves differently. Nitrous acid reacts with alkylamines to give a variety of products, including alkenes and alcohols; the proportion of the various products depends upon the alkyl group present in the amine. Because of the variety of products this reaction is not particularly useful as a method of synthesis.

Arylamines however are converted by nitrous acid to diazonium compounds, in a reaction known as **diazotization.**

An example of diazotization is the reaction of phenylamine with hydrochloric and nitrous acids to give benzenediazonium chloride:

COMMENT
This reaction is an important distinction between aryl and alkyl amines.

$$C_6H_5-NH_2 \xrightarrow[HNO_2]{HCl} C_6H_5-\overset{+}{N}\equiv N\ Cl^-$$

Diazonium compounds are important as **intermediates**, that is, they are readily converted into one of a number of other classes of compounds. For example their reaction with phenols gives azo-dyes, in a reaction known as **coupling**.

An example of coupling is the reaction between benzenediazonium chloride and naphthalen-2-ol.

$$C_6H_5\text{-}\overset{+}{N}\equiv N + \text{naphthalen-2-ol} \longrightarrow C_6H_5\text{-}N=N\text{-}(naphthalenyl\text{-}OH) + H^+$$

naphthalen-2-ol

The two delocalized ring systems are now linked through the **azo group**, —N=N—. Delocalized systems like this absorb light in the blue region of the spectrum and form the basis of the orange-red azo dyes. The colour tone can be varied by altering the substituents on the benzene rings.

EXPERIMENT 17.4a The diazotization and coupling reactions

Three reaction mixtures are to be prepared. Butylamine, $CH_3(CH_2)_3NH_2$, will be used as an example of an alkylamine; ethyl 4-aminobenzoate as an example of an arylamine; the third mixture is a blank for comparison.

Naphthalen-2-ol will be used as the phenol in the coupling reaction.

Procedure

1 The diazotization reaction

To about 25 cm^3 of a crushed ice–water mixture at 5–10 °C in a 250 cm^3 beaker, add a solution of 0.5 cm^3 of butylamine dissolved in 10 cm^3 of 2 M hydrochloric acid.

Prepare a similar solution using 0.5 g of ethyl 4-aminobenzoate in place of the butylamine.

Prepare a blank solution using the inorganic reagents only and omitting any amine.

Also prepare a solution of 1.5 g of sodium nitrite in 30 cm^3 of water and add 10 cm^3 to each of the three solutions you have just prepared.

Allow the three reaction mixtures to stand for five minutes (but no longer).

2 The coupling reaction

In the meantime, prepare a solution of 3 g of naphthalen-2-ol in 20 cm^3 of 2 M sodium hydroxide. Warm, if necessary, to dissolve the solid.

Divide the solution into three equal portions, and dilute each with 50 cm^3 of cool water. At the end of five minutes, add small portions of the three reaction mixtures to the separate portions of naphthalen-2-ol solution.

Finally, add all the reaction mixtures to the naphthalen-2-ol solutions.

■ Does it look as if the amines have given distinctive reaction products? What does the blank mixture tell you?

> **SAFETY** ⚠
> Butylamine is irritant and flammable; sodium nitrite is toxic; naphthalen-2-ol is harmful. Wear protective gloves for this experiment.

Interpretation of the experiment

The mechanism of diazotization is complicated, but involves the following stages. Sodium nitrite reacts with hydrochloric acid to produce the weak acid nitrous acid, HNO_2.

$$NaNO_2(aq) + HCl(aq) \longrightarrow HNO_2(aq) + NaCl(aq)$$

In the presence of an excess of acid, nitrous acid forms the ion $\overset{+}{N}O$. This then joins to the nitrogen of the amine group using the lone pair of electrons on this amine group, and eventually the $-\overset{+}{N}\equiv N$ ion is formed.

In the coupling reaction the diazonium compound is acting as an electrophile in an electrophilic substitution on the phenol, in this case naphthalen-2-ol.

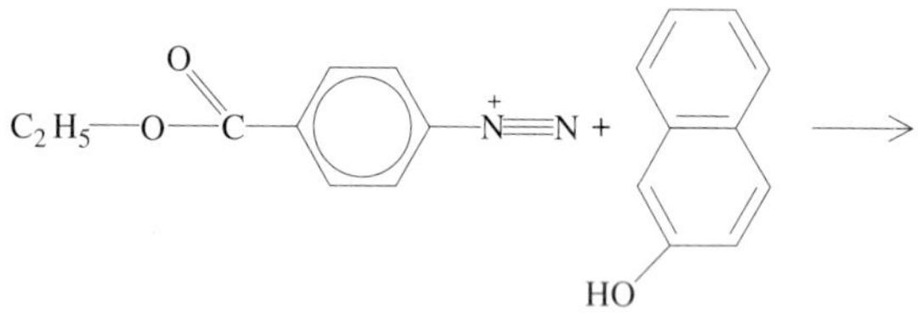

naphthalen-2-ol

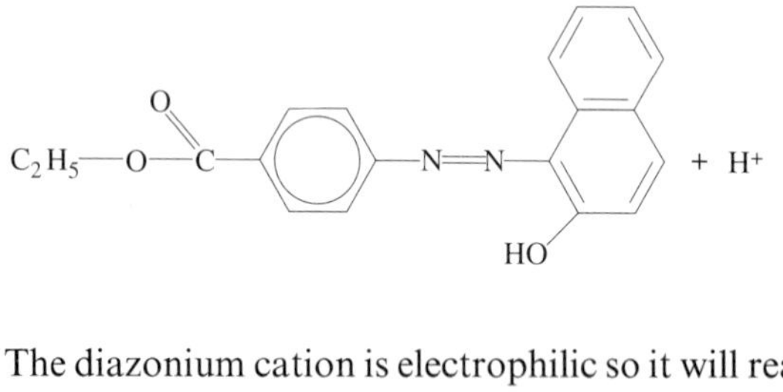

The diazonium cation is electrophilic so it will react with benzene rings which are activated to electrophiles. The diazonium compound will decompose if a low temperature is not maintained.

The diazonium compounds formed by alkylamines, such as butylamine, usually decompose without forming useful products even at low temperature.

Two further examples of coupling reactions are:

1 Coupling with phenol

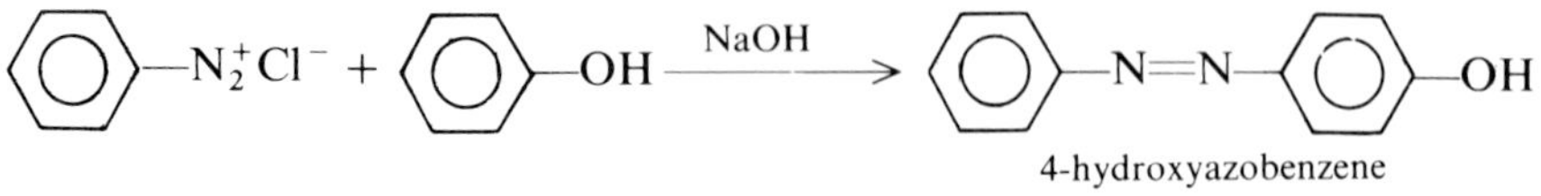

4-hydroxyazobenzene
(orange)

2 Coupling with N,N-dimethylphenylamine

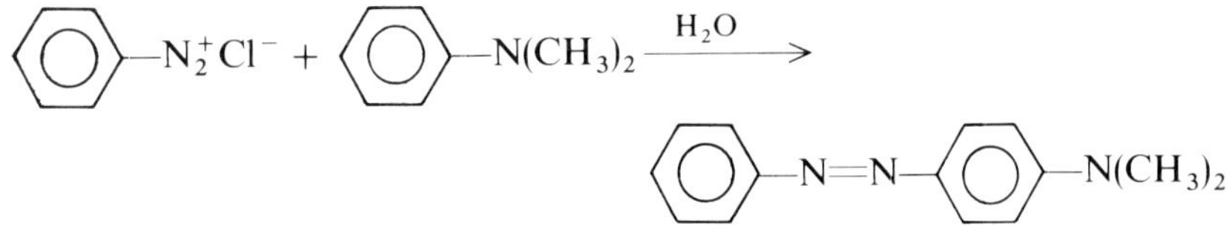

$$\text{C}_6\text{H}_5\text{-N}_2^+\text{Cl}^- + \text{C}_6\text{H}_5\text{-N(CH}_3)_2 \xrightarrow{\text{H}_2\text{O}} \text{C}_6\text{H}_5\text{-N=N-C}_6\text{H}_4\text{-N(CH}_3)_2$$

4-dimethylaminoazobenzene
(yellow)

The dyeing of different fabrics

A dye is a coloured compound which is capable of attaching itself firmly to fabrics. Once attached, it must be able to resist removal by water, soap and cleaning fluids, and it must not be subject to atmospheric oxidation. The best way for a dye to be attached to a fabric is by some form of chemical bonding to reactive groups on the molecules of the fabric. Because the molecules of different fabrics are of quite different types, it is not surprising to find that different fabrics require dyes of quite different chemical structure.

EXPERIMENT 17.4b   The dyeing of different fabrics

You will be provided with a mixture of three dyes: Direct Red 23, Disperse Yellow 7, and Acid Blue 40.

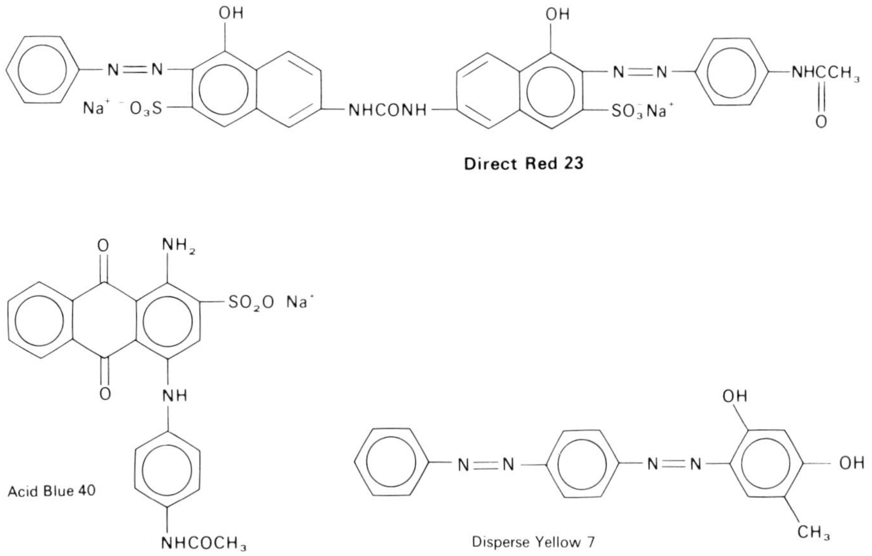

Figure 17.10 The structural formulae of Direct Red, and Acid Blue 40 and Disperse Yellow 7

Direct Red 23

Acid Blue 40

Disperse Yellow 7

Procedure

Dissolve 0.05 g of the dye mixture in 200 cm^3 of hot water and add to the dyebath 25 cm^2 pieces of various fabrics. A good choice would be a piece of cotton, a piece of nylon, and a piece of cellulose acetate or polyester. Try to avoid fabrics that have a mixed composition, and record the weave and surface texture of your fabrics so that you can identify them after dyeing.

> **SAFETY** ⚠
> Wear protective gloves for this experiment.

Allow the pieces of fabric to boil gently in the dyebath for about 5–10 minutes, remove with a pair of tongs, and rinse under the tap.

- What colours have your various fabrics been dyed?

Interpretation of the experiment

Direct Red 23 dyes by hydrogen bonding to the compound making up a fabric; Acid Blue 40 dyes by ionic attraction of its sulphonic acid group to ionizable groups in a fabric; and Disperse Yellow 7 has a small molecule that will form a solid–solid solution with a fabric, using dipole–dipole forces of attraction.

Cotton consists essentially of cellulose, a polysaccharide based on glucose; nylon is a polyamide; cellulose ethanoate is chemically treated cellulose in which most of the hydroxyl groups have been ethanoylated; polyester contains only ester functional groups.

QUESTION
Which dyes have dyed which fabrics and by what method?

17.5 Study task: Dyestuffs – the origins of the modern organic chemical industry

QUESTIONS
Read the passage below and answer the questions based on it.
1 Suggest five possible structural formulae for Perkin's arylamine, $C_{10}H_{13}N$.
2 What is the molecular formula of Perkin's mauve?
3 What structural knowledge would Perkin have needed to understand the structural formula of mauve?

The eighteenth century philosopher and statesman Edmund Burke said, 'People will not look forward to posterity, who never look backward to their ancestors'. We should certainly not neglect the tremendous contributions to the present state of our science which were made by those individuals who effectively laid the foundations of the modern organic chemical industry.

In the years before 1850 the organic chemical industry scarcely existed, and nothing in the progress of industrial chemistry has been more spectacular than its emergence, involving as it does the manufacture of thousands of complex substances, including dyes, drugs, explosives, plastics, man-made fibres, fuels, plant protection chemicals, insecticides, and a host of others.

There is a marked difference in character between the inorganic and organic chemical industrial scenes. In the former, developed primarily in Britain and France during the first half of the nineteenth century, the chemist devised processes for the manufacture of chemicals such as iron, steel, sulphuric acid, sodium hydroxide, and ammonia in large quantities. These processes, once developed, could then be carried on by trained workers for many years, and the chemist himself would be involved mainly in a trouble shooting role. The organic chemical industry, however, is one that changes so rapidly in character,

17 Nitrogen compounds

with the frequent discovery of new compounds and new synthetic routes, that the chemist is continuously involved. The methods of the organic chemical industry were laid down during the establishment of the synthetic dyestuff and drug industries in the latter half of the nineteenth century.

Let us consider some aspects of the story of the dyestuff industry, because the principles of working which governed it are still basic to the philosophy of our modern organic chemical industry. We can begin with the discovery, in 1856, by an eighteen-year-old student, W. H. Perkin, of the first synthetic colouring material, a purple dye known as aniline purple or mauve. Perkin was a student of the German chemist, A. W. von Hofmann (then Professor of Chemistry at the Royal College of Chemistry in London). Hofmann suggested that the drug quinine might be synthesized from arylamines derived from coal tar, and Perkin, on his own initiative, set out to attempt this. At that time the structural formulae of organic compounds had not been worked out, and chemists knew only the molecular formulae; for quinine, this was $C_{20}H_{24}N_2O_2$. Starting with an arylamine, whose empirical formula was $C_{10}H_{13}N$, Perkin attempted his synthesis on the basis of the proposed reaction

$$2C_{10}H_{13}N + 3[O] \longrightarrow C_{20}H_{24}N_2O_2 + H_2O$$

He obtained, not the quinine he sought, but a dirty brown precipitate.

So he decided to investigate the oxidation of the simpler amine, phenylamine, C_6H_7N. After preparing the sulphate of phenylamine, he oxidized it with potassium dichromate(VI) and obtained a black precipitate which, after drying and extraction with ethanol, yielded a brilliant purple solution. This product was rapidly accepted by British and French dyers. It is interesting to note that the dye was as costly as platinum and that by the time theoretical knowledge had progressed far enough to work out its structure in 1888, the dye had fallen out of general use. The dye, known as Perkin's mauve, found its last major application in 1881 for the printing of lilac coloured 1d postage stamps; a colour plate showing the stamp is to be found in the *Book of data*.

Figure 17.11 A self-portrait photograph of William Perkin at the age of fourteen.

mauve

Hofmann forecast that Britain would become the chief dye manufacturing and exporting country in the World because of the ready availability of the starting material for dyestuffs, coal tar. This did not prove to be an accurate forecast.

In the first twenty years of the dye industry, the inventive genius in synthesizing new phenylamine-based dyes lay almost wholly in Britain and France. German chemists sought experience in Britain, and many involved in this 'brain drain' made enormous contributions to the field. However, the vast

British commitment in the textile, coal, and iron industries of the Industrial Revolution overshadowed the growth of the dyestuffs industry which began to take firm roots in Germany. When the First World War broke out, Britain was importing a large proportion of its dyes from Germany. With the cessation of imports in 1914, British dyers were so deprived that they had insufficient dyestuffs to dye the uniforms of the troops who were to fight the Germans. So acute was the shortage that Royal Warrants were isued to permit trading with the enemy and dyes for a while were purchased from Germany by way of the Netherlands.

The colour of dyes

The colour of compounds is due to their absorption of visible light but the compounds have the colour of the light **reflected** from the compound, not the light the compound absorbs. Thus a yellow colour is due to the absorption of blue light and a purple colour is due to the absorption of green light.

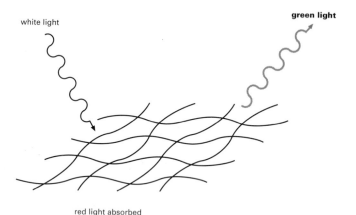

Figure 17.12 What is the real colour of a dyed fabric?

The absorption of light by a compound occurs when the compound can use the energy absorbed to change from its normal energy state to a higher energy state, from its 'ground' state to an 'excited' state. Such changes are associated with changes in structure and in the case of carbon compounds, the absorption of light energy is particularly associated with delocalized double bond systems. The delocalized double bond system of benzene has its longest wavelength absorption in the ultra-violet, at 260 nm, with maximum absorption at 204 nm; however, substituents which can form delocalized systems with the benzene ring can shift the wavelength of maximum absorption into the visible region of the spectrum. Thus the ion of 4-nitrophenol has its maximum absorption at 400 nm and is yellowish in colour.

The diazo group is an especially useful one because it forms a delocalized system linking two benzene rings. Not all dyestuffs are azo-dyes with a diazo group, of course. Two examples of azo-types are methyl orange and 'Dispersol' Fast Yellow G.

$$Na^+ \; ^-O_3S-\langle\bigcirc\rangle-N=N-\langle\bigcirc\rangle-N(CH_3)_2$$

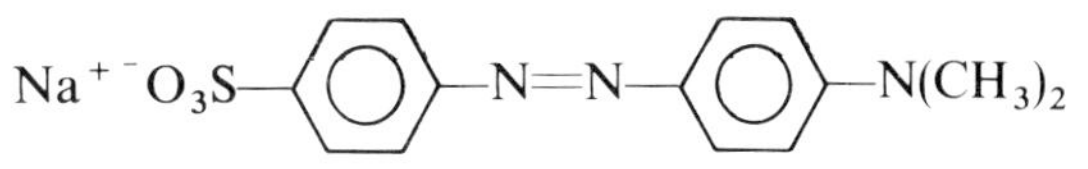

sodium salt of methyl orange

Methyl orange is not a very good dye, but is a useful indicator.

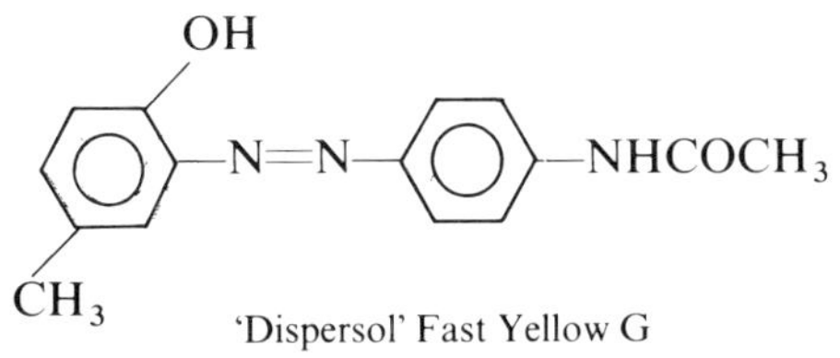

'Dispersol' Fast Yellow G (structure with OH, CH$_3$, N=N, NHCOCH$_3$ groups)

The point to be borne in mind here is that there is a link between colour and constitution, between the structure of a molecule and the properties we desire. This seeking for, and exploitation of, the relationships between structure and properties is fundamental to the operation of the organic chemical industry.

A large chemical company is likely to synthesize thousands of new compounds every year in the search for better or cheaper dyes. Without continuous research a chemical company could find that its products soon became uncompetitive.

INVESTIGATION 17.5 Nitrates in nature

Plan and, if you have time, carry out one of these suggestions:
1 Study the concentration of nitrate in natural waters: test strips are available that can be used to determine the approximate concentration of nitrate in water.
2 Study the effect of nitrate concentration on the growth of radishes: only two weeks are needed to grow the radishes sufficiently for valid results.

Summary

At the end of this Topic you should be able to:

a demonstrate understanding of the chemistry of the compounds of non-metallic elements, when provided with information, using ideas such as
   electronic structure
   enthalpy changes
   acid–base reactions
   redox reactions
   equilibrium reactions
   entropy changes
   Gibbs free energy changes

b recall the diazotization reaction of arylamines and the coupling reactions of diazonium compounds

c evaluate information by extraction from text and the *Book of data* about nitrogen compounds.

Review questions

*Indicates that the *Book of data* is needed.

17.1 The following is an extract from a textbook of structural chemistry.
'In all its compounds, nitrogen has four pairs of electrons in its valency shell. According to the numbers of the lone pairs, there are the five possibilities exemplified by the series

A	B	C	D	E
NH_4^+	NH_3	NH_2^-	NH^{2-}	N^{3-}
ammonium ion	ammonia	amide ion	imide ion	nitride ion

The last three, the NH_2^-, NH^{2-} and N^{3-} ions, are found in the salt-like amides, imides, and nitrides of the most electropositive metals.'

  a Draw dot-and-cross diagrams of the structures A to E.
  b Sketch the shapes you would expect NH_4^+ ions and NH_3 molecules to have. Explain the differences, if any, in the H—N—H bond angles in NH_4^+ and NH_3.
  c Show by means of a sketch the shape you would expect the amide ion to have. Make an estimate of the likely values of the bond angles.
  d The ammonia molecule, NH_3, can form the positive ion NH_4^+. Would you expect methane, CH_4, to form an ion CH_5^+? Give reasons for your answer.

17.2 Diazonium salts are important starting materials for a wide variety of synthetic procedures which include the production of azo dyes.

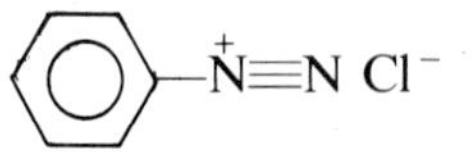

$$C_6H_5-\overset{+}{N}\equiv N \ \ Cl^-$$

  a What experimental conditions are necessary to make an aqueous solution of the diazonium ion from phenylamine?

b   Suggest reasons for the following:
i   Solid diazonium salts are often explosive (and therefore rarely isolated).
ii Stable solutions of salts are only obtained from primary amines containing a benzene ring.
c   The diazonium ion reacts with phenol in alkaline solution to form an azo dye.

$$C_6H_5\overset{+}{N}\equiv N + C_6H_5-O^- \longrightarrow C_6H_5-N=N-C_6H_4-OH$$

(4-hydroxyphenyl) azobenzene

i   Why is the phenate ion shown in the equation rather than phenol?
ii What kind of reagent is the diazonium ion in this reaction?
iii What kind of attack has taken place on the phenate ion?
d   Benzene rings are usually attacked by electrophilic reagents, but if an attached group is sufficiently electron withdrawing, electrophilic attack may be discouraged to the point where nucleophilic attack is preferred.
i   Give two reasons why you might expect diazonium salts to react with nucleophilic reagents.
ii Which atom is normally substituted in arene reactions? Which atom, or group of atoms, would you expect to be substituted when diazonium salts react? Give reasons.
iii Account for the formation of phenol when solutions of diazonium salts are boiled.
iv What do you think might be formed if a solution of a diazonium salt were warmed with potassium iodide solution?

Examination questions

17.3   The information in this table may be used to predict the likely course of some reactions of nitrous acid, HNO_2.

Half reaction	$E^\ominus/V$
$I_2(aq) + 2e^- \rightleftharpoons 2I^-(aq)$	+0.54
$NO_3^-(aq) + 3H^+(aq) + 2e^- \rightleftharpoons HNO_2(aq) + H_2O$	+0.94
$HNO_2(aq) + H^+(aq) + e^- \rightleftharpoons NO(g) + H_2O$	+0.99
$Br_2(aq) + 2e^- \rightleftharpoons 2Br^-(aq)$	+1.09

Nitrous acid is normally made freshly when it is required by adding ice cold aqueous sodium nitrite, $NaNO_2$, to dilute hydrochloric acid.

a   Explain why you would not expect nitrous acid to be stable.
b   What are the likely products of reaction between nitrous acid and a solution containing iodine and iodide ions?
c   What are the likely products of reaction between nitrous acid and a solution containing bromine and bromide ions?
d   What type of reactant is nitrous acid in the reaction mentioned in **b** and **c**?
e i What are the likely products of the reaction between nitrous acid and a solution containing bromide ions and iodide ions?

ii What would you expect to see in the organic layer if the resulting mixture were shaken with a few drops of hydrocarbon solvent?

17.4 This question concerns hydrazine, which has the molecular structure:

$$\begin{array}{cc} H & H \\ \diagdown & \diagdown \\ N\!\!-\!\!N \\ \diagup & \diagup \\ H & H \end{array}$$

Some data for hydrazine can be found in table 5.3 of the *Book of data*; note that hydrazine is a liquid at room temperature.

a The enthalpy change of formation for gaseous hydrazine at 298 K is $+95$ kJ mol^{-1}. Using bond energy data and an appropriate energy cycle, check that this value is consistent with the bond energy data.

b Because of its structure, the properties of hydrazine are likely to be similar to those of ammonia. Describe at least some of the major physical and chemical properties you would expect hydrazine to have on the basis of this similarity to ammonia.

17.5 In the laboratory, the adsorption of sulphur dioxide may be demonstrated by passing sulphur dioxide through active charcoal, but this is not a practical method in industry. One way of tackling the acid rain problem, used in some German power stations, is to pass the waste gases containing sulphur dioxide through an aqueous suspension of limestone.

The overall reaction is

$$2SO_2(g) + 2CaCO_3(s) + O_2(g) \longrightarrow 2CaSO_4(aq) + 2CO_2(aq)$$

Gypsum is then crystallized out as $CaSO_4.2H_2O(s)$. 1.2 million tonnes of gypsum are produced per year in West Germany by this method.

a Describe the changes you would expect to *see* when sulphur dioxide is passed through an aqueous suspension of limestone.

b Sulphur dioxide can be detected by the reduction of $Cr_2O_7^{2-}$ ions to Cr^{3+} ions.

i Describe the changes you would expect to see when sulphur dioxide is passed through a solution of $Cr_2O_7^{2-}$ ions.

ii What is the oxidation number of chromium in the ion, $Cr_2O_7^{2-}$?

iii In the reaction, the oxidation number of sulphur increases from $+4$ to $+6$. Suggest the likely product of the oxidation of sulphur dioxide and hence deduce the equation for this reaction.

c The amount of sulphur dioxide adsorbed by active charcoal can be determined by a titration method.

10 g of active charcoal containing adsorbed sulphur dioxide was added to 1000 cm^3 of iodine solution of concentration 0.00500 mol of I_2 per dm^3.

20.0 cm^3 portions of this solution were then titrated with 0.0100 mol dm^3 sodium thiosulphate solution: 11.6 cm^3 were required for complete reaction.

The relevant equations are:

$$SO_2(g) + I_2(aq) + 2H_2O(l) \longrightarrow 2I^-(aq) + SO_4^{2-}(aq) + 4H^+(aq)$$

$$I_2(aq) + 2S_2O_3^{2-}(aq) \longrightarrow 2I^-(aq) + S_4O_6^{2-}(aq)$$

    i   Calculate the number of moles of sodium thiosulphate in 11.6 cm^3 of its solution.
    ii   Calculate the total number of moles of iodine, I_2, which reacted with the sodium thiosulphate.
    iii Deduce the total number of moles of iodine which reacted with the sulphur dioxide.
    iv Hence calculate the number of moles of sulphur dioxide present in 10 g of active charcoal.
    v   What does the term **adsorption** mean?
d A power station produces 55 000 tonnes of gypsum per year. (1 tonne = 10^3 kg)
    i   How many moles of gypsum are produced per year? (Molar masses: H = 1, O = 16, S = 32, Ca = 40 g mol^{-1})
    ii   What volume of sulphur dioxide was absorbed in the production of 55 000 tonnes of gypsum? (1 mol of sulphur dioxide at this temperature has a volume of 24 dm^3)
e Suggest a use for the gypsum produced by this method.

17.6 When the elements of Group 1 are heated in oxygen, the type of oxide formed varies from metal to metal. Lithium oxide, Li_2O, sodium peroxide, Na_2O_2, and potassium superoxide, KO_2, illustrate this variation.

a i   In which one of the above three compounds do **both** elements have their usual oxidation numbers?
    ii   Draw a dot-and-cross electron diagram for the peroxide ion, O_2^{2-}, showing only the outermost electrons.
    iii What is unusual about the electron arrangement in the superoxide ion, O_2^-?
b Sodium peroxide reacts exothermically with cold water:

$$2Na_2O_2(s) + 2H_2O(l) \longrightarrow 4NaOH(aq) + O_2(g)$$

    i   Which element is both oxidized and reduced? Give the changes in oxidation number.
    ii   Why should sodium peroxide be kept out of contact with material such as paper or organic liquids?
c Sodium peroxide is used to absorb carbon dioxide and release oxygen in a closed environment such as inside a submarine, but lithium peroxide, Li_2O_2, is preferred in space capsules.
    i   Write an equation for the reaction of lithium peroxide with carbon dioxide.
    ii   Suggest why lithium peroxide is preferred to sodium peroxide in space capsules.
    iii Suggest three factors which must be taken into account in calculating an appropriate mass of lithium peroxide to be carried on a space voyage.

17.7 The yellow dye, Disperse Yellow 3, can be prepared using the reaction sequence shown:

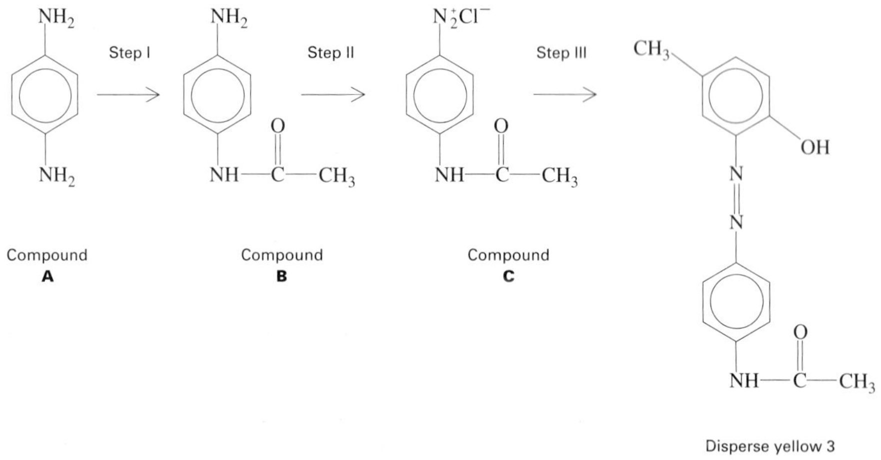

[Reaction scheme: Compound A (1,4-diaminobenzene, with NH$_2$ groups at 1 and 4 positions) → Step I → Compound B (4-aminophenyl with NH—C(=O)—CH$_3$ group) → Step II → Compound C (with N$_2^+$Cl$^-$ and NH—C(=O)—CH$_3$ groups) → Step III → Disperse yellow 3 (structure with CH$_3$, OH, N=N linkage and NH—C(=O)—CH$_3$ group)]

a i Give the name and structural formula of the reagent needed to carry out step I.
ii Give the names of the reagents and the reaction conditions needed to carry out step II.
iii Give the name of the functional group —N$_2^+$ in compound C.
iv Give the structural formula of the reagent used in step III.
b Suggest two important properties, apart from being coloured, which are essential for a compound used as a dye.

*17.8 Iodobenzene can be obtained in a yield of 75% by the following reactions:

$$C_6H_5NH_2(aq) + NaNO_2(aq) + 2HCl(aq) \xrightarrow{5°C} C_6H_5N_2^+Cl^-(aq) + NaCl(aq) + 2H_2O(l)$$

and then

$$C_6H_5N_2^+Cl^-(aq) + KI(aq) \xrightarrow{heat} C_6H_5I(l) + N_2(g) + KCl(aq)$$

You are asked to prepare 20 g of pure iodobenzene using this method. Plan a procedure in sufficient detail that a student with your own experience could follow your instructions. Include the calculation of suitable quantities of all reagents, the apparatus and reaction conditions to be used and the method for separating and purifying the iodobenzene.

The *Book of data* provides some information about the physical properties of the substances involved.

17.9 Describe the main reactions of amines, explaining any significant differences between the behaviour of alkylamines (such as butylamine) and arylamines (such as phenylamine). Quote formulae and equations wherever appropriate.

Show how the reactions of amines are used to produce at least two industrially important products, such as polymers and dyestuffs.

17 Nitrogen compounds

17.10 Reduction can be defined as:
  removal of oxygen
  or addition of hydrogen
  or addition of electrons
  or reduction in oxidation number

Select one example in each case to illustrate these four definitions. Your four examples should involve four different elements and be drawn from organic and inorganic chemistry and from both laboratory and industrial situations.

For each example write a balanced equation and give full details of the reaction conditions. Why do chemists use several different definitions of reduction?

17.11 When the chemical industry wishes to manufacture a desired chemical product, chemists have to solve two basic problems:

- Is there a chemical reaction which will form that product to an appreciable extent from available starting materials?
- Is the rate of reaction adequate?

Discuss how these problems have been investigated and overcome by chemists in the case of the synthesis of ammonia. A further problem is the separation and collection of the desired product. Suggest ways in which ammonia could be separated from excess starting materials: justify your suggestions.

17.12 Give an account of the **safety procedures** that are used to reduce the risks involved in performing experiments that require the use of hazardous chemicals. Suitable chemicals to discuss might include chlorine, sodium dichromate(VI), sulphuric acid, cyclohexene and ethanoyl chloride.
Your answer should refer to at least **four experiments** involving hazardous chemicals, mentioning the hazards involved and the procedures and apparatus that you would use to perform the experiments safely.

READING TASK 6

NITRATES IN AGRICULTURE

Plants require a number of nutrients for growth but the three major constituents of fertilizers are nitrogen, phosphorus and potassium (N, P and K). However, not all the nutrients applied as fertilizers are used by the crops and the unused portions represent a financial loss to the farmer and sometimes a hazard to the environment. Fertilizers must therefore be used as efficiently as possible.

The commonly used fertilizers in the nineteenth century were sodium nitrate (Chile saltpetre), ammonium sulphate (a by-product from coal gas manufacture) and guano from South America. But the introduction of the Haber process meant that by the mid 1930s over 80% of the nitrogen fertilizer used in the UK was ammonium sulphate prepared from synthetic ammonia. Since then ammonium nitrate, ammonium phosphates and urea have increased in importance. All these are water soluble and plants can make use of both ammonium and nitrate ions. In neutral temperate soils ammonium ions are oxidized to nitrate ions which crops prefer.

Since 1960 the use of nitrogenous fertilizers has increased rapidly, contributing to increased agricultural production. However there are problems associated with larger amounts of N in circulation. The nitrate concentration found in rivers in both cultivated (arable) and grassland areas of England has been increasing. The activity of bacteria in the soil leads to the release of nitrogen compounds and may result in an increase in the oxides of nitrogen in the upper atmosphere, affecting the ozone layer.

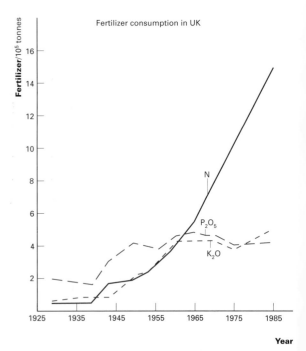

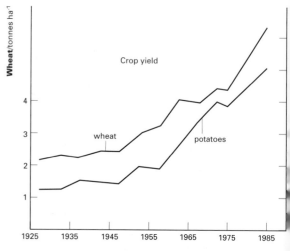

Figure R6.1 Increasing use of fertilizers and increasing yields of crops.

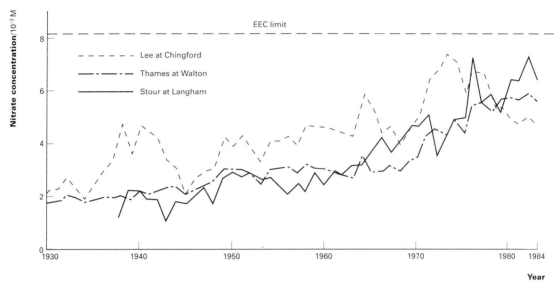

Figure R6.2 Increasing nitrate concentration in rivers.

The European Community Directive on the quality of water for human consumption now restricts the maximum acceptable concentration of nitrate in drinking water to 50 mg dm^{-3}. Too much nitrate in water can cause depletion of oxygen through excessive plant growth and in drinking water can cause blue-baby syndrome in infants. Water is a valuable resource. It should be used efficiently, and we want our water – whether for drinking, washing, or swimming – to be uncontaminated. So how are the excessive concentrations of nutrients such as nitrate to be avoided?

The nitrogen cycle

Nitrogen circulates in the environment and the nitrogen cycle is an attempt to describe the varied processes involved. In the soil the processes depend on groups of bacteria. One group oxidizes nitrogen compounds (ammonia to nitrite, to nitrate – nitrification) to obtain energy. Other bacteria are responsible for releasing the nitrogen from dead matter (mineralization); while some nitrogen will be built into the amino acids of the bacteria (immobilization). For all agricultural systems the nitrogen cycle is 'leaky', with losses occurring both to the atmosphere (NH_3, NO, N_2O) and to ground and surface waters (NO_3^-, NO_2^-, NH_4^+).

The movement of N through the nitrogen cycle can be followed by isotopic labelling. This is done by adding extra ^{15}N to the system. ^{15}N is the heavy, non-radioactive isotope of nitrogen. It occurs naturally in the air at an abundance of 0.3663%. It can be concentrated by various techniques and measured by mass spectrometry, and thus used to trace the movement of the N in the cycle.

Large amounts of fertilizer nitrogen are used each year by farmers to increase yields of crops. However, these crops also use N that is mineralized from soil organic matter together with any atmospheric inputs. By using ^{15}N to label the fertilizer the amount of nitrogen in the harvested crop that is derived from the applied fertilizer can be measured.

By sampling the soil at harvest it is possible to see how much of the labelled N remains in the soil unused and whether it is present as organic N, or as inorganic N which may be lost by leaching. By measuring the labelled N in both the crop and the soil, the amount of N that is lost from the system can be calculated. This loss may be by denitrification, leaching or possibly ammonia volatilization. In addition, the fate of any N residues remaining in the soil can be followed in subsequent years.

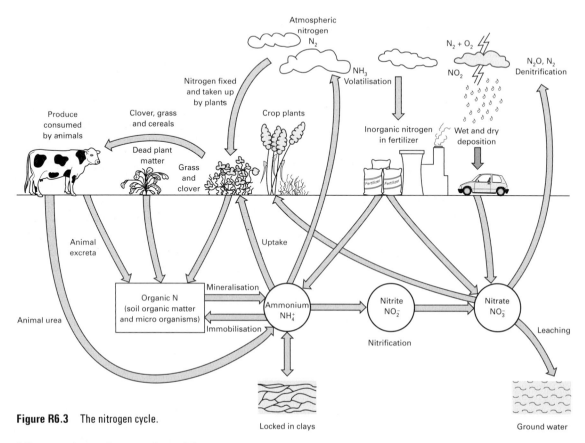

Figure R6.3 The nitrogen cycle.

Measuring nitrate leaching

There are three principal methods that can be used for directly estimating the amount of nitrate leached from soil – lysimeters, porous ceramic suction cups and soil sampling.

Lysimeters are large undisturbed blocks of soil, cut from the ground and encased so that all of the water draining from them can be collected and analysed.

Porous ceramic suction cups consist of a hollow ceramic head which can be inserted into the soil at any depth and connected to the surface by using narrow bore tubing. By creating a vacuum, soil solution is drawn into the cup and then extracted for analysis. This gives a direct measure of nitrate concentration in the soil solution near the cup. However, an estimate of drainage must also be made to convert concentrations into amounts leached. Ceramic cups are cheap and easy to use, and thousands of them are in use in leaching studies.

The simplest method, and the cheapest for a single measurement, is to take a sample of soil from whatever depth is required, cut it into sections and extract the nitrate and ammonium from these, usually with 2 M potassium chloride. This gives the mineral nitrogen profile of the soil. The amount of nitrate leached is calculated from the difference between successive profiles and an estimate of drainage, but this is very imprecise. Processes other than leaching cause nitrate to disappear, for example crop uptake.

Finally, samples can be taken from field drains, ditches, streams and boreholes. These again show directly the concentrations of nitrate leached from fields, farms or larger catchments. But unless the drains or streams can be related to the area of land, or water catchment area, that they drain it is impossible to estimate the quantities of nitrate leached on an area basis.

Figure R6.4 Lysimeter in use: studying the effects of water-logging on crop growth.

Possible sources of leached nitrogen

Before solutions to the problem of nitrate leaching can be suggested, all of the sources of nitrate in soil need to be identified. For arable farming these are – apart from a very small amount in the seed – soil organic matter (humus), fertilizers and manure, and the atmosphere; the growing crop does not discriminate between the various sources.

Soil can contain many tonnes per hectare of organic N. The amount mineralized to ammonium and then nitrified to nitrate each year depends on the soil type, the weather and probably the forms of organic N, although this is not well understood. Less than 1% of the organic N is mineralized each year, but this still represents a large amount of nitrate. A small variation in mineralization (caused by climate or farm practice) makes a big difference to the nitrate content of the soil. Mineralization occurs mainly in the autumn and spring when the soil is moist and warm.

Fertilizers have been suggested as the main source of the nitrate leached into waters. Increases in nitrogen fertilizer usage seem to be linked to increased nitrate in surface and ground waters. Unfortunately the chemistry is not as simple as the statistics suggest. Experiments have shown that very little fertilizer nitrogen is leached directly when it is applied correctly. However, the farmers who apply more than the recommended dose as an insurance, do risk leaching.

Organic manures such as farmyard manure used to be an important source of nutrients on farms. The poorly-timed application of such manures in early autumn can lead to enormous losses of nitrate by leaching. The disposal to land of increasing amounts of sewage sludge in coming years is likely to make this problem worse. Land at Rothamsted experimental farm regularly manured with farmyard manure in the autumn loses four times the amount of N by leaching and by denitrification as that lost from land receiving approximately the same amount of N as inorganic fertilizer applied in the spring.

N is also deposited on the land in rain: measurements began at Rothamsted in 1853. Measurements show that the total atmospheric input, much of which comes from vehicles and industry, is equivalent to one quarter of the average N application to intensively farmed land. We cannot label and trace atmospheric N but computer models can be used to study its fate. Almost 30% of the N deposited from the atmosphere onto arable land will be leached, because much falls when the crop cannot use it.

The fixation of atmospheric nitrogen is an important source of nitrate in farming systems where clover or other legumes are grown. When the clover is killed, usually by ploughing, several hundred kilograms of nitrate may be released, but at a rate which is usually uncontrolled and depends on the soil and the weather. If this nitrate is not used it may be leached.

In terms of the amount of nitrogen contributed to arable agriculture each year, the order of importance of the sources is

fertilizers/manures > mineralization > atmosphere.

In terms of a direct contribution to nitrate leaching in the year of application of fertilizers or ploughing of grassland, the order is

mineralization (of humus/manure) > inorganic fertilizer = atmospheric inputs.

In the long-term, on soils in which organic residues have been built up through repeated fertilizer application, the order of contribution to nitrate leaching may well change to

fertilizer ≥ mineralisation of humus > atmospheric inputs.

Reducing nitrate leaching

Leaching of nitrate cannot be prevented wherever crops are grown. But our understanding of the nitrogen cycle suggests that there are ways to minimize nitrate leaching:
- Reduction in arable farming would certainly reduce nitrate losses, but nitrate is also lost from grassland.
- Using less nitrogen fertilizer makes little difference to nitrate leaching in the short term but will reduce leaching in the long term.
- Better use of fertilizers and manure offers the best hope in terms of reduced leaching from conventional, intensive farming.

Nitrate leaching is not an easy problem to solve. There are real challenges in improving farming methods to reduce leaching losses. Chemistry and chemists will be at the forefront of such research.

Questions

Reread the passage and answer these questions:

1 What is meant by the terms leaching, nitrification, mineralization and denitrification?

2 Plot a graph of fertilizer consumption against nitrate concentration in rivers. What other changes in farming practice might account for the increasing nitrate concentration in rivers?

3 What are the advantages and disadvantages of the different methods of investigating nitrate leaching? How should the evidence be collected so that it is reliable?

4 There is a strong movement in favour of organic farming in which the use of inorganic fertilizers, pesticides and other chemical is avoided. What are the arguments for and against organic farming?

TOPIC 18

Instrumental methods

The chemical and physical properties of materials are strongly influenced by their structure at the atomic level. Therefore to understand the properties of materials it is necessary to understand their structures. This applies as much to naturally occurring substances such as rocks and minerals, and the constituents of living organisms such as cells, muscles, and bone, as it does to substances such as semiconductors for computer chips and polymers for drip-dry fabrics.

A wide range of methods is available for obtaining information about the structures of substances; in fact, any property of a material can be made to produce evidence of its structure. The most useful investigations are those in which:

1 Properties of solutions of substances are measured (e.m.f., conductivity, colour, temperature changes).
2 Matter interacts with an electric or a magnetic field (mass spectrometry, dipole moments, see Topic 9).
3 Electromagnetic radiation is emitted or absorbed by matter, giving rise to emission or absorption spectra (infra-red absorption, see Topics 7 and 12, nuclear magnetic resonance).
4 Electromagnetic radiation interacts with matter to give diffraction patterns (X-ray diffraction, see Topics 3 and 6).

Investigations in each of these areas gives different information about a substance, and when the evidence from several of them is added together it is often possible to obtain a detailed knowledge of its structure.

Descriptions of some of the most important instruments used to obtain this information now follow. You are not expected to memorize the details of the instruments; look at the **Summary** at the end of the Topic to appreciate what you are expected to learn when you study this Topic.

18.1 Using a pH meter

The hydrogen ion concentration of a solution is usually expressed as a pH number, where

$$\text{pH} = -\lg [\text{H}^+(\text{aq})]$$

In principle, the simplest method for measuring hydrogen ion concentration is to use a hydrogen electrode (see Topic 13) in the solution of unknown $\text{H}^+(\text{aq})$ concentration. This can then be combined with a standard electrode to form a complete cell, the e.m.f. of which can be measured. The value of $[\text{H}^+(\text{aq})]$ and hence the pH can then be calculated from the e.m.f.

Figure 18.1 A pH meter and electrode.

Figure 18.2 A 'stick' pH meter

For example, if the cell

Pt[H_2(g)] | 2H$^+$(aq) ⦀ Cu^{2+}(aq, M) | Cu(s)
(Concentration unknown)

is set up and the e.m.f. is found to be 0.43 V, rather than the standard e.m.f., $E^\ominus = 0.34$ V, then the value of the pH can be calculated and would be 1.5.

However, the hydrogen electrode is not easy to use. It is bulky when the hydrogen generator is taken into account, slow to reach equilibrium, and rather easily 'poisoned' by impurities. The **glass electrode** is an alternative electrode in common use.

The glass electrode consists of a thin-walled bulb blown from special glass of low melting point. A solution of constant pH (a buffer solution) is placed inside the bulb with a platinum wire dipping into it. If the bulb is now immersed in a solution of unknown pH a potential is developed on the platinum wire and the whole arrangement can be used as a half-cell. When it is combined with a suitable reference electrode it is possible to make e.m.f. measurements. The resistance of the glass bulb is high (10^7–10^8 ohm) and a very sensitive solid-state voltmeter must be used to measure the e.m.f. The reference electrode is usually a **silver/silver chloride electrode** (a silver wire coated with silver chloride dipping into saturated potassium chloride solution).

The theory of the glass electrode is complicated but an arrangement such as

Pt | solution A of known pH | glass bulb | solution B of unknown pH ⦀ reference electrode

can be attached to a voltmeter calibrated in pH units. An instrument designed on this basis is called a **pH meter**. In commercial pH meters the glass electrode and the reference electrode are often combined in one unit which can be dipped into the solution under investigation. Simple, reliable and robust pH meters are available which are useful for field work or when comparative rather than exact readings are all you need.

A pH meter should first be checked using solutions of reliable pH. Buffer solutions of pH 4 and pH 9 are suitable for checking the acid–base range of a meter. You will not get reliable readings of pH 7 in pure water because the concentration of ions is too low.

When measuring the pH of solutions the pH electrode must always be rinsed in a large volume of pure water before and between readings. It should never be used in solutions of concentrated acid or alkali.

The composition of buffers can be found in table 6.7 in the *Book of data*.

18.2 Using a conductivity meter

The measurement of the electrical conductivity of solutions enables their behaviour to be investigated at very low concentrations. The actual measurement is of **resistance** using an alternating current of at least 1000 cycles per second to avoid electrolysis of the solution. For the same reason electrodes coated with platinum 'black' must be used.

To use a conductivity meter to make comparative studies of compounds or to detect the end-point of reactions is simple. Meter readings are usually in siemens (ohm^{-1}) and can be used without calculation to compare the degree of

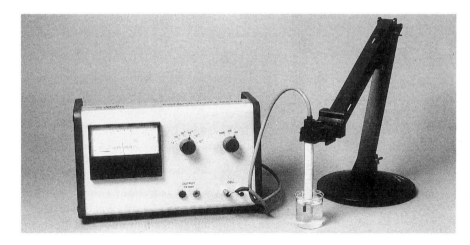

Figure 18.3 A conductivity bridge and dip cell.

The molar conductivities of ions are listed in table 6.4 of the *Book of data* and will enable you to see the variation in conductivity of ions.

ionization of compounds, for example to look for weak acids and bases. Provided there is a significant change in ionic conductivity during a reaction the end point of a titration can be detected by plotting the meter readings against volume of added reagent. The method can also be used in studies of rates of reaction where covalent reactants are forming ionic products.

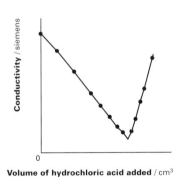

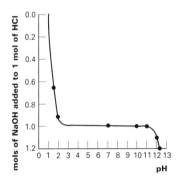

Figure 18.4 Strong acid–strong base titration with a conductivity meter and a pH meter.

Changes in electric conductance during the titration of a solution of NaOH by HCl.

Changes in pH in the titration of a solution of HCl with NaOH.

To make quantitative determinations of the conductivity of ions requires special care. Pure water of laboratory quality still contains too many ions and must be further purified, for example by being passed through a mixed cation-anion exchange resin. All solutions must be prepared with this very pure water and stored in plastic bottles: some types of glass dissolve sufficiently to interfere with the measurements. Temperature control is also important as the viscosity of water varies with temperature and has a significant effect on the results.

18.3 Using a colorimeter

Using a colorimeter is a quick and easy method of determining the concentrations of solutions which are coloured. Basically the instrument consists of the components shown in figure 18.5.

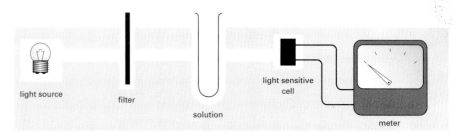

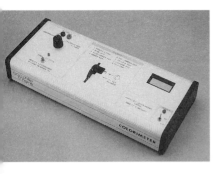

Figure 18.5 A colorimeter.

The light-sensitive cell is commonly a silicon photodiode which gives an increase in current proportional to the intensity of the light falling on it. A meter reading gives an indication of the intensity of light emerging from the solution.

Before a colorimeter can be used, the most suitable filter must be chosen. This must let through that band of wavelengths of light which are most strongly **absorbed by the solution**.

The instrument must also be **calibrated** because the meter reading may not be accurately proportional to the intensity of light. This can be done by making a set of solutions of known concentration by accurate dilution of a standard solution, and then taking readings of the intensity of light passing through the solutions and filter.

Measurements are made by first inserting a tube of pure solvent and adjusting the intensity of the incident light by means of the shutter so that the meter reading is at its maximum for transmission. A tube of solution is then put in the colorimeter and the new meter reading taken. It is most important that this procedure is adopted for all readings if possible. The meter readings are then plotted against concentration.

> **COMMENT**
> Remember that a **blue solution** is absorbing red light, not blue light, so the correct filter to use with a blue solution is a **red filter**.

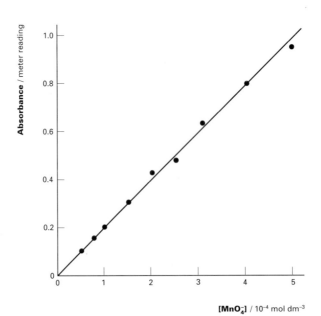

Figure 18.6 A calibration curve for manganate(VII) solutions.

[MnO_4^-] / 10^{-4} mol dm^{-3}

To carry out an experiment you will require a test-tube for the pure solvent and an optically matched one for your solution. To take a reading, the meter is adjusted to maximum for transmission with pure solvent in place, then the reading is obtained with solution in place. It is best then to check again with solvent in place. The value of the meter reading obtained can then be turned into a molarity using the calibration curve.

18.4 Using an electrical compensation calorimeter

Good values for the enthalpy changes of a number of reactions in solution can be obtained using an electrical compensation calorimeter. This apparatus consists of a vacuum flask in which are placed a small electric heater, and a thermometer which can be read to the nearest $0.1\ °C$.

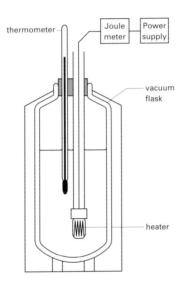

Figure 18.7 An electrical compensation calorimeter.

The reaction is carried out in the vacuum flask and the temperature change is noted. Electrical energy is then supplied:
- In exothermic reactions to produce a further rise in temperature by the same number of degrees celsius
- In endothermic reactions, to take the temperature back to its original value.

The electrical energy that is supplied can be measured directly, using a joulemeter. Alternatively, it can be calculated from readings obtained from a voltmeter, an ammeter, and a clock, and then using the relationship

$$\text{joules} = \text{volts} \times \text{amperes} \times \text{seconds}$$

This electrical energy is equal to the enthalpy change of the chemical reaction taking place in the vacuum flask. Provided the number of moles taking part is known, the enthalpy change per mole can be calculated.

To use the calorimeter the reaction is first carried out in the usual manner: mix measured amounts of the reactants, and note the initial and final temperatures. Next note the reading on the joulemeter and switch on the electric heater. Watch the thermometer carefully and switch off the heater as soon as the temperature of the contents of the flask have risen by the same number of degrees as the change which took place in the first part of the experiment.

Record the new reading on the joulemeter, and work out the number of joules of electrical energy that have been supplied. The enthalpy change per mole of reaction can then be calculated from the amounts of reactants used.

18.5 The mass spectrometer

The most accurate method of determining molar masses of atoms and molecules is by use of the mass spectrometer. Figure 18.8 shows how it works.

First let us consider the determination of the molar mass of an element. A stream of the vaporized element enters the main apparatus, which is maintained under high vacuum. The atoms of the element are bombarded by a stream of high-energy electrons which, on collision with the atoms, knock electrons out of them and produce positive ions. In most cases single electrons are removed from atoms of the element

$$E(g) \longrightarrow E^+(g) + e^-$$

although in some cases more electrons may be removed.

The positive ion stream passes through holes in two parallel plates to which a known electric field is applied, and the ions are accelerated by this field. They then enter a region to which a magnetic field is applied, and they are deflected by it.

Five main operations are performed by the mass spectrometer:
1 The sample is vaporized.
2 Positive ions are produced from the vapour.
3 The positive ions are accelerated by a known electric field.
4 The ions are then deflected by a known magnetic field.
5 The ions are then detected.

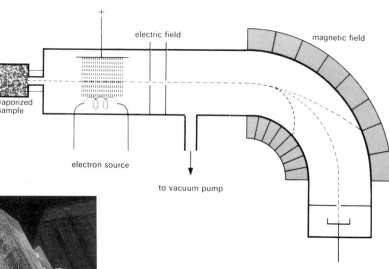

Figure 18.8 The mass spectrometer and how it works.

For given electric and magnetic fields only ions with the same charge and mass reach the detector at the end of the apparatus: all other ions hit the side walls of the instrument. By gradually increasing the strength of the magnetic field, ions of increasing masses are brought successively to the detector. A computer is used to calculate their masses from measurements of the strength of the applied fields; and their relative abundance is found from the relative magnitudes of the current produced in the detector.

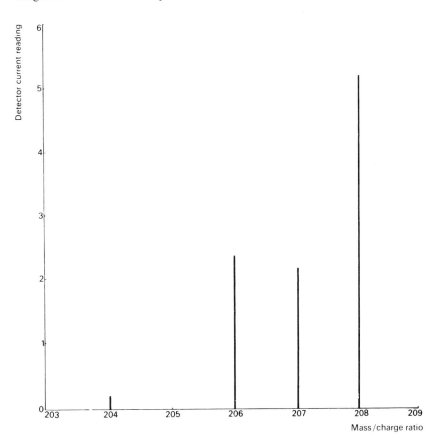

Figure 18.9 Mass spectrum for naturally occurring lead.

Figure 18.9 shows a mass spectrum for naturally occurring lead, as an example of the type of spectrum which is obtained. It will be seen that the mass spectrometer gives the relative abundance of the various isotopes of the element.

Note that the horizontal axis is labelled 'Mass/charge ratio'. This is because ions of the same mass but with different charges give separate traces in the mass spectrum. For ions carrying a single charge the mass/charge ratio is equal to the isotopic mass.

From figure 18.9 the relative abundances can be seen to be:

Isotopic mass	Relative abundance	% relative abundance
204.0	0.2	2
206.0	2.4	24
207.0	2.2	22
208.0	5.2	52
	10.0	100

From these values the molar mass of naturally occurring lead can be worked out as follows.

In 100 atoms of naturally occurring lead there will be, on average, 2 atoms of isotopic mass 204.0, 24 of 206.0, 22 of 207.0, and 52 of 208.0. If we find the total mass of all these 100 atoms we may find the average mass by dividing by 100.

Molar mass $(g\ mol^{-1}) =$
a.m.u. $(g) \times L\ (mol^{-1})$
(L = Avogadro constant)
1 a.m.u. = 1.661×10^{24}
$L = 6.002 \times 10^{23}\ mol^{-1}$

Isotopic mass	Number of atoms in 100 atoms of mixture	Mass of isotopes in 100 atoms of mixture
204.0	2	408
206.0	24	4944
207.0	22	4554
208.0	52	10816
		20722

Average mass of 1 atom $= \dfrac{20\,722}{100} = 207.2$ atomic mass units (a.m.u.)

STUDY TASK

1 Use table 2.2 in the *Book of data* to work out the molar masses to 3 SF of naturally occurring lithium and of iron. Record the working, and the result, in your notes.

2 Examine the abundance of the isotopes of tellurium, $_{52}$Te, and the molar mass of tellurium. Compare this with the molar mass of $_{53}$I and comment on their positions in the Periodic Table and their relative molar masses. Do the same for $_{18}$Ar and $_{19}$K.

We will now consider the determination of molar masses of compounds, particularly volatile compounds such as most carbon compounds.

For the determination of its molar mass, the compound under investigation is injected into the instrument as a vapour. It must, of course, be stable at whatever temperature is needed to turn it to a vapour at the low pressure inside the instrument. High velocity electrons then bombard the molecules and produce a variety of positively charged ions.

The ion detected to have the highest mass, the 'parent ion', normally indicates the molar mass of the compound. From this, and from a knowledge of the elements present, some idea of the molecular formula can be obtained by reference to tables of masses which have been compiled for the purpose. Having found the molecular formula, some idea of the structure of the compound can be obtained from the ions of smaller mass, caused by the break-up of some of the original molecules under high velocity electron bombardment.

Once the molecular formula has been found, the molecular structure can be determined. This can be done by examination of, for example, the infra-red spectrum or nuclear magnetic resonance spectrum, as will be explained in the next sections.

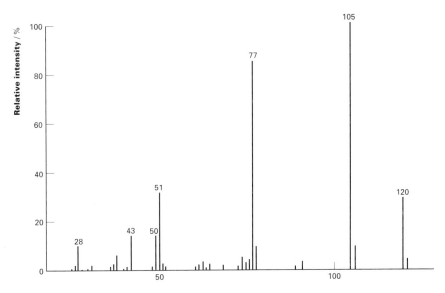

Figure 18.10 Mass spectrum of the compound v C_8H_8O.

> ### STUDY TASK
> The mass spectrum of the compound v, C_8H_8O, has a parent ion peak at 120, corresponding to the $C_8H_8O^+$ ion. The molecule also breaks into smaller fragments, such as the peaks at 105 and 77 (see figure 18.10).
>
> Try to work out a structural formula for the compound.
> How many isomers can you find with the formula C_8H_8O?
> What atoms might correspond to the loss of mass 120 $\longrightarrow$ 105 $\longrightarrow$ 77?
> Which isomers are unlikely to produce fragments of mass/charge ratio 105 and 77?
> What structure do you suggest?
> Compare your answer with the deductions that can be made from the infra-red spectrum shown in figure 18.15.

18.6 X-ray diffraction

As an example of the interaction of electromagnetic radiation with matter to give diffraction patterns, we shall consider X-ray diffraction, for this is the most precise and versatile method available for the determination of structure in solids.

X-rays have a wavelength in the region of 10^{-9} m, and are to be found in the electromagnetic spectrum beyond the far ultraviolet. When a beam of X-rays of one particular wavelength, that is, a monochromatic beam, falls on a crystalline solid the X-rays are scattered in an orderly manner. This scattering is known as **diffraction** and gives rise to a **diffraction pattern**, which can be recorded electronically. A typical X-ray diffraction pattern is shown in figure 18.11a.

Figure 18.11 **a** An X-ray diffraction photograph of a single crystal of urea. **b** Electron density map of urea. Contours are at electrons per 10^{-30} m³. **c** A model of the structure of urea.

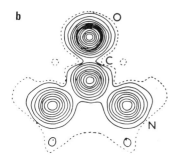

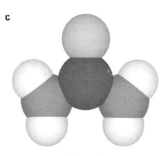

Figure 18.12 Professor Dorothy Hodgkin who won a Nobel prize for her work on X-ray cryslallography.

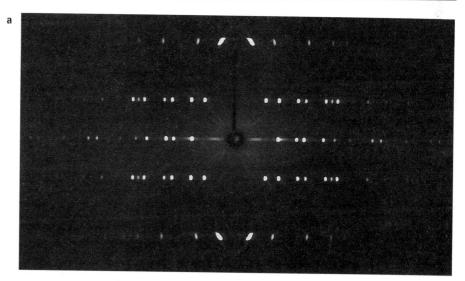

The X-ray diffraction pattern is related to the pattern of the electrons in the solid.

By analysis of the X-ray pattern it is possible to deduce the pattern of the electrons in the solid, and thus the identity of the atoms and their relative positions. The electron pattern is usually presented as an electron density map, such as figure 18.11b.

From these maps it is possible to construct a model of the structure of the solid to a high degree of precision; bond lengths, for example, can be determined to an accuracy of better than one per cent. The position of hydrogen atoms in molecules is not usually determined by X-ray diffraction, because the hydrogen atom has a low electron density which is not easily detected by X-rays. The position of the hydrogen atoms cannot be seen in figure 18.11b for this reason.

Even for a relatively simple structure the calculations which are involved in translating a diffraction pattern into a crystal structure model can be very complex. For structures such as those of proteins and of DNA the quantity of calculation involved is immense and requires huge computing capacity. Such calculations are leading to increasing knowledge of the immensely complicated molecules upon which the processes of life depend.

18.7 Infra-red spectrometry

In order to record the infra-red spectrum of a compound, an infra-red spectrophotometer is used. The intensity of a beam of infra-red radiation passed through a sample is compared with the intensity of a reference beam. The source of infra-red radiation is usually a ceramic rod heated to around 1500 °C, which emits infra-red radiation covering the whole of the required range of wavenumbers (200 to 4000 cm^{-1}).

A liquid sample is usually held as a thin film between two sodium chloride discs; a solid sample is powdered, mixed with potassium bromide, and crushed

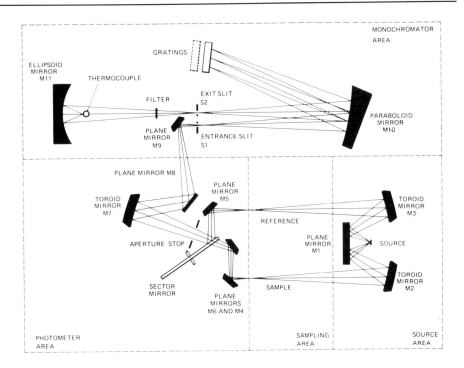

Figure 18.13 Layout of a typical spectrophotometer.

under considerable pressure to form a disc. The use of alkali halides such as sodium chloride and potassium bromide is necessary because they are transparent to infra-red radiation, whereas glass discs would absorb most of the radiation. This means, of course, that all samples have to be completely dry before their spectra can be recorded.

The level of intensity of the sample beam is compared with that of the reference beam for a range of wavenumbers. A recorder plots a graph, or **spectrum of percentage transmission against wavenumber**. Percentage transmission is the proportion of the original beam that passes through the sample without being absorbed. The sample beam is typically 90 per cent of the intensity of the reference beam when no absorption is occurring, but it can drop to 50 per cent or less at wavenumbers where absorption is occurring. Absorptions are seen as inverted peaks, or troughs, in recordings of infra-red spectra.

The absorption of infra-red radiation by molecular vibrations

A covalent molecule such as methane is not a rigid structure: the electron cloud which binds the atomic nuclei of carbon and hydrogen together allows the nuclei to move as if connected by springs.

If you hold a spring-connected model of a molecule of methane by a ball representing hydrogen and shake it, you should be able to observe some of the patterns of vibration that can occur in the model.

In the molecule of methane, the two types of vibration that take place in a C—H bond are shown in figure 18.14. Have you been able to observe these in the model?

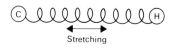
Stretching

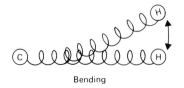
Bending

Figure 18.14 Stretching and bending of the C—H bond.

A chemical bond not only has particular patterns of vibration, it also has a natural frequency of vibration just as a spring has a natural frequency of vibration.

The significance of molecular vibrations is that a molecule will absorb electromagnetic radiation whose frequency is the same as any of the vibrations in the molecule. The vibrations in the molecule will increase in amplitude as a result of the energy absorbed from the radiation. However, there is an important restriction to this behaviour: the absorption of radiation only occurs if the vibration is accompanied by a change of dipole in the molecule. A dipole is formed by two electric charges of equal magnitude but opposite sign, a small distance apart. Thus, molecules such as hydrogen and chlorine will not absorb radiation as a result of their molecular vibration, but polar molecules such as hydrogen chloride will absorb radiation.

The radiation that is absorbed as a result of molecular vibrations lies in the infra-red region of the electromagnetic spectrum. Particular vibrations in particular bonds give rise to absorption in a particular part of this region.

In recordings of infra-red spectra (figure 18.15) these absorptions are seen as deep troughs (inverted peaks).

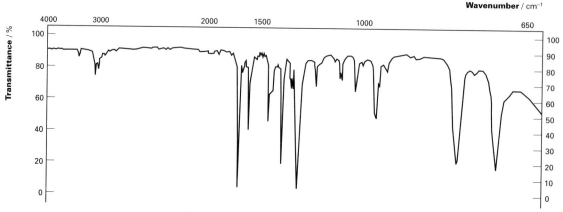

Figure 18.15 The infra-red spectrum of the compound v, C_8H_8O.

When describing infra-red spectra, the position of peaks is referred not to the frequency of the radiation but to the wavenumber, measured in cm^{-1}.

Individual wavenumbers in an infra-red spectrum are useful because each peak is characteristic of a vibration of a particular molecular structure. Thus the C—H stretching vibration absorbs at about 2900 cm^{-1} in alkanes but at about 3050 cm^{-1} in alkenes. C=C vibrations in the plane of a benzene ring absorb at both 1600 cm^{-1} and 1500 cm^{-1}.

The characteristic absorptions that are useful for the identification of particular groups of atoms in molecules of organic carbon compounds are found in the region from 200 cm^{-1} to 4000 cm^{-1}. Details are given in table 3.3 of the *Book of data*.

STUDY TASK
Identify the bonds responsible for the peaks in the spectrum of the compound v, C_8H_8O, shown in figure 18.15. Hence predict possible structures for the compound and compare your answer with the deductions that can be made from the mass spectrum shown in figure 18.10.

18.8 Nuclear magnetic resonance spectroscopy

Nuclear magnetic resonance (NMR) spectrometers are designed to investigate the absorption of radio waves by nuclei when they are exposed to a powerful magnetic field. NMR spectroscopy is one of the most valuable methods available to chemists for the determination of the structure of carbon compounds, in part because NMR enables the position of hydrogen atoms to be identified (which is not normally possible by X-ray crystallography).

Hydrogen nuclei (protons) behave like small magnets when placed in a magnetic field: the hydrogen 'magnets' become aligned either in the same direction as the applied magnetic field or in the opposite direction. The directions of alignment differ in energy. Absorption of energy from electromagnetic radiation of the appropriate frequency will cause a hydrogen nucleus to 'flip' from the lower energy alignment to the higher one. For this to happen the energy of the electromagnetic radiation must exactly match the energy difference between the two alignments – a resonance state. An appropriate frequency is 60 MHz, in the radio wave region of the electromagnetic spectrum.

However the behaviour of a hydrogen nucleus is affected by its bonding electrons. Bonding electrons shield the hydrogen nucleus to some extent from the applied magnetic field and the effect depends on the electron density of the bond: the shorter the bond the greater the shielding. So the behaviour of the hydrogen nucleus in H—C, H—O, and H—N is different and can be detected.

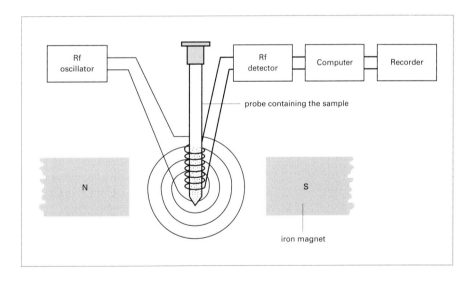

Figure 18.16 An NMR spectrometer.

COMMENT
The procedure is to record the strength of the applied magnetic field at which resonance absorption occurs for a fixed frequency radio wave; this is reported as the difference, or **shift** from the strength of magnetic field at which a reference standard substance, tetramethylsilane $Si(CH_3)_4$, showed resonance absorption. Chemical shifts are listed in table 3.4 of the *Book of data*.

Using a high resolution NMR spectrometer more information about the molecular environment of protons can be obtained. For example, the magnetic field experienced by protons in a methyl group is influenced by any proton on an adjacent atom.

As an example consider the protons in ethanol.

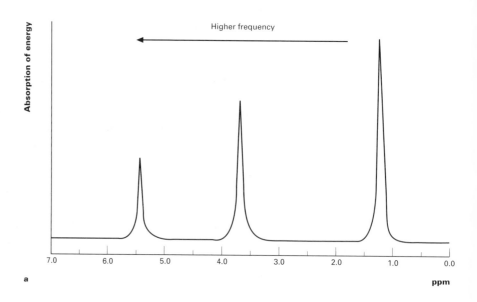

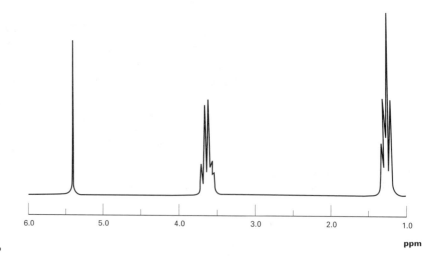

Figure 18.17 NMR spectra of ethanol **a** low resolution, **b** high resolution.

In figure 18.17a, the low resolution spectrum has peaks which correspond to the single hydrogen nucleus in the hydroxyl group, H—O, and higher peaks corresponding to the two hydrogen nuclei in the —CH_2— group and the three methyl hydrogen nuclei. But the magnetic field experienced by the methyl protons is affected by the alignment of the magnetic field of the adjacent —CH_2— hydrogen nuclei. In the high resolution spectrum (figure 18.17b) it can be seen that the single peak at low resolution is actually made up of three peaks.

As well as the nuclei of hydrogen atoms, a number of other atomic nuclei behave like small magnets: examples include ^{13}C, ^{19}F and ^{31}P. NMR spectroscopy can therefore be extended to find the positions of these atoms as well and proceeding in this way, the molecular structure of an organic compound can be studied in great detail.

Fullerenes

The use of NMR and mass spectrometry is not limited to the determination of the structure of compounds, it can also alert chemists to the presence of new substances. The new group of carbon allotropes, the **fullerenes**, was first noticed and eventually identified through their mass spectra and NMR.

In 1970 a Japanese chemist suggested on the basis of theoretical calculations that a C_{60} structure based on hexagonal and pentagonal faces (like a football) might be stable enough to exist. A Russian group made a similar suggestion but both proposals went largely unnoticed. Then in 1984 an American group identified C_{60} and C_{70} clusters as products from the vaporization of graphite by a high powered laser, but it was British and American co-workers who first prepared C_{60} and suggested that the molecule might have a football shaped structure.

Finally a joint German–American group were able to prepare C_{60} and C_{70} by an electric arc process that produced enough material for purification by solvent extraction and chromatography. This enabled proof of their structures to be obtained.

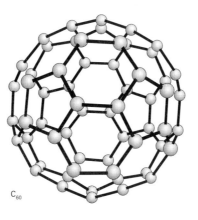
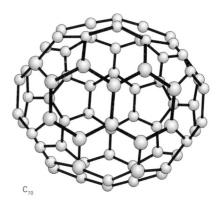

Figure 18.18 Buckminsterfullerene (C_{60}) and the C_{70} fullerene.

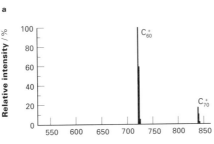
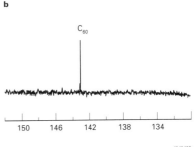

Figure 18.19 Buckminsterfullerene C_{60} **a** the mass spectrum **b** NMR spectrum.

Summary

At the end of this Topic you should be able to:
- **a** extract information from text and the *Book of data* about the application and techniques of mass spectrometry, infra-red spectrometry, X-ray diffraction and NMR
- **b** deduce empirical formulae, molecular formulae and structural formulae from data drawn from elemental percentage composition, infra-red spectra, mass spectra, and physical and chemical properties (see Topic 16)
- **c** calculate molar masses and relative isotope abundances from appropriate mass spectra
- **d** deduce the groups in organic compounds from infra-red spectra using simple correlation tables. (See Topics 7 and 12.)

APPENDIX 1

Help with mathematics

Problems with chemical calculations

The problem with chemical calculations is usually knowing how to arrange the numbers and it is often not easy to get a feel for the magnitude of the answer. There are some basic mathematical procedures we would like to remind you about.

Most of the relationships you will use in chemistry are like the relationship of the number of moles of a substance used in an experiment to the mass that was used. When you want to *double* the number of moles you have to *double* the mass you use. When you want to *treble* the number of moles used you have to *treble* the mass you use. This is an example of **direct proportionality** and the symbol $\propto$ has the meaning 'is proportional to', so

number of moles $\propto$ mass of substance

Whatever we increase or decrease the number of moles by, the mass will increase or decrease by the same multiplier.

If the number of moles = 2 mol when the mass = 80 g
then $\qquad\qquad$ 3 × 2 mol $\qquad\qquad$ 3 × 80 g
$\qquad\qquad\qquad$ 0.2 × 2 mol $\qquad\qquad$ 0.2 × 80 g

The other type of proportionality is **inverse proportionality**. An example is the time taken for sulphur to precipitate from sodium thiosulphate solution when acid is added. When the concentration of thiosulphate is increased the time for precipitation decreases.

Increasing the thiosulphate concentration × 3 reduces the time to $\frac{1}{3}$. So for this reaction, time is inversely proportional to concentration:

$$\text{time} \propto \frac{1}{\text{concentration}}$$

The multiplier rule still applies but we multiply by the reciprocal.

If a concentration = 2 M gives a time = 50 seconds
then $\qquad\qquad$ 3 × 2 M $\qquad\qquad\quad$ $\frac{1}{3}$ × 50 seconds
$\qquad\qquad\quad$ 0.2 × 2 M $\qquad\qquad\quad$ $\frac{1}{0.2}$ × 50 seconds

One approach to a calculation is to do it in stages using what is known as the **unitary method**. For example, if a measurement shows that 1.8 cm^3 of mercury has a mass of 24.5 g, it is sensible to work out the mass of 1 cm^3 of mercury as a first stage. Then you can work out the mass of mercury in any other volume, such as 500 cm^3.

As the mass and volume are in direct proportion to one another
1.8 cm³ of mercury has a mass of 24.5 g
you divide both quantities by 1.8 to find the mass of 1 cm³ of mercury:
1 cm³ has a mass of 13.61 g
and multiply both by 500 to find the mass of 500 cm³:
500 cm³ has a mass of 6800 g, to 2 SF

To get a feel for the answer to calculations that involve large numbers of figures you can do the calculation with easier numbers first. For example, you can make a rough estimate of the answer to the mercury problem by working out the mass of 500 cm³ on the basis of 25 g in 2 cm³. You can see quite readily that the volume has increased $\times 250$, so the mass will also increase by $\times 250$. If you then write down the procedure you used for this easier example you can use the same procedure for the actual problem:
(final volume ÷ initial volume) × mass = $(500 \div 1.8) \times 24.5 = 6800$ g, to 2 SF.

Significant figures (SF) are explained in the next section.

QUESTIONS

1 4.63 g of copper are precipitated by 4.07 g of iron. What mass would be precipitated by 55.85 g of iron (1 mol)? Are the masses directly or inversely proportional to each other? What is your first estimate of the answer?

2 100.2 g of cyclohexanol (1 mol) should form 82.1 g of cyclohexene (1 mol). How much should you obtain from 15.0 g?

3 The density of cyclohexanol is 0.962 g cm⁻³. What is the volume of 100.2 g (1 mol)?

4 The mass of 1 mol of sodium hydroxide is 40 g. What is the mass of
   i 2 mol ii 0.1 mol iii 0.02 mol iv 0.17 mol?

5 What is the mass of sodium hydroxide dissolved in
   i 1 dm³ of a 2 M solution ii 100 cm³ of a 0.1 M solution iii 35 cm³ of a 0.02 M solution iv 182 cm³ of a 0.17 M solution?

ANSWERS
1 63.5 g to 3 SF
2 12.3 g to 3 SF
3 104 cm³ to 3 SF
4 i 80 g ii 4 g iii 0.8 g iv 6.8 g
5 i 80 g ii 0.4 g iii 0.028 g
   iv 1.2 g

Significant figures

In a number such as 126.9045, which is the most accurate value we have for the molar mass of iodine, I,
the **first** figure 1**26.9045** occupies the highest place value and is called the most significant figure, abbreviated to SF, or the first significant figure, 1 SF
the **second** most significant figure is 12**6**.9045
the **third** most significant figure is 126.**9**045, and so on.

When you do not need such an accurate value you can **round off** the figures. You can round off to one significant figure: this means rounding off to the most significant figure
126.9045 ⟶ **100** for **1** SF
or you can round off to two significant figures:
126.9045 ⟶ **130** for **2** SF
or you can round off to three significant figures:
126.9045 ⟶ **127** for **3** SF.

To round off a number you have to know whether to round up or down:

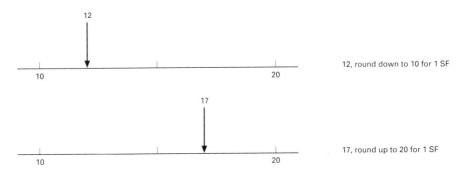

The rule is that when the next significant figure is less than five (4, 3, 2, or 1) round down, when it is five or more (5, 6, 7, 8 or 9) round up. Note particularly how the rule applies to the figure 5.

You can now see how we rounded off the molar mass of iodine.

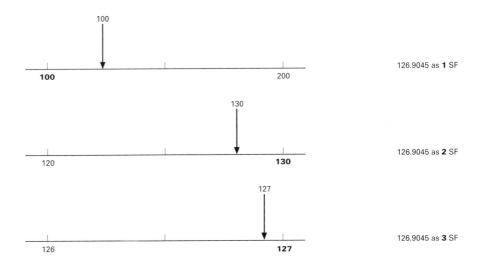

However you will often use molar masses as exact numbers and should not adjust your answers for their apparent number of significant figures. Thus molar masses are usually quoted as $H = 1$, $C = 12$ g mol^{-1} but these numbers should not be treated as having only 1 SF and 2 SF.

In chemistry, another rule you will need is:

Zeros at the end of a number are significant figures when they occur after a decimal point.

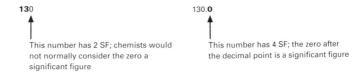

You are most likely to need this rule when doing calculations involving the concentrations of solutions. The zeros are used to describe the accuracy with which the solution was prepared: for example 5.844 g of sodium chloride (4 SF) in 100.0 cm^3 of solution (4 SF) would be described as 1.000 M (4 SF).

0.02 M is only accurate to 1 SF
0.020 M is accurate to 2 SF
0.0200 M is accurate to 3 SF

When reporting the results of calculations you need to think about how many significant figures there should be in the answer. You need to identify the quantity used in the calculation with the **least number** of significant figures; your final answer should have the **same number** of significant figures.

After a titration you might need to calculate the mass of dissolved reactant. If you used a pipette to measure 25.0 cm^3 of 0.0200 M acid and found that the end-point was 21.3 cm^3 your numbers all have 3 SF so you should state your answer to 3 SF.

But if the concentration of acid was 0.020 M your answer can only have 2 SF. And if you used a measuring cylinder rather than a pipette you would have to decide how to record the volume of acid: 25 cm^3 to 2 SF would probably be appropriate, with your answer restricted to the same number of significant figures.

We should mention one other guideline. You should only adjust your **final** answer to an appropriate number of significant figures and not the starting quantities or intermediate answers. Otherwise you may not arrive at the best answer:

$$27 \times 16 \times 2 = 858$$
$$= 900 \text{ to 1 SF}$$

but adjusting the **starting** numbers to 1 SF:

$$30 \times 20 \times 2 = 1200$$

QUESTIONS

1. Round off the molar mass of iron 55.847 g mol^{-1} to **i** 1 SF **ii** 2 SF **iii** 3 SF **iv** 4 SF
2. Round off the Faraday constant 96 484.6 C mol^{-1} to **i** 1 SF **ii** 2 SF **iii** 3 SF **iv** 4 SF
3. Round off the solubility of nickel bromide 0.516 mol/100 g to **i** 1 SF **ii** 2 SF
4. Round off the solubility of iron(II) sulphate 0.103 mol/100 g to **i** 1 SF **ii** 2 SF
5. Round off the Avogadro constant 6.02204531×10^{23} mol^{-1} to **i** 1 SF **ii** 2 SF **iii** 3 SF **iv** 4 SF **v** 5 SF **vi** 6 SF **vii** 7 SF.

Indices

A common method of writing down very large and very small numbers is to use **indices** (singular: index).

Numbers can be written down as 'powers of a base number' such as 10^{23} or 2^3 or e^5 or 10^{-9}, where 10, 2, e are the base numbers and the indices are 23, 3, 5 and -9.

When the base number is 10 the form of expression is called **scientific notation**.

There are four rules that you need to learn in order to deal with indices correctly when you meet them in calculations:

Rule 1 A positive index means $\quad n^3 = n \times n \times n$

Rule 2 A negative index means $\quad n^{-3} = 1/n^3 = \dfrac{1}{n \times n \times n}$

Rule 3 A fractional index means $\quad n^{\frac{1}{3}} = \sqrt[3]{n}$

Rule 4 A zero index means $\quad n^0 = 1$

From these rules we can derive some guidelines for working out the answers to problems involving indices:

1 **To multipy add the indices** $\qquad\qquad 10^5 \times 10^3 = 10^8$
2 **To divide subtract the indices** $\qquad\quad 10^5 \div 10^3 = 10^2$
3 **To work out the value of fractional indices take the root**
$$10^{\frac{1}{3}} = \sqrt[3]{10} = 2.15 \text{ (to 3 SF)}$$

QUESTIONS

Work out the answers to these problems, using the values in the *Book of data*:
1 Multiply the elementary charge on the electron by the Avogadro constant. Which other constant have you calculated?
2 Multiply the Boltzmann constant by the Avogadro constant. Which other constant have you calculated?
3 Divide the rest mass of the proton by the rest mass of the electron.

Logarithms

The difficulties associated with the division and multiplication of large numbers were made easier by the invention of logarithms in 1614 by a Scotsman, John Napier. The effect is to reduce division and multiplication to the much simpler processes of subtraction or addition.

Logarithms are a special case of indices, called common logarithms, abbreviated to lg, when the base number is 10

$$10^2 = 100 \qquad \Rightarrow \log_{10} 100 = 2$$

Some more examples

$10^4 = 10000 \Rightarrow \log 10\,000 = 4$

$10^{-2} = 0.01 \Rightarrow \log 0.01 = -2$

$10^{-4} = 0.0001 \Rightarrow \log 0.0001 = -4$

Converting numbers to their common logarithms

Example 1 For the logarithm of 123.4 on most calculators you can enter 123.4 followed by the log button, producing the answer 2.091315.

$$\log(123.4) = \log(1.234 \times 10^2)$$
$$= 2.091 \text{ to 4 SF}$$

which is the equivalent of

$$123.4 = 10^{2.091}$$

Example 2 For the logarithm of 0.009876, the answer from your calculator should be -2.0054.

$$\log(0.009876) = \log(9.876 \times 10^{-3})$$
$$= -2.005 \text{ to 4 SF}$$

which is the equivalent of

$$0.009876 = 10^{-2.005}$$

Converting logarithms to their numbers

Example To find log 3.5 on most calculators you can enter 3.5 followed by the 10^x button

$$x = \log 3.5$$
$$= 10^{0.5} \times 10^3$$
$$= 3200$$

Example For log $x = -1.5$ the answer from your calculator should be 0.0316

$$x = 10^{0.5} \times 10^{-2}$$
$$= 0.032$$

QUESTIONS

1 Convert these numbers to logarithms

   75, 7.5, 0.75, 0.00075, 7.5×10⁴, 7.5×10⁻⁴

2 Convert your answers to **1** back to numbers.

3 Write down the exact procedure you used on your calculator to get the correct answers to **1** and **2**.

4 Use the relationship

   pH = −log [H⁺(aq)]

   to calculate the pH of solutions in which [H⁺(aq)] is
   i 0.5 mol dm⁻³ **ii** 0.0005 mol dm⁻³ **iii** 5×10⁻⁶ mol dm⁻³
   iv 5×10⁻¹² mol dm⁻³.

5 Convert your answers to **4** back to [H⁺(aq)].

APPENDIX 2

Laboratory safety

You are more likely to suffer a minor injury – a cut, burn or scald – in a kitchen than in a laboratory. The reason is quite simple; we know there are hazards involved in working in a laboratory so risk assessments are made for every experiment and then protective measures taken to control those risks. In chemistry lessons most risks arise from the use of chemicals but some other practical activities have associated hazards (for example, micro-organisms or electricity).

Laboratory safety is about minimizing exposure to risk as well as protecting yourself from the results of mishaps. When you do an experiment that involves the production of an unpleasant gas like nitrogen dioxide, you should think about using small quantities and containing the gas in a fume cupboard as well as thinking about what you might need to do if the gas is produced unexpectedly fast.

All the experiments in this course have been checked for safety but when you are going to do an investigation of your own you will be expected to carry out a risk assessment (and have it checked before starting any practical work).

Figure A2.1

RISK ASSESSMENT FORM

Title of the experiment			
Outline of the procedures			
Hazardous substances being used or made	Nature of the hazards (e.g. toxic, flammable)	Quantities being used or made	Control measures and precautions
Any non-chemical hazards and precautions to be taken			Signed:
Disposal of residues			Date:

Risk assessments are most conveniently made on a standard form and an example is shown here. The procedure is straightforward when you realize that the intention is to protect you from any risks. The steps in making an assessment are:

1 **Write down the procedures** you will be using (chemicals used or made, quantities, techniques; any non-chemical hazards)

2 Use reference sources to **identify any hazardous chemicals** you are planning to use or make. The appropriate warning symbol should be on reagent bottles and in suppliers' catalogues.

3 **Record the nature of the hazards** involved and the way you might be exposed to the hazard. There are standard reference sources with this information such as the 'Hazcards' published by CLEAPSS.

4 **Decide what protective or control measures to take** so that you can carry out your practical work in safety.

5 Find out how to dispose of any hazardous residues from your practical work.

The protective measures you need to take will depend on your laboratory as well as your experiment. The experiments in this course have been assessed for use in a well lit, well ventilated and uncrowded laboratory. Where conditions are different additional protective measures may be necessary.

As well as the specific protective measures to be taken when hazardous chemicals are being used there are also general procedures to be observed in all laboratories at all times.

- Long hair should be tied back, and you should not wear wet look hair preparations, which can make hair unusually flammable. Do not let ties, scarves and cardigans hang freely, where they could be a fire hazard. We strongly recommend the wearing of laboratory coats.

- Eating, drinking and chewing are not permitted in laboratories. It is in fact contrary to the COSHH Regulations to permit eating, drinking or indeed smoking or the application of cosmetics in any area which could be contaminated with hazardous chemicals.

- **Eye protection should be worn whenever a Risk Assessment requires it**, or whenever there is any risk to your eyes. This includes, for example, washing up at the end of the lesson and even when you have finished practical work, as long as other students are still working.

- You should find that the chemicals that you are going to use are in clearly labelled stock bottles, with the name of the chemical, any hazards, and the date of acquisition or preparation. In taking liquids from a bottle, remove the stopper with one hand, and keep the stopper in your hand whilst pouring from the bottle. This way, the stopper is likely to be replaced at once, and remain uncontaminated. Pour liquids from the opposite side to the label, so that it does not become damaged by corrosive chemicals.

- Study carefully the best techniques for safely heating chemicals. Small quantities of solid can be heated in test-tubes; liquids present greater problems, because of the risk of 'bumping' and 'spitting'. Boiling tubes are safer than test-tubes (because of their greater volume), but should be

less than one-fifth full. You are fairly likely to point test-tubes away from your own face, but do remember the need to do the same for your neighbours. **Use a water bath to heat flammable liquids, NEVER use a naked flame**.

- When testing for the odour of gases the gas should be contained in a test-tube (not a larger vessel) and the test-tube held about 10–15 cm from your face, pointing away. Fill your lungs with air by breathing in, and then cautiously sniff the contents of the test-tube, by using a hand to waft the vapours cautiously to your nose. Slowly bring the test-tube nearer, if necessary. If you are asthmatic you should not smell gases without a report from other students because gases such as chlorine are harmful.

- **You must always clear up chemical spillages straight away**. Whilst a few spills may need chemical neutralization or similar treatment most minor spills can be dealt with by a damp cloth. (Don't forget to rinse it afterwards.)

- In the event of getting a chemical in your eye, or on your skin, **flood the area with large quantities of water**. Keep the water running for at least ten minutes. Rubber tubing on a tap is the most convenient way of doing this. Even if the chemical reacts exothermically with water, provided a large quantity of water is used, the heating effect will be negligible.

- A heat burn from apparatus, scalding liquids or steam is treated by **immersing the area in cool water** for at least ten minutes. Preferably use running water from rubber tubing, fixed to a tap.

- **Report all accidents at once**.

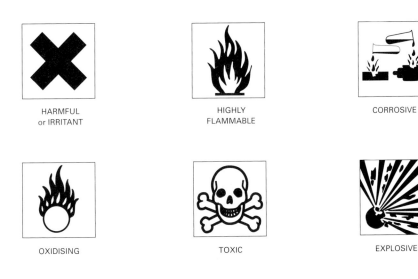

Figure A2.2 Warning symbols

Acknowledgements

Photographic acknowledgements

Allsport/Shaun Botterill: page 317
American Chemical Society: page 30 bottom
Apple Computer UK Ltd: page 436 top left
BP Chemicals: pages 191, 323
The Bancroft Library: page 71 top
Bayer: pages 29, 348
Bayer Diagnostics: page 434
C Blackie: page 462
Boltzmann Society: page 284
Paul Brierley: page 68
British Library: page 23
Bruce Coleman/John Cancalosi: page 414
Bruce Coleman/Hans Reinhard: page 358
Bruce Coleman/Leonard Lee Rue: page 31
Bruce Coleman/Kim Taylor: page 256
Bruce Coleman/Michael Viard: page 32 left
Bruce Coleman/Gunter Ziesler: page 104
De Beers: page 281
Mary Evans Picture Library: pages 73, 88
European Vinyls Corporation UK Ltd: page 212 bottom
Ronald Grant Archive: page 245
Robert Harding Picture Library: pages 4 bottom, 161 right
Philip Harris: pages 538 top, 539
Harvard University Archives: page 474
Dorothy Hodgkin: page 546 bottom
Michael Holford: pages 46, 436 bottom left
Hulton Deutsch: page 276
Hulton Deutsch/Bettman Archive: page 433
Hydro Agri Persgrunn: page 289
ICI Chlor Chemicals: pages 71 bottom, 175
ICI Paints: page 435
Johnson Matthey: page 508
Professor A. Keller, University of Bristol: page 249
Keystone: page 164 bottom
Frank Lane Picture Agency: page 279
Dame Kathleen Lonsdale: page 546 top
Mansell Collection: page 22
Museum of London: page 340
Dr RB Moyes, The University of Hull: page 542
National Medical Slide Bank: page 121
National Portrait Gallery: page 523
Natural History Museum: page 7
Natural History Photographic Agency/GJ Cambridge: page 105
Natural History Photographic Agency/Jane Gifford: page 313 left
Natural History Photographic Agency/Michael Leach: page 313 right
Natural History Photographic Agency/David Woodfall: page 341
Norfolk Lavender Ltd: page 32 right
Norwegian Road Research Laboratory: page 30 top
Perkin-Elmer: page 547
Planet Earth/W.R. Whiteway: page 465
Prestige Group: page 212 top
Dr J.P.G. Richards: page 132
Rothamsted Experimental Station: pages 78, 535
Royal Mail: page 56
Science Photo Library: pages 9, 280, 378, 396, 474 left
Science Photo Library/Martin Dohrn: page 436 right
Science Photo Library/Vaughan Fleming: page 85
Science Photo Library/Richard Folwell: page 272
Science Photo Library/Simon Fraser: page 299
Science Photo Library/François Gohier: page 4 top
Science Photo Library/Adam Hart-Davis: page 404
Science Photo Library/Harvey Pincis: page 4 centre
Shell: page 171
Solexpress: page 538 bottom
TWI: page 13
University of Bristol: page 416
University of Pennsylvania, Edgar Fahs Smith Collection: page 24 top
Michael Vokins: pages 12, 24 bottom, 118, 103, 161, 170, 180, 349, 436
N. Uyeda: page 57 bottom

Examination questions

The Trust and the Publishers are very grateful to the University of London Examinations and Assessment Council for permission to reproduce questions from past Examination papers.

Other acknowledgements

The poems *Iron* (page 3) and *Nitrogen* (page 507) are by Roger McGough. They are reprinted by permission of the Peters Fraser and Dunlop Group Ltd.
Reading task 1 and figures in it (pages 52 to 55) are based on Scurlock, J. and Hall, D. 'The carbon cycle'. Inside Science Number 51. *New Scientist*, 1991.
Reading task 2 and figures in it (pages 128 to 131) are based on Cox, T. 'Origin of the chemical elements'. Inside Science Number 29. *New Scientist*, 1990.
Reading task 3 (pages 240 to 244) is based on Emsley, J. 'Photochemistry' Inside Science Number 58. *New Scientist*, 1993.
Figures R3.1 to R3.4 are based on Folkins, I. and Brasseur, G. 'The chemical mechanisms behind ozone depletion'. *Chemistry and Industry*, 8, 1992.
Reading task 4 (pages 431 to 434). We are very grateful for the help of the British Diabetic Association.
Figure R4.1, after Hernandez-Pacheco, F. *Bulletin de Real Sociedad Espagnola*, Madrid, 1921.
Figures R5.2, R5.4 and R5.5 are from Snyder, Solomon. *Drugs and the brain*. Copyright © 1986 by Scientific American Books, Inc. Reprinted with permission of W.H. Freeman and Company.
Reading task 6 (pages 532 to 536) is based on Goulding, K. and Poulton, P. 'Unwanted nitrate'. *Chemistry in Britain*, 28, 1992.
Figure R6.1, data from the Fertillizer Manufacturers Association.
The study task (pages 347 to 350) is based on Tolhurst, T. 'Production of citric acid'. *NCSB Newsletter*, 6, 1989.
Figure 3.2 is from Perrin, J. *Les Atomes*. Presses Universitaires de France, 1970.
Figures 3.17, 6.3 and 15.4 are after Witte, H. and Wofel, E. *Reviews of Modern Physics*, 30, 1958.
Figure 6.1 is after Coulson, C. A. *Proceedings of the Cambridge Philosophical Society*, 34, 1938.
Figure 7.30a is after Cox, E.G., Cruickshank, D.W.J. and Smith, J.A.S. 'Crystal structure' *Proceedings of the Royal Society*, 1958.
Figure 9.18 is after Dickerson, R.E. (ed. Neurath, H.) *The proteins*. Academic Press, 1964.
Figure 14.16 is after Thompson, E.O.P. 'The Insulin molecule'. *Scientific American*. Copyright 1955 by Scientific American, Inc. All rights reserved.
Figure 18.11b is after Vaughan, P. and Donohoe, J. *Acta crystallographica*, 5, 1952.

Index

absolute potentials 362
absolute zero of temperature 88, 281
acetic acid *see* ethanoic acid
Acetobacter 312
achiral compounds 393
acid amides *see* amides
acid anhydrides 329
    reactions 330–31, 333, 335
acids and bases 72–4
    acid–base reactions 276
        entropy changes 283
        equilibria 299–306
        in living materials 311–14
        see also neutralization reactions
    acid–base titrations 307–8
    strengths 74, 303–6
    see also dissociation constants
acrylamide *see* propenamide
'acrylic' fibres 421
actinides 19
activated complex 227
activation energy 225–6
    effect of catalyst 227, 228
activity series of metals 359–60
acyl chlorides 329
    reactions 330–32
addiction 506
addition polymerization 192, 414
addition reactions 204
    of alkenes 188–9, 190–91, 192
    of benzene 203
adiabatic changes 315
adsorption 229–30
air
    acid constituents 314
    see also atmosphere
alanine 394
'alcohol' *see* ethanol
alcohols
    bond energies 138–40
    manufacture 191
    naming 36, 41
    preparation 183, 329, 331–2
    reactions
        dehydration 37–8, 40, 43–4
        ester formation 325, 327–8, 331, 332
        hydroxyl group replacement 178, 179, 184
        oxidation 37, 38–40
            products 45–6
        with sodium 37, 39
    see also primary alcohols; secondary alcohols; tertiary alcohols
aldehydes 45
    oxidation 45–6
    preparation 39–40
    reduction 329
aldohexoses 399
alkali metal halides
    lattice energies 97
    reactions 114–15

alkali metals *see* Group 1 elements
alkaline earth metals *see* Group 2 elements
alkaline manganese cell 383–4, 385
alkaloids 313, 503–4, 506
alkanes 29, 160–64
    boiling points 163, 247
    naming 35, 157–8
    preparation 191
    reactions 164–71
    see also cycloalkanes
alkenes 40, 44, 184
    isomerism 185–6, 187
    in petrol 174
    reactions 187–9
    synthesis 183
alkoxides 39
alkyl groups, naming 158
alkylation
    of arenes *see* Friedel–Crafts reactions
    of petrochemicals 173
Alsophila pometaria (moth), 'antifreeze' in 41
aluminium
    electron density map 437
    reactions
        with acids 374
        with iron(III) oxide *see* Thermit reaction
amides 329
    preparation 331, 475
    reactions 330, 331
amines 343
    preparation 183, 476
    reactions 344–7
        with acid anhydrides and acyl chlorides 330, 331
        with nitrous acid 518–21
amino acids 403–8
    metabolism 312, 433
    pH of solutions 410
    separation 405–6
ammonia
    –ammonium chloride buffer solution 310, 512
    electronic structure 133–5, 508, 509
    elimination from living organism 312
    ligand 449
    manufacture 289, 515–18
    reactions 344–5, 513
        with acid anhydrides and acyl chlorides 330, 331
        catalytic oxidation 513
        with halogenoalkanes 183, 346
        with hydrogen chloride 72
            enthalpy change 91
        with hydrogen halides 115
        with water 301, 512
    stability 509–10
    synthesis 227
        catalysts for 230–31

ammonium chloride
    –ammonia buffer solution 310, 512
    acidic solution 301
    crystals 67
    reaction with barium hydroxide
        entropy change 276, 283
        free energy change 380
ammonium ion 512
    Brønsted–Lowry acid 301
    electronic structure 143–4
ammonium iron(II) sulphate, preparation 15–16
ammonium iron(III) sulphate, preparation 17
amorphous polymers 418–19
amounts of substance 9–10
amphetamines 504, 506
α-amylase 409–10
anaemia 464–5, 466
analgesic drugs 503–4
aniline *see* phenylamine
animals
    acid–base reactions in 311–12
    micronutrients 464–6
anisole *see* methoxybenzene
antifreeze 41, 191
antiseptics 336
archaeological fraud 7
arenes
    naming 197
    production in catalytic reforming 173, 174
    reactions 198–203
argon, van der Waals forces in 247, 248
aroma compounds 391–2, 482–3
aromatic hydrocarbons *see* arenes
Arrhenius equation 226
Arrhenius, Svante 52, 73
    definitions of acids and bases 73
ascorbic acid oxidase 465
aspartame 431, 432
aspartic acid, isomer conversion 214
Aspergillus niger (mould) 312, 346–7
aspirin 337–8, 341–3
    infra-red spectrum 339
Atacama salt lake 104
atactic polymers 420
atmosphere
    carbon dioxide levels 52, 53, 54–5
    nitrogen deposition from 536
atomic number 19
atomic radii 65, 66
    of transition elements 461
    see also covalent radii; van der Waals radii
atomic volume 20
atomization *see* enthalpy changes of atomization
atoms
    arrangement of electrons in 61–6
    models 56–7

structure 20
attacking groups 181–2
Avogadro, Amedeo 9
Avogadro constant 9, IBC
azo-dyes 518–25

Bacon, Francis 345
bacteria, metabolism 312
Bakelite 423
balanced equations 14, 117
    redox 373–4
Banting, Frederick 433
barium hydroxide
    reaction with ammonium chloride
        entropy change 276, 283
        free energy change 380
bases *see* acids and bases
batteries 382–5
BCC (body-centred cubic) crystal
    structure 438, 441
Benedict's solution 39, 46, 397, 398, 433
benzene 35
    crystal structure 259
    molecular structure 149, 194–7
    reactions
        addition 203
        substitution 202–3, 480
    toxicity 197–8
benzenediazonium chloride 518–19, 520–21
benzenesulphonic acid, preparation 202
benzocaine 504, 505
benzoic acid
    reaction with methanol 327–8, 484
    reduction 329
Best, Charles 433
bidentate ligands 451, 453
'big bang theory' 128–9
Bill of Mortality (London, 1665) 340, 341
biodegradability 402
biuret test 405
'bleach' *see* sodium chlorate(I)
blood
    glucose levels in 432, 434
    pH 312
body-centred cubic (BCC) crystal
    structure 438, 441
boiling points
    of alkanes 163, 247
    of alkenes 184
    effects of intermolecular forces 246–7
    of halogens 246
    of hydrogen halides 252–3
    of noble gases 252
Boltzmann, Ludwig 276–7, 284
Boltzmann's constant 276, 277
bond angles 135, 147, 148, 162
bond breaking 167, 204
bond energies 136–43
    in alkanes 162
    in alkenes 186
    in compounds of carbon and
        silicon 156
    in halogenoalkanes 176, 181, 182
    of hydrogen bonds 252
    in organic compounds 474
bond lengths
    ethane 162
    methanoic acid and methanoates 147–8
    multiple bonds 142–3
    nitric acid and nitrates 147, 149
bond polarization 145–6, 176, 182
Born–Haber cycle 94–5, 97
brain 501–2
bromine
    in human metabolism 122
    occurrence and extraction 104, 105
    reactions
        with arenes 199–200, 202
        with halide ions 106–7
        with phenols 333, 335
        with sodium thiosulphate 117
        with unsaturated compounds 32–3, 188, 189–90, 199
bromoalkanes, hydrolysis 221–4
bromobenzene, preparation 202
1-bromobutane, hydrolysis 222
1-bromo-1-chloroethane, isomerism 392, 393
bromoethane, preparation 179, 184
2-bromo-2-methylpropane,
    hydrolysis 223
bromophenol blue 308
1-bromopropane, preparation 191
2-bromopropane, preparation 191
bromothymol blue 308
Brønsted, J. N. 74
bronze 436
Brownian motion 57, 284
Buchner flask 111–12
buckminsterfullerene 551
buffer solutions 309–10, 327, 512, 513
    in living materials 311, 313
butane 33–4, 35, 169
butanoic acid 312
trans-butenedioic acid, enzyme catalysed
    reaction 409
button cells 384
butylamine 343
    reactions 344–5

calcium
    plant nutrient 514
    role in agriculture 78–9
calcium carbonate
    reaction with hydrochloric acid, kinetic
        study 215–16
    see also limestone
calcium fluoride, electron density map 68
calcium hydroxide
    application to soil 79
    solubility 77
caliche 104
calorimeters
    combustion 139
    electrical compensation 541–2
cancer, skin 243
candle 1
cannabis 504, 506
caprolactam, manufacture 244
carbocations 189
carbohydrates 397–400
carbon
    carbon–carbon bonds 135, 185
        free rotation at 157, 162
        reactions forming 475–6
    carbon–hydrogen bonds,
        vibrations 547–8
    enthalpy change of graphite–diamond
        conversion 93
    ionization energies 96
    origin in Universe 130
    see also fullerenes
carbon compounds 155–7
    see also organic compounds
carbon cycle 52–5
carbon dioxide 314
    elimination from living organisms 311
    see also carbon cycle
carbon monoxide 170
    electronic structure 144
carbonic acid–hydrogencarbonate buffer
    system 311
carbonic anhydrase 311
carbonyl compounds *see* aldehydes;
    ketones
carboxylic acids 323–9
    naturally occurring 400–403
    preparation 39, 40, 46, 475, 476
    see also fatty acids
Carothers, W. H. 414
castor oil, hydrolysis 401
catalysis 227–9, 374
    by transition metals 454–6, 462
    heterogeneous 227, 229–31
    homogeneous 227, 228, 455
    see also enzymes
catalytic cracking 172–3
    of alkanes 165–6, 170–71, 184
catalytic reforming 173, 174
cell diagrams 362, 369, 370
cellulose 258, 400
    see also cotton
cellulose ethanoate, dyeing 521–2
CFCs *see* chlorofluorocarbons
chain reactions 168–9, 170, 204
chance, operation in chemistry 273–9
chemical calculations 553–8
chemisorption 229, 230–31
'chemists' toolkit'
    amounts of substance 9–10
    balancing redox equations 117
    electronegativity 145–6
    formulae and equations 13–14
    molar masses of organic compounds 42
    naming organic compounds 157–60
    oxidation numbers 107–9
    solution concentration 75–6
    standard electrode potentials 371–4
charge, elementary IBC
chiral compounds 393–4, 399
    amino acids 403, 405
chitin 400
chloride, ligand 449
chlorides, periodicity of properties 120
chlorine
    in human metabolism 122
    occurrence and extraction 104, 105
    oxidation numbers 110
    reactions
        with alkanes 166–9, 170
        with halide ions 106–7
        with iron 457–8
        with sodium hydroxide 110–11
chloroalkanes, bond energies 142
1-chlorobutane, mass spectrum 177
4-chloro-3,5-dimethylphenol *see* Dettol

Index

chloroethane, reaction with ammonia and amines 346
chloroethene, polymerization 193, 212
chlorofluorocarbons (CFCs) 241–2
chloromethane 175
2-chloro-2-methylbutane, preparation 179
2-chloro-2-methylpropane
   hydrolysis 222–3
   preparation 179
   reactions 180–83
chlorophyll 240, 241, 464
chlorpromazine 503
cholesterol 244, 394
cholesteryl benzoate, liquid crystals 394
chromate–dichromate equilibrium 292
chromatography
   gas, purification of organic compounds 487
   paper, separation of amino acids 405–6
   thin layer, of nitrobenzoates 485
chromium, crystal structure 438
chromium(II) ethanoate, preparation 451
cis–trans isomerism 185–6
citric acid 29, 313
   manufacture 312, 347–9
   reaction with sodium hydrogencarbonate 86
   uses 349–50
Clausius, Rudolf 276, 278, 280
cobalt
   cobalt-60 436
   micronutrient 466
cocaine 504–6
codeine 503–4
collision theory (reaction kinetics) 225–6
colorimetry 539–41
   in blood sugar measurements 434
   in kinetic studies 217
combustion
   of alkanes 164–5, 169–70, 174
   of alkenes 187–8
   of arenes 198
   of halogenoalkanes 180
   of magnesium, entropy changes 275, 280, 281–2
   of phenols 333
   see also enthalpy changes of combustion
combustion analysis 486–8
combustion calorimeter 139
complex ions
   containing iron 8, 18, 290, 292
   determination of formula 454
   ligand substitution reactions, entropy changes 453
   shapes 448
   stability constants 448–50
   of transition metals 448–54, 462
concentration 75–6
   effect on reaction rate 213–14
condensation polymerization 414–15, 423
conductance (solutions) 66
conductivity measurement, in kinetic studies 217, 222–3
conductivity meter 66, 217, 538–9
conservation of energy 89
copper
   copper–tin alloy *see* bronze
   crystal structure 438

micronutrient 465–6
zinc–copper cell 365–6
   see also Daniell cell
copper(II) complexes 448–51
copper(II) sulphate
   migration of ions 67
   reactions
     with amines 345, 346
     with iron 15
     with zinc 86, 360
corrosion 358
   of iron (rusting) 18
cortisone 474, 475, 476
cotton 258, 400
   dyeing 521–2
Coulson, C. A. 132
coupling reactions 518–21
covalent bonds 133–6
   bond energies 136–43
   dative 143–4
covalent radii 247, 248
Crab Nebula 131
crude oil 160–62
   fractionation 172
crystal structures
   ionic compounds 67–9
   metals 438–41
crystalline polymers 419–21
curare 501
cyclamate 431, 432
cycloalkanes, naming 35, 159
cycloalkenes, naming 187
cyclohexane 32–3, 35
   in caprolactam production 244
   fire 164
cyclohexanol, dehydration 43–4
cyclohexanone, oxidation 329
cyclohexene 32–3
   preparation 43–4
   reactions 187–9
cysteine 312

Daniell cell 362, 365–6, 380, 382
dative covalency 143–4
debye (unit) 176
detergents 402
Dettol 336
deuterium 129
Devarda's alloy 514, 515
diabetes 431–4
1,2-diaminoethane, ligand 453
diazotization 518–20
1,2-dibromopropane, preparation 190
dichromate–chromate equilibrium 292
di(dodecanoyl) peroxide, polymerization initiator 188
diethylamine
   infra-red spectrum 344
   preparation 346
diffusion, in living organisms 311
dihalogenoalkanes, preparation 190
dihydrofolate reductase, analysis 412
1,2-dihydroxybenzene, ligand 449, 450, 451
dilatometry, in kinetic studies 217
4-dimethylaminoazobenzene, preparation 521
N,N-dimethylphenylamine, coupling reactions 520

diols 41
   preparation 191
dipole moments 250–51
   of halogenoalkanes 176, 181
dipole–dipole interactions 246
dipoles
   permanent 249–50
   see also polar molecules
direct proportionality 553–4
disaccharides 400
'Dispersol' Fast Yellow G 525
displayed formulae 34
disproportionation reactions 111
dissociation constants 304–6, 513
distillation 43–4, 514
   see also steam distillation
DNA 257
dopamine 503
dot-and-cross diagrams 70, 71, 133–4
double bonds 135–6, 143, 185
double salts 15–17
drawing (polymer film) 420–21
drugs 501–6
dyes 518–25
dynamic equilibrium 291, 293

Earth, composition 3, 128
edta, ligand 449, 450, 451, 453
Einstein, Albert 284
elastomers 422–3
electric eel 358
electrical compensation calorimeter 541–2
electrochemical cells 361, 382–5
   cell diagrams 362, 369, 370
   Daniell cell 362, 365–6, 380, 382
   entropy changes in 375–7
   zinc–carbon 359
   see also electromotive force
electrode potentials 366, 511
   measuring 369–70
   see also standard electrode potentials
electromotive force (e.m.f.) 361, 380–81
   calculating 374
   contributions of electrode systems 363–4
   measuring 365–6
electron affinity 92–3, 95, 96
electron density maps
   aluminium 437
   benzene 195
   covalent compounds 132–3
   ionic compounds 68, 69, 133
   urea 546
electron microscopy 57
electronegativity 145–6
electrons 20
   delocalization 147–9
     in azo-compounds 519, 524–5
     in benzene ring 195
     in carboxylate group 148–9, 324
     in metals 438
     in nitrate ion 147, 509
     in phenoxide ion 334–5
   energy levels 59–64
   in ionic compounds 69–71
   lone pairs 135, 251–2, 448, 508
   rate of transfer 214
   rest mass IBC
   transfer in redox reactions 107, 360–61

electrophiles 189, 204
electrophilic addition reactions 189, 190
electrophilic substitution reactions 201–3
elements
   atomic volumes 20
   classification *see* Periodic Table
   molar masses 9, 19
   origin 128–31
   standard enthalpy changes of atomization 92
   standard enthalpy changes of formation 93, 94–5
   standard free energies 379
   symbols 13
elimination reactions 40, 182–3, 204
Ellingham diagrams 463
emission spectra 58–60, 128
empirical formula 487
endothermic reactions 12, 275–6, 318
   enthalpy changes 87, 88
   entropy changes 278, 283
energy changes of reactions
   measuring 85–8
   see also enthalpy changes
energy levels *see under* electrons
energy sharing 273–4
enthalpy changes of atomization 142
   standard 92, 94–5
enthalpy changes of combustion
   measuring 139–40
   standard 136–8
enthalpy changes of formation
   standard 88, 89–90, 91, 137–8
      of elements 93
      of solutions 264
enthalpy changes of hydration 265
enthalpy changes of reactions 87–90
   measuring 541–2
   standard 88, 91
   see also Hess's Law
enthalpy changes of solution 263–4
enthalpy changes of sublimation 246
enthalpy changes of vaporization 253–5
entity (chemical) 9–10
entropy 276–9
   changes in electrochemical cells 375–7, 377–8
   changes in ligand replacement reactions 453
   in elastomers 422
   and equilibrium reactions 314–18, 516–17
   measuring 279–84
   standard 280–81
enzymes 256–7
   copper-containing 465–6
   in fermentation 46
   molybdenum-containing 466
   see also names of specific enzymes
ephedrine 501
equations 14
   ionic 292
   redox 117, 373–4
equilibrium constant 294–8, 380–81
   for ionization of water 301–2
   see also dissociation constants; stability constants
Equilibrium Law 294–8
equilibrium state 291–3

   and entropy 314–18
ester exchange 261
esters
   formation 325, 327–8, 330, 331–2, 333
   hydrolysis 227, 295, 296, 331
   see also pheromones
ethanal, catalytic decomposition 228
ethanamide 329, 330, 409
ethane 33–4, 35, 162
   bond energies 141
   electronic structure 136
ethanedioic acid, reaction with potassium manganate(VII) 454–5
ethane-1,2-diol 41
   preparation 191
ethanenitrile, preparation 475
ethanoic acid
   –ethanoate buffer system 309–10, 327
   determination in vinegar 310
   glacial 324
   manufacture 323
   reactions 324–5
      with ethanol 295, 296, 325
ethanoic anhydride 329
   reactions 330, 333, 335, 337–8, 342
ethanol 36
   hydroxyl group replacement 179, 184
   infra-red spectrum 325, 326
   manufacture 191
   NMR spectra 550–51
   physiological effects 506
   production in fermentation 46–8
   reactions 37
      with acid anhydrides and acyl chlorides 330
      with ethanoic acid 295, 296, 325
ethanoyl chloride 329
   reactions 330, 346, 347
2-ethanoylaminobenzoic acid 345
ethene 185
   electronic structure 136
   manufacture 171
   reactions 191–2
ethyl 4-aminobenzoate
   reactions 344–5
      diazotization/coupling 519–20
ethyl ethanoate
   preparation 325
   reaction with water 227, 295, 296
ethylamine 346
ethylbenzene, preparation 202, 475
ethylene *see* ethene
ethylenediaminetetra-acetic acid *see* edta
exothermic reactions 12, 275–6, 318
   enthalpy changes 87, 88
   entropy changes 278, 280, 281–2

face-centred cubic (FCC) crystal structure 438, 440, 441
Faraday, Michael 1
Faraday constant IBC
fats and oils, hydrogenation 191
fatty acids 402, 432–3
FCC (face-centred cubic) crystal structure 438, 440, 441
Fehling's solution 39
fermentation 46–8, 347–8
fertilizers 79, 532–4, 535
   nitrogen determination in 513–15

fibres 421–2
flame colours 58–9
flame spectra *see* emission spectra
flash photolysis 241
floating corks experiment 245
floating needles experiment 246
flow sheets (organic syntheses) 478–80
fluorine
   in human metabolism 121–2
   occurrence and extraction 104
Fluothane 175
forests, destruction 54, 55
formulae 13–14, 33–4, 487
fossil fuels 52, 53, 54
Fourier, Jean-Baptiste 52
Francis, A. W. 190
free energy *see* Gibbs free energy
free radicals 204
   in caprolactam synthesis 244
   in combustion in car engine 174
   in ethanal decomposition 228
   in photochemical reactions 167–9, 170, 203
   in polymerization reactions 192
   in stratosphere 241–2
Friedel–Crafts reactions 199, 202, 475
fructose, reactions 397–9
fuels
   alkanes 169–70
   see also fossil fuels
fullerenes 551
fumarase 409
functional groups 36, 159–60, 474

gas chromatography *see under* chromatography
gas constant IBC
gastric juice 312
geometric isomers, alkenes 185–6
germanium, existence predicted 25
Gibbs free energy 377–81, 382, 460, 509–10
Gibbs, Willard 280, 284, 378
ginger beer 47–8
glacial ethanoic acid 324
glass, photochromic 461–2
glass electrode 538
glass temperature
   elastomers 422
   thermoplastics 418
glucose
   metabolism in diabetics 432, 433, 434
   reactions 397–9
   solubility 257, 259–60
glucose oxidase 434
glutamic acid 404–5
glycerol *see* propane-1,2,3-triol
glycine 403, 404
glycol *see* ethane-1,2-diol
graphite–diamond conversion 93
graphs, straight line 225
'greenhouse effect' 52, 54–5
Group 1 elements, flame colours 58–9
Group 2 elements, flame colours 58–9
gutta-percha 186

Haber process *see* ammonia, manufacture
haemocyanin 465
haemoglobin 411, 464, 465

Index

half-cells 363, 369
half-reactions 107, 360
hallucinogenic drugs 504
halogenoalkanes 175–8
  preparation 178–9, 184
  reactions 180–83
    with amines 346
    with cyanides 475, 476
    hydrolysis 221–4
halogens 103
  boiling points 246
  electronegativities 145
  in human metabolism 121–2
  occurrence and extraction 104–6
  oxidation numbers 110
  properties 119–21
  reactions
    with alkalis 110–11
    with alkanes 166–9, 170
    with halide ions 106–7
  see also individual halogens
hazard warning symbols 180, 561
hazards 560
heat capacity 85
HCP (hexagonal close-packed) crystal structure 438, 440, 441
heating safely 560–61
helium
  discovery 59
  intermolecular forces in 246
  origin in Universe 128–9
heroin 503, 504
Hess's Law 89–90, 91, 264
heterogeneous catalysis 227, 229–31
heterolytic fission 167, 204
hexadentate ligand 451, 453
hexagonal close-packed (HCP) crystal structure 438, 440, 441
hexane
  properties 164–5, 170
  as solvent 259–60
hexanedioic acid, preparation 329
Hippocrates 342
Hodgkin, Dorothy 546
Hofmann, A. W. von 523
Hofmann, F. 342
homogeneous catalysis 227, 228, 455
homolytic fission 167, 204
hydration (ions) 263–5
hydration energy 264
hydrazine 509–10
hydrazoic acid 509–10
hydrocarbons
  aromatic *see* arenes
  saturated 157, 160
  unsaturated *see* alkenes; cycloalkenes
hydrochloric acid
  in gastric juice 312
  reactions
    with calcium carbonate, kinetic study 215–16
    with sodium thiosulphate, kinetic study 224–5
hydrogen
  chemisorption 229
  electron density for H_2 132
  emission spectrum 59
  energy levels 59–60
  origin in Universe 128–9

reactions
  with alkenes 191
  with nitrogen *see* ammonia, manufacture
hydrogen bonds 251–3
  in carboxylic acids and alcohols 325, 326
  in living organisms 256–8
  in nylon 421
  in protein molecules 412
  in water 251, 252, 255–7, 260
hydrogen bromide, preparation and properties 115
hydrogen chloride
  electronic structure 133–5, 146
  occurrence in volcanic gases 105
  preparation and properties 115
  reactions
    with ammonia 72
      enthalpy change 91
    with iron 458–9
    with water 300–301, 303
  see also hydrochloric acid
hydrogen electrode 363–4, 537–8
  standard 364, 366–7
hydrogen fluoride, occurrence in volcanic gases 105
hydrogen halides
  addition to alkenes 191
  hydrogen bonding in 252–3
hydrogen iodide, preparation and properties 115
hydrogen ions 73, 300–301
  see also hydroxonium ions; pH scale
hydrogen peroxide 105
  catalytic decomposition 228
  reaction with Rochelle salt 228
hydrogen sulphide, reaction with potassium manganate(VII) 373–4
hydrogencarbonate–carbonic acid buffer system 311
hydroxonium ions (H_3O^+) 74, 143, 301
4-hydroxyazobenzene, preparation 520
2-hydroxybenzoate ion, ligand 449, 450, 451
2-hydroxybenzoic acid 342
  infra-red spectrum 339
  preparations using 337–8, 342

ice, hydrogen bonding in 255–6
ideal gas, molar volume IBC
indicators
  acid–base 72, 300, 307–8
  corrosion 18
  for thiocyanate titrations 297–8
indices 557
infra-red spectra 489, 491, 492, 546–8
  of alkanes 163
  of alkenes 185–6
  of amines 343–4
  of benzene 196
  of carboxylic acids 325, 326
  of ethanol 326
  of halogenoalkanes 177
  of propanone 327
initiators (polymerization) 188, 192, 193
insects, 'antifreeze' in 41
insulin 33
  molecular structure 411, 412, 413

role in metabolism 432–3
intermolecular forces 245–58
  and solubility 259–65
  see also hydrogen bonds
International Union of Pure and Applied Chemistry (IUPAC) 157, 362
inverse proportionality 553
invert sugar 399
iodide
  reactions
    with iron(III) 368–9, 372, 373
    with persulphate ion 456
iodine
  crystal structure 259
  in human metabolism 122
  micronutrient 464
  occurrence and extraction 104, 105–6
  partition between solvents 290
  reactions
    with acid and base 290
    with alkali 111–12, 292
    with halide ions 106–7
    with propanone, kinetic study 214, 218–20, 221
    with sodium thiosulphate 112–13
  solid–solution equilibrium 291
  solubility 259
ionic compounds 66–71
  electron densities 133
  formulae 14
  oxidation numbers in 108
  solubility 260, 263–5
ionic equations 292
ionic radii 65, 66
ionization energies 60, 61–3, 65–6, 93, 94
  and ion formation 95–6
  of transition elements 461
ions
  electronic structure 69–70
  formation 95–6
  migration 67
  polarization 97–8
  shape 69
  solvation (hydration) 263–5
iron 3–4
  chemistry 5–8
  colour reaction with thiocyanate 6, 8, 290, 292
  corrosion (rusting) 18
  crystal structures 441
  micronutrient 464–5
  origin in Universe 130
  reactions
    with copper(II) sulphate 15
    of iron(III) with iodide 368–9, 372, 373
    redox 442–4
    Ag^+/Fe^+ equilibrium 297–8
iron alum, preparation 17
iron(III) chloride
  hydrated 264–5
  reactions
    with phenols 333
    with potassium ethanedioate 18
iron chlorides
  preparation 456–9
  properties 459–60
iron(II) compounds 5
  detection 6, 8

Index

oxidation 6, 7
iron(III) compounds 5
  detection 6, 8
  reduction 6, 8
iron(II) nitrate, thermal decomposition 8
iron(III) oxide, reaction with aluminium *see* Thermit reaction
iron(II) sulphate
  reactions
    with potassium halates(V) 118
    with sodium chlorate(I) 116
  thermal decomposition 8
'iron tablets', analysis 445
isomerization (petrochemicals) 173, 174
isomers 34, 247
  geometric (alkenes) 185–6
  optical (mirror-image) 392–3, 399
  structural 187
isotactic polymers 420, 421
isotope tracer studies
  esterification reaction 328
  nitrogen cycle 533
  solid–solution equilibria 291
isotopes 19, 543–4
IUPAC *see* International Union of Pure and Applied Chemistry

jet, deflection in electrostatic field 250
joulemeter 87

Kekulé, F. A. 194–5
Kelvin, Lord 88, 284
kelvin (unit) 88
ketohexoses 399
ketones 45
  detection in urine 434
  oxidation 45–6
  reduction 329
'knocking' (car engines) 172, 174
Kolbe synthesis 342

laboratory safety 2, 559–61
lactase 466
lactic (2-hydroxypropanoic) acid 312
lanthanides 19
latent heat of vaporization *see* enthalpy changes of vaporization
lattice energies 94–7, 263
laundry bags, soluble 261–2
lavender oil 32
Lavoisier, Antoine 22–3
Laws of Thermodynamics
  First 89
  Second 277
  Third 281
Le Châtelier, H. L. 284
Le Châtelier's Principle 318, 517–18
lead, mass spectrum 543–4
leaving groups 181
Leclanché cell 383–4, 385
  see also zinc–carbon cell
Lewis, G. N. 71
Liebig, Justus von 52
ligands 346, 448, 449, 450–51
limestone 53, 54
  application to soil 79
limonene 31–3
line emission spectra 58–60
lipids 400–403

liquid crystals 394–6
Lister, Joseph 336
lithium cells 384, 385
lithium chloride, electronic structure 70
lithium halides, ionic polarization in 97–8
lithium oxide, electronic structure 70
lithium tetrahydridoaluminate reductions 329, 331–2, 476
local anaesthetics 504–5
logarithms 557–8
Loschmidt, Josef 9, 195
Lowry, T. M. 74
lysimeter 534, 535

macromolecules 391
macronutrients 463
magnesium
  combustion, entropy changes 275, 280, 281–2
  ionization energies 96
  magnesium–copper cell 365
  magnesium–lead cell, e.m.f. 374
  magnesium–zinc cell 365
  plant nutrient 514
  reactions
    with acid, kinetic study 232
    with zinc sulphate 360, 361
magnesium fluoride, electronic structure 70
malic acid 29, 313
manganese, micronutrient 465
manure 535
margarine 191
marijuana *see* cannabis
Marker, Russell 30
mass number 19
mass spectrometry 19, 328, 542–5
  for determining molar mass 487, 489, 490, 492, 542–4
  of fullerene 551
  of halogenoalkanes 177–8
  in protein analysis 411–12
mathematics 553–8
mauve (dye) 523
melamine 423
melamine–methanal polymer 423
melanoma 243
melting points
  determination 485–6
  of hydrides 252
Mendeleev, Dmitri 22, 24–5
metal/metal ion systems 359–67, 371
  entropy changes in 375–6
metallic bonds 437–8
metals
  activity series 359–60
  corrosion 358
  crystal structures 438–41
methanal polymers 423
methane 29–30, 33–4, 35, 349
  bond energies 141
  combustion 169
  electronic structure 133–5
  reaction with chlorine 166–9
  standard enthalpy change of formation 137–8
methanoate ion, electronic structure 147–8

methanoic acid, electronic structure 147–8
methanol 36
  reactions
    with 2-hydroxybenzoic acid 337, 338
    with benzoic acid 327–8, 484
  as solvent 259–60
methoxybenzene 197–8
  reactions 198–9, 200–201
4-methoxybenzoic acid, electron density map 132
methyl benzoate
  nitration 484–6
  preparation 327–8, 484
methyl 2-hydroxybenzoate *see* oil of wintergreen
methyl 4-hydroxybenzoate, reactions 332–4
methyl 2-methylpropenoate (methacrylate) 188
methyl 3-nitrobenzoate, preparation 484–6
methyl orange 300, 308, 525
methylbenzene 197–8
  reactions 198, 200
  as solvent 259–60
2-methylbutan-2-ol, reaction with hydrochloric acid 179
2-methylpropane 34
2-methylpropan-2-ol, reaction with hydrochloric acid 179
methylpropene, synthesis 183
methylurea 409
micronutrients 463–6
models
  atoms 56–7
  carbonyl compounds 45
  metallic structures 438–41
  organic molecules 392–3
  wax molecule 2
Mohr's salt, preparation 15–16
molar masses 9–10, 19, 555
  determination 487, 489, 490, 492, 542–4
  of organic compounds 42, 487
  molar volume, ideal gas IBC
mole 9
molecular compounds
  electron densities 133
  oxidation numbers in 108
  shapes of molecules 133–5
molecular crystals, van der Waals forces in 247
molecular formulae 34, 487
molybdenum, micronutrient 464, 466
monoamine oxidase 503
monodentate ligands 450
monosaccharides 400
morphine alkaloids 503–4, 506
multiple bonds 135–6
  lengths and energies 142–3
myoglobin 258, 411

naphthalen-2-ol, coupling reactions 519, 520
Napier, John 557
natural product chemistry 29–33, 391–3, 397–413
neon, electronic structure 64
neurotransmitters 502–3

Index

neutralization reactions 72, 73, 76–7, 301
  energy change 86
neutrons 20, 128–9, 130
  rest mass IBC
nickel 436
  hydrogen chemisorption 229
  Raney 191
nickel complexes 453
nicotine 29
ninhydrin test 404, 405, 406
nitrate ion
  electronic structure 147, 149, 508–9
  levels in water 525
nitrates
  in agriculture 532–6
  effects on plant growth 525
nitration reactions 199, 202–3, 485–6
nitric acid
  electronic structure 144, 147, 149
  manufacture 508
  properties 510–12
  in rainwater 314
  reactions
    with arenes 199, 202–3
    with phenols 334, 335–6
nitriles
  preparation 475
  reactions 475–6
nitrite ion, levels in water 525
nitrobenzene, preparation 202–3
nitrogen
  chemisorption 230–31
  determination in fertilizer 513–15
  plant nutrient 514
  reaction with hydrogen see ammonia, manufacture
nitrogen compounds 508–25
  structure and energetics 508–10
nitrogen cycle 533–4, 535
nitrogen oxides 511
  NO_2–N_2O_4 equilibrium 295, 297, 315, 316
2-nitrophenol 336
4-nitrophenol 336
  colour of ion 524
nitrosyl chloride 244
nitrous acid 511, 512
  reactions with amines 518–21
NMR see nuclear magnetic resonance spectroscopy
noble gas structures 69–71, 133
noble gases 64
  boiling points 252
nomenclature
  of alkenes 187
  of arenes 197
  of carbon compounds 33–6
  of halogenoalkanes 175–6
  of organic compounds 157–60
  use of oxidation numbers 109
Nomex 419
norepinephrine 503
nuclear fusion reactions
  after 'big bang' 129
  in stars 129–30
nuclear magnetic resonance spectroscopy 549–51
nucleophiles 182, 183, 204
nucleophilic substitution reactions 182, 183, 184, 330–31, 475
  S_N1 and S_N2 222, 224
nylon 29, 414, 422
  dyeing 521–2
  effect of drawing 420
  fibres 420, 422
'nylon rope trick' 416

ocean, role in carbon cycle 52, 53, 54
octane number 172–4
oct-1-ene, infra-red spectrum 186–7
odour, testing for 561
ohm (unit) 66
oil of wintergreen
  infra-red spectrum 340
  preparation 337, 338
optical activity 399
optical isomers 392–3, 399
oranges, extraction of limonene from peel 31
orbitals 62, 64
order of reaction 213
  determination 215–20
organic compounds 29–30
  identification 486–92
  naming 33–6, 157–60, 175–6, 187, 197
organic synthesis 473–92
ornithine cycle 312
oxalic acid see ethanedioic acid
oxidation 107
  of alcohols 37, 38–40, 46
  of alkanes 165
  of alkenes 188, 191
  of aluminium see Thermit reaction
  of ammonia 513
  by electron transfer 360–61
  of carbonyl compounds 45–6
  of cyclohexanone 329
  of ethanedioic acid 454–5
  of hydrogen sulphide 373–4
  of iron(II) compounds 6, 7
  of unsaturated compounds 33, 188, 191
oxidation numbers 6, 107–9
  and balancing equations 117
  and cell diagrams 369
  fractional 113
  variable 110, 120, 442–8, 461–2
oxide ion, enthalpy change of formation 96
oxidizing agents 109
  determination 112–13
oxygen
  in Cambrian atmosphere 241
  origin in Universe 130
'ozone layer' 240, 241–2, 243, 532

paper 400
paper chromatography see under chromatography
paracetamol 343
Pauling, Linus 145
pentanenitrile, preparation 475
peptide group 403
perfumery see aroma compounds
Periodic Table IFC, 19–20, 64–5
  history 22–5
periodicity 120
  atomic volumes 20
  electronegativities 145
  ionization energies 63
  properties of halogen compounds 120, 145
Perkin, W. H. 523
peroxidase 434
peroxides, polymerization initiators 188, 192, 193
Perrin, Jean Baptiste 284
Perspex, preparation 188
persulphate, reaction with iodide ion 456
petrochemicals 29, 170–71, 184, 402
petrol hydrocarbons 172–4
petroleum see crude oil
pH
  of amino acid solutions 409
  of blood 312
  change during acid–base titrations 307
  effects on enzymes 409–10
  from dissociation constant data 305–6
  of gastric juice 312
  measurement 299–300
    see also pH meters
  of soil 78–9
  of solutions of acids and bases 73, 74
  of solutions of iron compounds 5
pH meters 303, 537–8
pH scale 302–3
phenelzine 503
phenol, coupling reaction 520
phenol–methanal polymer see Bakelite
phenolphthalein 300, 308
phenols 332–6
phenyl ethanoate, preparation 335
phenylamine 346
  diazotization 518
  oxidation (Perkin's synthesis) 523
  reaction with acyl chlorides 347
phenylmethanol, preparation 329, 331
phenylmethylamine, preparation 476
pheromones 331, 481
phosphoric acid, reactions with halides 115
phosphorus, plant nutrient 514
photochemical reactions 240–44
  halogen–alkane 166–9, 170
photochromic glass 461–2
photographic film 114, 240
photons 167, 240
photosynthesis 241, 313
physical constants IBC
π-bonds (pi bonds) 135, 185
picric acid see 2,4,6-trinitrophenol
pigments 8
Piltdown fraud 7
pining (disorder of sheep) 466
Planck constant IBC
plane-polarized light 393, 395–6, 397
plants
  acid–base reactions in 313–14
  chemicals from 30–33
  effect of nitrates on growth 525
  micronutrients 463–6
  mineral requirements 514
  photosynthesis 241, 313
  preferred soil pH 78–9
  role in carbon cycle 52, 53
platforming see catalytic reforming
polar molecules 146, 250, 300
polarimeter 393, 398

polarization
  of bonds 145–6, 176
  of ions 97–8
  of molecules *see* polar molecules
polyamide *see* nylon
poly(chloroethene) 175, 413–14
  manufacture 193
polydentate ligands 450
poly(epoxyethene) 422
polyester 421
  crystallinity 420
  dyeing 521–2
  preparation of resin 416
poly(ethene) 192, 193–4, 414
  crystals 249
  high- and low-density 420
  properties 164–5, 417, 420
poly(ethenol)
  crosslinking 417
  'slime' 417, 423
  soluble laundry bags 261–2
polymerization 204, 414–17
  of alkenes 188, 192, 193–4
  initiators 188, 192, 193
polymers 413–22
poly(methyl 2-methylpropenoate) *see* Perspex
polypeptides 403
  see also proteins
poly(propenamide) 415
poly(propene)
  fibres 421
  isotactic and atactic 420
poly(propenenitrile), fibres 421
polysaccharides 400
polystyrene 30
polyvinyl alcohol *see* poly(ethenol)
Porter, George 241
potassium
  ionization energies 62–3
  plant nutrient 514
potassium ethanedioate, reaction with iron(III) chloride 18
potassium halates(V), reactions 118
potassium hexacyanoferrate(III) test 6, 8, 368
potassium hydrogencarbonate, enthalpy change of decomposition 90
potassium hydroxide
  reactions
    with halogenoalkanes 180, 181, 182–3
    with iodine 111–12
potassium iodate(V)
  determination 113–14
  preparation 111–12
potassium iodide
  reactions
    with potassium halates(V) 118
    with sodium chlorate(I) 116
potassium manganate(VII)
  migration of ions 67
  reactions
    with cyclohexanone 329
    with ethanedioic acid 454–5
    with hydrogen sulphide 373–4
    with unsaturated compounds 33, 188, 191
  titrations using 445

potassium nitrate
  crystals 67
  solid–solution equilibrium 291, 292–3
potassium peroxodisulphate(VI), polymerization initiator 193
potassium sodium tartrate, reaction with hydrogen peroxide 228
potassium thiocyanate, colour reaction with iron 6, 8, 290, 292
potassium triethanedioatoferrate(III), preparation 18
precipitation reactions, iron compounds 5, 7
pressure, effect on entropy change 318, 517–18
primary alcohols 41
  dehydration 37–8, 40
  hydroxyl group replacement 178, 179, 184
  oxidation 37, 38–9, 39–40, 46
  reaction with sodium 37, 39
primary electrochemical cells 382, 385
primary halogenoalkanes 178
propanal 45
  preparation 40
  reactions 45–6
propane 33–4, 35, 169
propane-1,2,3-triol 41
  esters *see* fats and oils
propanoic acid, preparation 40
propan-1-ol 36
  dehydration 37–8, 40
  oxidation 38–40
propanone 45
  detection in urine 434
  dipole 249
  reactions 45–6
    with iodine, kinetic study 214, 218–20, 221
propenamide, polymerization 415
propene
  preparation 40
  reactions 190–91
proportionality 553–4
propylamine, solubility 260
propylene *see* propene
proteins 403–5, 410–11
  amino acid composition 411–12
  hydrogen bonding in 256–7
  metabolism 312
  shapes of molecules 257, 258, 412–13
  see also enzymes
protons 20, 128–9
  mass IBC
  see also hydrogen ions; nuclear magnetic resonance spectroscopy
psoriasis 243
psychedelic drugs 504
PVC *see* poly(chloroethene)

quantum shells 62, 64

r-process (element formation) 131
Ramón y Cajal, Santiago 501
Raney nickel 191
rate constant 214
rate-determining step 220
  in ammonia synthesis 230
rate equations 213–14, 216

rates of reaction 212
  effect of concentration 213–14
  effect of temperature 224–6,
  measuring 214–19
  and reaction mechanism 219–24
  see also catalysis
Rauwolfia 341
reaction kinetics *see* rates of reaction
reaction mechanism 219–24
reactivity series 11–12
red giants (stars) 130
redox equations 117, 373–4
redox reactions 358–87, 461–2
  energy changes in 463
  entropy changes in 376–7
  and oxidation numbers 106–9
  of transition elements 6, 7–8, 11, 442–5, 446–8
reducing agents 109
reducing sugars 398, 433
reduction 107
  of alkenes 191
  by electron transfer 360–61
  of carboxylic acids and derivatives 329, 331–2
  of iron(III) compounds 6, 8
  of iron(III) oxide *see* Thermit reaction
  of nitriles 476
reflux apparatus 38
Reinitzer, Friedrich 395
reserpine 341
resistance (solutions) 66
reversible reactions 289–98, 316
  see also acid–base reactions
risk assessment 559–60
Roberts, John 328
Robinson, Robert 474
Rochelle salt *see* potassium sodium tartrate
rock salt 105
rocks, role in carbon cycle 53, 54
rounding off 554–6
rubber 186, 422
rusting 18

s-process (element formation) 130
saccharin 431, 432
safety in laboratory 2, 559–61
salicin 341
salicylic acid *see* 2-hydroxybenzoic acid
salt bridge 365, 382
Sanger, Frederick 411, 412
saturated compounds 33
saturated fatty acids 402
saturated hydrocarbons 157, 160
schizophrenia 503
scientific notation 557
seaweeds, iodine in 105–6
secondary alcohols 41
  hydroxyl group replacement 184
  oxidation 46
secondary electrochemical cells 382, 385
secondary halogenoalkanes 178
separating funnel 43–4
serine 403
SI units 9, IBC
siemens (unit) 66
s-bonds (sigma bonds) 135, 185
significant figures 554–6

Index

silicon compounds 155–6
silver, Ag^+/Fe^{2+} equilibrium 297–8
silver halides 114
  lattice energies 97
silver/silver chloride electrode 538
skin, photochemical reactions in 242–3
slaked lime *see* calcium hydroxide
'slime' 417
S_N1 and S_N2 reaction mechanisms 222, 224
soaps 401–2
sodium
  electronic structure 64
  ionization energies 61–2
  reactions
    with alcohols 37, 39
    with phenol 332–3
sodium carbonate, reactions with phenols 333
sodium chlorate(I) (sodium hypochlorite) 110–11
  reactions 116
sodium chloride
  crystals 68
  electron density map 68, 132
  lattice energy 94–5
  see also rock salt
sodium chromate(VI), reactions 290
sodium dichromate(VI)
  reactions 290, 452
    with alcohols 37, 38–9, 40
    with carbonyl compounds 45
sodium hydrogencarbonate, reaction with citric acid 86
sodium hydroxide
  reactions
    with ethanamide 330
    with halogens 110–11, 292
    with phenols 333, 334
sodium phenoxide, preparation 334
sodium thiosulphate
  reactions
    with bromine 118
    with hydrochloric acid, kinetic study 224–5
    with iodine 112–13
soil
  acid–base reactions in 313–14
  nitrate leaching from 534–6
  pH 78–9, 313
  role in carbon cycle 53
Solar System, element abundances in 128, 129
solid–solution equilibria 291, 292–3
solubility
  of alcohols 37, 39
  of amines 344, 345
  of calcium hydroxide 77
  of carboxylic acids 324, 325
  effect of temperature 77
  of ionic compounds 263–5
  of iron compounds 5, 264–5
  of molecular compounds 259–62
  of organic compounds in water 257, 260
  of phenols 332
solute 5
solutions
  concentration 75–6
  electrical resistance (conductance) 66

enthalpy changes of formation 264
  see also solid–solution equilibria
solvation (ions) 263
solvent 5
solvent extraction 105–6, 338
spectator ions 72, 292
spontaneity of change 272–3, 275–80, 282–4, 314, 317
stability constants 448–51
standard electrode potentials 364, 366, 370–74, 442–4, 446–7, 455–6
standard entropies 280–81
standard free energies, of elements 379
standard hydrogen electrode 364, 366–7
starch 400
  enzyme-catalysed hydrolysis 409–10
stars 129–31
state symbols 14
Staudinger, Hermann 414
steam distillation 31–2
steric factors (reaction kinetics) 226
Stock notation 109
Stone, Edmund 342
straight line graphs 225
structural formulae 34
structural isomers 187
sublimation *see* enthalpy changes of sublimation
substitution reactions 182–3, 200, 204
sucrose 431
  acid hydrolysis 304, 398
  reactions 397–9
sugar 257
sulphamic acid, hydrolysis, kinetic study 229
sulphonation reactions 199, 202
sulphur, plant nutrient 514
sulphur dioxide, in atmosphere 314
sulphur trioxide, manufacture 71
sulphuric acid
  effect on alkanes 165
  reactions
    with alkenes 188, 191
    with arenes 199, 202
    with halides 115
Sun
  helium discovered in 59
  spectrum of sunlight 128
sunbathing 242–3
supernovae 130–31
surroundings 87, 277–9
  entropy change in 279–80, 282, 316
swayback (disorder of sheep) 466
sweeteners 431–2
symbols
  elements 13
  hazard warnings 180, 561
  state 14
system, closed 277–8
  enthalpy changes 87–8
  entropy changes 281–2, 315
  required for equilibrium 293

target molecule, defined 476
TCP 33, 336
tea plant, fluorine in 105
teeth, fluorides in 121–2
temperature
  absolute zero 88, 281

effects on enzymes 409
  and entropy changes 279–80, 316–17
tertiary alcohols 41
  hydroxyl group replacement 178, 179, 184
tertiary halogenoalkanes 178
  hydrolysis 222–4
tetraethyl-lead 174
thalidomide 394, 478
thermal decomposition
  of iron compounds 6, 8
  of potassium halates(V) 118
  of potassium hydrogencarbonate, enthalpy change 90
  of zinc carbonate, entropy changes 276, 283–4, 317
Thermit reaction 11–13, 463
thermodynamics *see* Laws of Thermodynamics
thermoplastics 418–22
thermosetting polymers 423
Thiele melting-point apparatus 486
thin layer chromatography *see under* chromatography
thyroxine 122, 175
tin–copper alloy (bronze) 436
titanium oxide 435
titrations
  acid–base 307–8
    conductimetric 539
    in solubility measurement 77
  back 513, 515
  in kinetic studies 217, 218
  using potassium manganate(VII) 445
  using potassium thiocyanate 297–8
  using sodium thiosulphate 112–14, 455
toluene *see* methylbenzene
trace elements (plant nutrition) *see* micronutrients
transition elements 19, 435–66
triboluminescence 345
2,4,6-tribromophenol, preparation 335
2,4,6-trichlorophenol *see* TCP
triethylamine, preparation 346
2,4,6-trinitrophenol, preparation 336
triols 41
trypsin 411
Tyndall, John 52
Tyrian purple 105, 175

unit cell 441
unitary method (proportionality calculations) 553–4
units 9, IBC
  amount of substance 9
  dipole moment 176
  e.m.f. 361
  entropy 277
  equilibrium constant 296
  molar mass 9
  resistance and conductance 66
  volume 16
unsaturated compounds 33
  see also alkenes
unsaturated fatty acids 402
urea 312, 546
  enzyme-catalysed hydrolysis 229, 408–9
urease 229, 408–9, 411

Urey, Harold 328
urine
   glucose in 433
   ketones in 434
   pH 312

valine 403
van der Waals forces 246–8, 252, 259
van der Waals radii 247–8
vanadium 436
   redox reactions 446–7
van't Hoff, Jacobus 284
vinegar, ethanoic acid in 310, 312
vitamin B12 466
vitamin D, production 244
volcanic gases, hydrogen halides in 105
volt (unit) 361
volume
   of solution 75
   units for 16

water
   action on chlorides 120
   behaviour of acids in 72, 73, 74
   chlorination 122
   electronic structure 133–5, 143
   enthalpy change of vaporization 253–5
   fluoridation 121–2
   'hard' 314
   hydrogen bonding in 251, 252, 255–7, 260
   ionization 301–2
   nitrate/nitrite levels in natural 525, 533
   as solvent 259–62
   see also hydration
water of crystallization (in formulae) 14
weed killers 118
welding, Thermit 12–13
white dwarfs (stars) 130
willow extract (herbal remedy) 341–2
Woodward, Robert Burns 474, 475

X-ray diffraction 545–6
xanthine oxidase 466

yeast 46–7
yield of reaction 44, 477–8

zero order reactions 220–21
Ziegler, Karl 193
zinc
   crystal structure 438
   micronutrient 465
   reaction with copper(II) sulphate 360, 361, 380
      energy change 86
   zinc–carbon cell 359, 383
   zinc–copper cell 365–6
      see also Daniell cell
   zinc–mercury(II) oxide cell 384, 385
   zinc–silver oxide cell 384, 385
zinc carbonate, thermal decomposition, entropy changes 276, 283–4, 317
zinc sulphate, reaction with magnesium 360, 361